T0335651

Optical Interconnects for Data Centers

Related titles

Optical Fiber Telecommunications Volume VIA, 6th ed, Academic Press
(ISBN 978-0-12-396958-3)

Polymer Optical Fibres
(ISBN 978-0-08-100039-7)

Subsea optics and imaging
(ISBN 978-0-85709-341-7)

Woodhead Publishing Series in Electronic and Optical Materials: Number 90

Optical Interconnects for Data Centers

Edited by

Tolga Tekin

Richard Pitwon

Andreas Håkansson

Nikos Pleros

AMSTERDAM • BOSTON • HEIDELBERG • LONDON
NEW YORK • OXFORD • PARIS • SAN DIEGO
SAN FRANCISCO • SINGAPORE • SYDNEY • TOKYO
Woodhead Publishing is an imprint of Elsevier

Woodhead Publishing is an imprint of Elsevier
The Officers' Mess Business Centre, Royston Road, Duxford, CB22 4QH, United Kingdom
50 Hampshire Street, 5th Floor, Cambridge, MA 02139, United States
The Boulevard, Langford Lane, Kidlington, OX5 1GB, United Kingdom

British Library Cataloguing-in-Publication Data
A catalogue record for this book is available from the British Library

Library of Congress Cataloging-in-Publication Data
A catalog record for this book is available from the Library of Congress

ISBN: 978-0-08-100512-5 (print)
ISBN: 978-0-08-100513-2 (online)

For information on all Woodhead Publishing visit our website at https://www.elsevier.com

Publisher: Matthew Deans
Acquisition Editor: Kayla Dos Santos
Editorial Project Manager: Alex White
Production Project Manager: Debasish Ghosh
Cover Designer: Maria Ines Cruz

Typeset by MPS Limited, Chennai, India

Contents

List of contributors ix
Biography xiii
Preface xv
Woodhead publishing series in electronic and optical materials xvii

Part I Introduction

1 Data center architectures **3**
C. DeCusatis
1.1 Introduction 3
1.2 Data center environment considerations 4
1.3 Data center classifications 9
1.4 Application architectures 10
1.5 Cloud data center architectures 14
1.6 Physical architecture 17
1.7 Data center design considerations 21
1.8 Next generation data center architectures 32
1.9 Optical interconnects for data centers 36
References 39

2 Optical interconnects: fundamentals **43**
D. Tsiokos and G.T. Kanellos
2.1 Optical interconnects: the driver behind future data centers 43
2.2 Classes of optical interconnects in data centers 46
2.3 Current status and future trends of optical interconnects systems 51
2.4 Overview of photonic key enabling technologies 56
2.5 Summary and practical conclusions 66
References 67

3 Key requirements for optical interconnects within data centers **75**
M. Duranton, D. Dutoit and S. Menezo
3.1 An explosion of data 75
3.2 What are data centers? 78
3.3 Data communication requirements 82
3.4 Optical interconnect: a solution to energy and bandwidth requirements? 83

3.5 Conclusion 90
References 92

Part II Materials and Components

4 Indium phosphide (InP) for optical interconnects 97
M. Lebby, S. Ristic, N. Calabretta and R. Stabile
4.1 Introduction 97
4.2 InP photonic integration platforms 98
4.3 State-of-the-art in InP photonic integrated circuits (PICs) for data centers 100
4.4 Future trends 114
References 118

5 Photonic crystal cavities for optical interconnects 121
Liam O'Faolain
5.1 Photonic crystal background 121
5.2 Mass production 127
5.3 Light emission 131
5.4 The fiber coupling problem and its solution 137
5.5 Optical modulation 142
5.6 Photo-detection 149
5.7 Outlook 152
References 153

6 Types and performance of high performing multi-mode polymer waveguides for optical interconnects 157
M. Singh and K. Weidner
6.1 Introduction 157
6.2 Polynorbornene 159
6.3 Silicones 163
6.4 Connectors and coupling 166
6.5 Conclusions 169
References 169

7 Design and fabrication of multimode polymer waveguides for optical interconnects 171
T. Ishigure and M. Immonen
7.1 Introduction 171
7.2 Structure of multimode polymer optical waveguide 172
7.3 Fabrication method 174
7.4 Characterization 179
7.5 Polymer optical waveguide circuit for optical PCB 192
7.6 Summary 193
References 194

8 **Silicon photonics for multi-mode transmission** **197**
K. Kurata, Y. Suzuki, M. Tokushima and K. Takemura
8.1 Expectations for optical interconnection 197
8.2 Multi-mode wiring for silicon photonics technology 198
8.3 Chip-scale silicon photonics transceiver "optical I/O core" 201
8.4 Evaluation 216
8.5 Application 219
References 221

9 **Scalable three-dimensional optical interconnects for data centers** **223**
R. Morris, A.K. Kodi and A. Louri
9.1 Introduction 223
9.2 Photonic and three-dimensional interconnects 224
9.3 Optical architecture: three-dimensional-NoC 225
9.4 Reconfiguration 232
9.5 Performance evaluation 236
9.6 Conclusions and future directions 244
References 244

10 **Electronic drivers/TIAs for optical interconnects** **247**
J. Bauwelinck and X. Yin
10.1 Co-design and co-simulation of electronics and photonics 250
10.2 Electronic drivers 252
10.3 Transimpedance amplifiers 256
References 262

Part III Circuit Boards

11 **Electrical and photonic off-chip interconnection and system integration** **265**
M. Zia, C. Wan, Y. Zhang and M. Bakir
11.1 Introduction 265
11.2 Emerging electrical and photonic interconnects 267
11.3 Large-scale interconnected system using a "silicon bridging" concept 272
11.4 Conclusion 282
References 283

12 **Electro-optical circuit boards with single- or multi-mode optical interconnects** **287**
L. Brusberg, M. Immonen and T. Lamprecht
12.1 Motivation and classification of optical interconnects at the board level 287
12.2 Manufacturing of integrated planar polymer waveguides 289
12.3 Integrated glass waveguide based EOCBs 297

12.4 Mass production and reliability 304
References 305

13 International and industrial standardization of optical circuit board technologies 309
R. Pitwon, M. Immonen, J. Wu, L. Brusberg, H. Itoh, T. Shioda, S. Dorresteiner, H.J. Yan, H. Schröder, D. Manessis P. Schneider and K. Wang
13.1 Introduction 309
13.2 Industrial manufacturing processes for OPCBs 312
13.3 International standardization of OPCBs 324
13.4 OPCB measurement 326
13.5 Conclusion 339
References 340

14 Requirements for process automation of optical interconnect technologies 343
I. Piacentini and T. Vahrenkamp
14.1 An introduction and a list of issues 343
14.2 Positional accuracy and the debate on passive/active alignment 351
14.3 Machine "ingredients" and machine technologies 354
14.4 Machine vision 363
14.5 The role of software and HMI in the optimization of new processes 367
14.6 Test and measurement instrumentation 369
14.7 Design for automated assembly/testing and standardization 370
14.8 Automated testing 371
14.9 Conclusions 371
References 372

Part IV Using Optical Interconnects to Improve Network Architectures in Data Centers

15 The role of optical interconnects in the design of data center architectures 375
A. Siokis, K. Christodoulopoulos and E. Varvarigos
15.1 Introduction 375
15.2 Overview of optical interconnects technologies 376
15.3 Applications of optical interconnects in the individual layers of the packaging hierarchy 376
15.4 On-optical printed circuit boards (OPCB) layout strategies 378
15.5 Conclusion 390
Acknowledgments 390
References 391

Index 395

List of contributors

M. Bakir Georgia Institute of Technology, Atlanta, GA, United States

J. Bauwelinck Ghent University - iMinds - imec, Ghent, Belgium

L. Brusberg Fraunhofer Institute for Reliability and Microintegration (IZM), Berlin, Germany

N. Calabretta Eindhoven University of Technology, Eindhoven, The Netherlands

K. Christodoulopoulos University of Patras, Patras, Greece; Computer Technology Institute and Press – Diophantus Patras, Greece

C. DeCusatis Marist College, Poughkeepsie, NY, United States

S. Dorresteiner TE Connectivity, s'Hertogenbosch, The Netherlands

M. Duranton Commissariat à l'énergie atomique et aux énergies alternatives, Gif-sur-Yvette Cedex, France

D. Dutoit Commissariat à l'énergie atomique et aux énergies alternatives, Grenoble Cedex, France

M. Immonen Meadville Aspocomp International Limited, TTM Technologies, Inc., Salo, Finland; TTM Technologies Inc. Turku Area, Finland

T. Ishigure Keio University, Yokohama, Japan

H. Itoh Industrial Technology Center of Tochigi Prefecture, Tochigi, Japan

G.T. Kanellos Aristotle University of Thessaloniki, Thessaloniki, Greece; Center for Research and Technology Hellas, Thessaloniki, Greece

A.K. Kodi Ohio University, Athens, OH, United States

K. Kurata Photonics Electronics Technology Research Association (PETRA), Tsukuba, Japan

T. Lamprecht Vario-optics AG, Heiden, Switzerland

M. Lebby OneChip Photonics Inc., Ottawa, ON, Canada; Glyndwr University, Wales, United Kingdom; Lightwave Logic Inc., Longmont, CO, United States; University of Southern California, Los Angeles, CA, United States

A. Louri George Washington University, Washington, DC, United States

D. Manessis Fraunhofer Institute for Reliability and Microintegration, Berlin, Germany

S. Menezo Commissariat à l'énergie atomique et aux énergies alternatives, Grenoble Cedex, France

R. Morris Ohio University, Athens, OH, United States

Liam O'Faolain SUPA, School of Physics and Astronomy, St Andrews, KY, Scotland

I. Piacentini ficonTEC Service GmbH, Achim, Germany

R. Pitwon Seagate, Langstone Road, Havant, Hampshire, United Kingdom; University of St Andrews, St Andrews, KY, Scotland

S. Ristic McGill University, Montreal, QC, Canada

P. Schneider TE Connectivity, s'Hertogenbosch, The Netherlands

H. Schröder Fraunhofer Institute for Reliability and Microintegration (IZM), Berlin, Germany

T. Shioda Mitsui Chemicals, Inc., Chiba, Japan

M. Singh Sumitomo Bakelite, Tokyo, Japan

A. Siokis University of Patras, Patras, Greece; Computer Technology Institute and Press – Diophantus, Patras, Greece

R. Stabile Eindhoven University of Technology, Eindhoven, The Netherlands

Y. Suzuki Photonics Electronics Technology Research Association (PETRA), Tsukuba, Japan

K. Takemura Photonics Electronics Technology Research Association (PETRA), Tsukuba, Japan

M. Tokushima Photonics Electronics Technology Research Association (PETRA), Tsukuba, Japan

D. Tsiokos Aristotle University of Thessaloniki, Thessaloniki, Greece; Center for Research and Technology Hellas, Thessaloniki, Greece

T. Vahrenkamp ficonTEC Service GmbH, Achim, Germany

E. Varvarigos University of Patras, Patras, Greece; Computer Technology Institute and Press – Diophantus, Patras, Greece

C. Wan Georgia Institute of Technology, Atlanta, GA, United States

K. Wang Seagate, Langstone Road, Havant, Hampshire, United Kingdom

K. Weidner Dow Corning, Auburn, MI, United States

J. Wu TTM Technologies, Shanghai, P.R. China

H.J. Yan TTM Technologies, Shanghai, P.R. China

X. Yin Ghent University - iMinds - imec, Ghent, Belgium

Y. Zhang Georgia Institute of Technology, Atlanta, GA, United States

M. Zia Georgia Institute of Technology, Atlanta, GA, United States

Biography

Tolga Tekin received the PhD degree in electrical engineering and computer science from the Technical University of Berlin, Berlin, Germany, in 2004. He was a Research Scientist with the Optical Signal Processing Department, Fraunhofer HHI, where he was engaged in advanced research on optical signal processing, 3R-regeneration, all-optical switching, clock recovery, and integrated optics. He was a Postdoctoral Researcher on components for O-CDMA and terabit routers with the University of California. He worked at Teles AG on phased-array antennas and their components for skyDSL. At the Fraunhofer Institute for Reliability and Microintegration (IZM) and at Technical University of Berlin, he then led projects on microsystems, 3-D heterogeneous integration, optical interconnects and silicon photonics packaging. He is engaged in photonic integrated system-in-package, millimeter-wave photonics, photonic interconnects and 5G research activities. He is group manager of Photonics and Plasmonics Systems in the System Integration and Interconnection Technologies Department at Fraunhofer IZM. He is coordinator of European flagship project "PhoxTroT" on optical interconnects for data centers. He is Senior Member of IEEE and co-chair of "Photonics-Communication, Sensing, Lighting" Technical Committee in the IEEE Components, Packaging and Manufacturing Technology Society.

Nikos Pleros (b. 1976) joined the Department of Informatics, Aristotle University of Thessaloniki, Greece, in 2007, and serves currently as an Assistant Professor in the field of Optical Communications. He obtained the Diploma and the PhD Degree in Electrical & Computer Engineering from the National Technical University of Athens (NTUA) in 2000 and 2004, respectively. His research interests include all aspects of optical communications and photonic systems spanning from integrated photonic devices to optical networking concepts. He has been actively engaged in research on optical interconnects for data centers and computing, on optical RAM memories and optical buffering, on optical cache architectures, on silicon photonics and plasmonics, on multi-wavelength cw and pulsed laser sources for WDM/OTDM, on all-optical signal processing and digital logic modules, on optical packet/burst/label switching systems and architectures, semiconductor-based switching devices, optical wireless access and radio-over-fiber systems towards 5G networks, and in the fields of biophotonics and biosensing. Dr. Pleros has published more than 190 articles in scientific journals and international conferences including several invited contributions, while his work has received more than 2000 citations (h-index=25). He has participated in several research projects (often as the project coordinator) funded by the European Commission within the

FP6, FP7, and H2020 framework programs, including MUFINS, PLATON, PhoxTrot, MIRAGE, ICT-STREAMS, PlasmoFab, L3Matrix, etc. He is currently coordinating the FP7 project COMANDER, and the H2020 projects ICT-STREAMS and PlasmoFab. Dr. Pleros is a recipient of the 2003 IEEE/LEOS Graduate Student Fellowship that is annually granted to 12 PhD candidates worldwide. Dr. Pleros has held positions of responsibility at several top-tier conferences in the field, and has been the (co)-organizer of several workshops taking place in major conferences, as well as of the 1st European Summerschool on Optical Interconnects.

Richard Pitwon holds the BSc (Hons) degree in Physics from the University of St Andrews and MSc in Computer Science. He leads the photonics research and development group at the Seagate Systems UK and has over 15 years experience in the design and development of high speed photonic interconnect technologies for data storage and communication systems including passive and active optical connectors, optical printed circuit boards, optical interconnect interfaces and transceivers. He holds 46 patents in the field of embedded optical interconnect and has authored numerous publications in this area including journal and conference papers, book chapters, white papers, online articles, and international standards. He is a Chartered Engineer (CEng) and serves as secretary of the International Electrotechnical Commission (IEC) standards group for optical circuit board and is also the permanent UK expert member on IEC standards groups for photonic integrated circuits, optical connectors and optical cable technologies.

Dr. Andreas Håkansson (M) received an MS in applied physics and electrical engineering from Linköping Technical University, Sweden, in 2002, and a PhD degree in electrical engineering from the Polytechnic University of Valencia in 2006, for which he received the universities extraordinary thesis award. In 2006–07 he continued his academic work at the International Centre for Young Scientists at National Institute for Material Science in Tsukuba, Japan. There he led various international collaboration projects related to the optimized design of microscaled optical elements. In December 2007, he joined DAS Photonics S.L. as a senior researcher and PM, where he was managing various European projects related to plasmonics and silicon photonics integration and optoelectronics. In 2014 he joined Fraunhofer as a PM for the FP7-PHOXTROT project, where he is currently active as the project Quality Manager. Since 2015 he has been coordinator of the H2020 project L3Matrix on silicon photonics integration with CMOS for high radix switches in data centers.

Preface

The burden of our rapid growth as an information affluent society is our increasingly ravenous demand for digital information. Over the past few years, the rise of mobile computing systems has seen the location of our information move from local computers to the Cloud, with most information predicted to reside in remote data centers by 2020. However, the exponential rise in data consumption is now pushing modern communications systems and infrastructures beyond their limits, thus fueling the migration of optical interconnect solutions in modern data center environments from the external optical fiber fabric right down into the systems themselves.

Indeed, this migration is already strongly reflected in the research, development, and strategic activities of mainstream organizations across the world, many of whom have contributed to this book.

In this book we introduce the latest advances in high-performance, low-energy, and low-cost optical interconnects in data center environments, guiding the reader through the fundamentals of data centers and optical interconnects. In particular, we consider system embedded photonic interconnect technologies from optical circuit boards based on polymer or glass waveguides to photonic integrated circuits including silicon photonics, III − V photonics, photonic nanocrystals, and CMOS electronics, and how to design data centers to best leverage these new technologies.

This book will serve as a valuable reference for researchers, scientists, and engineers from academia, research, and industry alike with an interest in this exciting and rapidly advancing field, and we hope it will guide you, the reader, to innovate, disrupt, and tear down boundaries toward a world more connected than ever before.

Finally, we would like to thank all the authors whose contributions made this book possible, and our partners and close friends on the European Union-funded flagship project, PhoxTroT.

Berlin, Germany — Tolga Tekin

Fareham, United Kingdom — Richard Pitwon

Thessaloniki, Greece — Nikos Pleros

Berlin, Germany — Andreas Håkansson

Woodhead publishing series in electronic and optical materials

1. Circuit analysis
   J. E. Whitehouse
2. Signal processing in electronic communications: For engineers and mathematicians
   M. J. Chapman, D. P. Goodalli, and N. C. Steele
3. Pattern recognition and image processing
   D. Luo
4. Digital filters and signal processing in electronic engineering: Theory, applications, architecture, code
   S. M. Bozic and R. J. Chance
5. Cable engineering for local area networks
   B. J. Elliott
6. Designing a structured cabling system to ISO 11801: Cross-referenced to European CENELEC and American Standards
   Second edition
   B. J. Elliott
7. Microscopy techniques for materials science
   A. Clarke and C. Eberhardt
8. Materials for energy conversion devices
   Edited by C. C. Sorrell, J. Nowotny, and S. Sugihara
9. Digital image processing: Mathematical and computational methods
   Second edition
   J. M. Blackledge
10. Nanolithography and patterning techniques in microelectronics
    Edited by D. Bucknall
11. Digital signal processing: Mathematical and computational methods, software development and applications
    Second edition
    J. M. Blackledge

12. Handbook of advanced dielectric, piezoelectric and ferroelectric materials: Synthesis, properties and applications
Edited by Z.-G. Ye

13. Materials for fuel cells
Edited by M. Gasik

14. Solid-state hydrogen storage: Materials and chemistry
Edited by G. Walker

15. Laser cooling of solids
S. V. Petrushkin and V. V. Samartsev

16. Polymer electrolytes: Fundamentals and applications
Edited by C. A. C. Sequeira and D. A. F. Santos

17. Advanced piezoelectric materials: Science and technology
Edited by K. Uchino

18. Optical switches: Materials and design
Edited by S. J. Chua and B. Li

19. Advanced adhesives in electronics: Materials, properties and applications
Edited by M. O. Alam and C. Bailey

20. Thin film growth: Physics, materials science and applications
Edited by Z. Cao

21. Electromigration in thin films and electronic devices: Materials and reliability
Edited by C.-U. Kim

22. *In situ* characterization of thin film growth
Edited by G. Koster and G. Rijnders

23. Silicon – germanium (SiGe) nanostructures: Production, properties and applications in electronics
Edited by Y. Shiraki and N. Usami

24. High-temperature superconductors
Edited by X. G. Qiu

25. Introduction to the physics of nanoelectronics
S. G. Tan and M. B. A. Jalil

26. Printed films: Materials science and applications in sensors, electronics and photonics
Edited by M. Prudenziati and J. Hormadaly

27. Laser growth and processing of photonic devices
Edited by N. A. Vainos

28. Quantum optics with semiconductor nanostructures
Edited by F. Jahnke

29. Ultrasonic transducers: Materials and design for sensors, actuators and medical applications
Edited by K. Nakamura

30. Waste electrical and electronic equipment (WEEE) handbook
Edited by V. Goodship and A. Stevels

31. Applications of ATILA FEM software to smart materials: Case studies in designing devices
Edited by K. Uchino and J.-C. Debus

32. MEMS for automotive and aerospace applications
Edited by M. Kraft and N. M. White

33. Semiconductor lasers: Fundamentals and applications
Edited by A. Baranov and E. Tournie

34. Handbook of terahertz technology for imaging, sensing and communications
Edited by D. Saeedkia

35. Handbook of solid-state lasers: Materials, systems and applications
Edited by B. Denker and E. Shklovsky

36. Organic light-emitting diodes (OLEDs): Materials, devices and applications
Edited by A. Buckley

37. Lasers for medical applications: Diagnostics, therapy and surgery
Edited by H. Jelínková

38. Semiconductor gas sensors
Edited by R. Jaaniso and O. K. Tan

39. Handbook of organic materials for optical and (opto)electronic devices: Properties and applications
Edited by O. Ostroverkhova

40. Metallic films for electronic, optical and magnetic applications: Structure, processing and properties
Edited by K. Barmak and K. Coffey

41. Handbook of laser welding technologies
Edited by S. Katayama

42. Nanolithography: The art of fabricating nanoelectronic and nanophotonic devices and systems
Edited by M. Feldman

43. Laser spectroscopy for sensing: Fundamentals, techniques and applications
Edited by M. Baudelet

44. Chalcogenide glasses: Preparation, properties and applications
Edited by J.-L. Adam and X. Zhang

45. Handbook of MEMS for wireless and mobile applications
Edited by D. Uttamchandani

46. Subsea optics and imaging
Edited by J. Watson and O. Zielinski

47. Carbon nanotubes and graphene for photonic applications
Edited by S. Yamashita, Y. Saito and J. H. Choi

48. Optical biomimetics: Materials and applications
Edited by M. Large

49. Optical thin films and coatings
Edited by A. Piegari and F. Flory

50. Computer design of diffractive optics
Edited by V. A. Soifer

51. Smart sensors and MEMS: Intelligent devices and microsystems for industrial applications
Edited by S. Nihtianov and A. Luque

52. Fundamentals of femtosecond optics
S. A. Kozlov and V. V. Samartsev

53. Nanostructured semiconductor oxides for the next generation of electronics and functional devices: Properties and applications
S. Zhuiykov

54. Nitride semiconductor light-emitting diodes (LEDs): Materials, technologies and applications
Edited by J. J. Huang, H. C. Kuo and S. C. Shen

55. Sensor technologies for civil infrastructures
Volume 1: Sensing hardware and data collection methods for performance assessment
Edited by M. Wang, J. Lynch and H. Sohn

56. Sensor technologies for civil infrastructures
Volume 2: Applications in structural health monitoring
Edited by M. Wang, J. Lynch and H. Sohn

57. Graphene: Properties, preparation, characterisation and devices
Edited by V. Skákalová and A. B. Kaiser

58. Silicon-on-insulator (SOI) technology
Edited by O. Kononchuk and B.-Y. Nguyen

59. Biological identification: DNA amplification and sequencing, optical sensing, lab-on-chip and portable systems
Edited by R. P. Schaudies

60. High performance silicon imaging: Fundamentals and applications of CMOS and CCD sensors
Edited by D. Durini

61. Nanosensors for chemical and biological applications: Sensing with nanotubes, nanowires and nanoparticles
Edited by K. C. Honeychurch

62. Composite magnetoelectrics: Materials, structures, and applications
G. Srinivasan, S. Priya and N. Sun

63. Quantum information processing with diamond: Principles and applications
Edited by S. Prawer and I. Aharonovich

64. Advances in non-volatile memory and storage technology
Edited by Y. Nishi

65. Laser surface engineering: Processes and applications
Edited by J. Lawrence, C. Dowding, D. Waugh and J. Griffiths

66. Power ultrasonics: Applications of high-intensity ultrasound
Edited by J. A. Gallego-Juárez and K. F. Graff

67. Advances in delay-tolerant networks (DTNs): Architectures, routing and challenges
Edited by J. J. P. C. Rodrigues

68. Handbook of flexible organic electronics: Materials, manufacturing and applications
Edited by S. Logothetidis

69. Machine-to-machine (M2M) communications: Architecture, performance and applications
Edited by C. Anton-Haro and M. Dohler

70. Ecological design of smart home networks: Technologies, social impact and sustainability
Edited by N. Saito and D. Menga

71. Industrial tomography: Systems and applications
Edited by M. Wang

72. Vehicular communications and networks: Architectures, protocols, operation and deployment
Edited by W. Chen

73. Modeling, characterization and production of nanomaterials: Electronics, photonics and energy applications
Edited by V. Tewary and Y. Zhang

74. Reliability characterisation of electrical and electronic systems
Edited by J. Swingler

75. Industrial wireless sensor networks: Monitoring, control and automation
Edited by R. Budampati and S. Kolavennu

76. Epitaxial growth of complex metal oxides
Edited by G. Koster, M. Huijben and G. Rijnders

77. Semiconductor nanowires: Materials, synthesis, characterization and applications
Edited by J. Arbiol and Q. Xiong

78. Superconductors in the Power Grid
Edited by C. Rey

79. Optofluidics, sensors and actuators in microstructured optical fibres
Edited by S. Pissadakis

80. Magnetic Nano- and Microwires: Design, Synthesis, Properties and Applications
Edited by M. Vázquez

81. Robust Design of Microelectronic Assemblies Against Mechanical Shock, Temperature and Moisture
E-H. Wong and Y-W. Mai

82. Biomimetic technologies: Principles and Applications
Edited by T. D. Ngo

83. Directed Self-assembly of Block Co-polymers for Nano-manufacturing
Edited by R. Gronheid and P. Nealey

84. Photodetectors
Edited by B. Nabet

85. Fundamentals and Applications of Nanophotonics
Edited by J. Haus

86. Advances in Chemical Mechanical Planarization (CMP)
Edited by S. Babu

87. Rare Earth and Transition Metal Doping of Semiconductor Materials: Synthesis, Magnetic Properties and Room Temperature Spintronics
Edited by V. Dierolf, I. Ferguson and J. M. Zavada

88. Materials Characterization Using Non-Destructive Evaluation (NDE) Methods
Edited by G. Huebschen, I. Altpeter, R. Tschuncky and H.-G. Herrmann

89. Polymer Optical Fibres: Fibre Types, Materials, Fabrication and Applications
Edited by Christian-Alexander Bunge, Markus Beckers and Thomas Gries

90. Optical Interconnects for Data Centers
Edited by Tolga Tekin, Richard Pitwon, Andreas Håkansson and Nikos Pleros

Part I

Introduction

Data center architectures

1

C. DeCusatis
Marist College, Poughkeepsie, NY, United States

1.1 Introduction

Data centers house the computational power, storage, networking, and software applications that form the basis of most modern business, academic, and government institutions. It is difficult to overstate the importance of information technology (IT) in our daily lives. Data centers support the hardware and software infrastructure which is critical to both Fortune 500 companies and startups. It is difficult to find a market related to business, finance, transportation, health care, education, entertainment, or government which has not been disrupted by IT [1,2]. The widespread availability of data on demand anytime, anywhere, has caused new applications to emerge, including cloud computing, big data analytics, real-time stock trading, and more. Workloads have evolved from a predominantly static environment into one which changes over time in response to user demands, often as part of a highly virtualized, multitenant data center. In response to these new requirements, data center architectures are also undergoing significant changes. This chapter provides an overview of data center fundamentals, with particular emphasis on the role of optical data networking.

It can be difficult to describe a "typical" data center, since there are many types of servers, storage devices, and networking equipment which can be architected in different ways. Historically, large centralized mainframe computers were first developed in the 1950s and 1960s, at a time when computer networking was extremely limited (peripheral devices such as printers could be placed no more than 400 feet away from the mainframe, connected by inch-thick copper cables [2]). This led to most of the computer equipment being kept in a large room or "glass house," the forerunners of contemporary data centers. In this environment, computer security was equivalent to physical security, since few users had access to the expensive computer hardware required. Over time, mainframes evolved to become significantly more cost-efficient and powerful, setting the standard for so-called "enterprise class computing" (analogous with "carrier class" networking in the central office of large telecom providers). An enterprise class data center provided very high levels of reliability, availability, and scalability (RAS), as well as large amounts of centralized processing power. Over time, the description for enterprise computing came to be associated with platforms other than the mainframe, as x86 based servers gradually became capable of offering similar performance and reliability.

Optical Interconnects for Data Centers. DOI: http://dx.doi.org/10.1016/B978-0-08-100512-5.00001-2

Figure 1.1 Rack mounted server.

Mainframes continue to be used today by Fortune 500 companies worldwide due to their extremely high performance and RAS characteristics,[1] although other platforms comprise most of the server volumes in data centers and cloud computing environments. While mainframe architectures derive their increased processing power by scaling up the capacity of their processors in each successive technology generation, other platforms achieve high performance by scaling out (adding more processors to form a cluster or server farm). The predominant server platform is currently based on the Intel x86 microprocessor, packaged into servers which mount into standardized 19-inch wide equipment racks (see Fig. 1.1). Servers with other processors are available (including AMD and IBM Power microprocessors) in different form factors, including cabinets and towers. To improve the density of servers per rack, as well as potentially reducing power consumption and heat generation, other form factors such as blade servers can also be used (see Fig. 1.2). Enterprise data centers can become quite large; e.g., a typical data center used by Microsoft could easily be the size of a warehouse (25–30,000 m^2), consume 20–30 MW of electricity, and cost over $500 M to construct [3].

Enterprise class servers were the first to take advantage of server virtualization as early as the 1960s, using software to create multiple virtual machines (VMs) hosted on a single physical machine. Virtualization technology was not widely adopted on x86 servers until much later, but has since become important for modern data centers. Virtualization software helps improve utilization of physical servers, and allows multiple VMs running different operating systems to be installed on a single processor. However, there are still many data centers consisting of lightly utilized servers running a "bare metal" operating system, or a hypervisor with a small number of VMs.

1.2 Data center environment considerations

Designing a data center is far more complicated than simply assembling enough servers, storage, and networking in a large building. Considerations ranging from

[1] The IBM System Z enterprise servers, for example, can support well over 10 billion instructions per second, using custom microprocessors with over 18 billion transistors per chip, and offer "continuous availability" through features such as concurrent software upgrades and hardware replacement [2].

Figure 1.2 Blade server.

the data center geographic location, proximity to inexpensive electrical power sources, disaster recovery planning, and much more, can require large investments of time and money, even for a moderately sized facility. In this section, we provide a brief overview of some important data center design features.

The location of a data center and its backup facilities is typically influenced by factors such as safety (proximity to earthquake fault lines, flood plains, etc.), access to high-speed wide area networking, and cost of construction and real estate. As with any large facility, data centers require physical security such as video monitoring systems, restricted access using badge locks, and similar precautions. Often designed to operate continuously, some facilities will use three or four shifts of employees managing and servicing equipment.

Data centers often have unique architectural requirements, the most prominent being raised flooring which provides space for cabling and air cooling under the servers (see Fig. 1.3). A typical raised floor may be 80–100 cm deep, and covered with a grid of removable 60 cm square floor tiles. Some tiles may be perforated to facilitate air flow, since the underfloor area provides a plenum for air circulation that helps cool the equipment racks as part of the air conditioning system. There are several general types of raised floors, which are documented in detail as part of various industry standards, originally developed for telecom environments compliant with the Network Equipment Building Standard (NEBS) [4,5] and other environments [6]. For example, one type consists of vertical steel pedestals called stringers, which are mechanically fastened to a concrete floor. The pedestal height can be adjusted to allow for different raised floor heights. Stringered raised floors provide good structural integrity, including support for lateral loads, through mechanical attachments between the pedestal heads. Equipment racks are generally attached to the floor using toggle bars or some other form of bracing. This approach is

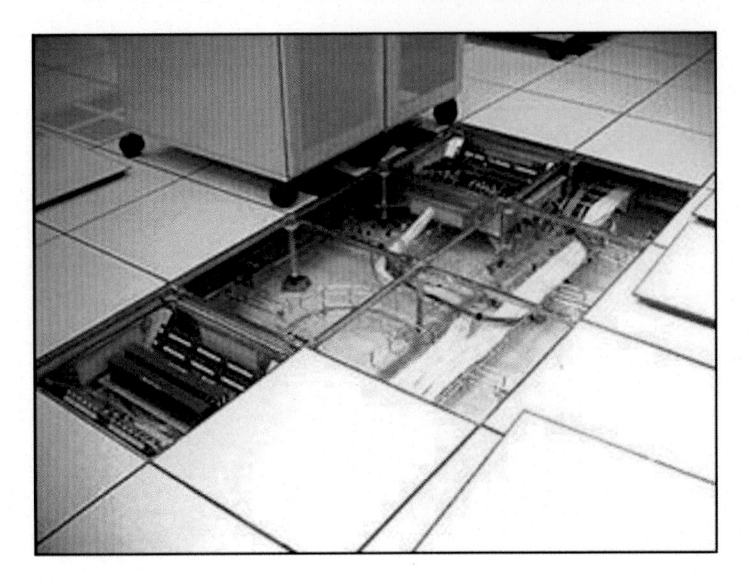

Figure 1.3 Data center raised floor.

commonly used in areas where the floor will bear large amounts of equipment or be exposed to earthquakes. Structured platforms can also be constructed of welded or bolted steel frames, sturdy enough to permit equipment to be fastened directly to the platform. For areas which require less mechanical integrity, stringerless raised floors may be used. This method is significantly weaker, since it only provides pedestals to support floor panels at their corners, but it also provides fewer obstructions to access the space under the raised floor.

The data center must also be protected by a fire suppression system, preferably using a chemical agent which will not damage the servers (unlike plain water). Safety regulations affect the different types of optical and copper cables which can be installed in a data center. In particular, there are several classifications for optical fiber cables, as established by the National Fire Code [7] and National Electrical Code [8], which are enforced by groups such as the Underwriters Laboratories [9]. These include riser, plenum, low halogen, and data processing environments. Riser rated cable (UL 1666, type OFNR) is intended for use in vertical building cable plants, but provides only nominal fire protection. An alternative cable type is plenum rated (UL 910, type OFNP) which is designed not to burn unless extremely high temperatures are reached. Plenum rated cable is required for installation in air ducts by the 1993 National Fire Code 770-53, although there is an exception for raised floor and data processing environments which may be interpreted to include subfloor cables. There is also an exception in the National Electrical Code that allows for some cables installed within a data center to be rated "DP" (data processing) rather than plenum (see the "Information technology equipment" section of the code, Article 645-5 (d), exception 3). Some types of plenum cable are also qualified as "limited combustibility" by the National Electrical Code. There are two basic types of plenum cable, manufactured with either a Teflon- or

PVC-based jacket. Although they are functionally equivalent, the Teflon-based jackets tend to be stiffer and less flexible, which can affect installation. Outside North America, another standard known as low halogen cable is widely used; this burns at a lower temperature than plenum, but does not give off toxic fumes. Yet another variant is low smoke/zero halogen, in which the cable jacket is free from toxic chemicals including chlorine, fluorine, and bromides. It remains challenging to find a single cable type which meets all installation requirements worldwide. Since the requirements change frequently and are subject to interpretation by local fire marshals and insurance carriers, network designers should consult the relevant building code standards prior to installing any new cables within a data center.

In recent years, the cost of energy to power data center equipment and provide cooling/heating of the building have become the driving factors in determining the location of large data centers. In fact, the cost of power may soon exceed the cost of the original capital equipment investment at some facilities. Some recent studies [10] have suggested that a single large data center such as those supporting Amazon, Google, or Facebook can easily consume as much power as a small city (in the 60−70 MW range). Data centers worldwide currently consume around 30 billion watts of electricity, the equivalent of about 30 nuclear power plants [11]. Thus, it makes sense to locate data centers near hydroelectric dams or similar facilities. Most data centers also provide redundant power supplies to maintain continuous 24/ 7 operation in the event of a power outage. This may include racks of batteries, so-called "uninterruptable power supplies" (UPS), or diesel-fueled onsite generators. Power supplies are usually built to provide N + 1 redundancy throughout the data center. There have been many efforts to reduce energy consumption in large data centers (so-called "green" data centers), by taking advantage of economies of scale and frequently upgrading to more energy efficient equipment [12]. Some large data centers operate in a "lights-out" mode, having virtually eliminated the need for onsite personnel during normal operations by using remote management systems. Data centers also consume significant energy in cooling/heating the server equipment. Most servers are air-cooled, and data centers are designed with server racks oriented such that all the servers intake cooler air from one aisle, and expel heated air into an adjacent aisle, to optimize heating and cooling of the overall building. Fortunately, due to extensive server virtualization and workload consolidation, the amount of energy consumed per unit of computing is actually declining [13]. New standards and best practices are emerging for improving power usage effectiveness (PUE), which is the ratio of energy imported by a facility to the amount of power consumed by IT resources. Networking equipment is not exempt from this trend. Use of relatively thin, flexible fiber optic cables to replace copper can provide improved air flow and more efficient cooling. Recent enhancements to copper and backplane links (such as the IEEE P802.3az standard, known as Energy Efficient Ethernet [14]) have attempted to create a more energy efficient interconnect by allowing the links to idle during periods of low activity. These factors help contribute to modern data centers achieving a PUE close to unity. Network virtualization technologies have also been shown to reduce energy consumption for switches and routers by up to 25%, depending on the type of workload [12]. Modern data centers must comply with various energy consumption regulations, which vary in different

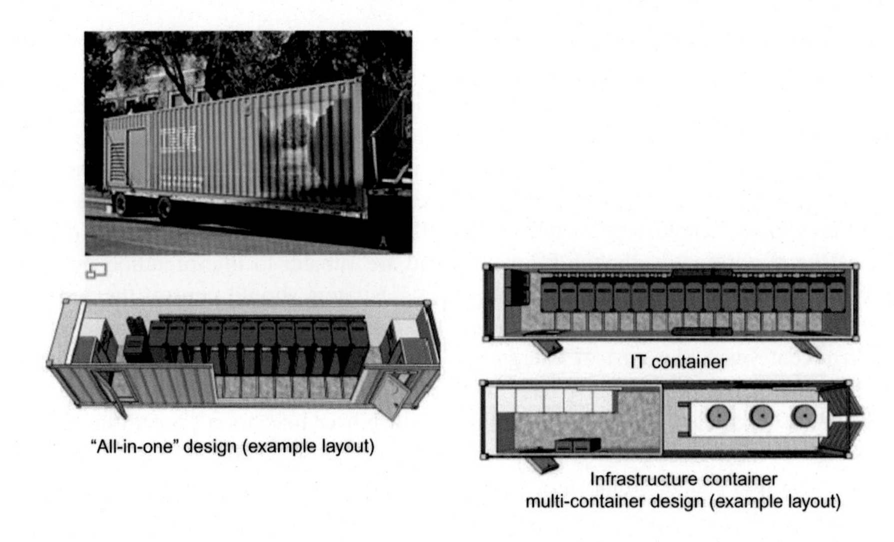

Figure 1.4 A 40-foot IBM Portable Modular Data Center, exterior and interior views [18].
Source: From Handbook of Fiber Optic Data Communication, Chapter 12, Figure 12.3.

countries. For example, the US Environmental Protection Agency and US Department of Energy have defined an Energy Star rating for standalone or large data centers [15]. The European Union also has a similar initiative, the Data Centre Energy Efficiency Code of Conduct [16].

Not all data centers are housed in large, special purpose buildings. Following a natural disaster, it may be necessary to quickly deploy a mobile data center solution which can temporarily replace lost data processing capacity. One common approach is to package servers, networking, and storage within a standard sized shipping container, which can be delivered to the disaster site and quickly be made operational. Systems which can be used for this purpose may contain 1000 or more servers each, often utilize 1 or 10 Gbit/second Ethernet switches to interconnect several racks worth of data processing equipment, and interconnect this equipment with an outside network. There are many examples of commercially-available container data centers from companies such as Cisco, HP, Sun, Google, and others. For instance, the IBM Portable Modular Data Center (PMDC) [17] shown in Fig. 1.4 is built into a standard 20-, 40-, or 53-foot shipping container, which can be transported using standard shipping methods. The unit is weather resistant and insulated for use in extreme heat or cold, and includes its own power management and cooling systems. In some cases, large cloud computing data centers are constructed with modular container-like building blocks within a large warehouse, as a semipermanent installation. This makes it possible to swap out entire containers when a component fails, rather than isolating and removing a single piece of equipment. Load balancing, security, and other functions can be organized around these container-like modules. The size and capacity of such modules can vary widely depending on the application.

1.3 Data center classifications

Data centers are governed by many industry standards and operational metrics; a common approach is to classify data centers based on their calculated availability to end users. The Telecommunications Industry Association (TIA) is a trade association accredited by ANSI (American National Standards Institute). In 2005 they published ANSI/TIA-942, Telecommunications Infrastructure Standard for Data Centers [6], which was amended in 2008 and again in 2010. This standard defined four levels (called tiers) of data centers. The simplest is a Tier 1 data center, a basic server room which follows guidelines for the installation of computer equipment. The most stringent level is a Tier 4 data center, which is designed to host mission critical computer systems, with fully redundant subsystems and compartmentalized security zones controlled by biometric access controls methods.

The requirements for a Tier 1 data center include the following:

- Single nonredundant distribution path serving the IT equipment;
- Nonredundant capacity components;
- Basic site infrastructure with expected availability of 99.671%.

The requirements for a Tier 2 data center include all of the Tier 1 requirements, plus a redundant site infrastructure capacity component with expected availability of 99.741%. A Tier 3 data center meets or exceeds all Tier 1 and 2 requirements, and in addition meets the following requirements:

- Multiple independent distribution paths serving the IT equipment;
- All IT equipment must be dual-powered and fully compatible with the topology of a site's architecture;
- Concurrently maintainable site infrastructure with expected availability of 99.982%.

Finally, a Tier 4 data center meets or exceeds all requirements for the previous three tiers, and in addition:

- All cooling equipment is independently dual-powered, including chillers and heating, ventilating, and air-conditioning (HVAC) systems;
- Fault-tolerant site infrastructure with electrical power storage and distribution facilities with expected availability of 99.995%.

The difference between 99.982% and 99.995% is only 0.013%; however, this can be quite significant depending on the application. For example, consider a data center operating for 1 year or 525,600 minutes; we expect that a Tier 3 data center will be unavailable 94.608 minutes/year, whereas a Tier 4 data center will only be unavailable 26.28 minutes/year. Therefore, each year a Tier 4 data center would be expected to be available for 68.328 minutes more than a Tier 3 data center. Similarly, a Tier 3 data center would be expected to be available for 22.6 hours longer than a Tier 2 data center. The higher the availability needs of a data center, the higher the capital and operational costs of building and managing it. Business needs should dictate the level of availability required. In other words, an appropriate level of availability should be addressed by the data center design criteria, in order to avoid financial and operational risks as a result of downtime. If the estimated cost of downtime per day/week/year exceeds the amortized capital costs and operational expenses, a higher level of availability should be factored into the data center design. If the cost of avoiding downtime greatly exceeds the cost of downtime itself, a lower level of availability should be factored into the design. Some data centers claim to exceed even Tier 4 requirements, such as the Stone Mountain data-plex which is housed underground inside a former limestone mine [19]. Independent from the ANSI/TIA-942 standard, the Uptime Institute (later acquired by the 451 Group [20]) defined its own four levels of data center design. These levels also describe the availability of data, with higher tiers providing greater availability. Other efforts have also been proposed, including a five level certification process for data center criticality based on the German Datacenter star audit program (which is no longer in service).

1.4 Application architectures

Within the data center, an application architecture describes how functions are spread across servers in order to deliver end services to data center users. Historically, in the early 1960s when compute power was expensive, applications processing was done on a large centralized computer; users interacted with this system using so-called "dumb terminals" (keyboards and displays with no processing power of their own). Data networking was limited to modems which communicated between dumb terminals and large computers over the public telephone network, often at very low speeds (perhaps 10–56 kbits/second). With the development of microprocessors, the first personal computers appeared in the mid-1970s. Although these could be configured to emulate a terminal connection to a larger computer, many new applications emerged which were supported on desktop or laptop computers. As more data became distributed across individual user computers, the need for an efficient file sharing mechanism quickly became apparent. The process of copying files to disk, manually carrying the disk to another computer, and reinstalling the files became far too cumbersome for most users. This led to the development of local area networks (LANs) in the late 1980s and early 1990s, which enabled file sharing between computers.

LANs enabled a new type of architecture called client – server computing, as shown in Fig. 1.5. Processing work could be divided between the personal computer client and a larger, centralized server. Generally, the client performance was less than the server for most applications, although the steady improvements in client performance with each technology generation shifted the nature of applications that would preferably run on the server. Client – server architectures were widely adopted during and after the 1990s, and are still used today by many large enterprises. In a traditional client – server design, the client may be a personal computer while the server is a mainframe or enterprise-class computer. The server-centric approach offered all the benefits of centralized processing and control. Since data

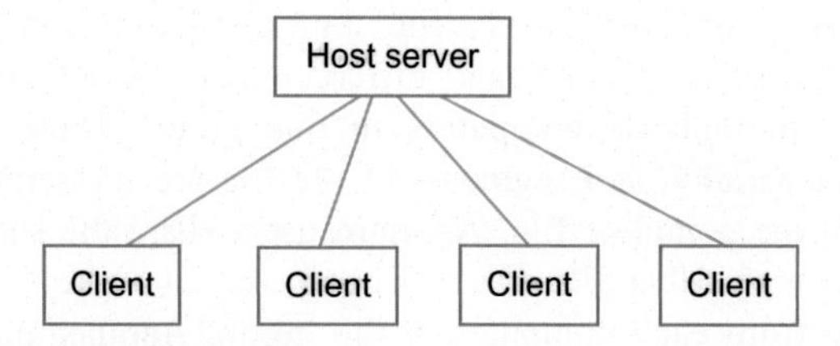

Figure 1.5 Client – server architecture.

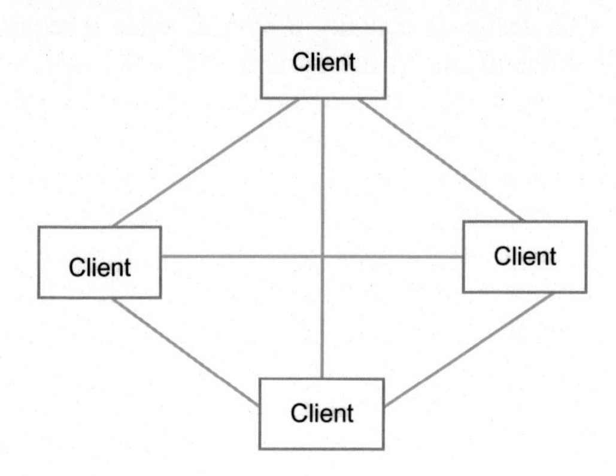

Figure 1.6 Peer-to-peer architecture.

communication flowed through the central server, implementation of policy-based control and security was simplified. Some enterprise architectures attempted to deploy "thin clients" (essentially dumb terminals) to reduce cost. More recently, with the increasing demands of a mobile and telecommuting workforce and the falling cost of computing hardware, many users needed the flexibility of a more powerful client device. In early client – server deployments, a more powerful client was often under-utilized, wasting storage, processing power, and bandwidth. The modern client – server model has evolved to leverage Internet connectivity into cloud computing or application as a service model. The role of clients has expanded to include smart phones and other mobile devices, which are served by VMs within warehouse-scale cloud data centers worldwide.

An alternative design is the peer-to-peer system depicted in Fig. 1.6. Servers work directly with each other as peers to accomplish all or part of the workload, without the assistance of a central server. This is possible because of the increased processing power of low cost, commodity architecture servers, whose capacity may be under-utilized in a client – server design. Well known examples of peer-to-peer architectures include file sharing service BitTorrent, Skype's voice over IP system, and distributed processing applications including SETI@Home [21]. Similar designs have also begun to see adoption in large business environments. In a file sharing program such as BitTorrent, there is no concept of downloading files from a central server to a client. Instead, each computer hosts a client program, and a set of files. When one computer requests a file, a group of computers which contain all or parts of the file are assembled, and different parts of the file are downloaded simultaneously from multiple computers in this group. These downloads occur simultaneously, in parallel, and segments of the file are reassembled on the target computer to form the complete file. As more users share the same file, download speeds will increase for that file, since a computer can take smaller and smaller pieces of the file from each computer in the group. Another popular example of peer-to-peer architecture (with one exception that is technically a centralized server) is the Skype voice over IP system, which offers free or inexpensive phone calls over the Internet. This design is actually a hybrid, since it requires a central login server where users authenticate to the system. All other operations of Skype are done using a peer-to-peer design. To look up the name and address of someone you wish to call, Skype uses a directory search process. Compute nodes on the Skype network can be promoted to act as "super nodes" if they have enough memory, bandwidth, and processor capacity. The Skype directory search (as well as other signaling functions) is a peer-to-peer process performed on the super nodes. Transport of the call data is done by routing voice packets between the two host computers at either end of the call.

In addition to the hardware architecture, data centers also employ various software architectures. A detailed discussion of software architecture is beyond the scope of this chapter, although we will mention a few examples for the sake of completeness. Modern cloud computing systems can employ a service oriented architecture, which is actually an application development methodology which creates solutions by integrating one or more web services [3]. Each web service is treated as a subroutine or function call designed to accomplish a specific task (e.g., processing a credit card or checking airline departure times). These web services are treated as remote procedure calls or application program interfaces (APIs). Cloud computing environments may develop software-as-a-service or platform-as-a-service offerings based on these approaches. Software architectures may include agile development methodologies, particularly the collaborative approach between developers, IT professionals, and quality assurance known as DevOps. These approaches emphasize principles, such as fail fast, and borrow from other continuous improvement development processes such as the Deming Cycle (also known as Plan, Do, Check, Act, or PDCA) [22].

The most common general purpose application architecture for enterprise data centers is a multitier model, as shown in Fig. 1.7. Based on the layered designs supporting enterprise resource planning and content resource management solutions, this design includes tiers of servers hosting web, application, and database systems. Multitier server farms provide improved resiliency by running redundant processes on separate servers within the same application tier. In this way, one server can be taken out of service without interrupting execution of the process. Resiliency is also encouraged by load balancing between the tiers. Using server virtualization, the web and application servers can be implemented as VMs on a common physical server (assuming this meets resiliency goals). Conventional database servers deploy separate physical machines rather than using virtualization, due to performance

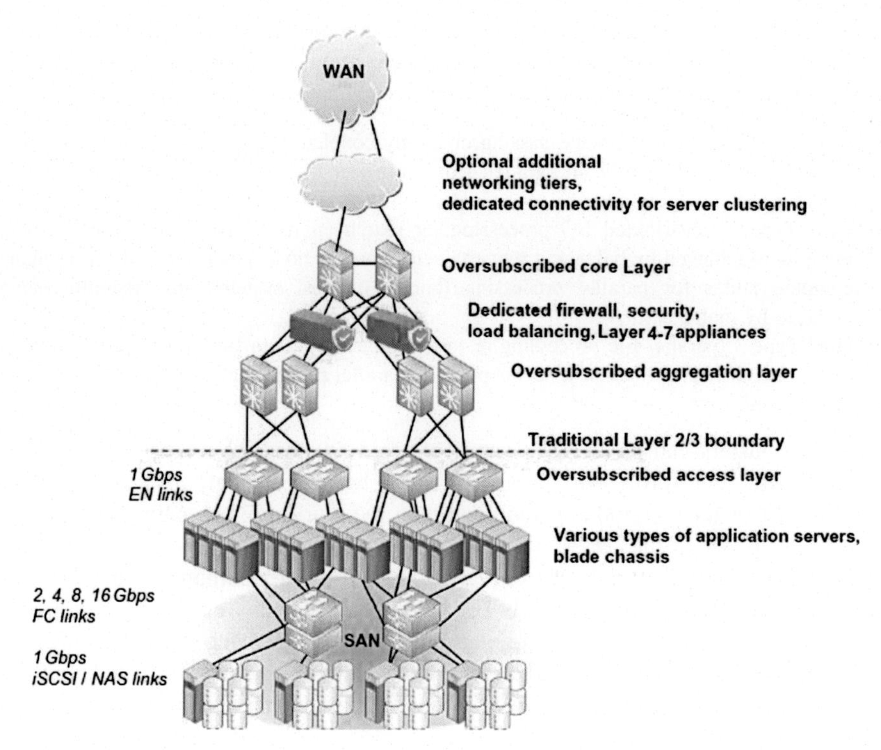

Figure 1.7 Traditional data center network architecture.
Source: From Handbook of Fiber Optic Data Communication, Chapter 1, Figure 1.1. After InfiniBand Trade Association [Online], http://www.infinibandta.org/content/pages.php?pg = technology_overview [accessed 29.01.13].

concerns. The data center tiers can be segregated from each other by using different physical routers within each tier, or by provisioning virtual local area networks (VLANs). Often VLAN-aware firewalls and load balancers will also be employed between tiers. Physically separate networks may achieve better performance, with the tradeoff of higher deployment cost and more devices to manage. The main advantages of VLANs are the reduced complexity and cost. System performance requirements and traffic patterns will often help determine whether physical or virtual network segmentation is preferred for a given design.

The multitier approach creates clusters of servers, storage, and networking equipment which are used to achieve high availability, load balancing, and improved security. Resource clustering is a general principle which can be applied to other computing applications. For example, there is a particular class of high performance computing (HPC) applications which combine multiple processors to form a unified, high performance system, leveraging special software and high-speed networking. Examples of HPC clusters can be found in scientific and technical research (including meteorology, seismology, and aerodynamics), real-time financial trading analytics, rendering of high resolution graphics, and many other fields. HPCs are available in many different types, using both commodity, off-the-shelf hardware, and custom designed processors. There are three main categories of HPC which are generally recognized by the industry:

- HPC Type I (parallel message passing or tightly coupled): Applications run on all compute nodes simultaneously, in parallel, while a master node determines workload allocations for each compute node.
- HPC Type 2 (Distributed I/O processing, or search engines): Rapid response to client requests is achieved by balancing requests across master nodes, then sprayed across many compute nodes for parallel processing (current unicast systems are gradually being replaced by multicast).
- HPC Type 3 (parallel file processing or loosely coupled): Data files are divided into segments and distributed across a server pool for parallel processing; partial results are later recombined.

These clusters can be large or small (up to 1000 servers), organized into subgroups with inter-processor communication between subgroups. A currently updated list of the top 500 supercomputers in the world [23] is maintained to provide an overview of the server, storage, and network interconnects currently in use. Most cluster networks are based on variations of Ethernet, although other proprietary network fabrics are also used. Topologies can include variations on a hypercube, torus, hypertree, and full or partial meshes (to provide equal latency and shortest paths to all compute nodes). HPC networks may include four-way or eight-way Equal Cost Multi-Pathing (ECMP), and distributed forwarding based on both Layer 3 and Layer 4 port hashing.

1.5 Cloud data center architectures

Cloud computing is a method for delivering computing services from a large, highly virtualized data center to many independent end users, using shared applications and pooled resources. While there are many different definitions for cloud computing [2,3,24], it is typically distinguished by the following attributes: On-Demand Self-Service, Broad Network Access, Resource Pooling, Rapid and Elastic Resource Provisioning, and Metered Service at Various Quality Levels. Implementation of these attributes as part of a large, enterprise-class cloud computing service which provides continuous availability to a large number of users typically requires significantly more server, networking, and storage resources than conventional data centers (up to an order of magnitude more in many cases). This is only achievable in a cost-effective way through extensive use of virtualization. In recent years, many equipment vendors have contributed to the hardware and software infrastructure which has made enterprise-class virtualization widely available on all major compute platforms and storage systems (network virtualization remains an emerging technology, but appears to be gaining significant support across the industry). This, in turn, enables new architectures for cloud computing data centers, including hosting multiple independent tenants on a shared infrastructure, rapid and dynamic provisioning of new features, and implementing advanced load balancing, security, and business continuity functions. Modern cloud data centers employ resource pooling to make more efficient use of data center appliances and to enable dynamic reprovisioning in response to changing application needs. Examples of this include elastic workloads where application components are added, removed, or resized based on the traffic load; mobile applications relocating to different hosts based on distance from the host or hardware availability; and proactive disaster recovery solutions which relocate applications in response to a planned site shutdown or a natural disaster.

It has been shown [25] that highly virtualized servers place unique requirements on the data center network. Cloud data center networks must contend with huge numbers of attached devices (both physical and virtual), large numbers of isolated, independent subnetworks, multitenancy (application components belonging to different tenants are collocated on a single host), and automated creation, deletion, and migration of VMs (facilitated by large Layer 2 network domains). Further, many cloud data centers now contain clusters or pods of servers, storage, and networking, configured so that the vast majority of traffic (80–90% in some cases) flows between adjacent servers within a pod (so-called east – west traffic). This is a very different traffic pattern from conventional data center networks, which supported higher levels of traffic between server racks or pods (so-called north – south traffic).

Traditionally, data centers were built for the sole use of one company. This had the benefit of giving the company complete control over their systems, allowing them to tweak the architecture for higher performance. However, this model has become much less cost competitive with the advent of technologies that enable renting or sharing of data center resources from a cloud service provider. The economics are analogous to renting a hotel room or apartment; when you travel to a distant location, especially for a short stay, it is not cost-effective to build a second house [26]. Yet this is precisely what most private companies have done for many years when their current data center capacity is exhausted. Modern cloud providers offer infrastructure-as-a-service, meaning that the cloud provider makes data center resources available on request, and charges only for the time those resources are in use. Similarly, it is possible to do application development in the cloud without owning the underlying development platform (platform-as-a-service), or to lease entire applications from the cloud provider, such as a remote desktop (application-as-a-service). The ability to rent or lease capacity from a cloud provider, sometimes on an hourly basis, has changed the economics of data center design. About 30% of the server, storage, and network equipment manufactured today are sold to cloud providers rather than private data centers [27], and this percentage is expected to steadily increase for the foreseeable future.

Some application workloads are better suited to cloud computing than others. For example, workloads for batch processing or similar applications that require periods of work interspersed with long periods of inactivity are well suited to the dynamic nature of cloud computing. Applications which are expected to grow or scale dramatically over time are also well suited to cloud computing, since it is possible to quickly add data center resources and avoid processing bottlenecks which could hinder growth. Applications with traffic bursts are good candidates, for the same reason. Traffic bursts can either be predictable (high volume of online retail sales near major holidays), or unpredictable (natural disasters). In either case, the cloud provider service level agreement and quality of service guarantees must be considered carefully in order to take full advantage of the cloud data center's flexibility.

A key distinguishing feature of cloud computing architectures is that physical server and storage resources may be located anywhere in the world (subject to performance restrictions and legal requirements for certain types of data that must be stored within their country of origin). This allows many enterprises to employ duplicate or redundant resources at a remote site for failover and load balancing, an approach known as colocation. This approach has also been used by telecommunication providers, where they are known as carrier hotels, since many different telecom carriers will house their equipment in the same facility. In telecom applications, these facilities enable interconnection of carriers, known as telecom exchanges, and act as regional fiber hubs serving local businesses, in addition to hosting carrier equipment. Similar designs have been adopted by enterprise and cloud service providers, where the hotel analogy continues to be relevant (compute resources can be rented from a cloud service provider, just like rooms in a hotel, and many enterprises often share the same cloud data center). Colocation makes the data center less susceptible to downtime caused by natural or man-made disasters, since they can use redundant systems to continue critical operations. Colocation may also improve application performance through distributed workload balancing across sites, and due to their high traffic volumes may be a near term application for multiterabit optical communication networks.

Modern colocation centers (also called colos or colocs) provide rental space for housing data center equipment in standard 19-inch racks, including facilities such as power, cooling, fire protection, and physical security. The colo may be served by one or more network service providers, and carrier neutrality in this environment has been an important issue for many colo clients. The colo provider realizes economies of scale by renting space to several clients in a large facility (perhaps 50,000–100,000 feet^2). Colos must often comply with independent audits to meet industry standards and reliability levels. The most common standards for a colo include the tier system from the Uptime Institute discussed previously. Various other compliance standards deal with specific industries. For example, colos for financial or banking applications need to demonstrate compliance with Sarbanes – Oxley requirements; a key standard in this area is SSAE 16 SOC Type I and II (Statement on Standards for Attestation Engagements) [28], which supersedes the former SAS 70 regulations. This is a highly detailed and comprehensive standard, requiring a description of the data center environment and its various control systems. As another example, colos supporting health care providers must demonstrate compliance with the Health Insurance Portability and Accountability Act (HIPPA) [29].

1.6 Physical architecture

Many corporations, standards, bodies, and industry consortiums have proposed data center architectures. These include proposals such as the Open Data Center Alliance [30], the Open Compute Project [31], and the Open Data Center Interoperable Network (ODIN) [32]. While it is difficult to define a "typical" data center physical architecture, some of the most common features are illustrated in Fig. 1.7. This figure illustrates some of the common practices deployed in enterprise class networks and Fortune 1000 companies which own and maintain their own data centers, as well as some cloud computing providers. There are many variations on this approach, and the nature of data center architectures is currently changing in response to new applications such as social networks, mobile applications, and big data/analytics. We will discuss some of these variants in later sections of this chapter, but first we will cover the traditional multitier data center architecture.

Large data centers would not be possible without advanced networks interconnecting racks of servers and storage devices. These networks are based on the design of telecommunication or campus level Ethernet data networks, which form a hierarchical tree structure characterized by access, aggregation, and core layers. Data traffic is aggregated at each layer of this design, making it possible to scale the number of endpoints (servers or storage devices) by adding more switches at the lower layers. Lower layer switches tend to have fewer ports, use lower data rates, and are thus less expensive, while switches further up the hierarchy need to carry higher data rates and tend to be larger, more expensive devices. Some large networks may have three, four, or more tiers of switching. Data traffic flows from the endpoints up through the access, aggregation, and core tiers, traveling as far as required until a path is found to the desired destination endpoint. Traffic then flows back down the hierarchy until it reaches its destination. Since the basic switch technology at each tier is the same, the TCP/IP stack is usually processed multiple times at each successive tier of the network. To reduce cost and promote scaling, oversubscription is typically used for all tiers of the network. Layer 2 and 3 functions are generally separated within the access layer of the network. Services dedicated to each application (firewalls, load balancers, etc.) are placed in vertical silos dedicated to a group of application servers.

Due to concerns that early Ethernet LANs would be overwhelmed by large file transfers, some enterprises created dedicated storage area networks (SANs). Different protocols were developed for the SAN in order to provide functionality which was not present in Ethernet, such as guaranteed lossless delivery of data frames in the proper order. The Fibre Channel protocol [33], originally developed by ANSI, continues to be used today in many large enterprises, which rely on Fibre Channel switches and a dedicated storage infrastructure separate from the Ethernet network. Other special purpose networking protocols have been developed, with their own dedicated switch networks, such as InfiniBand [34], which was proposed by the InfiniBand Trade Association (IBTA) for high performance server clustering applications. Each of these protocols has its own physical layer, specifying cable/connector types, data rates, and optical transceiver form factors. Most of these networks are switch fabrics within a single data center, or provide limited connectivity across dedicated links between redundant data centers. Since Ethernet is more widely adopted, and may serve as a convergence point for other networks in the future, we will concentrate on Ethernet networking for the remainder of this chapter.

A typical Ethernet architecture places a rack-mounted access switch or router at the top of an equipment rack to handle connectivity between all the equipment inside the rack, and to connect the rack with nearby racks through an aggregation or core switch located elsewhere in the data center. This is called a top of rack (TOR) design, and the switch is often called a TOR switch, even though the switch may actually reside in the middle or bottom of the physical rack; even a blade form factor switch can be configured to act as a TOR. TORs are typically smaller switches (up to 64 ports) which serve as the access layer for the data center network. They can interconnect with aggregation or core layer switches, sometimes called end of row switches because of their position in the physical data center. A variation on this approach places the aggregation or core switches in the middle of several equipment racks (sometimes called middle of row or MOR) to reduce the longest path lengths to nearby servers.

As noted previously, many data center networks are based on well-established approaches developed for telecommunication networks and later transplanted into the data center or cloud computing environment. For example, telecom networks have historically faced the problems associated with building very large crossbar switches for the network core, where traffic density is highest. When physical circuit switching requirements exceeded the capacity of the largest practical switch, a multistage switch fabric was introduced. It was shown that the number of cross points required in a multistage fabric was significantly less than if the same fabric was implemented with a single large switch. By using at least three cascaded crossbar switches, or any odd number of switching stages, it was possible to build economical nonblocking switch fabrics (known as Clos networks, since they were first formalized by Charles Clos in 1952 [18]). Variations of this approach, including folded Clos designs, Banyan networks [35], and rearrangable nonblocking Benes networks [36], are still used today as the building blocks for large data communication networks. Many data centers will also employ a "fat tree" design, developed by Charles Leiserson of MIT [37], in which link bandwidth increases closer to the network core (see Fig. 1.8). This allows the network to make efficient use of any bandwidth links that are available, as opposed to other network topologies which are based on strict mathematical relationships and thus cannot be tailored easily to a specific application. Some applications of these principles are fairly specialized; e.g., all optical cross-connects have recently found applications in reducing overhead associated with persistent traffic flows in cloud computing data centers.

1.6.1 Layer 2 and Layer 3 network architectures

A network tree is subject to bottlenecks if links between layers are oversubscribed. An alternative approach is the leaf-spine architecture (Fig. 1.9), in which the access layer is formed by a series of leaf switches fully meshed to a series of spine

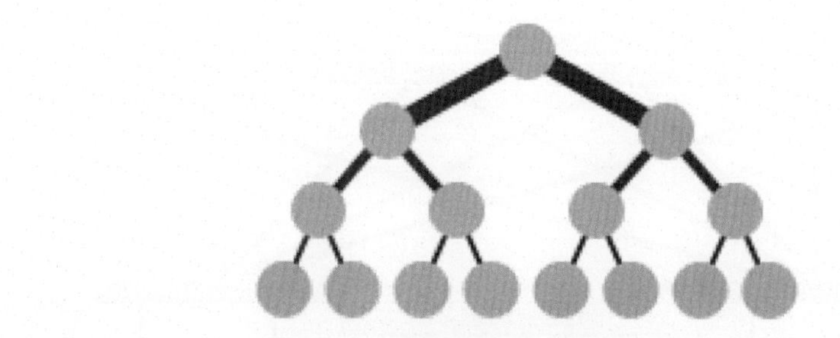

Figure 1.8 Example of a fat tree data network.

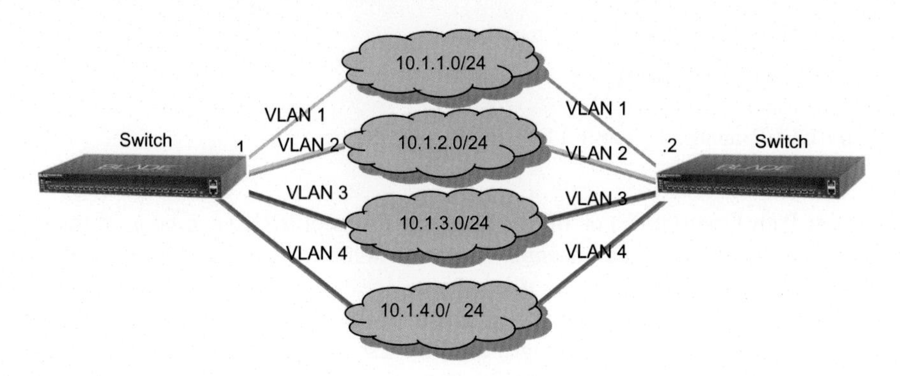

Figure 1.9 Example of a four-way Layer 3 ECMP design.

switches. The mesh ensures that all the leaf switches are no more than one hop away from each other, minimizing latency links within a leaf-spine can be either switches (Layer 2) or routed (Layer 3). Many data center network will employ a multitier architecture based on a Layer 3 fat tree design or Clos network using ECMP. As shown in Fig. 1.10, a Layer 3 ECMP design creates multiple load balanced paths between nodes. The number of paths is variable, and bandwidth can be adjusted by adding or removing paths up to the maximum allowed number of links. Unlike a Layer 2 STP network, no links are blocked with this approach. Broadcast loops are avoided by using different VLANs, and the network can route around link failures. Typically, all attached servers are dual homed (each server has two connections to the first network switch using active–active network interface card (NIC) teaming). Using a two tier design with a reasonably sized (48 port) leaf and spine switch, and relatively low oversubscription (say 4:1), it is possible to scale this network up to around 1000–2000 or more physical ports.

If devices attached to the network support Link Aggregation Control Protocol (LACP) it becomes possible to logically aggregate multiple connections to the same device under a common Virtual Link Aggregation Group (VLAG). It is also possible to use VLAG inter-switch links (ISLs) combined with Virtual Router Redundancy Protocol (VRRP) to interconnect switches at the same tier of the network. VRRP supports IP forwarding between subnets, and protocols such as Open

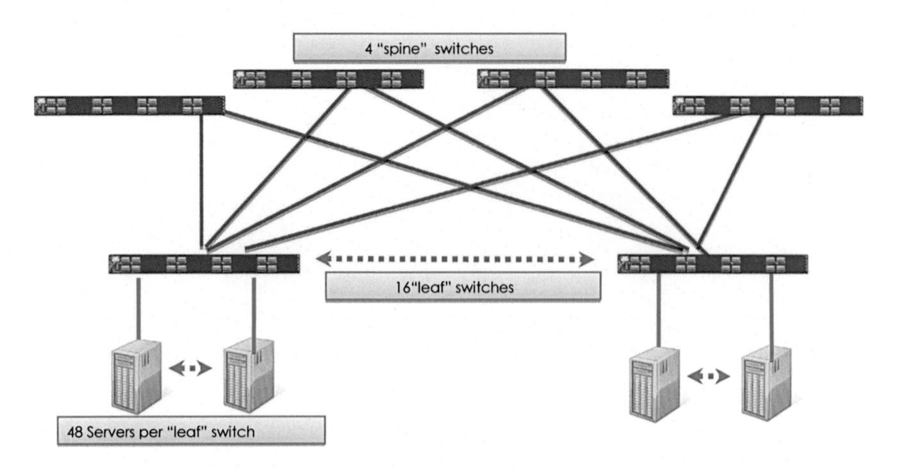

Figure 1.10 Example of a Layer 3 ECMP leaf-spine design.

Shortest Path First (OSPF) or Border Gateway Protocol (BGP) can be used to route around link failures. Virtual machine migration is limited to servers within a VLAG subnetwork.

Layer 3 ECMP designs offer several advantages. They are based on proven, standardized technology which leverages smaller, less expensive, rack or blade chassis switches (virtual soft switches typically do not provide Layer 3 functions and would not participate in an ECMP network). The control plane is distributed, and smaller isolated domains may be created.

There are also some tradeoffs when using a Layer 3 ECMP design. The native Layer 2 domains are relatively small, which limits the ability to perform live VM migrations from any server to any other server. Each individual domain must be managed as a separate entity. Such designs can be fairly complex, requiring expertise in IP routing to set up and manage the network, and presenting complications with multicast domains. Scaling is affected by the control plane, which can become unstable under some conditions (e.g., if all the servers attached to a leaf switch boot up at once, the switch's ability to process Address Resolution Protocol (ARP) and Dynamic Host Configuration Protocol (DHCP) relay requests will be a bottleneck in overall performance). In a Layer 3 design, the size of the ARP table supported by the switches can become a limiting factor in scaling the design, even if the media access control (MAC) address tables are quite large. Finally, complications may result from the use of different hashing algorithms on the spine and leaf switches.

New protocols are being proposed to address the limitations of a Layer 3 ECMP design, while at the same time overcoming the limitations of Layer 2 designs based on STP or MC-LAG. All of these approaches involve some implementation of multipath routing, which allows for a more flexible network topology than STP (see Fig. 1.11). In a Layer 2 leaf-spine, loop prevention protocols such as STP are often replaced by other protocols, such as transparent interconnect for lots of links (TRILL), shortest path bridging (SPB), or various vendor proprietary protocols

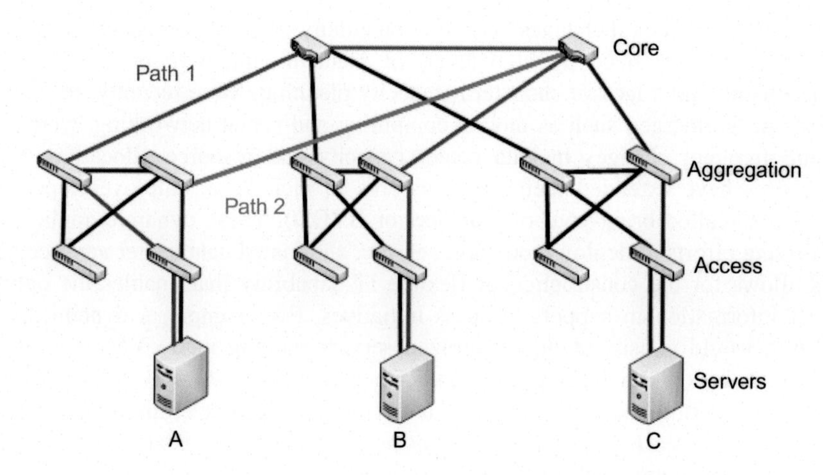

Figure 1.11 Example of multipath routing between three servers A, B, and C.

including Qfabric, Fabric Path, and VCS [2]. In a Layer 3 leaf-spine, OSPF protocols may be used for this purpose. There are also variations on the network physical architecture, including networks which compress the leaf-spine into a single network with an MOR switch (sometimes called a spline), a Layer 2 design with prevents blocking by using versions of the MLAG (multiple link aggregation) protocol, or even overlay networks which impose Layer 2 over Layer 3 using tunneling protocols such as VXLAN, NV-GRE, or NV03/DOVE [2]. Some of these variants are illustrated in Fig. 1.10.

1.7 Data center design considerations

The following sections describe important data center design requirements. There are many ways to implement the physical and virtual architecture within a data center, but a sound design should take the following aspects into consideration. Proper planning for the data center infrastructure is critical, including consideration of such factors as latency and performance, cost-effective scaling, resilience or high availability, rapid deployment of new resources, virtualization, and unified management. Many IT organizations are redesigning their existing data centers as they attempt to reduce cost (both capital and operating expense), while simultaneously implementing the ability to support an increasingly dynamic (and in some cases highly virtualized) environment.

1.7.1 Agility and elasticity

Conventional data centers were designed with the implicit assumption that most of their applications would not change much over time. Examples of static applications include payroll processing, which typically will not change very much from

1 week to the next, and changes very little on a daily or hourly basis. These applications were characterized by high levels of manually intensive provisioning and management, and a lack of short-term capacity planning. More recently, new applications have emerged such as mobile computing and social networking, which can require frequent changes in data center capacity and resource allocation. Some enterprises have extended their data networks to include an employee's personal devices (so-called bring-your-own-device, or BYOD). These dynamic applications are driving efforts to deploy more agile, elastic, automated data center architectures. This allows for the construction of flexible IT capability that enables the optimal use of information to support business initiatives. For example, a dynamic infrastructure would consist of highly utilized servers running many VMs per server, using high bandwidth links to communicate with virtual storage and virtual networks, both within and across multiple data centers. The accelerating pace of innovation in this area has also led to many new proposals for next generation data centers to be implemented within the next decade.

An agile data center architecture should ideally enable users to treat computing, storage, and network resources as fully fungible pools that can be dynamically and rapidly partitioned. While this concept of a federated data center is typically associated with cloud computing environments, it also has applications to enterprise data centers and other common use cases. Consider the concept of a multitenant cloud data center, in which the tenants represent clients sharing a common application space. This may include sharing data center resources among different divisions of a company (accounting, marketing, research), stock brokerages sharing a real-time trading engine, government researchers sharing VMs on a supercomputer, or clients sharing video streaming from a content provider. Some designs make an effort to decouple the application requirements from the infrastructure, in order to simplify the data center and improve efficiency. Cloud computing designs have implemented orchestration software to automatically provision data center resources based on application policies.

Specifically, an agile data center architecture should provide connectivity between all available data center resources with no apparent limitations due to the network. An ideal network would offer infinite bandwidth and zero latency, be available at all times, and be free to all users. Of course, in practice there will be certain unavoidable practical considerations; each port on the network has a fixed upper limit on bandwidth or line rate and minimal, nonzero transit latency, and there are a limited number of ports in the network for a given level of subscription. Still, a well-designed network will minimize these impacts or make appropriate tradeoffs between them in order to make the network a seamless, transparent part of the data processing environment.

1.7.2 Flattened, converged networks

Classic Ethernet networks are hierarchical, with three, four, or more tiers (such as the access, aggregation, and core switch layers). Each tier has specific design considerations, and data movement between these layers is known as multitiering. The movement of traffic between switches or routers is commonly referred to as "hops" in the network (there are actually more precise technical definitions of what constitutes a "hop" in different types of networks, which are beyond the scope of this chapter). In order for data to flow between racks of servers and storage, data traffic needs to travel up and down a logical tree structure. This adds latency and potentially creates congestion on ISLs. Network loops are prevented by using STP, which allows only one active path between any two switches. This means that ISL bandwidth is limited to a single logical connection, since multiple connections are prohibited. To overcome this, link aggregation groups (LAGs) were standardized, so that multiple links between switches could be treated as a single logical connection without forming loops. However, LAGs have their own limitations, e.g., they must be manually configured on each switch port.

Some data centers are seeking a flattened network that clusters a set of switches into one large (virtual) switch fabric. The definition of a fabric as opposed to a network is somewhat vague, although it is generally agreed that such interconnects offer at least some of the following distinguishing characteristics:

- Fabrics are typically flatter than existing networks, eliminating the need for STP while still providing backwards compatibility with existing Ethernet networks.
- Fabrics can be architected in any topology to better meet the needs of a variety of workloads.
- Fabrics use multiple least-cost paths for high performance and reliability, and are more elastic (scaling up or down as required). In principle, this provides improvements in performance, utilization, availability, and simplicity of design.
- More advanced fabrics are self-configuring and function as a single logical entity (or perhaps several redundant entities for high availability configurations), in which all switch elements are aware of each other and aware of all connected physical and logical devices. Management of a fabric can thus be domain-based rather than device-based, and defined by an overall fabric policy.

Fabrics are intended to improve VM mobility and automated network management, and lend themselves to a flattened, two tier design as shown in Fig. 1.12. Note that when blade server chassis are used instead of rack mounted servers, the switch access layer can reside on a switch blade in the same chassis as the servers. Topologically, a "flat" network architecture implies removing tiers from a traditional hierarchical data center network such that it collapses into a two tier network (access or TOR switches and core switches). Most architectures agree that a flat fabric implies that connected devices can communicate with each other without using an intermediate router. A flat network also implies creating larger Layer 2 domains (connectivity between such domains will still require some Layer 3 functionality). This flat connectivity simplifies the writing of applications since there is no need to worry about the performance hierarchy of communication paths inside the data center. Flat fabrics have been touted as a way to prevent resources in a data center from becoming stranded and not efficiently usable, although this is a common issue with all types of network designs and can also be addressed with more conventional networks. Some flat fabrics include elimination of STP and LAG. Replacing the STP protocol allows the network to support a fabric topology (tree, ring, mesh, or core/edge) while avoiding ISL bottlenecks, since more ISLs

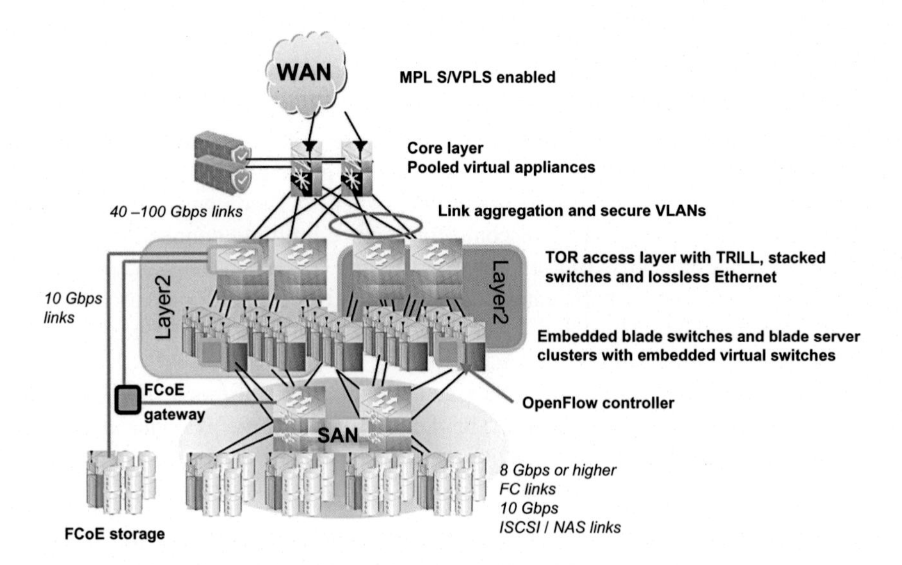

Figure 1.12 Two tier data center network architecture.
Source: From Handbook of Fiber Optic Data Communication, Chapter 1, Figure 1.2.

become active as traffic volume grows. Self-aggregating ISL connections replace manually configured LAGs.

As noted previously, data centers have long sought the means to converge multiple protocol networks (Ethernet, Fibre Channel, Infiniband) into a single unified network. Recently, Ethernet has emerged as a possible candidate for network convergence, due to its low cost and widespread adoption. Enhancements to the Ethernet protocol have been standardized in an effort to provide features supported by other protocols, such as guaranteed packet delivery. This approach is known as Lossless Ethernet [38], or Converged Enhanced Ethernet (CEE). Traditional Ethernet is a lossy protocol; i.e., data frames can be dropped or delivered out of order during normal operation (this is handled by upper layer protocols such as TCP/IP). In an effort to improve the performance of Ethernet, the IEEE has developed a new standard known as Lossless Ethernet or Data Center Bridging (DCB). Historically, this work grew out of a series of proposals made by a consortium called the Converged Enhanced Ethernet (CEE) Author's Group; for this reason, the resulting standard is also sometimes known as CEE. There are three key components which are required to implement lossless Ethernet, namely Priority-based Flow Control (PFC), Enhanced Transmission Selection (ETS), and Data Center Bridging Exchange protocol (DCBx). A fourth, optional component of the standard is congestion notification (QCN).

Incremental steps towards network convergence include migration from Fibre Channel to Fibre Channel over Ethernet (FCoE), and the adoption of RDMA over converged Ethernet standards (RoCE) for high performance, low latency clustering applications. Despite significant interest in this area, available data suggests that network convergence will take time, and alternate protocol networks will likely persist for the foreseeable future. As Ethernet data rates increase, this technology holds the potential to disrupt various aspects of the data center design. For example, data rates of 10 Gbit/second are sufficient to encourage the convergence of storage networking with Ethernet. Data rates of 40 Gbit/second and above will begin to disrupt the server clustering market and potentially impact technologies such as InfiniBand. Eventually, data rates in the 100–400 Gbit/second range or higher (available in perhaps 3–5 years) may disrupt the fundamental server and I/O market dynamics.

Flattening and converging the network reduces capital expense through the elimination of dedicated storage, cluster and management adapters, and their associated switches, and the elimination of traditional networking tiers. Operating expense is also reduced through management simplification by enabling a single console to manage the resulting converged fabric. Note that as a practical consideration, storage traffic should not be significantly over-subscribed, in contrast to conventional Ethernet design practices. The use of line rate, nonblocking switches is also important in a converged storage network, as well as providing a forward migration path for legacy storage. Converging and flattening the network also leads to simplified physical network management. While conventional data centers use several tools to manage their server, storage, network, and hypervisor elements, best practices in the future will provide a common management architecture that streamlines the discovery, management, provisioning, change/configuration management, problem resolution and reporting of servers, networking, and storage resources across the enterprise. Converged and flattened data centers may require new switch and routing architectures to increase the capacity, resiliency, and scalability of very large Layer 2 network domains.

1.7.3 Virtualization and latency

In a data center, virtualization (or logical partitioning) refers to the number of VMs which are supported in a given data center environment, but it also has important implications on the network. All elements of the network can be virtualized, including server hypervisors, NICs, converged network adapters, switches, and routers. Furthermore, virtual Ethernet Layer 2 and 3 switches with a common control plane can be implemented to support multitier applications. Full mobility of VMs from any interface to any other interface should be supported without compromising any of the other network properties.

Latency refers to the total end-to-end delay within the network, due to the combination of time-of-flight and processing delays within the network adapters and switches. For some applications, such as HPC, both the magnitude and consistency of the latency are important (consistency, or variation in packet arrival times, is also known as jitter). Low latency is critical to high performance, especially for modern applications where the ratio of communication to computation is relatively high compared to legacy applications. For example financial applications are especially sensitive to latency; a difference of microseconds or less can mean millions of dollars in lost revenue. High latency translates directly to lower performance because applications stall or idle when they are waiting for a response over the network. Further, some types of network traffic are particularly sensitive to latency, including VM migration and synchronous storage. Consistency of network latency is affected by many factors, including the number of switch chips and internal design of TOR and core switches.

Today there is a tradeoff between virtualization and latency, so that applications with very low latency requirements typically do not make use of virtualization. Some applications might prefer a virtual Ethernet bridge (VEB) or virtual switch (vSwitch) model, with VMs located on the same server. In the long term, increased speeds of multicore processors and better software are expected to reduce the latency overhead associated with virtualization, making this tradeoff less important.

While latency sensitive applications will continue to keep VMs on the same physical platform as discussed earlier, there are other business applications which are interested in dynamically moving VMs and data to under-utilized servers. In the past, physical servers typically hosted static workloads, which required a minimal amount of network state management (e.g., creation/movement of VLAN IDs, ACLs, etc.). Modern virtualized systems host an increasing number of VMs, and clients demand the ability to automate the management of network state associated with these VMs. Tactical solutions use a Layer 2 virtual switch in the server, preferably with a standards-based approach for coordinating switch and hypervisor network state management. A more disruptive approach is emerging in which flow control is handled through a separate fabric controller, thus reducing and simplifying physical network management.

VM migration is currently limited on many switch platforms by low bandwidth links, Layer 2 adjacency requirements, and manual network state virtualization. The IEEE 802.1Qbg standard [39] is used to facilitate virtual environment partitioning and VM mobility. The state of the network and storage attributes must be enabled to move with the VMs. This addresses the previous concerns with high capital and operating expense by balancing the processor workload across a wide number of servers. This approach also reduces the amount of data associated with a VM through VM delta-cloning technologies, which allow clients to create VM clones from a base VM and the base VM's data/files, thereby minimizing the amount of data that is created for each VM and which moves when the VM moves. This type of automatic VM migration capability requires coordination and linkages between management tools, hypervisor, server, storage, and network resources. It also requires high bandwidth links to quickly create and deploy VMs and their associated data, and to move those VMs from highly utilized servers to under-utilized servers (a form of load balancing). This environment will require service management tools that can extend all the way to the hardware layer to manage IT and network resources. Virtualization should also be automated and optimized using a common set of standards-based tools.

1.7.4 Scalability

The number of interfaces at the network edge, N, is typically defined as the scale of the network (although scale may be limited by many thresholds other than physical ports, including the number of VLANs, MAC addresses, or ARP requests). The fundamental property of scalability is defined as the ability to maintain a set of defining characteristics as the network grows in size from small values of N to large values of N. For example, one defining characteristic may be the cost per port of the network. While it is certainly possible to scale any network to very large sizes, this requires a brute force approach of adding an increasing number of network ports, aggregation switches, and core switches (with the associated latency and performance issues). A well designed network will scale to large numbers of ports while controlling the cost per port; this is the concept of cost-efficient scaling. It should be apparent that there are other defining characteristics of the network which are affected by scalability, including performance and reliability. In this way, scalability is related to many of the other fundamental network properties. It is desirable for networks to scale linearly or sublinearly with respect to these characteristics; however, in many traditional networks, power and cost scale much faster than this. Scalability can be facilitated by designing the network with a set of modular hardware and software components; ideally this permits increasing or decreasing the network scale while traffic is running (sometimes called dynamic scalability).

Classic Ethernet allows only a single logical path between switches, which must be manually configured in the case of LAGs. As the fabric scales and new switches are added, it becomes increasingly more complex to manually configure multiple LAG connections. Ethernet fabrics overcome this limitation by automatically detecting when a new switch is added, and learning about all other switches and devices connected to the fabric. Logical ISLs can be formed which consist of multiple physical links (sometimes called VLAN aggregation or trunking) to provide sufficient bandwidth. Traffic within a trunk may be load balanced so that if one link is disabled, traffic on the remaining links is unaffected and incoming data is redistributed on the remaining links.

Scalability is a prerequisite for achieving better performance and economics in data center networks. Many modern data centers are preparing to deploy an order of magnitude more server infrastructure than they had considered only a few years ago. At the same time, many organizations are striving to improve the utilization of their IT infrastructure and reduce inefficiencies. This implies avoiding large disruptions in the existing network infrastructure where possible. Server rack density and power density per rack are a few of the common metrics applied in this context. Economies of scale (both capex and opex) apply to the network infrastructure; small-scale data centers cannot be made as efficient as large-scale ones, provided that resources are kept fully fungible. For most applications the raw performance of a set of tightly coupled computing elements in a single large data center is significantly better than the collective performance of these same elements distributed over a number of smaller data centers. This performance difference has to do with the inherently lower latency and higher bandwidth of inter-processor communication in a single data center.

1.7.5 Network subscription level

The difference between the input and output bandwidth for each layer of switching in the network (or the difference between the number of downlinks and uplinks from a switch) is defined as the subscription rate. This concept was first applied to conventional multitier networks with significant amounts of north – south traffic. In a fully subscribed network, each switch (or layer of switching) will have the same bandwidth provisioned for downlinks (to servers, storage, or other switches below it in the network hierarchy) and for uplinks (to switches above it in the network hierarchy). An oversubscribed switch will have more downlink bandwidth than uplink bandwidth, and an undersubscribed switch will have more uplink bandwidth than downlink bandwidth. Oversubscription is commonly used to take advantage of network traffic patterns which are shared intermittently across multiple servers or storage devices. It is a cost-effective way to attach more devices to the network, provided that the application can tolerate the risk of losing packets or occasionally having the network unavailable; this risk gets progressively worse as the level of oversubscription is increased. Note that in networks with a large amount of east – west traffic, oversubscription can still impact multiple switches on the same tier of the network. For conventional Ethernet applications, oversubscription of the network at all layers is common practice; care must be taken in migrating to a converged storage network, in which certain applications (such as Fibre Channel or FCoE storage) will not tolerate dropped or out of order packets. Oversubscription is not recommended at the server access switch due to the nature of traffic patterns in traditional Ethernet networks. A non-oversubscribed network is recommended for converged storage networking using FCoE.

Traditionally, enterprise data center networks were designed with enough raw bandwidth to meet peak traffic requirements, which left the networks over-provisioned at lower traffic levels. In many cases, this meant that the network was over-provisioned most of the time. This approach provided an acceptable user experience, but it does not scale in a cost-effective manner. New approaches to network subscription levels must be considered. Further, with the introduction of new applications, it is becoming increasingly difficult to predict traffic patterns or future bandwidth consumption.

Scalability and subscription rate collectively impact the network's ability to transmit data with no restrictions or preplanning, sometimes known as any-to-any connectivity. This includes the ability to absorb rapid changes to the rate of transmission or to the number of active senders and receivers. The network's ability to equally share its full available bandwidth across all contending interfaces is known as fairness. The network should also be nonblocking, which means that the only apparent congestion is due to the limited bandwidth of ingress and egress interfaces and any congestion of egress interfaces does not affect ingress interfaces sending to noncongested interfaces. An ideal network provides any-to-any connectivity, fairness, and nonblocking behavior (this is very expensive to achieve in practice, and most network architects have concluded this is not strictly necessary to achieve good performance). Further, when carrying storage traffic, the network should not drop packets under congestion (which occurs when the instantaneous rate of packets coming in exceeds the instantaneous rate at which they are going out). To achieve this, the network should throttle the sending data rate in some fashion when network interfaces are nearing congestion. The ability to not drop packets when congestion occurs is critical to efficiently transporting "bursty" server-to-disk traffic. Applications assume that reads and writes to disk will succeed. Packet drops due to congestion break this assumption and force the application to handle packet loss as an error, resulting in a drastic reduction in performance or availability given the relative frequency of congestion events, if not in the failure of the application or the operating system. These considerations are important when flattening the data center and pooling the computing and storage resources, as well as supporting multitenancy and multiple applications within a tenant.

1.7.6 Availability and reliability

Although ensuring that the network is available for use at all times seems fairly self-explanatory, this property is closely related to other attributes such as redundancy, reliability, and serviceability of network components. Designing a network from modular components which are distributed and federated can provide high levels of availability; it can also promote scalability, as noted earlier.

Reliability can also be enhanced through simplicity of design, while high availability is often achieved by duplicating network components. A modular implementation where the modules are kept independent through the use of physical or logical separation means that failures in either the hardware or software are unable to compromise the entire system. The same principles apply to network management complexity, which should be kept as simple as possible. One of the challenges associated with the redesign of data center networks is that a combination of server consolidation, virtualization, and storage convergence tends to create systems which encompass a high level of functionality in a single physical package, which becomes a potential single point of failure.

One approach to increasing the availability of a data center network is to use a combination of redundant subsystems including a combination of link- and device-level redundancy (TOR and core switches in conjunction with redundant network designs that feature multiple links between devices), as is typically done today in a high availability enterprise data center. Redundancy and nondisruptive replacement of switch subsystems such as power supplies and cooling elements is expected in a high availability network device. Converged networks may also reduce the complexity of the infrastructure. The network should be designed where possible to automatically recover from switch and link failures, avoiding perceptible service interruptions. It typically requires significant effort to create solutions that are as resilient as possible while maintaining the performance, flexibility, and scalability expected by users. Additionally both business users and consumers have increasing 24/7/365 "always on" service level expectations that are also being driven by extended supply chains and increasing globalization (end users spanning multiple time zones). These reduce the opportunity to negotiate planned data center outages for platform upgrades, making network resilience, redundancy, quality of service, and concurrent upgrade capabilities increasingly critical.

Data centers should have a business continuity plan which includes policies for backup and recovery of data to a remote location in the event of a natural or man-made disaster. While data backup can be as straightforward as loading physical media onto a truck and driving to a remote location, modern data centers use either the Internet or dedicated private networks for backup to redundant data centers or the cloud. Long distance server connectivity may use optical fiber and wavelength division multiplexing (WDM) systems connected to remote systems, anywhere from a few miles away to the other side of the world. Of course, unlike relatively low bandwidth telephone signals, performance of many data communication protocols begins to suffer with increasing latency (the time delay incurred to complete transfer of data from storage to the processor). While it is easy to place a long distance phone call from New York to San Francisco (about 42 ms round trip latency in a straight line, longer for a more realistic route) it is impossible to run a synchronous computer architecture over such distances. Further compounding the problem, some data communication protocols were never designed to work efficiently over long distances. They required the computer to send overhead messages to perform functions such as initializing the communication path, verifying it was secure, and confirming error-free transmission for every byte of data. This meant that perhaps a half dozen control messages had to pass back and forth between the computer and storage unit for every block of data, while the computer processor sat idle. The performance of any duplex data link begins to fall off when the time required for the optical signal to make one round trip equals the time required to transmit all the data in the transceiver memory buffer. Beyond this point, the attached processors and storage need to wait for the arrival of data in transit on the link, and this latency reduces the overall system performance and the effective data rate. Efforts to design lower latency networks and protocols with less overhead have led to some improvements in this area.

The design of redundant data handling systems is a large and complex topic; this chapter will only briefly comment on some highlights in this field. There have been several proposed models for data consistency, both within data centers and between multiple data centers. The more lenient this requirement, i.e., different end users can be returned different versions of the same data (e.g., a stock price or a web page), then the easier it is to implement an availability solution. There are several different taxonomies for data consistency. At one extreme, representing the minimal acceptable requirements for data consistency, is an approach called BASE (Basically Available, Soft State, Eventual Consistency) proposed by Fox et al. [40]. At the other extreme is a much stricter set of requirements for reliable processing of database requests (called transactions), which was first developed by J. Gray and later became known by the acronym ACID (Atomicity, Consistency, Isolation, Durability) [41].

BASE availability solutions provide the following key features:

1. Basically Available: Data returned to applications can be stale if a system is not up.
2. Soft State: Data can be lost if a server fails.
3. Eventual Consistency: The data returned to an application may not have had the most recent changes applied to it.

On the other hand, ACID availability solutions provide key features including:

1. Atomicity: If one part of a transaction fails, then the entire transaction fails and the data is left unchanged. An atomic system must guarantee atomicity in each and every situation, including power failures, errors, and crashes.

2. Consistency: The system will ensure that data changed by a transaction is left in a consistent state, across all replicas of that data during applications access.
3. Isolation: Concurrent changes of the data by separate transactions will result in changes to the data that appear serially ordered.
4. Durability: Once a transaction commits any changes to data, the change is permanent, even if the system crashes immediately after the transaction commits.

Data networking equipment needs to support the infrastructure required for both ACID and BASE methodologies, as well as a spectrum of different alternatives between these two extremes.

Wide area networks (WANs) are used to interconnect multiple data centers for workload sharing, business continuity, and backup/recovery applications. In recent years WAN traffic volumes have been increasing by about 30% per year, and traffic volume is expected to increase even further with the advent of larger cloud data centers. Modern highly virtualized data centers may require hundreds of virtual servers on a single domain; if this is extended over distance, it would require a huge amount of WAN bandwidth (otherwise, it might take a very long time to move a VM and its associated data). Higher data rates on the WAN are related to correspondingly higher aggregate data rates within the data center. Multisite connectivity can be implemented in a number of ways. Leased line data services are available from service providers which include options for private management of point-to-point networks (known as private circuits or Layer 2 VPN) or full mesh connectivity (Layer 3 VPN). In areas where leased optical fiber (or "dark fiber") is available, it is often cost-effective for larger enterprises to use dedicated optical WDM solutions. For storage applications, dark fiber WDM solutions are currently preferred, and may be supplemented with Fibre Channel over IP (FC-IP) solutions [42].

1.7.7 Network security

Security for data centers and networks is an extensive topic, well beyond the scope of this chapter. We will mention only a few key considerations in this area, and encourage the reader to review supplemental references on this topic [43]. Security policies not only protect the client, but also ensure compliance with industry and government regulations. Best practices include dividing the network into security zones ranging from trusted (critical information within the organization) to untrusted (not supervised by the organization). In many environments, front end and back end traffic must be physically separated. In a converged environment this can be done logically but not physically, and thus would be inappropriate for many sensitive applications. In multitenant cloud data centers, the concept of a demilitarized zone (DMZ) between secure and untrusted assets is breaking down, as the cloud becomes the new network perimeter. This has led to significant security research, including concepts such as micro-segmentation and zero-trust fabrics [43]. Built-in network access control (NAC) and intrusion detection and prevention systems (IDS/IPS) offer additional layers of protection that supplement existing endpoint and server security solutions. LAN security should be based on MAC address, not switch port, providing specific endpoint and user policy enforcement. Post admission control should be based upon Access Control Lists, not simply VLAN assignment, providing fine-grained controls. Note that modern switches are not susceptible to VLAN-hopping techniques, so it is considered secure to provide separate VLANs for isolation of network traffic.

There are potential security exposures between the aggregation and core layers of a conventional network. Not only may perimeter protection be insufficient, but also it may not be enough to guard against problems associated with cross-site scripting, buffer overflows, Spear Phishing, DDoS, URL tampering, or SQL injection (to name only a few known issues). Further, there is always the possibility of human error in configuring the network security policy, leaving network assets vulnerable to a security breach. Emerging standards related to software-based flow control in network devices can help reduce some of these risks.

As the number of different types of malicious attacks on the network increases (including viruses, phishing, spam, spyware, malware, denial of service, hacking, and more) the network architecture can no longer afford to focus on protection against a single type of threat. This goes beyond conventional network monitoring, authentication, and data logging features. Network security should be part of a unified threat management policy, which addresses a wide range of incursions from both inside and outside the data center (securing client to server transactions also limits a client's liability for attacks which originate from within their IP address space). This may include intrusion prevention systems, which use a variety of techniques including firewalls, access control lists, data encryption, deep packet inspection, application level stateful inspection, anomaly detection, zoning of storage networks, segregated VLANs, web filtering, and more. Traffic policing on the control plane helps prevent certain attacks such as denial of service. Layer 2 authentication of users, endpoint integrity, and network information is part of this approach. Further, the network architecture should be designed to prevent rogue switches from being attached to a trusted network (either accidentally or maliciously). This should be part of a unified access control policy which is designed to restrict access to sensitive resources or protect the data center infrastructure. Consideration should be given to logical segmentation of the network (how many networks share the same LAN).

In a long distance network, WDM systems can be used to isolate traffic to a subset of wavelengths, detect intrusion at the physical layer, and provide additional data in-flight encryption. Remote connectivity solutions also require IPSec VPN tunnels with sufficient capacity for simultaneous tunnels. Encrypted throughput including 3DES, AES, key exchange, and user authentication options (Layer 2 tunneling protocol) are important features in this application. Other WAN networking options will also be discussed later in this book, including encapsulation of storage data into IP networks [21].

1.8 Next generation data center architectures

As noted previously, current data center network architectures have evolved from classic telecom or campus networks. However, this approach was never intended to meet the traffic patterns or performance requirements of a modern data center. For example, the network consists of many tiers, where each layer duplicates many of the IP/Ethernet packet analysis and forwarding functions. This adds cumulative end-to-end latency (each network tier can contribute anywhere from 2 to 25 μs) and requires significant amounts of processing and memory. Oversubscription, in an effort to reduce latency and promote cost-effective scaling, can lead to lost data and is not suitable for storage traffic, which cannot tolerate missing or out-of-order data frames. Further, most of the traffic in modern data centers flows between servers which are on the same architectural tier (so-called east – west traffic). Classic tree networks are designed for traffic patterns which flow predominantly north – south, with all traffic eventually flowing through the network core. This means that excess latency is incurred whenever two servers on the same tier are communicating. For this reason, some designers have advocated a flatter network design.

Conventional networks do not scale in a cost-effective or performance effective manner. Scaling requires adding more tiers to the network, more physical switches, and more physical service appliances. Management functions also do not scale well, and IPv4 addresses may become exhausted as the network grows. Network topologies based on STP can be restrictive for modern applications, and may prevent full utilization of the available network bandwidth. The physical network must be manually rewired to handle changes in the application workloads, and the need to manually configure features such as security access makes these processes prone to operator error. Further, conventional networks are not optimized for new features and functions. There are unique problems associated with network virtualization (significantly more servers can be dynamically created, modified, or destroyed, which is difficult to manage with existing tools). Conventional networks do not easily provide for VM migration (which would promote high availability and better server utilization), nor do they provide for cloud computing applications such as multitenancy within the data center.

Attempting to redesign a data center network with larger Layer 2 domains (in an effort to facilitate VM mobility) can lead to various problems, including the well-known "traffic trombone" effect. This term describes a situation in which data traffic is forced to traverse the network core and back again, similar to the movement of a slide trombone, resulting in increased latency and lower performance. In some cases, traffic may have to traverse the network core multiple times, or over extended distances, further worsening the effect. In a conventional data center with small Layer 2 domains in the access layer and a core IP network, north – south traffic will be bridged across a Layer 2 subnet between the access and core layers, and the core traffic will be routed east – west, so that packets normally traverse the core only once. In some more modern network designs, a single Layer 2 domain is stretched across the network core, so the first-hop router may be far away from the host sending the packet. In this case, the packet travels across the core before reaching the first-hop router, then back again, increasing the latency as well as the east – west traffic load. If Layer 3 forwarding is implemented using VMs, packets may have to traverse the network core multiple times before reaching their destination. Thus, inter-VLAN traffic flows with stretched or overlapping Layer 2 domains can experience performance degradation due to this effect. Further, if Layer 2 domains are extended across multiple data centers and the network is not properly designed, traffic flows between sources and destinations in the same data center may have to travel across the long distance link between data centers multiple times.

Installation and maintenance of this physical compute model requires both high capital expense and high operating expense. The high capital expense is due to the large number of under-utilized servers and multiple interconnect networks. Capital expense is also driven by multitier IP networks, and the use of multiple networks for storage, IP, and other applications. High operational expense is driven by high maintenance and energy consumption of poorly utilized servers, high levels of manual network and systems administration, and the use of many different management tools for different parts of the data center. As a result, the management tasks have been focused on maintaining the infrastructure, and not on enhancing the services that are provided by the infrastructure to add business value.

Virtualization and abstraction have played a role in server and data center architectures since the use of virtual logical partitions on mainframe servers in the 1950s. However, in order to manage a server, storage device, or network equipment, it has been necessary to use a control interface hosted on each separate device (which was often vendor specific and interoperated poorly with equipment from other vendors). This often led to a distributed management system, particularly for networking devices which were not virtualized to the same degree as servers or storage. Distributed management architectures offer distinct advantages, such as robustness, the Internet and public telephone system are good examples of a distributed system which can reroute operations around a failed part of the network. This approach also suffers from some disadvantages, not unlike the distinction between peer-to-peer and client-server architectures discussed previously. A distributed architecture takes longer to provision new features and services, since each device must be managed separately. It is also difficult to implement end-to-end service quality and security, since there is no central point of control which has visibility to the entire data path. Some designers have attempted to address this issue by standardizing "affinities", or software constructs that link application requirements with infrastructure provisioning. While progress has been made in this area, implementing affinities benefits from centralized management and control architectures, as opposed to the traditional distributed network architectures widely used today.

Recent data center designs have begun to realize the benefits of separating the device control and management plane from the data handling plane, and centralizing management functions for the entire data center into a controller. A prominent example of this has been software defined networking (SDN), in which networking equipment is managed by an SDN controller (software residing on a server that is connected with each network device). The flow tables which control data routing are no longer stored in each device on the network; instead, a single copy of the routing information resides in the SDN controller. An industry standard protocol (such as OpenFlow) forms a control plane between the SDN controller and all the network devices. The controller also contains rules describing what actions should be taken when certain types of data packets flow through the network. In this way, it is possible to create policies which manage common network conditions, and to automate many aspects of network management. The physical infrastructure can be wired once, then dynamically reconfigured using a software-defined control plane. Similar benefits can be realized by using centralized provisioning of server and storage resources. The combination of these management efforts results in a software defined data center (SDDC) architecture.

One example of the SDDC is shown in the software stack of Fig. 1.13 [44]. Physical and virtual data center resources at the bottom of the stack are managed through REST APIs; a popular example shown here is OpenStack [53], an open source middleware project [4]. OpenStack provides southbound APIs with different names to manage and control servers (Nova), storage (Cinder), and networking (Neutron) resources. Other APIs provide a management dashboard and handle data center policy or user authentication services. The northbound APIs from OpenStack can interface with application aware orchestration software, which uses service profiles to translate application requirements into data center provisioning. For example, an application such as financial trading might require low latency (fewer hops to cross the network), high security (virtual firewalls and encryption for all data in flight and at rest), and high performance processors to compute analytics which respond to changing market conditions in real time. These requirements would be composed onto pools of available data center resources, along with performance, service, and reliability guarantees. Different service profiles corresponding to different applications can be customized to adjust the control of data center resources as needed. While this approach is still emerging and much work remains to be done, SDDC represents a significant shift in the way future data centers may be designed.

Another emerging architectural trend is the so-called disaggregated data center. Conventional data center architectures use servers as building blocks: processors, memory, and storage are organized and prepackaged into a server hardware

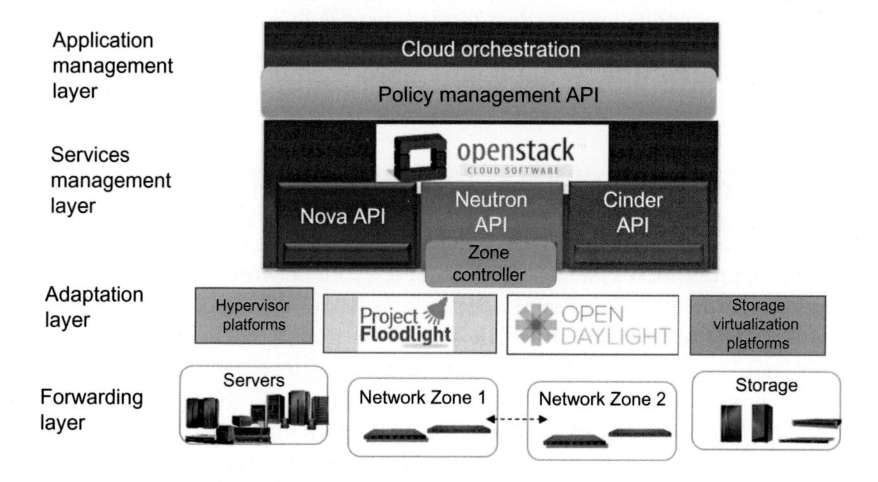

Figure 1.13 Example software stack for a software defined data center.

platform. However, if an improved processor or memory becomes available, the entire server must be replaced in order to take advantage of this improvement. Since software development takes place much more rapidly than hardware development, it is challenging to predict future hardware requirements. Alternatively, a disaggregated architecture makes it possible to upgrade individual components without disturbing the rest of the system by packaging processors, memory, and other key components onto modular hardware units, interconnected by a high-speed electrical or optical interconnect. By disaggregating or reaggregating the server in this manner, the processor, memory, and other resources are treated as pools of fungible resources that can be reconfigured as needed. The hardware is decoupled from the software, which in turn is agnostic to the underlying optical switching elements. In principle, this approach frees up capacity that is stranded on the server boards and results in a more modular, flexible architecture. This is an extension of the workload optimization approaches being used in many data centers. By creating subsystems that can be independently upgraded, this approach may promote open innovation from multiple third-party vendors. This approach also provides the option of placing low cost, efficient photonic devices closer to the processor chips. Such approaches are in the early stages of development by consortia such as the Open Compute Project.

Emerging applications may also have a significant impact on data center architectures. The so-called Internet of Things (IoT) has received considerable attention lately. Some analyses project that within the next several years, the number of devices connected to the Internet will grow to over 50 billion [45]. Such a drastic increase in connected devices may drive the adoption of multitier networks, rather than flattened network fabrics. Further, the traffic pattern for the IoT will drive high volumes of data from large numbers of devices connected at the network edge, which is very different from conventional networks that have concentrated high volume traffic in the core among relatively few devices. These are only some of the challenges which the IoT will impose on next generation data center designs; we can expect other applications to drive further changes to next generation data center architectures.

1.9 Optical interconnects for data centers

Optical networks for data center environments have unique requirements compared with telecommunication systems. Traditional telecom networks invest more heavily in the link end points to maximize spectral efficiency of long distance links, in an environment where fiber resources are scarce. In a data center, fiber is relatively plentiful and low cost; spectral efficiency of the optoelectronics tends to be traded for lower power, cost effective transceivers, which leads to networks with low latency and high path diversity. Data networks have other unique requirements, including lower bit error rates (10e-12 to 10e-15 compared with 10e-9 for telecom networks), higher use of multimode fiber for limited distances (and many more transceivers per kilo meter of fiber distance), interoperability for many different brands of networking equipment, and limitations of class 1 laser eye safety [46].

The industry's first roadmaps for optical communication in scale-out data centers based on quantitative, system level metrics were created by the Optoelectronics Industry Association (OIDA) [47]. In collaboration with the US National Science Foundation (NSF), the Optical Society of America (OSA), and the Center for Integrated Access Networks (CIAN), the OIDA helps formulate industry-wide and government research requirements and standard setting activities. The current draft version of key OIDA metrics and roadmaps is summarized in Fig. 1.14.

There has also been considerable discussion about whether these metrics are achievable, and in what time frame. For example, cost per gigabit for optical components is heavily influenced by component volumes; recent trends such as lossless Ethernet disrupting storage and RDMA applications, and the significant growth of warehouse-scale data centers, appear to be driving increasing volumes of Ethernet components. Network virtualization is also expected to drive higher volumes of networking equipment, in much the same way that server virtualization drove increased server volumes in accordance with Jevon's Law [48]. However, virtualizing a network (making many independent, physically distributed switches behave as if they were one large, logical switch) presents unique challenges compared with virtualizing a server or storage device (making a single physical device behave as if it was many independent logical devices). Network complexity may prevent cost per bit from decreasing as the network scales. As networks become larger, the

Metric	2012	2017	2022
Link speed (interconnect)	10-100 Gbps	100 Gbps to 1 Tbps	250 Gbps to 2.5 Tbps
Cost per bandwidth (interconnect)	1-5 $/Gbps	0.50 $/Gbps	0.025 to 0.15 $/Gbps
Energy per bit (rack-rack or greater) 1 pJ/bit = 1 mW/Gbps	20-100 pJ/bit	10-50 pJ/bit	10-20 pJ/bit
Energy per bit (chip/board level link) 1 pJ/bit = 1 mW/Gbps	2-10 pJ/bit	1-5 pJ/bit	1-2 pJ/bit
Optical switch port count Ref: CIAN	200x200	400x400	1000x1000
Optical switching speed (fiber & wavelength) Ref: CIAN	10 msec (circuit reconfiguration)	100 microsecond (per-connection reconfiguration)	100 ps (packet-scale reconfiguration)

Figure 1.14 Example of 2014 OIDA proposed data center metrics.
Source: OIDA example; table created with data taken from OSA Industry Development Associates (OIDA) *Roadmap Report: Photonics for Disaggregated Data Centers* (2015).

complexity of pairwise interactions and any-to-any communication requires more advanced management, forecasting, and control mechanisms. Without these mechanisms, it becomes necessary to over-provision the network in an effort to handle worst-case traffic patterns without congestion or packet loss. Enhanced routing mechanisms also need to enable scheduling of network resources and flow optimization based on application requirements. Some of these features may be addressed by SDN and related network virtualization techniques, thus making it possible to achieve larger scale and lower costs, as well as addressing the faster than Moore's Law growth of data center traffic. Future work in this area will include further interpretation and refinement of these metrics, product commercialization based on the metrics, and reference models based on test beds such as the CIAN environment [49].

Optical interconnects for data centers are preferably based on open industry standards and multisource agreements (MSAs), which lower the investment cost and risk for optical component suppliers and promote interoperability at the physical and link layer. Standardization also encourages future-proofing of the network and helps promote buying confidence. Some examples of MSAs for optical interconnects include the Quad Small Form Factor Pluggable (QSFP), C Form Factor Pluggable (the "C" stands for the Latin numeral centum, used to express the number 100), parallel optical interconnects (typically implemented as multiples of 12 fibers, which is the standard size for multifiber connectors such as the multifiber push-on (MPO) connector), and others.

Active optical cables function like a conventional transceiver/cable pair, but without a detachable optical cable. Available in several different form factors, the optoelectronics are integrated with the cable endpoints and permanently attached to the fiber cable, so that the cable connectors are electrical while the transmission media is optical. There are manufacturing and testing cost savings associated with this form factor; since both ends of the optical link are always associated with the same transceiver/fiber pairing, it becomes possible to select these components to optimize the end-to-end optical link. Active optical cables have the potential to reach cost/gigabit levels sufficient to displace copper cables for short distances (tens of meters) at data rates of 10–40 Gbit/second and higher.

Although short-term requirements are focused on optical links for multirack level interconnects, many researchers feel that rack, board, and even chip level optical links are inevitable. While these interconnects remain highly speculative, they hold the potential to enable entirely new system level architectures. Recent proposals in this area include a photonic reference architecture proposed by Intel and Facebook [50], which provides rack level optical interconnects for disaggregated, large scale-out data centers, and the recently announced Consortium for Onboard Optics (COBO) [51]. The integration of conventional silicon and photonics technologies, loosely known as Silicon Photonics or Nanophotonics, extends the use of optical interconnects to the board or chip level, and includes integration of photonic and electronic components in a single device.

Optical cross-connects provide an alternative to full function Layer 2/3 switch interconnects in applications which can benefit from high throughput and low latency with relatively long reconfiguration time. Research into networking for analytics of large data sets and Hadoop processing suggests that an optical cross-connect might help alleviate bottlenecks at the top of an equipment rack, in cases where high bandwidth interconnects must be shared by the application [52].

References

[1] Satell G. Every business will be disrupted by open technology, Forbes online, http://www.forbes.com/sites/gregsatell/2013/12/14/in-2014-every-business-will-be-disrupted-by-open-technology/ [last accessed 21.06.15].

[2] DeCusatis C, editor. Handbook of fiber optic data communications. 4th ed. Boston: Academic Press/Elsevier; 2013.

[3] Jamsa K. Cloud computing. New York, NY: Jones & Bartlett; 2013.

[4] Telcordia GR-3160. NEBS Requirements for Telecommunications Data Center Equipment and Spaces, provides guidelines for data center spaces within telecommunications networks, and environmental requirements for the equipment intended for installation in those spaces, http://telecom-info.telcordia.com/site-cgi/ido/docs.cgi?ID=SEARCH&DOCUMENT=GR-3160& [last accessed 21.06.15].

[5] Telcordia GR-2930. NEBS: Raised Floor Generic Requirements for Network and Data Centers presents generic engineering requirements for raised floors that fall within the strict NEBS guidelines, http://telecom-info.telcordia.com/site-cgi/ido/docs.cgi?ID=SEARCH&DOCUMENT=GR-2930& [last accessed 21.06.15].

[6] ANSI/TIA-942 Telecommunications Infrastructure Standard for Data Centers, specifies the minimum requirements for telecommunications infrastructure of data centers and computer rooms including single tenant enterprise data centers and multi-tenant Internet hosting data centers. The topology proposed in this document is intended to be applicable to any size data center, http://www.tiaonline.org/standards/ [last accessed 21.06.15].

[7] National Fire Code as provided by the National Fire Protection Association (NFPA), http://www.nfpa.org/index.asp?cookie_test=1 [last accessed 21.06.15].

[8] National Electrical Code, http://www.necplus.org/Pages/Default.aspx?sso=0 [last accessed 21.06.15].

[9] Underwriters Laboratories, http://www.ul.com/global/eng/pages/offerings/industries/buildingmaterials/fire/communication/wire/ [last accessed 21.06.15].

[10] Wall Street. Jounal article on energy efficiency, http://www.nytimes.com/2012/09/23/technology/data-centers-waste-vast-amounts-of-energy-belying-industry-image.html?pagewanted=1&_r=1 [last accessed 21.06.15].

[11] Liu H, Jin H, Xu CZ, Liao X. Performance and energy modeling for live migration of virtual machines. Cluster Comput 2013;Vol. 16(No. 2):249–64. Springer.

[12] DeCusatis C, Cannistra R, Carle B. A demonstration of energy efficient optical networks for cloud computing using software-defined provisioning. In: Proc. IEEE GreenCom (November 12–14, 2014), http://www.ieee-onlinegreencomm.org/program.html [last accessed 21.06.15].

[13] IEEE response to energy efficienty article, http://www.informationweek.com/cloud-computing/infrastructure/ny-times-data-center-indictment-misses-b/240007880?pgno=1 [last accessed 21.06.15].

[14] IEEE Energy Efficient Ethernet standard, http://www.ieee802.org/3/az/index.html [last accessed 21.06.15].
[15] Energy Star program, http://www.energystar.gov/index.cfm?c=prod_development.data_center_efficiency_info [last accessed 21.06.15].
[16] European Union Code of Conduct for Energy Efficient Data Centers, http://iet.jrc.ec.europa.eu/energyefficiency/ict-codes-conduct/data-centres-energy-efficiency [last accessed 21.06.15].
[17] Delivering rapid deployment of a complete, turnkey modular data center to support your unique business objectives, IBM GTS white paper, July 2009, http://www-935.ibm.com/services/us/its/pdf/sff03002-usen-00_hr.pdf [last accessed 21.06.15].
[18] Clos C. A study of non-blocking switching networks (PDF). Bell Syst Tech J 1953;32 (2):406–24. <http://dx.doi.org/10.1002/j.1538-7305.1953.tb01433.x>. ISSN 0005-8580. Retrieved 22 March 2011.
[19] Stone Mountain Dataplex, http://erecordssite.com/SMD//index.htm [last accessed 21.06.15].
[20] The 451 Group, https://451research.com/ [last accessed 21.06.15].
[21] Panko R, Panko J. Business data networks and security. 10th ed. New York, NY: Pearson; 2015.
[22] The W. Edwards Deming Institute, https://www.deming.org/theman/theories/pdsacycle [last accessed 21.06.15].
[23] Top 500 supercomputer list, http://www.top500.org/ [last accessed 21.06.15].
[24] DeCusatis CJS, Carranza A, DeCusatis C. Communication within clouds: open standards and proprietary protocols for data center networking. IEEE Commun Mag 2012;50(9):26–34.
[25] Barabesh K, Cohen R, Hadas D, Jain V, Recio R, Rochwerger B. A case for overlays in DCN virtualization. In: Proc. 2011 IEEE DC CAVES workshop, collocated with the 22nd international tele-traffic Congress (ITC 22).
[26] Marsh D. Analytic solutions for cloud computing. In: Proc. NSF 7th annual enterprise compute conference, June 14 – 17, Marist College, Poughkeepsie (NY); 2015.
[27] Roadmap Report: Photonics for Disaggregated Data Centers, from OSA Industry Development Association (OIDA) workshop held March 22, 2015, collocated with OFC 2015 Annual Meeting, Los Angeles, CA (2015); see also Proc. Majorca at MIT, July 27–30, 2015, Cambridge (MA) (2015).
[28] SSAE 16 standard, http://ssae16.com/SSAE16_overview.html [last accessed 21.06.15].
[29] Summary of the HIPPA privacy rule, http://www.hhs.gov/ocr/privacy/hipaa/understanding/summary/ [last accessed 21.06.15].
[30] Open Data Center Alliance, http://www.opendatacenteralliance.org/ [last accessed 21.06.15].
[31] Open Compute Project http://www.opencompute.org/projects/data-center/ [last accessed 21.06.15].
[32] DeCusatis C. Software-defined networking for the Open Datacenter Interoperable Network (ODIN), Open Network Exchange 2013 (presented by Network World), University of Chicago Gleacher Center, Chicago (IL) (May 14, 2013) https://www.etouches.com/ehome/53138.
[33] ANSI Fibre Channel Standard, http://www.ansi.org [last accessed 21.06.15].
[34] InfiniBand Trade Association [Online], http://www.infinibandta.org/content/pages.php?pg = technology_overview [accessed 29.01.13].

[35] Pattavina A. Switching theory: architecture and performance in broadband ATM networks. Chichester: John Wiley & Sons Ltd; 1998, ISBNs: 0-471-96338-0 (Hardback); 0-470-84191-5 (Electronic) (Banyan networks).
[36] Beneš VE. Mathematical Theory of Connecting Networks and Telephone Traffic. New York, NY: Academic Press; 1965.
[37] Charles E. Leiserson Fat-trees: universal networks for hardware-efficient supercomputing. IEEE Trans Comput 1985;34(10):892–901.
[38] DeCusatis C. Optical interconnect networks for data communications. IEEE J Lightwave Technol 2014;32(4):544–52.
[39] IEEE 802.1Qbg—Edge Virtual Bridging standard, http://www.ieee802.org/1/pages/802.1bg.html [last accessed 21.06.15].
[40] Gray J, Reuter A. Transaction processing. New York, NY: Morgan Kauffman; 1993.
[41] Pritchett D. Base: an ACID alternative. Queue 2008;6(13):48.
[42] Benson T, et.al. Network traffic characteristics of data centers in the wild. In: Proc. IMC conference, published by the ACM (2010) [Online] http://pages.cs.wisc.edu/~tbenson/papers/imc192.pdf [accessed 05.05.13].
[43] Smith R. Elementary information security. 2nd ed. New York, NY: Jones & Bartlett; 2015.
[44] Ahmed R, Boutaba R. Design considerations for managing wide area software defined networks. IEEE Commun Mag 2014;5(7):116–23.
[45] Lopez Research Report. An introduction to the internet of things, http://www.cisco.com/web/solutions/trends/iot/introduction_to_IoT_november.pdf (November 2013) [last accessed 21.06.15].
[46] IEC 60825-1laser safety standard, published by the International Electrotechnical Commission; see also U.S. National Regulation 21 CFR Chapter 1 Subchapter J, of the Center for Device and Radiological Health (CDRH), a subgroup of the Department of Health and Human Services (DHHS) of the Occupational Safety and Health Administration (OSHA) and the U.S. Food and Drug Administration (FDA), also published by the American National Standards Institute (ANSI).
[47] Proc. OIDA Workshop on Data Center Metrics, concurrently with OFC/NFOEC 2013 (Anaheim, CA) [Online], http://www.oida.org/home/ [accessed 05.05.13].
[48] Jevon's Paradox [Online], http://www.vattenfall.com/en/jevons-paradox.htm [accessed 05.05.13].
[49] CIAN [Online], http://www.cian-erc.org/testbed.cfm [accessed 05.05.13].
[50] Intel Photonic Reference Architecture v 1.0 [Online], http://goparallel.sourceforge.net/facebook-redefines-data-centers-parallel-with-intel-photonics/ [accessed 29.01.13].
[51] Consortium for Onboard Optics (COBO) announcement, http://thestack.com/microsoft-cobo-cisco-optical-data-center-240315 [last accessed 21.06.15].
[52] Want G, Eugene Ng TS, Shaikh A. Programming your network at run-time for big data applications. In: Proc. HotSDN 12, Helsinki, Finland (August 13, 2012) [Online], http://www.cs.rice.edu/~eugeneng/papers/HotSDN12.pdf [accessed 05.05.13].
[53] OpenStack: open source software for building public and private clouds, www.openstack.org [last accessed 09.10.14].

Optical interconnects: fundamentals

2

D. Tsiokos[1,2] *and G.T. Kanellos*[1,2]
[1]Aristotle University of Thessaloniki, Thessaloniki, Greece, [2]Center for Research and Technology Hellas, Thessaloniki, Greece

2.1 Optical interconnects: the driver behind future data centers

The immense growth of social networking and data sharing, via the cloud and smart mobile handsets, has been the main driver for optical interconnects. At first glance, this statement sounds unjustified; however, it is a plain way of describing the reality behind the need for high-performance optical interconnects. This will become apparent after we take a closer look at a group of very interesting Internet statistics. There are 1.65 billion active mobile social accounts globally and most of them, apart from texting, share photos, videos, and other personal and business data. Four and a half billion likes are generated on a daily basis in Facebook which, by 2014, has served more video views (12.3 billion) than YouTube (11.3 billion). Viber and Whatsapp have 800 million users communicating daily, 500 million tweets are coming out every day, while the +1 button is hit 5 billion times per day in Google+. Seventy million photos and videos are sent daily at Instagram, while 200 billion emails are sent or received every day. Those numbers are constantly increasing as more applications are released and faster communication and wireless networks become available. Alongside the data transfers created by people, a further 3.9 exabytes of data will be generated by Internet of Things applications by 2017, bringing the total stored data to an overwhelming level of a few zetabytes (10^{21} bytes). All this excess of new data has to be stored, processed, and shared between a disproportionately small number of mega data centers located across the world. It is now safe to conclude that without the data center, the cloud would not have been implemented.

Over the past decades, photonics have been leveraged to boost telecommunication technology for the transfer of massive amounts of data around the globe, enabling the Internet and changing the very nature of communications. During the last decade scientists and engineers have been trying to mimic the paradigm of telecommunication networks. They are still intensely seeking ways to adapt optical systems in shorter range data communication environments for computer interconnection in order to address massive data transfer demands already building up inside data centers. In this context, power consumption and size appear to be the main barriers to next-generation data center (DC) and High-Performance Computing (HPC)

Optical Interconnects for Data Centers. DOI: http://dx.doi.org/10.1016/B978-0-08-100512-5.00002-4

environments. The limited capacity and physical quantity of electrically wired interconnects is in contrast to the increased clock speeds and wiring densities inside machines, forming the main source of bottlenecks in information exchange across all hierarchical communication levels: intra-chip, chip-to-chip (on-board), board-to-board, and rack-to-rack (machine-to-machine).

Optical technology is already used in optical interconnection cables called Active Optical Cables (AOCs) for rack-to-rack communication, either in a short range between adjacent racks or between racks in different rooms of the data center, as illustrated in the indicative floor plan of Fig. 2.1. The floor plan shows how the data center can scale, depending on the service provider needs and the customer's requirements, for special conditions such as caged private areas for security or customized equipment cooling. Therefore, AOCs should be able to cover a few meters interconnection for close proximity servers (in the same row or room) up to hundreds of meters or even kilometer-range interconnection in mega-data centers for remote high-bandwidth interconnection between different collocation rooms. AOC technology is constantly advancing in order to address the ever increasing need for traffic transportation from one computing machine (rack or chassis) to another. In stark contrast, data moving across chips and from chip-to-chip within a single board of a computer or server still move electronically through tiny metal wires. But, as computers get faster and faster, manipulating ever more data, these wires are just not up to the task. Moving data to a computer's central processor from its memory, e.g., is already a notorious bottleneck; the wires simply cannot ship data from memory fast enough to keep the processor busy. And the problem is only going to get worse if optical interconnects technology does not reach all

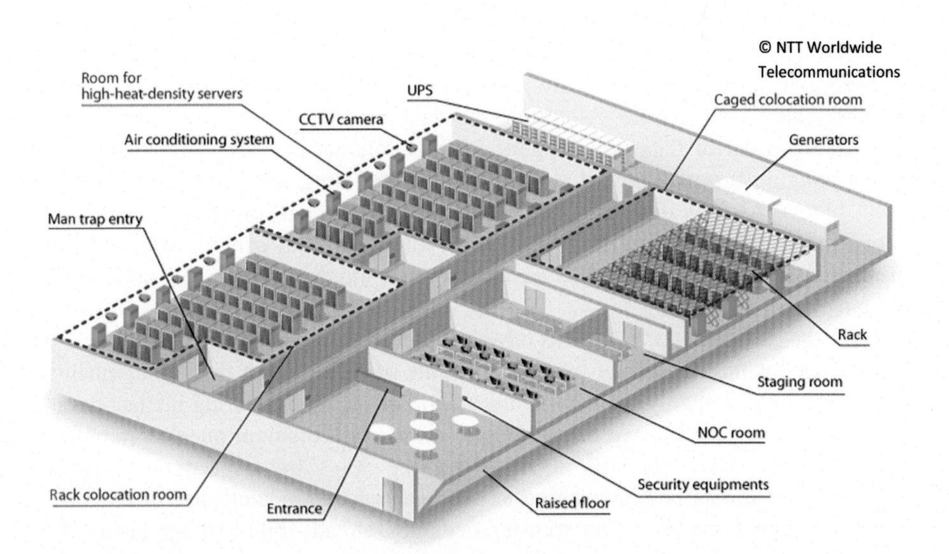

Figure 2.1 Indicative floor plan representation of a single storey data center. *Source*: Courtesy of NTT Worldwide Telecommunications.

hierarchy levels of interconnections in data centers and HPCs. Systems that replace wires with optics will ultimately move data faster and use less power. Optics is expected to effectively maximize the usage of available computing power by making data transfer within the network more than 50 times faster than is possible using copper connections.

But how will optical interconnects fill the gap left by electronic counterparts in the datacom world?

One of the problems an engineer has to deal with when interconnects have to be realized over long cables is signal attenuation. Such attenuation is unavoidable because of the resistance of the wires and other factors such as dielectric loss. Those losses are also frequency dependent, meaning that the high frequency components of a signal will have greater attenuation than the lower frequency parts, and will also have different phase shifts leading to distortion of signals. Advanced signal formats and/or signal processing circuits can be used to maximize the channel capacity up to the Shannon limit that is determined by the physical noise limitations. All of these interventions will, however, increase the complexity of the system, which in turn will be expected to increase the cost and power dissipation of the communication link. The simplest solution to increase the information capacity for a given wire is to increase its cross-sectional size, which effectively reduces the resistance yet increases cost. Once we get inside large complex systems like data centers, the cross-sectional size of wires becomes a problem in itself, limiting the density of wiring.

Why is this happening? In practice, there are two kinds of electrical wiring: resistive-capacitive (RC) lines and inductive-capacitive (LC) lines, primarily determined by the wire's cross-sectional size, as is clearly explained in Miller and Ozaktas [1] and Miller [2]. At the gigahertz frequency range, chip interconnects are primarily RC, limited by the bulk resistance of the metals, and lines off chips are primarily LC transmission lines, limited by the resistance through the *skin effect*. Skin effect is the tendency of an alternating electric current (AC) to become distributed within a conductor such that the current density is highest near the surface of the conductor, and decreases deeper in the conductor. The skin effect causes the effective resistance of the conductor to increase at higher frequencies where the skin depth is smaller, thus reducing the effective cross-section of the conductor.

In practice, neither the capacitance nor the resistance of wires scales well at smaller sizes. If the cross-sectional size of a wire is reduced, its resistance will go up. However, for the same case, the capacitance per unit length does not change since it depends on the shape of the cross-section. As Miller concludes in Miller [2], at least for on-off signaling with a simple model, the capacity of electrical lines from such resistive limits can be written approximately as:

$$B = B_0 \frac{A}{L^2}$$

where A is the cross-sectional area of the wiring, L is the length of the wires, and B_0 is a constant [1]. $B_0 \sim 1016$ bits/second for RC-limited lines that are typical on a chip, while a slightly smaller number holds for LC lines with resistive loss (RLC lines), and $B_0 \sim 1017$ bits/second for off-chip equalized RLC lines [2]. The ratio A/L^2 is dimensionless, making the capacity of a wiring system independent of the actual size of the whole system. This briefly explains why the capacity of the wiring cannot be increased further once system real estate is no longer available. The above argument becomes even more convincing if additional losses are present in the electrical link like surface scattering or dielectric loss, contributing to the link capacity degradation. Finally, the use of electrical interconnects also leads to several other problems in signal quality for dense, high-speed wiring like "cross-talk", signal reflections from connectors, and other discontinuities in the transmission lines.

All the above present themselves as serious barriers towards next generation, ultra high-speed, and small dimension interconnections, the demands for which are constantly increasing in data centers. In conclusion, it is interesting to look at the forecasts of chip interconnects given by par excellence industrial and standardization bodies. Specifically it is foreseen that more than 10,000 connections are required in semiconductor chips to enable the trends of Tb/second bandwidth interconnections [3]. At the same time, the International Technology Roadmap for Semiconductors (ITRS) has no roadmap for pin counts greater than 6500 [4]. The technological gap between ultra high-speed interconnectivity and lower pin/count requirements is therefore apparent, highlighting the need for new targets in transmission speed at low power and size.

2.2 Classes of optical interconnects in data centers

With big-data analytics turning into the cornerstone of modern business applications [5], the vast amount of new data being generated continues to grow at much higher rates than the development of infrastructures, a problem that is commonly referred to as the "data deluge problem". This brings current cloud- and big-data-driven computational machines in the struggle to exceed Exascale processing powers without leading to an explosion in energy and cost requirements. This ambitious target cannot anymore rely on simply increasing the number of low-power many-core microprocessors [6]. With processor chip size fragmentation having already reached its limits in trying to accommodate more cores [7], efforts for improving cost- and energy-efficiency in extreme-parallel computing are focusing on the next-level of hierarchy: chip-to-chip communication in server-board and intra-rack designs currently forms the hot-spot of the processor industry towards reducing physical space, network complexity, and resources (switches and cables), while enabling higher performance per watt. To put the situation in a practical perspective, and before we discuss the individual technologies that aim to address the above trends, we will briefly describe the three main application classes of optical interconnects in data centers.

2.2.1 On-board interconnects

Fig. 2.2 illustrates an intuitive schematic of how optical interconnects may be applied on a server board. A typical board will host separate chip processors (sockets) and memories for distributed data processing, a central router for routing data on the board between processors represented by the cloud in the schematic, as well as a series of I/Os for the communication of the board-processors with the rest of the data center network. This is either located into the same rack/different board, or to a different rack in the network. The router follows the network topology, which is in turn dictated by the executed application instructing data to move from one chip to a different chip on the same board (chip-to-chip interconnection), or redirecting data to the board IOs for inter-board or inter-rack communication.

The vision of optical interconnects entails the deployment of opto-electronic transceivers at the chip I/Os for generating optical data for transmission or for receiving optical data and converting them to electronic data for processing or storage in the chip. The transceiver includes optical source (either externally of directly modulated), photo-detectors as well as the associated driving electronics (laser or modulator drivers, trans-impedance amplifiers). The same requirement is imposed for the central router when the router will be solely implemented using electronics. Research efforts are being made into the development of high throughput, low latency, electro-optical switching matrices or optical matrices that will act as the switching fabric below higher level routers as the means to avoid this intermediate o/e/o conversion [8,9]. Such an approach will, however, require intense development efforts on the co-integration of photonic and electronic ICs and 3D integration processes. Finally, the board itself will have to accommodate optical waveguides as a replacement to copper wires, as well as optical interfaces

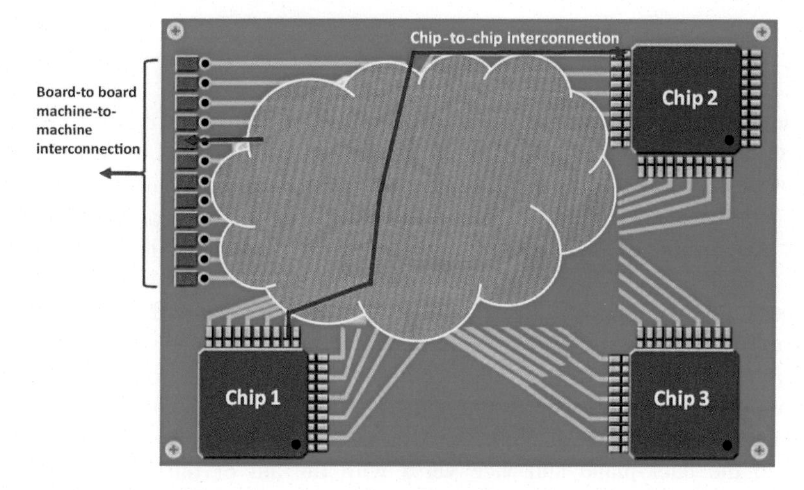

Figure 2.2 Schematic of board core elements and associated interconnects.

for counter propagating traffic traveling into and out of the board and towards the transceivers on each chip.

Having the above scheme in mind, it is now easier to understand technology limitations and future trends. In particular, the new server-board roadmap seeks eagerly for solutions that can bring more and more processors on the same board in order to increase performance at a reduced cost and energy envelope, allowing at the same time for almost linear reductions in physical space requirements and linear increases in throughput densities. According to an analysis of Intel [10], four-socket server-board designs can offer important advantages in high-volume virtualization deployment scenarios compared to single- or dual-socket layouts, almost doubling the aggregate performance per watt (1.92 × improvement), while reducing Total Cost of Ownership (TCO) expenses up to 35% [10]. The same analysis, in line with the industrial consensus, concludes that even greater performance, energy, and capacity gains are expected when scaling to 8-socket and beyond 8-socket server-board deployments [10,11], gradually moving towards "data-center-in-a-server" implementations [12].

However, current multi-socket server-boards continue to rely on four-socket implementations, with higher numbers of on-board directly-interconnected sockets being restricted by their high interconnect bandwidth requirements that are far beyond the capabilities of current electrical wiring [11]. "Glueless" socket architectures with direct point-to-point links between the sockets necessitate an up to 65% bandwidth waste for cache coherency purposes, usually enforcing bandwidth-hungry multi- and broadcasting schemes [11]. With processor capacity and memory bandwidth scaling faster than copper-based interconnect technologies over the last years, current point-to-point socket interconnects [13,14] fail to meet the needs of higher than eight-socket deployments: low-energy, high-bandwidth, low-latency, collision-free, and multi- and broadcast-friendly links, and all of them at once!

2.2.2 Board-to-board interconnects

The next level in the optical interconnects hierarchy involves board-level interconnection. This is the situation where data from on board need to be transmitted to another board on the same back-plane, as illustrated in the cartoon of Fig. 2.3, or on the same rack. In this case the loss budget of the link increases, as more losses are introduced from longer propagation distances, and from vertical coupling from the board to the optical waveguide of the back-plane and back to the board. In most board-to-board applications, designs require that the boards are pluggable. This allows for easy replacement in the event of board failure, more streamlined manufacturing for applications with multiple add-on board options, and ease of upgradeability or accommodation of future design changes. Therefore, in pluggable optical boards light will have to be coupled vertically from the board to the back-plane and vice versa with the aid of mirror structures and micro-lenses as described in the top-right hand side schematic of Fig. 2.3. For this purpose, low loss mirrors with high optical bandwidth will be necessary in order

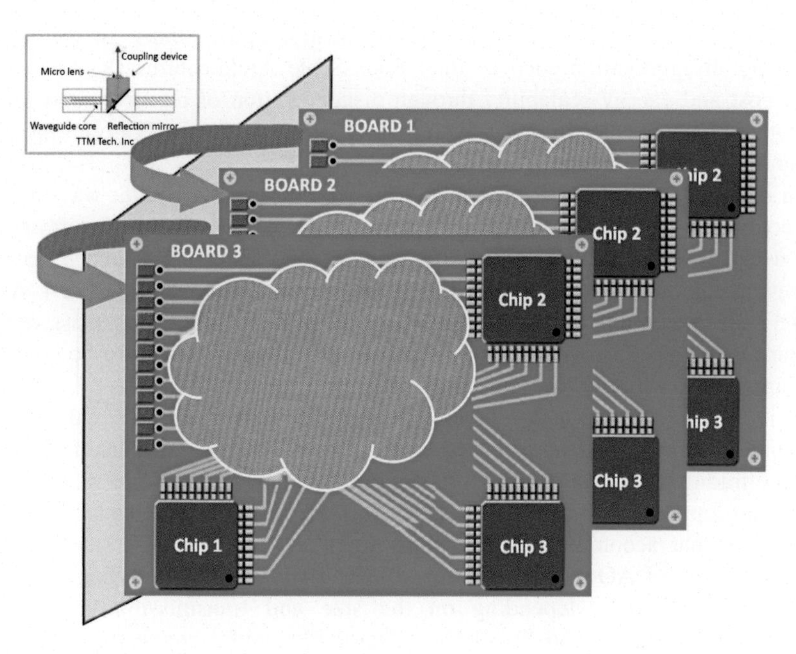

Figure 2.3 Schematic of a board-to-board interconnects scenario and vertical coupling concept. (TTM tech Inc.)

to accommodate strict power budgets and wavelength division multiplexed (WDM) optical data from different origins.

2.2.3 Rack-to-rack interconnects and AOCs

Different ICT requirements must be satisfied by different configurations of the modular data storage subsystems, which form the building blocks of modern data centers. These building blocks include, but are not limited to, data storage arrays, integrated application platforms, storage servers, switches, and high performance storage and computer subsystems. If such building blocks could be arranged to be truly modular, i.e., work independently of their location within the data center, and if interconnect length and bandwidth constraints between these subsystems could be disregarded, then this would allow a disaggregated architecture. In this case, the combination of subsystems required to satisfy a given set of ICT requirements may not be constrained to the same rack or cluster of racks, but they could be physically dispersed across the data center. Ideally, the user can be provided with a virtual data center solution with the optimum combination and amount of compute, memory, and storage, even though the corresponding hardware allocated could be dispersed. In order to satisfy these requirements without over-provisioning of hardware resources, one must have the capability to convey high bandwidth data over far longer distances than is typical or possible today between subsystems, and this can only be satisfied by low-cost high-bandwidth optical links. In this context, computer disaggregation such as Intel Rack-Scale Architectures (RSA) [15] promote cost and energy scalability through disaggregation of resources and modular hardware resource allocation, expecting significant savings in TCO during system upgrade.

In data centers, racks will need to be interconnected in a similar way that users are connected in a telecommunication network, following a network infrastructure and using data switches. This is achieved by using cables and switches that must, however, support long distances and high bandwidths, as shown in Fig. 2.4. As data centers are becoming larger and larger in order to accommodate massive cloud demands, distances between racks of up to a kilometer will have to be covered in extreme cases, while at the same time carrying 100 second of GBs of information (this is swiftly moving to the terabyte range). In parallel cabling real estate and weight is becoming a daunting overhead. AOCs have been introduced in order to replace traditional copper cables and overcome copper's traditional shortcomings. AOCs are optical fiber cables ending up in hot-pluggable high-end interfaces on both sides that accommodate both transmitter and receiver in a single compact housing. Sizes of AOC connectors have been standardized, including CXP, SFP, QSFP, and QSFP+ depending on the size and transmission capacity. For instance, the QSFP (Quad Small Form-factor Pluggable) connector specification accommodates Ethernet, Fibre Channel, InfiniBand, and SONET/SDH standards with different data rate options. QSFP+ transceivers are designed to carry Serial

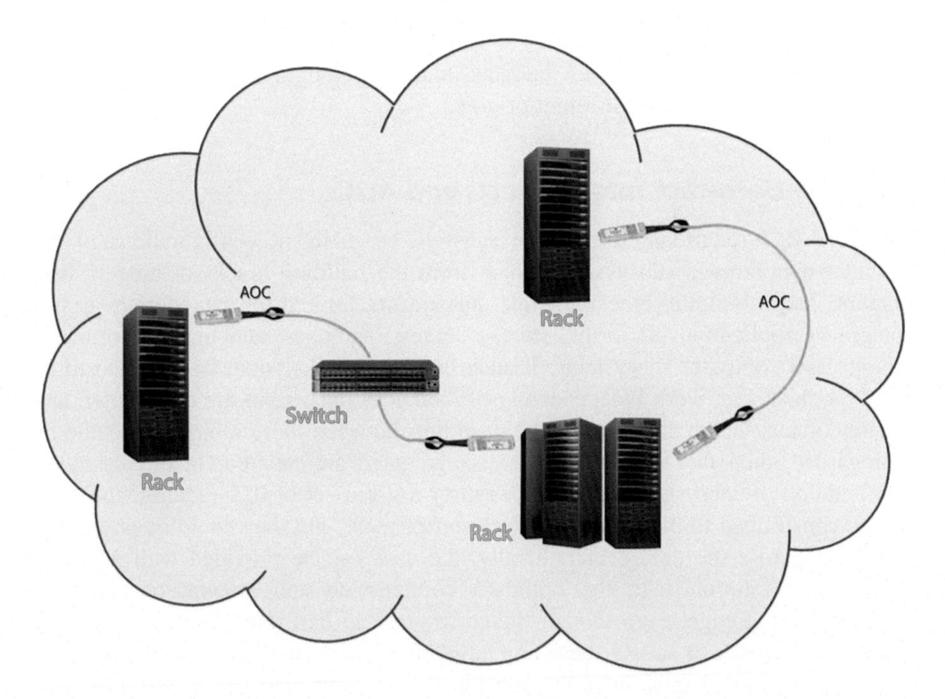

Figure 2.4 Rack-to-rack interconnection with the aid of switches and active optical cables.

Attached SCSI, 40G Ethernet, QDR (40G), FDR (56G) Infiniband, and other communications standards. QSFP modules increase the port-density by $3\times-4\times$ compared to SFP+ modules. Switches and AOCs supporting the abovementioned standards have been introduced in the market; however, the constant upgrade of the products alongside effective interconnection architectures are required to address mega-data centers with 100,000+ servers at ultra-high speed and ultra-low power consumption.

AOC lengths may vary from a few meters up to a kilometer addressing chassis-to-chassis on the same rack up to rack-rack within a data center warehouse while it can accommodate more than one optical fiber; but the cost of the fiber needs to be considered. If parallel fiber interfaces are not used, more efficient signaling schemes and/or WDM techniques optimized for short reach will be required. Some of the more efficient signaling schemes suitable for optical transmission are lower order m-PAM, PSK, or QAM [16]. As the baud rate increases, FEC may be required to close the link budget. The choice of the FEC must be considered carefully to address both latency and power concerns.

2.3 Current status and future trends of optical interconnects systems

Optical Interconnection Systems and fiber cabling solutions deployed in data centers today are designed to support data rate applications such as 100G Ethernet, Fibre Channel ≥32G, and InfiniBand ≥40G. To cope with the various demands for each application in terms of bandwidth, power consumption, cost, and reach distance, several forms of parallel optical interfaces have been adopted with varying channel rate, number of channels, and multi-mode/single-mode type of transmission. With respect to the number of channels, there are two forms of commercially available products for parallel optical interfaces. The first is a twelve-channel system consisting of an optical transmitter and an optical receiver. The second is a four-channel transceiver system that is capable of transmitting four channels and receiving four channels in one product.

Channel rates range from 10 to 25 Gb/second, as the industry currently has electrical interfaces for 10 Gb/second (OIF's CEI-11G), 25 Gb/second, and 28 Gb/second (OIF's CEI-25/28G) at its disposal, and work has started on 56 Gb/second (OIF's CEI-56G) to meet higher data rate needs. Infiniband roadmap was developed to keep the rate of performance increase in line with systems-level performance increases. Fig. 2.5 depicts the 2014 Infiniband roadmap that details 1×, 4×, and 12× port widths with bandwidths reaching 300 Gb/second data rate EDR this year, with 25 Gb/second single line rate, while 600 Gb/second data rate HDR is forecast for 2017.

Depending on the type of termination, Optical Interconnects are classified in two main categories: the Embedded Optic Modules (EOM) for on-board optical interconnection, and the AOCs as the pluggable form of optical interconnects.

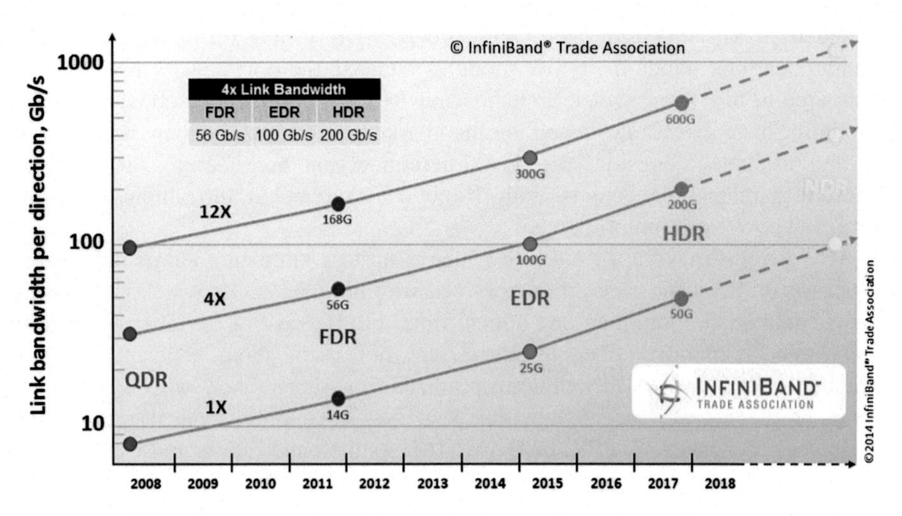

Figure 2.5 Infiniband roadmap 2014.
Source: Courtesy of IBTA.

In the following subsections we will review the state of the art for those two categories of optical interconnects.

2.3.1 Active optical cables

AOCs have been the cash cow of the optical interconnects business, offering a closed form of optical transmission, simple installation, long reach in harsh data center environments, and broad support to protocols (Fibre Channel, Infiniband, Ethernet). To this end, AOCs have been endorsed by all the big names in optical networking components (Finisar, Avago, Intel, Tyco, etc.) currently focusing on the data center market. The market for AOC is projected to surpass US$3.5 billion by 2022 according to an MSA report [15], driven by bandwidth demands. As the processing power of CPUs is increasing by the continuous progress of microelectronics with the 3TFlop/CPU benchmark on the corner [17], the bandwidth requirements for server interconnection in HPCs and data centers is expected to rise to a totally new level. In order to cope with these demands, AOCs focus on the dimension of parallelization for the throughput capacity increment. Typical examples of parallel AOCs are the Multiple-Fiber Push-On/Pull-Off (MPO) connector, allowing for simple installation of 12 fiber factory-terminated backbone/horizontal cabling into pre-terminated modules, panels, or harnesses, and the Quad Small Form-factor Pluggable (QSFP/QSFP+/QSFP28) hot-pluggable transceiver for four channel transmission. The 12 fiber connectivity is followed by Finisar that has packaged 12 transceivers, each operating at 14 Gb/second, at both ends of the cable for the demonstration of an 168 Gb/second AOC [18]. This device is based on vertically integrated Vertical Cavity Surface Emitting Laser (VCSEL) and photodetector (PD) arrays, consuming 23.3 mW/Gb per second power. A new product by Hitachi [19], based on the same architecture, has halved the power to 12 mW/Gb per second for the same capacity. Meanwhile, an approach by Fitel Photonics Laboratory aimed at power reduction by using highly-efficient 1060 nm VCSELs, achieved a power consumption value of 7 mW/Gb per s/link at 80°C for 10-Gb/second × 12-channel parallel-optical modules [20]. For distances exceeding the 100 m HPC and DC range enabling deployment in campus environments, Molex has demonstrated a solution formed with a DFB laser at 1490 nm and a Mach Zehnder Modulator targeting 40 Gb/second (4 × 10 Gb/second) interconnectivity with 18.75 mW/Gb per second power consumption [21]. Additionally, for 100 GbE connectivity, Molex has presented a QSFP + AOC based on 4 × 25 Gb/second transceivers, reaching distances up to 4 km and consuming 25 mW/Gb per second power [22]. In a different 16-channel bi-directional link CDFP configuration, TE has recently demonstrated a 400 Gb/second AOC based on 850 nm VCSEL technology with only 6 W of power consumption (15 pJ/bit) [23].

2.3.2 Mid-board optical engines

Chip-level optical interconnection can build on miniaturized optical engines, and CIR projects [24] optical engine revenues for chip-to-chip interconnection of $235 million by 2019, reaching around $775 million by 2020. Increasing equipment throughput in the network core requires new systems to support denser optical modules having 25–28 Gb/second interfaces, leading to today's current generation of 100G Optical Transceivers or QSFP28 with four channels at 28 Gb/second. The evolution to multi-terabit line cards will eventually require system interfaces for Very Short Reach Chip-to-Module Interfaces and Chip-to-Chip Interfaces to support rates up to 56 Gb/second, while the OIF-CEI-56G Close Proximity Chip-to-Chip Interface is currently being developed [25]. The Avago miniPOD series was the first commercially available transceiver for optical interconnects, with 12 channels operating at 8 Gb/second for compatibility with PCIe applications [26]. Meanwhile, the opto-module reported by IBM in a sub-100 nm complementary metal-oxide-semiconductor (CMOS) silicon-on-insulator (SOI) node, achieved a significant improvement in operational aspects, showing successful 24 × 20 Gb/second transmission while consuming 8.2 pJ/bit [27], based on 850 nm PD and VCSEL arrays flip-chip attached to the IC. Soon after, however, the need for cost-effective and power efficient optical transceivers at 25 Gb/second and beyond became imminent [28], leading to a series of mid-board transceivers operating at 25 Gb/second per lane [29–33]. Among them, TE Connectivity's mid-board optical module relied on 25 Gbp/second VCSEL and PIN devices, a TIA amplifier, and a driver IC to deliver 12 × 25.78 Gb/second while consuming less than 14.54 pJ/bit [24,30]. In an effort to further increase data rates, Molex and Finisar have already presented mid-board transceivers operating at 28 Gb/second using multi-mode VCSEL transmission, with 8 and 12 channels respectively [32,33]. Meanwhile, aiming for cost-effective, high-density single-mode optical fiber links, WDM has already been exploited by Kotura, delivering a 4 × 25 Gb/second QSFP package with only 3.5 pJ/bit energy consumption [34]. Moreover, IMEC has developed a 4×20 Gb/second hybrid CMOS silicon photonics transceiver, flip-chip integrated with a low-power 40 nm CMOS chip, aiming for high yield and low manufacturing cost [35] by exploiting small footprint, small capacitance, and low power consumption characteristics of four compact 25 Gb/second ring modulators, with their drivers being coupled to a common ring WDM bus, and an array of four 25 Gb/second Ge waveguide photo-detectors lying on the receiving end along with four 20 Gb/second trans-impedance amplifiers.

2.3.3 Techno-economic requirements of future optical interconnects

Next generation interconnection systems are being driven by the need to handle the increasing volume of data traffic in data centers. To this end, Infiniband forecasts HDR 56 Gb/second line rate and 12×600 Gb/second requirements by 2017. The IEEE 802.3 400 Gb Ethernet task force was set up in 2014 [146]. Among the task force objectives was to specify optional Energy Efficient Ethernet (EEE) capability for the 400 Gb/second physical layer, and support optional 400 Gb/second interfaces for chip-to-chip and chip-to-module applications. This was evident, as the next generation optical interconnects should not only keep track of the bandwidth demands, but also respect the constraints and limits imposed by power consumption, limits on the size of a system, and by the need to provide a cost-effective solution.

2.3.3.1 Energy efficiency

One of the biggest problems standing in the way of Exascale compute levels is power; not just the power required to run a task on a CPU, but the power required to share that data across the chip, node, and cluster. It is indicative that, considering the goal for 20 MW Exascale computing by 2018, today we consume 60% of the power (12 MW) for 1% (10 PF) of the processing performance [36].

Power efficiency, measured on a per-core basis, is expected to continue improving for multi-core and many-core architectures, but interconnect power consumption has not scaled nearly as well. This leads to a long-term problem: by 2018, it will cost more to move a FLOP off-die than to perform the calculation locally. According to Fig. 2.6, published in May 2013 at the Optical Interconnects Conference [37], data movement requires about 55 pJ/bit for off-chip DRAM access and falls below 10 pJ/bit only for the case of on-chip data movement. As energy efficiency performance needs improvement in all system components to reduce Exascale overall power consumption below the targeted 10 pJ/Op, a $\sim 60\times$ improvement [36] is also needed for off-chip optical interconnect communication, raising the goal to less than 1 pj/bit.

Besides progress in Si Photonics integration and throughput density increase that can help towards this direction, Software Defined Networking is another emerging concept that can significantly contribute to optical interconnects energy efficiency.

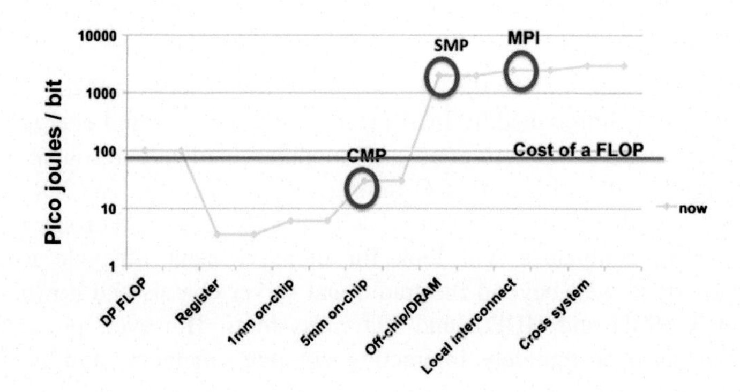

Figure 2.6 The cost of data movement.

Cloud-era technologies and today's volatile traffic patterns demand a network infrastructure that is more agile, more adaptable, and more in sync with the business it supports. SDN-capable optical engines [38] may transform hardware-intensive legacy networks into fully programmable networks that can automatically sense and respond to changing workload and bandwidth requirements.

2.3.3.2 Cost reduction

A target cost of $0.10 per Gb/second has been roughly set for the on-board optical interconnects for exascale systems, considering the optical equivalent of PCI-3 for multi-Tb/second I/O high-endchip performance [39]. Cost reduction may be achieved with the adoption of advanced system design concepts, as well as of fabrication processes. From the system perspective, throughput density increase is important for scaling cost, and WDM topologies can dramatically increase throughput density. In Ref. [40], Mellanox has released the wavelength specifications for the Open Compute Project, US Summit 2015, San Jose, CA, Mar. 9, 2015, revealing a 5× cost improvement when scaling CWDM optical engines to 1.6 Tb/second. Throughput density increase is also important for scaling cost through the real estate, and assembly of optical engines on electro-optical printed circuit boards (PCBs) with polymer waveguides certainly adds to that direction.

Improvement of fabrication processes may further contribute to cost efficiency. Assembly and packaging dominates the cost of short-reach optical interconnection. Simplification of assembly processes through passive alignment and self-alignment techniques [40] will relax mechanical placement alignment accuracy, enabling high volume manufacturing through high throughput assembly tools. Silicon Photonics integration and in-plane laser technology combined with the high bandwidth density of WDM can also partially compensate for cost. Moreover, in-plane lasers may help towards more reliable assembly processes with relaxed beam mechanical alignment requirements currently pursued in VCSELS-based products, a procedure that also makes them more fragile on real life operation.

2.3.3.3 Longer reach

The emerging trend of computer disaggregation, such as the Rack-Scale Architecture (RSA) introduced by Intel [41,42], promote cost and energy scalability through disaggregation of resources and modular hardware resource allocation, expecting significant savings in TCO during system upgrade. The IDC think tank predicts [43] that disaggregated systems will quickly gain market space as hyper-scale computing companies will look for more efficient lifecycle management options that extend well beyond the traditional server chassis and down into CPU, memory, disk (SSD and HDD), and I/O subsystems. However, pursuing performance advances over remotely interacting compute, memory, and storage pools requires enormous improvements in rack-scale interconnect fabrics. Once a PCB has adopted optical I/O at the chip package level, there are compelling reasons to extend it across the board. Because optical link performance and costs are largely independent of distance, optical buses within PCBs will provide architectural freedom for designers to place components and functions where other constraints are more easily managed. Multichannel, shared memory, e.g., can be placed where it can be more easily accessed by multiple sockets and remote processors. To support disaggregated computer architectures, future optical board interconnects should allow low-latency long-reach through single mode communication, and Intel is already turning towards silicon photonics for its next-generation intra-rack connectivity backbone [42].

2.4 Overview of photonic key enabling technologies

The looming necessity for optical interconnect systems with the techno-economic profile described above has attracted intense development efforts in many closely related underpinning technologies and indispensable components made from different materials and fabrication technologies. Recently, the research community has focused a lot of its efforts on silicon-based optical interconnects [44]. The development of a SOI platform for optical device development (a few hundred nanometers of single crystal silicon layer on top of a few micron thick buried oxide (BOX layer)) has triggered interest in silicon photonics technology, in an attempt to merge photonic mass production technology with established fabrication technology exploited for decades in electronics by CMOS foundries. Specifically, the high refractive index contrast between silicon ($nSi \sim 3.47$) and SiO_2 ($nSiO_2 \sim 1.45$), and their optical transparency in the 1.55 μm wavelength window enables ultra-compact device dimensions suitable for large-scale, high density integration on a chip. In addition, silicon photonics is expected to minimize time-to-market for innovative and disruptive photonic and opto-electronic devices, while maximizing return on investment for the industrial community and investors.

In this section, we aim to provide an overview of the current status in the underpinning technologies and components that aim to be deployed in future optical interconnect systems, while special attention is given to silicon photonics as the most tantalizing technology for next generation mass-produced photonics. Technologies and materials that are being researched as complementary technologies to silicon, i.e., active devices, are also discussed.

2.4.1 Photonic integrated circuit technologies

Photonics is one of the most vibrant areas of the global economy. The total world market of optical components and systems is currently estimated to be €15 billion. Given their pivotal importance across a wide range of industries and services, from telecommunications and information systems to health care, investment in generic photonic technologies can have a disproportionately large impact. In order to address this strong dependence on highly functional and low-cost photonic components, a strong momentum has been generated over recent years in photonic integrated circuits (PICs) technology following the electronics VLSI paradigm. Unlike electronic integration where silicon is the dominant material, PICs have been fabricated from a variety of material systems, including silica on silicon, silicon on insulator, various polymers, and semiconductor materials which are used to make semiconductor lasers such as GaAs and InP. Different material systems have been used because each provides different advantages and limitations, depending on the function to be integrated. For instance, silica (silicon dioxide)-based PICs have desirable properties for passive photonic circuits such as AWGs, due to their comparatively low losses and low thermal sensitivity, GaAs- or InP-based PICs allow direct integration of light sources, and silicon PICs enable co-integration of the photonics with transistor-based electronics. Such PIC fabrication technologies have been advanced by large research and industrial consortiums via multi-project wafer platforms including JEPPIX (http://www.jeppix.eu/) for III–V photonics, EPIXFAB (http://www.epixfab.eu/), and OPSIS (http://opsisfoundry.org/) for silicon photonics, and TRIPLEX (http://www.lionixbv.nl/triplexmpw) for silicon nitride photonics. Although III–V-based photonics is widely accepted as the most mature integration technology due to its capability to deliver high-performance active devices, it still remains an incompatible endeavor when compared to CMOS infrastructures and processes.

Amongst all candidate PIC technologies, silicon photonics are introduced as the most tantalizing next-generation technology that could dramatically reduce chip size and system power consumption, while it will keep development costs low by exploiting currently available CMOS infrastructures and volume manufacturing capabilities. Alongside silicon photonics (SiPh), Si_3N_4 PIC technology is also expected to grow in PICs for datacom, mainly due to the lower optical losses introduced and the inherent CMOS compatibility with electronics fabrication processes. SOI has become a material of choice for passive PICs including multi-mode interference (MMI)s and directional couplers [44], mode converters or in-plane couplers [44], and out-of-plane angled and completely vertical grating couplers [16,44], splitters, and optical filters [44]. Recent advances in the field have even pushed silicon beyond its passive niche, with demonstrations of high speed silicon-based modulators, silicon/germanium detectors, and even light sources when combined with III–V active regions via special processes. In a nutshell, the fact that today's commercial electronics are based on the same material can be used as leverage for the development of silicon photonics.

The unquestionable advantage of silicon photonics is its potential to combine itself with unavoidable electrical circuits on a single integration platform to reduce product size and development cost. Integration of optical and electrical devices can be realized via heterogeneous integration where optical and electrical devices use separate chips, or with monolithic integration [45,46] where optical and electrical devices share the same chip [47]. Heterogeneous integration has the advantage of an optimized fabrication process for best optical and electrical device performance [48]. Very recently a heterogeneously integrated 30 Gb/second chip-to-chip optical link with III–V/Si photonics and 32 nm CMOS electronics was demonstrated [49]. However, this strategy comes at the price of complicated packaging, limited density, and higher interconnect parasitic capacitance. Monolithic integration can address those shortcomings. Solutions rely either on SOI wafers [46,48,50], or on bulk silicon [45]. While the SOI approach is optimal for photonic waveguiding, high costs associated with inefficient use of die area, porting to high speed CMOS nodes (65–28 nm would be needed for higher data rates), and the increased complexity of the manufacturing process still act as retarding factors. Bulk silicon, on the other hand, allows for cost-aware high-volume applications, but lacks a buried oxide (BOX) layer for waveguiding.

Following these trends, the field of plasmonics emerged as a possible future intervention in data communications due to its metallic nature combining confined optical waveguiding and electrical circuitry [51,52]. At the same time, plasmonics when combined with silicon [53] or organic materials offers ultra high-speeds, since the plasmonic waveguides also serve as electrical contacts, thus making RC time constants very small [54].

2.4.2 III–V on SOI active devices

Optical sources are indispensable modules in all optical communication equipment and have therefore been absolutely essential for the optical interconnects industry to identify the most promising technologies for that purpose. In that respect, all integrated laser sources rely on III–V materials (InP, GaAs) for their active region, while the indirect recombination mechanisms of silicon have blocked its exploitation in the development of laser components. The material properties of silicon on one hand, and the incompatibility issues between III–V actives and silicon substrates on the other, have steered focus onto the prospect of having optical sources co-integrated with silicon-based components.

Despite those physical limitations, means to realize integrated lasers on silicon caught a lot of attention recently. Previous research attempts focused on integrating III–V material onto silicon by means of epitaxial growth, resulted in weak light emission due to poor material quality (lattice mismatch, misfit locations, etc.). More recently, there was a shift of research activities integrating III–V materials on silicon by means of bonding—either by using adhesives like BCB or by molecular bonding. Hybrid lasers developed by the University of California at Santa Barbara (UCSB) are a good example of silicon integrated sources based on molecular bonding. They consist of a III–V-based InAlGaAs quantum well structure molecularly bonded on top of a silicon waveguide [55,56]. As the main mode is located in the silicon, there is only little overlap of the optical mode with the III–V-based gain material, resulting in high threshold currents. In contrast, lasers based on adhesive bonding enable concentration on the light entirely in the III–V-based material, which results in lower threshold currents. However, as InGaAsP quantum wells in conjunction with poor thermally conducting BCB as adhesive are used, the thermal performance of these devices requires cooling by a power-hungry and costly thermo-electric cooler. Another popular type of laser source is the VCSEL which is a type of semiconductor laser diode with laser beam emission perpendicular from the top surface, contrary to conventional edge-emitting semiconductor lasers which emit in-plane from surfaces formed by cleaved facets. VCSELs are today the most efficient, lowest cost, and most widely used laser source for interconnects due to their low threshold current, small size, direct modulation capabilities, and their ability to be integrated in multi-component arrays. VCSELs have prevailed in short-reach interconnects using multi-mode 850 nm [57], yet VCSELs operating in the 1300 and 1550 nm wavelength regions have been produced for long-range interconnects [58]. Ultra high-speed long-wavelength VCSELs at 1550 nm have been reported modulated with data rates up to 35 Gb/second [59] or higher by using advanced modulation formats [60]. An inherent drawback of VCSELs is their incompatibility with SOI platforms and the vertical direction of the emitted beam, leaving the flip-chip bonding technique as the most appropriate option for hybrid integration with SOI based PICs.

Apart from optical sources, the issue of integrating optical amplifiers is a daunting question that sooner or later will have to be dealt with, compromising size, power consumption, and cost. The rapid development of PICs and tight power budgets imposed by modern standards has spawned the need for on-chip amplifying solutions in order to compensate for the inevitable optical losses, especially in complex circuits. Although Erbium Doped Fiber Amplifiers (EDFAs) have consistently been a solid amplification medium for optical systems over the years, only recently was an amplifier based on erbium-doped aluminum waveguides (Al2O3:Er3 +) developed, utilizing spiral channels [61]. The device exhibited a small-signal gain of 20 dB at 1532 nm for a waveguide length as large as 24.4 cm, relying on an optical pump at 976 nm. Significant research has been performed on the integration of III–V semiconductor materials on silicon, although hetero-epitaxial growth of InP on Si has proven difficult due to the large lattice and thermal expansion coefficient mismatch [62]. This has led to several alternative approaches, such as heterogeneous die-to-wafer bonding [63], thermo-compression bonding of pre-fabricated InP SOA arrays [64], or hybrid Si evanescent devices [65]. Photonic crystal-based structures have emerged as an alternative technology to potentially overcome integration density limits and create compact yet efficient PICs. In this regime, in Fang et al. [65] a theoretical approach is presented, proving the enhanced energy density provided by the slow-light mode of PhC waveguides, promising for hybrid

PhC-SOAs with significantly smaller size, higher efficiency, and better performance compared to conventional ones. In Ek et al. [66], an active PhC waveguide structure with gain is presented, while it is shown that the use of slow light may also reduce the power consumption of amplifiers. Although the aforementioned amplifier was based on optical pumping, recent work on PhC lasers has also demonstrated the feasibility of operation under electrical pumping [67,68].

2.4.3 Modulators

Next to the laser source, the modulator is the component required in all types of interconnects for transforming data from electrical into optical as an alternative to direct laser modulation. Optical designers and engineers need to optimize the optical and electrical power consumption for driving the modulator versus size, and speed of operation. High bandwidth data communications already call for modulation bandwidths that well exceed 40 Gb/second with low voltage, power consumption, and footprint requirements. As unstrained silicon does not possess an x2-nonlinearity, state-of-the art silicon photonic modulators mainly rely on plasma dispersion effects on doped silicon, III/V-on-SOI, or Silicon Organic Hybrid (SOH) approaches.

Plasma dispersion relying on pin and pn junctions has been extensively explored in MZIs as the mainstream solution since it requires only two standard processes, doping and etching. Most notably, vertical junctions have been employed in Liu et al. [69], exhibiting modulation data rates hardly up to 40 Gb/second. Rib SOI waveguides with horizontal 1 mm long lateral pn-junction driven by a 6.5 V peak-to-peak voltage to demonstrate a 50 Gb/second modulation [70]. Elongation of the lateral junction to 2 mm [71], 3.5 mm [72] or 4 mm [73] or 5 mm with careful design of the mm-long RF electrodes [74], provided only minor increases in speed and signal quality. A further increase of the waveguide height from 220 to 340 nm resulted in a 60 GHz modulation speed [75]. These demonstrations reveal that enhanced interaction of the light with the modulated carrier density is required for higher speeds, while exhaustive experimental investigations of the doping profiles [76] pinpointed the limitations imposed by impurity scattering losses in the mm-long pn-junctions. Even with a complex pipin junction, modulation speeds of 40 Gb/second were observed [77]. For enhanced phase modulation, carrier depletion has also been combined with slow light generated either by a 500 μm long corrugated waveguide [78] or by a 90 μm long photonic crystal waveguide [79] achieving 40 Gb/second; however, these techniques further exacerbate the power budget with additional 10 dB losses.

Resonant Si ring modulators have recently evolved as a promising candidate for compact on-chip integration as they confine light in high-quality factor (Q) devices that effectively increase the optical path length. A racetrack resonator with a $L = 300$ μm phase shifting section and advanced doping geometries in Rosenberg et al. [80] exhibited 40 Gb/second modulation with power consumption of 471 fJ/bit. Footprint reduction was achieved also in Li et al. [81], where a compact microring modulator with only 5 μm ring radius demonstrated 40 Gb/second modulation rate.

Although reversed biased pn-junctions are intrinsically faster, noticeably a forward biased pin-diode racetrack modulator achieved 50 Gb/second operation with engineering of a side-wall-grating section of $L = 15\ \mu m$ [82]. Further improvements in efficiencies with sub-fJ (0.9 fJ/bit) modulation were obtained using a compact microdisk with $R = 2.4\ \mu m$ radius, exhibiting simultaneously 25 Gb/second modulation, and low-voltage operation of 0.5 V. To achieve this result, a highly optimized vertical p-n junction [83] spanned the whole microdisk footprint supporting speeds up to 44 Gb/second.

A promising alternative solution suggests the use of *electro-absorption modulation* in III/V materials hybridly integrated on SOI waveguides [84]. In Feng et al. [85] a 45 μm long high-speed Ge electro-absorption modulator is buttcoupled with a deep-etched SOI waveguide to form a 30 Gb/second modulator. By further elongating the Ge active section to 100 μm, a 50 Gb/second traveling waveguide modulator was demonstrated with 9.8 dB modulation depth under a voltage of 2 V [86]. A 100 μm long phase shifter was also combined with asymmetric segmented electrode configuration for impedance matching purposes to achieve modulation speeds beyond 74 Gb/second in the 1.3 μm transmission window [87]. Generally, the electro-absorption effect in semiconductors is intrinsically an ultrafast process that requires short lengths, but hybrid III/V-on-SOI integration techniques are necessary and advanced modulation formats are difficult to support. For practical system application, symmetric MZI modulators offer several key properties, such as high thermal insensitivity, robustness against fabrication variations, and modulation format support.

The barrier of the 100 GHz modulation data rate was only recently broken by using *Silicon Organic Hybrid technology* [88]. Nanophotonic SOH modulators exploit the ultrafast Pockels effect of high-speed polymer materials deposited within a silicon photonic slot waveguide. An unprecedented 3 dB bandwidth of 100 GHz employing a phase shifter of 500 μm length and 2 dB insertion losses, highlighted the impact of narrow slot waveguide configurations [89]. However, the overall performance of the SOH MZI modulator is degraded by the additional 9 dB losses due to the strip-to-slot and slot-to-strip tapering mode converters, which also result in the length of 2.6 mm. It has been forecast for quite a while that plasmonics could overcome those tradeoffs [54]. Recently, a group from ETHZ in Switzerland has introduced plasmonic organic hybrid (POH) devices and was able to demonstrate a 40 Gbit/second plasmonic phase modulator of 29 μm length [90] and Mach-Zehnder modulators (MZMs) operating at speeds beyond 54 Gbit/second with a length of only 10 μm [91].

2.4.4 Photo-detectors

Moving from the transmitter to the receiver side, the main active component one will have to deal with is the photo-detector. Once more, the interconnection application will dictate the features of the receiver, i.e., a long-range interconnection will require higher speed operation arising from data aggregation, and higher receiver sensitivity stemming from higher interconnection losses. Responsivity and dark current are additional figures of merit used to quantify the operational efficiency of the photo-detection. Similarly to the other cases of active devices, the material that is required to function as a low-loss waveguide cannot simultaneously be used for PDs steering efforts towards III–V or germanium (Ge)-based detectors on silicon waveguides which may be directly epitaxially developed or bonded on the silicon chip.

Due to their small and direct bandgap properties, III–V compound semiconductors are inherently suitable for optical devices. However, the overall process cost is too high, and the integration of III–V materials with Si is rather complicated and expensive due to the inter-diffusion, surface polarity, and generally large lattice constant mismatch. On the other hand, germanium, a group IV material, is an attractive alternative for light detection at the C-band if it can be integrated on Si monolithically [92]. Although Ge is also an indirect bandgap material, its direct gap energy of 0.8 eV, which corresponds to 1500 nm, can be further reduced by tensile strain, and therefore, the absorption edge can be pushed further to cover most of the C-band [93]. Several discrete Ge-on-Si PDs have been illustrated, in spite of the inherent material difficulties in growing germanium on SOI [94,95], demonstrating bandwidths greater than 40 Gb/second or even integrating silicon-based high-speed modulators together with Ge-based detectors on a single SOI platform [96].

2.4.5 Optochips and 3D integration

The term optochip arose over recent years to highlight the fact that photonic technology is slowly crossing the boundaries of a chip-set so far dominated by electronics. The idea behind this vision is to either deploy photonics as the switching fabric, vertically controlled by electronics and higher layer routing functionalities in high throughput routers, or to integrate optical transceivers with supporting electronics and CMOS processors in next generation intra-chip interconnections, or both. Amazing progress has been witnessed in this respect by simultaneously advancing vertical integration techniques in the form of 3D integration (monolithic integration) or 3D stacking (hybrid integration) towards the vision of delivering multi-layer chip-sets accommodating both optics and electronics on the same socket [97]. For the first approach through-silicon-vias (TSV) may be used as "3D interconnects," providing all advantages of heterogeneous integration, yet on a common integration platform [98]. For the same purpose, electronics structures down to the transistor polysilicon layer have been combined with optical waveguides to demonstrate on-chip optical interconnects [45,46]. Again, a 3D stacking approach enables heterogeneous integration of different technologies that give multi-functionalities in a single chip [99]. Three-dimensional stacking photonic-electronic integration [147] can be exploited with current state-of-the-art III–V and SOI technologies in order to provide high performance opto-chips.

TSVs and hybrid integration have also been combined to embed the photonic layer into the last levels of metallization above the CMOS layer [100]. In addition, the backside process is common to electronics and photonics. In this approach, an external source is wafer bonded and coupled with a silicon wire circuit [101]. The solution takes advantage of the rear side of the electronics wafer. Integration of photonic layers at the backside of the CMOS wafer is performed by bonding and connecting a CMOS wafer and a photonic wafer. First, the CMOS wafer is processed up to the last metal layer, and the backside is thinned and fine polished to prime wafer surface quality. The photonic layer is then added by wafer-to-wafer bonding at low temperature. Afterward, electrical interconnects between the CMOS layer and the photonic layers are obtained using TSVs. Deep silicon etching is performed down to the metal layer of the photonic layers. Subsequent TSV isolation and metallization are deposited. Subsequent TSV isolation and metallization are deposited and structured. Finally, the substrate wafer is removed in order to release the photonic structures. Advantageously, in this approach, the IC and the photonic processing are rather independent and the packaging of such a double-sided chip may be developed for other applications. So far, optical TSVs in Si- [101,102] and glass-based [103] interposers have been demonstrated. Active photonic components have been integrated in a more complex chip in Wook Lee et al. [98], with a VCSEL and a PD flip-chip bonded on a 3D circuit.

Another method for optoelectronic assembly and vertical integration is flip-chip bonding. With this approach, contact bumps are required in order to connect the optoelectronic components with the substrate electrically, mechanically, and thermally. So far two methods for flip-chip bonding of optical components on SOI substrates have been reported: soldering [104,105], and thermocompression bonding [106,107]. Although this method imposes significant limitations in terms of alignment accuracy, process complexity, and mass production capabilities, it remains a promising assembly method for research purposes or small-scale customized solutions.

The optochip arena has been dominated so far by IBM, with the demonstration of two generations of Optobus program achieving initially a total throughput of 160 Gbps (16×10 Gbps) [108], and more recently 300 Gbps (24×12.5 Gbps) [27], consuming 8.2 mW/Gb per second. A newer approach was presented in Brusberg et al. [109], where 4×10 Gb/second transceivers were flip-chip bonded on a glass-formed PCB with 592 mW power consumption. Straight parallel polymer waveguides were fabricated with "waveguide-in-copper" technology in Nieweglowski et al. [110], with 4×10 Gbps transceivers flip-chip bonded with ceramic ball grid arrays on the PCB as well. Optochips that operate with VCSEL in the 1300 nm region for easier coupling to the waveguides due to smaller beam divergence were demonstrated in Ukaegbu et al.[111], with 4×10 Gb/second channels transmitted error-free to the corresponding receivers. All these optochip developments have, however, targeted solely transmission/reception functionalities without encompassing any routing capabilities.

2.4.6 Optical PCBs

The concept of chip-to-chip optical interconnects imposes the need for optical boards transmitting broadband light through embedded optical waveguides. Glass-based

[109,112] and polymer-based boards are credible technologies to address this task for either single mode or multimode links while bonding techniques between active photonics and optical printed circuit boards (OPCBs) are being developed. Specifically, the demonstration of adiabatic coupling [113] between silicon and polymer waveguides made it possible to interface silicon photonics with OPCBs, enabling coupling to/from a large number of waveguides with a single-step bonding process. Moreover, laser direct writing of polymer waveguides has also been demonstrated [114,115], and a connectivity solution has been realized [116] using passive alignment.

IBM has demonstrated the first optical PCBs with an MT pluggable optical connector for communication with other boards/instruments. The interconnection of two such boards via a ribbon cable was presented in Dangel et al.[117] with OPCBs exchanging data at 120 Gb/second (12 × 10 Gb/second) aggregate bit rate. Also, 12.5 Gb/second data were successfully transmitted via a 100 cm spiral shaped polymer waveguide (0.05 dB/cm losses @850 nm) formed on a PCB. Hitachi has presented a dual-layer embedded waveguide optical PCB with a total number of 48 waveguide channels; using a 4-transceiver optochip with each transceiver operating at 20 Gb/second and consuming 15.1 mW/Gb per second, this PCB has the potential for an aggregate capacity of 980 Gb/second [118]. In Takagi et al. [119], four channels at 10 Gbps were transmitted through 100 mm of polymer multimode waveguide. Fujitsu has also presented an optical PCB with 90 × 150 × 0.6 mm dimensions that incorporates three four-channel transceivers capable of operating error-free at bit rates up to 14 Gb/second, and the twelve channels are accessible through 2xMT 12-channel optical connectors [120]. For lower crosstalk between adjacent channels, an OPCB with 16 parallel Graded Index MM waveguides was presented in Ishigure and Niita[121] exhibiting more than 5 dB interchannel crosstalk reduction compared to step index ones, demonstrating transmission of one 12.5 Gb/second channel through a 5 cm waveguide with clear eye diagram at 850 nm wavelength. For optical backplanes that allow OPCB to OPCB interconnection, Xyratex presented a platform that relays 10.3 Gb/second optical data streams between four active connectors plugged to the optical mid-plane [122]. Additionally, a passive optical backplane that employs 100 embedded waveguides and is able to connect to 10 PCBs with a nonblocking architecture was presented in Beals et al.[123]. However, fully functional OPCB demonstrations equipped with routing elements have only been shown in Takagi et al. [119], where a hybrid network-on-chip with a 2 × 2 network interconnected via an electronic switch at 3.125 Gb/second line rate has been employed.

2.4.7 Optical memory elements

The prospect of having ultra-high bandwidth chip-to-chip interconnects on the motherboard by using WDM technology has spurred novel concepts on optical memories and how computing can be revolutionized by exploiting photonics. To make this argument clearer a brief analysis of the current computing challenges is required.

Current off-chip electronic random access memories (RAM) still exhibit insufficiently long access times and limited bandwidth to follow the progress of Chip-Multiprocessor technology (CMP) in processing speed, significantly decreasing their overall computational power [119]. To overcome the limited throughput between powerful CPUs and slow off-chip memory units, the so called Von Neumann bottleneck, on-chip cache memory schemes were introduced to allow memory operations at CPU frequency and in close proximity to processor cores [124,125]. These benefits, however, come at the expense of increased chip-footprint requirements, reaching up to 40% of the CMP chip real-estate spent for cache purposes [126]. As the processor chip size is fragmented due to cost and heat dissipation related issues [127,128], on-chip memory solutions restrict processor chips from accommodating more cores and thus increasing their processing power. A recent study [129] has shown that the way to relax the Von Neumann bottleneck and simultaneously release precious on-chip real-estate for processing purposes is to establish an alternative bit-level optical memory hardware that would allow the implementation of off-chip cache memories. Such optical technology would enable a high-throughput and energy-efficient interconnection system, but would require integration scale capabilities at least 14% that of electronics.

Optical interconnects have already enabled high bandwidth memory access to remote distances inspiring new computer architectures, such as computer disaggregation concepts [130], with significant performance and cost benefits in HPC and the data center industry. However, only when optical interconnects are combined with highly integrable optical RAM technology, will they have the potential to fully use the advantages of optical technology to revolutionize the computer industry with radically new computer architectures. In such a case, the fast response times of optical memories will meet with the WDM properties of optical transmission to enable external interconnection of CPU-memory with unprecedented speed and bandwidth optical links, while simultaneously avoiding costly EO conversions and additional critical time delays in a latency-sensitive environment.

Fortunately, several optical memory devices have been reported over the last few years exhibiting remarkable characteristics in terms of speed, footprint, and power consumption, underlining the potential of optical memory technology. Among them, devices that exploit light polarization bistability in VCSELs [131], integrated InP ring-lasers on SOI that exploit light direction bistability, coupled laser-based [132] and switch-based [133–135] "master-slave" configurations, reveal speed capabilities of optical memories in the 10–40 Gb/second regime and WDM characteristics. Very recently, a multilevel phase change memory has been presented [136,137] employing germanium-antimony-tellurium as the active material on top of a silicon waveguide. The device exhibited an eight-level operation, suitable for coding 3 bits simultaneously, with switching energies as low as 13.4 pJ at speeds approaching 1 GHz. Most importantly, the device underlines the ability of optical technology to deliver low power memories with reduced footprint, as the device active region was as low as $0.4 \times 0.4\ \mu m^2$.

On the other hand, photonic crystal nanocavity technology has also attracted much interest as their strong light-matter interaction and ultra-compact size raise expectation for devices with large-scale integration capabilities, bringing a whole new class of laser and switch devices [138–144]. Optical memory devices were among the first to take advantage of this technology, with the demonstration of devices based on the bistability of photonic crystal nanocavity lasers [140] or switches [141–144], reporting energy consumption of 30 nW bias power, 13 fJ switching energy, and $<10\ \mu m^2$ footprint [141]. PhC technology has also revealed dense integration potential, as an InP photonic crystal chip exhibiting 105 memory cells operating in 105 wavelengths with a total chip area of 1.07 mm was recently demonstrated [142]. Most importantly, optical RAM memory cells and optically connected CPU-Dynamic RAM (DRAM) technologies has already outlined the benefits in several experimental demonstrations so far [145].

2.5 Summary and practical conclusions

The purpose of this chapter was to convey to the reader the basic background and benefits of optical interconnects, and the motive behind their utilization as the means to grow data centers and other data demanding HPC systems. The interconnects classes have been presented in an hierarchical fashion, screening the needs for advanced optical technologies in each different level of the data center. In the same context, we went through a brief overview of the available and evolving photonic technology advances both in the component and the system level that aim to constitute the keystones of future optical interconnect scenarios before being explained in detail in the chapters to follow. In parallel, data center and related enabling technology trends have been pointed out in terms of current status and future requirements.

Although there has been much progress in the fabrication of photonic components targeting optical interconnects by using different material technologies, we have given special attention to silicon photonics due to its compatibility with CMOS technology and therefore its profound prospects for mass production. However, alongside silicon photonics, several newly emerging complementary technologies, such as plasmonic-based technology, Si:organic hybrid technology, and Si_3N_4-on-Si-based technology, all promise to positively impact the interconnects industry. It is expected that, once photonic development technologies align coherently with conventional electronics foundries, optical interconnect systems, as well as other application landscapes benefiting from photonics, will advance disproportionately.

In conclusion, the implementation of next generation optical interconnects poses specific technological challenges, especially in relation to power dissipation, limited I/O density, maximum channel data rate, and optimal reach. Strong system inter-dependencies such as higher integration, complex modulation schemes, chip break-out and routing, signal conditioning, thermal and power issues, package footprint, etc., need to be addressed in order to achieve a cost effective interconnect solution compatible with legacy data center standards that satisfies the power density (watt/meter^2) requirement.

References

[1] Miller DAB, Ozaktas HM. Limit to the bit-rate capacity of electrical interconnects from the aspect ratio of the system architecture. J Parallel Distrib Comput 1997;41:42–52.

[2] Miller DAB. Optical interconnects to electronic chips. Appl Opt 2010;49:F59–70.

[3] Polka LA. Package technology to address the memory bandwidth challenge for terascale computing. Intel Technol J 2007;11(3):197–204.

[4] ITRS. Overall roadmap technology characteristics tables. [Online]. Available: http://www.itrs.net/Links/2010ITRS/Home2010.htm; 2010.

[5] http://www.zdnet.com/article/intel-takes-its-next-step-towards-exascale-computing/.

[6] http://www.cresta-project.eu/the-exascale-challenge.html.

[7] Cunningham J.E. et al. Integration and packaging of a macrochip with silicon nanophotonic links. IEEE J Sel Top Quantum Electron 17(3):546–58.

[8] Dabos G, Bolten J, Prinzen A, Giesecke AL, Pleros N, Tsiokos D. Perfectly vertical and fully etched SOI grating couplers for TM polarization. Opt Commun 2015;350:124–7.

[9] Karinou F, Borkowski R, Zibar D, Roudas I, Vlachos KG, Tafur Monroy I. Advanced modulation techniques for high-performance computing optical interconnects. IEEE J Sel Top Quantum Electron 2013;19(2):324–37.

[10] http://www.intel.com/content/dam/doc/white-paper/virtualization-intel-it-xeon-7500-analyzing-the-virtualization-deployment-paper.pdf.

[11] http://www.bull.hu/adatok/bullion_an_efficient_server_architecture_for_virtualization_wp_en_2012_10.pdf.

[12] http://www.slideshare.net/LightCounting/ecoc-2012-market-focus-ao-cs-eoms-ecoc-2012-918.

[13] http://www.intel.com/content/www/us/en/io/quickpath-technology/quick-path-interconnect-introduction-paper.html.

[14] http://www.oracle.com/technetwork/server-storage/sun-sparc-enterprise/documentation/o13-024-m5-32-architecture-Server_Card_Specification_v0.5.pdf.

[15] http://www.prweb.com/releases/2014/06/prweb11976805.htm.

[16] Kanellos GT, Fitsios D, Alexoudi T, Vagionas C, Miliou A, Pleros N. IEEE/OSA J Lightwave Technol 2013;31(6):988–95.

[17] https://software.intel.com/sites/default/files/managed/e9/b5/Knights-Corner-is-your-path-to-Knights-Landing.pdf.

[18] https://www.finisar.com/active-optical-cables/fcbxd1xcd1cxx.

[19] Tokoro T, et al. Engineering a 150 Gbit/s optical active cable to meet the needs of the data center environment. In: Proc. OFC 2011; 6–10 Mar 2011.

[20] Uemura J, et al. 1060-nm 10-Gb/s × 12-channel parallel-optical modules for optical interconnects. In: Proc. IEEE CPMT symposium, Japan; 24–26 Aug 2011.

[21] http://www.molex.com/molex/products/family?key=quad_small_formfactor_pluggable_plus_qsfp_interconnect_solution&channel=products&chanName=family&pageTitle=Introduction.

[22] http://www.molex.com/molex/products/family?key=zqsfp_interconnect_system&channel=products&chanName=family&pageTitle=Introduction#overview.

[23] http://www.te.com/usa-en/products/fiber-optics/active-optics/active-optical-cable-assemblies.html?tab=pgp-story.

[24] http://cir-inc.com/news/cir-report-predicts-that-chip-level-optical-interconnect-market-will-genera.

[25] McDonough J. Next generation connectivity to drive optical networks beyond 100G. In: OIF workshop at LR POTE, New York City (NY); 17 May 2012.

[26] Fields MH et al. Transceivers and optical engines for computer and datacenter interconnects. In: Proc. OFC, OTuP1, Los Angeles (CA); 2011.
[27] Schow Clint L, et al. A 24-channel, 300 Gb/s, 8.2 pJ/bit, full-duplex fiber-coupled optical transceiver module based on a single "holey" CMOS IC. J. Lightwave Technol. 2011;29:542–53.
[28] Vlasov Y. Silicon CMOS-integrated nano-photonics for computer and data communications beyond 100G. IEEE Commun Mag 2012;50(2):s67–72.
[29] http://portal.fciconnect.com/Comergent/fci/documentation/datasheet/opticalinterconnect/oi_leap_obt.pdf.
[30] http://www.te.com/content/dam/te/global/english/industries/data-communications/active-optics/data-communications-coolbit-mbo-data-sheet.pdf.
[31] http://www.fujitsu.com/downloads/MAG/vol50-1/paper18.pdf.
[32] http://www.literature.molex.com/SQLImages/kelmscott/Molex/PDF_Images/987650-6751.PDF.
[33] https://www.finisar.com/optical-engines/fbotd25fl2c00.
[34] Jatar S. et al. Performance of parallel 4 × 25 Gbs transmitter and receiver fabricated on SOI platform. In: Group IV photonics, San Diego (CA); 2012.
[35] http://www2.imec.be/be_en/press/imec-news/imec-ugent-tyndall-silicon-photonics-transceiver.html.
[36] Sodani A. Race to exascale: opportunities and challenges. In: 44th annual IEEE/ACM international symposium on microarchitecture, Porto Alegre; 2011.
[37] Simon H. Why we need Exascale and why we wont geet there by 2020. In: Optical interconnects conference, Santa Fe (NM); 6 May 2013.
[38] Amaya N, et al. Software defined networking (SDN) over space division multiplexing (SDM) optical networks: features, benefits and experimental demonstration. Opt Express 2014;22:3638–47.
[39] https://mphotonics.mit.edu/docman/ctr/ctr-3/short-reach-interconnect-twg/722-on-board-optical-interconnection-digest-1/file.
[40] http://www.openopticsmsa.org/pdf/Open_Optics_Design_Guide.pdf.
[41] http://www.opencompute.org/assets/OCP-Summit-V-Slides/Mainstage/OCP-Intel-Hooper.pdf.
[42] http://www.intel.com/newsroom/kits/atom/c2000/pdfs/Architecting_for_Hyperscale_DC_Efficiency.pdf.
[43] Worldwide Server. Top 10 predictions: a time of transition. IDC #247001, IDC; February 2014.
[44] Subbaraman H, Xu X, Hosseini A, Zhang X, Zhang Y, Kwong D, et al. Recent advances in silicon-based passive and active optical interconnects. OSA Opt Express 2015;23(3):2487–511.
[45] Orcutt JS, Khilo A, Holzwarth CW, Popovic MA, Li H, Sun J, et al. Nanophotonics integration in state-of-art CMOS foundries. Opt Express 2011;19(3):2335–46.
[46] Orcutt JS, Moss B, Sun C, Leu J, Georgas M, Shainline J, et al. Open foundry platform for high performance electronic-photonic integration. Opt Express 2012; 20(11):12222–32.
[47] Lim Andy Eu-Jin, Song Junfeng, Fang Qing, Li Chao, Tu Xiaoguang, Duan Ning, et al. Review of silicon photonics foundry efforts. Invited, IEEE J Sel Top Quantum Elecrron 2014;20(4).
[48] Izhaky N, Morse MT, Koehl S, Cohen O, Rubin D, Barkai A, et al. Development of CMOS-compatible integrated silicon photonic devices. IEEE J Sel Top Quantum Electron 2006;12(6):1688–98.

[49] Dupuis N, Lee BG, Proesel JE, Rylyakov A, Rimolo-Donadio R, Baks CW, et al. 30-Gb/s optical link combining heterogeneously integrated III−V/Si photonics with 32-nm CMOS circuits. IEEE J Lightwave Technol 2015;vol. 33(Issue. 3):657−62.
[50] Yang X, Babakhani A. Optical waveguides and photodiodes in 0.18 μm CMOS SOI with no post processing. In: Proc. optical fiber communication conference; 2013, Paper OTu2C.6.
[51] Alam MZ, Aitchison JS, Mojahedi M. A marriage of convenience: Hybridization of surface plasmon and dielectric waveguide modes. Laser Photon Rev 2014;8(3):1863−8899.
[52] Kinsey N, Ferrera M, Naik GV, Babicheva VE, Shalaev VM, Boltasseva A. Experimental demonstration of titanium nitride plasmonic interconnects. Opt Express 2014;22(10):12238−47.
[53] Kumar A, Gosciniak J, Volkov VS, Papaioannou S, Kalavrouziotis D, Vyrsokinos K, et al. Dielectric-loaded plasmonic waveguide components: going practical. Laser Photon Rev 2013;1−14. Available from: http://dx.doi.org/10.1002/lpor.201200113.
[54] Melikyan A, Alloatti L, Muslija A, Hillerkuss D, Schindler PC, Li J, et al. High-speed plasmonic phase modulators. Nat Photon 2014;8(3):229−33.
[55] Zhang C, Srinivasan S, Tang Y, Heck MJR, Davenport ML, Bowers JE. Low threshold and high speed short cavity distributed feedback hybrid silicon lasers. J Appl Phys Opt Express 2014;22(9):10202−9.
[56] Jain SR, Tang Y, Chen H-W, Sysak MN, Bowers JE. Integrated hybrid silicon transmitter. J Lightwave Technol 2012;30(5):671−8.
[57] Westbergh P, et al. High-speed, low-current-density 850 nm VCSELs. IEEE J Sel Top Quantum Electron 2009;15(3):694−703.
[58] Hofmann WH, Moser P, Bimberg D. Energy-efficient VCSELs for interconnects. IEEE Photon J 2012;4(2):652−6.
[59] Muller M, Hofmann W, Grundl T, Horn M, Wolf P, Nagel RD, et al. 1550-nm high-speed short-cavity VCSELs. IEEE Sel Top Quantum Electron 2011;17(5):1158−66.
[60] Xie Chongjin, Spiga S, Dong Po, Winzer P, Bergmann M, Kögel, et al. 400-Gb/s PDM-4PAM WDM system using a monolithic 2 × 4 VCSEL array and coherent detection. J Lightwave Technol 2015;33(3):670−7.
[61] Vázquez-Córdova SA, et al. Erbium-doped spiral amplifiers with 20 dB of net gain on silicon. Opt Express 2014;22(21):25993−6004.
[62] Cheung Stanley, Kawakita Yasumasa, Shang Kuanping, Yoo S J Ben. Theory and design optimization of energy-efficient hydrophobic Wafer-bonded III−V/Si hybrid semiconductor optical amplifiers. J Lightwave Technol 2013;31(24):4057−66.
[63] Keyvaninia S, Roelkens G, Van Thourhout D, Fedeli J-M, Messaoudene S, Duan G-H, et al. A highly efficient electrically pumped optical amplifier integrated on a SOI waveguide circuit. In: IEEE group IV photonics conference; 2012.
[64] Fitsios D, et al. Dual SOA-MZI wavelength converters based on III-V hybrid integration on a um-scale Si platform. Photon Technol Lett 2014;26(6):560−3.
[65] Fang AW, Park H, Kuo Y-H, Jones R, Cohen O, Liang D, et al. Hybrid Si evanescent devices. Mater Today 2007;10(7−8).
[66] Ek S, Lunnemann P, Chen Y, Semenova E, Yvind K, Mork Jesper. Slow-light-enhanced gain in active photonic crystal waveguides. Nat Commun 2014;5:5039.
[67] Matsuo S, et al. Electrically-pumped photonic crystal lasers for optical communications. In: ECOC 2012, Amsterdam; 16−20 Sep 2012.
[68] Matsuo S, et al. 40 Gb/s direct modulation of membrane buried heterostructure DFB laser on SiO2/Si substrate. In: International semiconductor laser conference 2014, Mallorca; 7−10 Sep 2014.

[69] Liu A, Liao L, Rubin D, Nguygen H, Cifticioglu B, Chetrit Y, et al. High-speed optical modulation based on carrier depletion in a silicon waveguide. Opt Express 2007; 15(2):660–8.
[70] Thomson DJ, Gardes FY, Fedeli JM, Zlatanovic S, Hu Y, Kuo BPP, et al. 50-Gbs silicon optical modulator. IEEE Photon Technol Lett 2012;24(4):234–6.
[71] Dong P, Chen L, Chen Y. High-speed low-voltage single-drive push-pull silicon Mach-Zehnder modulators. Opt Express 2012;20(6):6163–9.
[72] Ding R, Liu Y, Ma Y, Yang Y, Li Q, Lim A Ej, et al. High-speed silicon modulator with slow-wave electrodes and fully independent differential drive. IEEE J Lightwave Technol 2014;32(12):2240–7.
[73] Tu X, Liow TY, Song J, Luo X, Fang Q, Yu M, et al. 50-Gb/s silicon optical modulator with traveling-wave electrodes. Opt Express 2013;21(10):12776–82.
[74] Jones TB, Ding R, Liu Y, Ayazi A, Pinguet T, Harris NC, et al. Ultralow drive voltage silicon traveling-wave modulator. Opt Express 2012;20(11):12014–20.
[75] Xiao X, Xu H, Li X, Li Z, Chu T, Yu Y, et al. High-speed, low-loss silicon Mach–Zehnder modulators with doping optimization. Opt Express 2013;21(4):4116–25.
[76] Yu H, Pantouvaki M, et al. Performance tradeoff between lateral and interdigitated doping patterns for high speed carrier depletion based silicon modulators. Opt Express 2012;20(12):12926–38.
[77] Ziebell M, Marris-Morini D, Rasigade G, Fédéli JM, Crozat P, Cassan E, et al. 40 Gbit/s low-loss silicon optical modulator based on a pipin diode. Opt Express 2012; 20(10):10591–6.
[78] Brimont A, Thomson DJ, Sanchis P, Herrera J, Gardes FY, Fedeli JM, et al. High speed silicon electro-optical modulators enhanced via slow light propagation. Opt Express 2011;19(21):20876–85.
[79] Nguyen HC, Hashimoto S, Shinkawa M, Baba T. Compact and fast photonic crystal silicon optical modulators. Opt Express 2012;20(20):22465–74.
[80] Rosenberg JC, Green WMJ, Assefa S, Gill DM, Barwicz T, Yang M, et al. A 25 Gbps silicon microring modulator based on an interleaved junction. Opt Express 2012; 20(24):26411–23.
[81] Li G, Zheng Z, Thacker H, Jin Y, Ying L, Shubin I, et al. 40 Gb/s thermally tunable CMOS ring modulator. In: Group IV photonics, San Diego (CA); 2012.
[82] Baba T, Akiyama S, Imai M, Hirayama N, Takahashi H, Noguchi Y, et al. 50-Gbs ring-resonator-based silicon modulator. Opt Express 2013;12(10):11869–76.
[83] Timurdogan E, Sorace-Agaskar CM, Sun J, Shah Hosseini E, Biberman A, Watts MR. An ultralow power athermal silicon modulator. Nat Commun 2014;5:4008.
[84] Liu J, Beals M, Pomerene A, Bernardis S, Sun R, Cheng J, et al. Waveguide-integrated, ultralow-energy GeSi electro-absorption modulators. Nat Photon 2008; 2:433–7.
[85] Feng NN, Feng D, Liao S, Wang X, Dong P, Liang H, et al. 30GHz Ge electro-absorption modulator integrated with 3 μm silicon-on-insulator waveguide. Opt Express 2011;19(8):7062–7.
[86] Tang Y, Chen HW, Jain S, Peters K, Westegren U, Bowers J. 50 Gb/s hybrid silicon traveling-wave electroabsorption modulator. Opt Express 2011;19(7):5811–16.
[87] Tang Y, Peters JD, Bowers JE. Over 67 GHz bandwidth hybrid silicon electroabsorption modulator with asymmetric segmented electrode for 1.3 μm transmission. Opt Express 2012;20:11529–35.
[88] Alloatti L, Palmer R, Diebold S, Pah KP, Chen B, Dinu R, et al. 100 GHz silicon–organic hybrid modulator. Nat Light Sci Appl 2014;3.

[89] Leuthold J, et al. Silicon-organic hybrid electro-optical devices. IEEE J Sel Top Quantum Electron 2013;19:114–26.
[90] Naik GV, Schroeder JL, Ni X, Kildishev AV, Sands TD, Boltasseva A. Titanium Nitride as a plasmonic material for visible and near-infrared wavelengths. Opt Mater Express 2012;2(4):478–89.
[91] Babicheva VE, Kinsey N, Naik GV, Ferrera M, Lavrinenko AV, Shalaev VM, et al. Towards CMOS compatible nanophotonics ultra compact modulators using alternative plasmonic materials. Opt Express 2013;21(22):27326–37.
[92] Kobayashi S-i, Nishi Y, Saraswat KC. Effect of isochronal hydrogen annealing on surface roughness and threading dislocation density of epitaxial Ge films grown on Si. Thin Solid Films 2010;518(6):S136–9.
[93] Nam Ju Hyung, Afshinmanesh Farzaneh, Nam Donguk, Jung Woo Shik, Kamins Theodore I, Brongersma Mark L, et al. Monolithic integration of germanium-oninsulator p-i-n photodetector on silicon. Opt Express 2015;23:12.
[94] Vivien L, Osmond J, Fédéli J-M, Marris-Morini D, Crozat P, Damlencourt J-F, et al. 42 GHz p.i.n germanium photodetector integrated in a silicon-on-insulator waveguide. Opt Express 2009;17(8):6252–7.
[95] Feng NN, Dong P, Zheng D, Liao S, Liang H, Shafiiha R, et al. Vertical p-i-n germanium photodetector with high external responsivity integrated with large core Si waveguides. Opt Express 2010;18(1):96–101.
[96] Liow T-Y, Ang K-W, Fang Q, Song J-F, Xiong Y-Z, Yu M-B, et al. Silicon modulators and germanium photodetectors on SOI: monolithic integration, compatibility, and performance optimization. IEEE J Sel Top Quantum Electron 2010;16(1):307–15.
[97] Chen C, Joshi A. Runtime management of laser power in silicon-photonic multibus NoC architecture. IEEE J Sel Top Quantum Electron 2013;19(2).
[98] Wook Lee K, et al. Three-dimensional hybrid integration technology of CMOS, MEMS, and photonics circuits for optoelectronic heterogeneous integrated systems. IEEE Trans Electron Dev 2011;58(3).
[99] Lu Jian-Qiang. 3-D hyperintegration and packaging technologies for micro-nano systems. Proc IEEE 2009;97(1):18–30.
[100] Kopp C, Bernabe S, Bakir BB, Fedeli J, Orobtchouk R, Schrank F, et al. Silicon photonic circuits: On-CMOS integration, fiber optical coupling, and packaging. IEEE J Sel Top Quantum Electron 2011;17(3):498–509.
[101] http://www.helios-project.eu/.
[102] Parekh MS, et al. Electrical, optical and fluidic through-silicon vias for silicon interposer applications. In: IEEE Proc. ECTC; 2011, Art. No. 5898790.
[103] Schroeder H, et al. glassPack—A 3D glass based interposer concept for SiP with integrated optical interconnects. In: IEEE Proc. ECTC; 2011, Art. No. 5490760.
[104] Heikkinen V, et al. Fiber-optic transceiver module for high-speed intrasatellite networks. J Lightwave Technol 2007;25(5):1213–23.
[105] Heikkinen H, et al. Indium-tin bump deposition for the hybridization of cdte sensors and readout chips. In: Proc. IEEE nuclear science symposium and medical imaging conference (NSS/MIC), USA; 2010, p. 3891–5.
[106] Reed JD, et al. High density interconnect at 10 um pitch with mechanically keyed Cu/Sn-Cu and Cu-cu bonding for 3-D integration. In: Proc. IEEE electronic components and technology conference (ECTC), USA; 2010, p. 846–52.
[107] Kapulainen M, et al. Hybrid integration of InP lasers with SOI waveguides using thermocompression bonding. In: Proc. group IV photonics 2008; 17–19 Sept. 2008.

[108] Doany FE, et al. 160 Gb/s bidirectional polymer-waveguide board-level optical interconnects using CMOS-based transceivers. IEEE Trans Adv Packag 2009;32(2).

[109] Brusberg L, et al. Glass carrier based packaging approach demonstrated on a parallel optoelectronic transceiver module for PCB assembling. In: IEEE Proc. ECTC; 2010, Art. No. 549096.

[110] Nieweglowski K, et al., Demonstration of board-level optical link with ceramic optoelectronic multi-chip module. In: IEEE Proc. ECTC; 2009, Art. No. 5074276.

[111] Ukaegbu I Augustine, et al. 2.5-Gb/s/ch long wavelength transmitter modules for chip-to-chip optical PCB applications. IEEE Photon Technol Lett 2011;23(19).

[112] Brusberg L, Manessis D, Neitz M, Schild B, Schroder H, Tekin T, et al. Development of an electro-optical circuit board technology with embedded single-mode glass waveguide layer. In: Electronics system-integration technology conference (ESTC); 2014, p. 1–5.

[113] Soganci IM, La Porta A, Offrein BJ. Multichannel optical coupling to a silicon photonics chip using a single-step bonding process. In: 2013 IEEE photonics conference (IPC); 2013, S. 115–6.

[114] Zgraggen E, Soganci IM, Horst F, La Porta A, Dangel R, Offrein BJ, et al. Laser direct writing of single-mode polysiloxane optical waveguides and devices. J Lightwave Technol 2014;3036–42. Bd. 32, Nr. 17, S.

[115] Zgraggen E. Fabrication and system integration of single-mode polymer optical waveguides. Zürich: ETH-Zürich; 2014.

[116] Krähenbühl R, Lamprecht T, Zgraggen E, Betschon F, Peterhans A. High-precision, self-aligned, optical fiber connectivity solution for single-mode waveguides embedded in optical PCBs. J. Lightwave Technol 2015;865–71. Bd. 33, Nr. 4, S.

[117] Dangel R, et al. Polymer-waveguide-based board-level optical interconnect technology for datacom applications. IEEE Trans Adv Packag 2008;31(4).

[118] Matsuoka Y, et al. 20-Gb/s/ch high-speed low-power 1-Tb/s multilayer optical printed circuit board with lens-integrated optical devices and CMOS IC. IEEE Photon Technol Lett 2011;23(18).

[119] Takagi Y, et al. 4-Ch × 10-Gb/s chip-to-chip optical interconnections with optoelectronic packages and optical waveguide separated from PCB. In: IEEE CPMT symposium; 2010, Art. No. 5680206.

[120] Shiraishi T, et al. Cost-effective on-board optical interconnection using waveguide sheet with flexible printed circuit optical engine. In: OFC; 2011, Paper OTu5.

[121] Ishigure T, Niita Y. On-board fabrication of multi-channel polymer optical waveguide with graded-index cores by soft-lithography. In: IEEE Proc. ECTC; 2010, Art. No. 5490962.

[122] Pitwon R. Embedded optical interconnect for use in data-storage systems. Proc. SPIE 2009;. Available from: http://dx.doi.org/10.1117/2.1200912.002528.

[123] Beals IV J, et al. Terabit capacity passive polymer optical backplane. In: CLEO; 2008, Art. No. 4551179.

[124] McKee. Reflections on the memory wall. In: Proc. 1st Conf. on Comp. frontiers, Ischia; Apr. 2004.

[125] Giles M. Future of HPC (trends, opportunities and challenges). In: HPC symposium, Bath; 4 June 2013.

[126] Conway P, et al. Cache hierarchy and memory subsystem of the AMD opteron procesor. IEEE Micro 2010;30(2):16–29.

[127] Shekhar Borkar, Andrew A. Chien Communications of the ACM, Vol. 54 No. 5, Pages 67–77.

[128] Moore G. No exponential is forever. In Proc. IEEE Int. Solid-State Circuits Conf. Feb. 2003, p. 20–3.
[129] Cunningham JE, et al. Integration and packaging of a macrochip with silicon nanophotonic links. IEEE J Sel Top Quantum Electron 2011;Volume:17(Issue: 3):546–58.
[130] Maniotis P, Fitsios D, Kanellos GT, Pleros N. Optical buffering for chip multiprocessors: a 16GHz optical cache memory architecture. J Lightwave Technol 2013; 31(24):4175–91.
[131] Han S, Egi N, Panda A, Ratnasamy S, Shi G, Shenker S. Network support for resource disaggregation in next-generation datacenters. In: ACM SIGCOM, Hotnets–XII; 2013.
[132] Liu L, Kumar R, Huybrechts K, Spuesens T, Roelkens G, Geluk EJ, et al. An ultra-small, low-power, all-optical flip-flop memory on a silicon chip. Nature 2010; 268:1–6.
[133] Hill MT, Dorren HJS, de Vries T, Leijtens XJM, Hendrik den Besten J, Smalbrugge B, et al. A fast low-power optical memory based on coupled micro-ring lasers. Nature 2004;432:206–9.
[134] Liu Y, Mcdougall R, Hill MT, Maxwell G, Zhang S, Harmon R, et al. Packaged and hybrid integrated all-optical flip-flop memory. Electron Lett 2006;42:1399–400.
[135] Wang J, Meloni G, Berrettini G, Poti L, Bogoni A. All-optical clocked flip-flops and binary counting operation using SOA-based SR latch and logic gates. J. Sel Top Quantum Electron 2010;16(5):1486–94.
[136] Pleros N, Apostolopoulos D, Petrantonakis D, Stamatiadis C, Avramopoulos H. Optical static RAM cell. IEEE Photon Technol Lett 2009;21:73–5.
[137] Wright CD, Hosseini P, Diosdado JAV. Beyond von-Neumann computing with nano-scale phase-change memory devices. Adv Funct Mater 2013;23:2248–54.
[138] Ríos Carlos, et al. Integrated all-photonic non-volatile multi-level memory. Nat Photon 2015;9:725–32.
[139] Zhou Y, et al. An upconverted photonic nonvolatile memory. Nat Commun 2014; 5:4720.
[140] Lengle K, et al. Fast all-optical 10 Gb/s NRZ wavelength conversion and power limiting function using hybrid InP on SOI nanocavity. In Proc. ECOC, We2E5, Amsterdam; 2012.
[141] Chen C, Matsuo S, Nozaki K, Shinya A, Sato T, Kawaguchi Y, et al. All-optical memory based on injection-locking bistability in photonic crystal lasers. Opt Express 2011;19(4):3387–95.
[142] Nozaki K, Shinya A, Matsuo S, Suzaki Y, Segawa T, Sato T, et al. Ultralow-power all-optical RAM based on nanocavities. Nat Photon 2012;6:248–52.
[143] Kuramochi Eiichi, Nozaki Kengo, Shinya Akihiko, Takeda Koji, Sato Tomonari, Matsuo Shinji, et al. Large-scale integration of wavelength-addressable all-optical memories on a photonic crystal chip. Nat Photon 2014;8:474–81.
[144] Tanabe T, Notomi M, Mitsugi S, Shinya A, Kuramochi E. Fast bistable all-optical switch and memory on a silicon photonic crystal on-chip. Opt Lett 2005;30:2575–7.
[145] Shinya A, et al. All-optical flip-flop circuit composed of coupled two-port resonant tunneling filter in two-dimensional photonic crystal slab. Opt Express 2006; 14:1230–5.
[146] IEEE P802.3bs 200 Gb/s and 400 Gb/s Ethernet Task Force. Official web site: http://www.ieee802.org/3/bs/index.html. Last Update: 16- May-2016.
[147] Tekin T. Review of packaging of optoelectronic, photonic, and MEMS components. IEEE J Sel Top Quantum Electron 2011;17(3):704–19.

Key requirements for optical interconnects within data centers

M. Duranton[1], D. Dutoit[2] and S. Menezo[2]
[1]Commissariat à l'énergie atomique et aux énergies alternatives, Gif-sur-Yvette Cedex, France, [2]Commissariat à l'énergie atomique et aux énergies alternatives, Grenoble Cedex, France

3.1 An explosion of data

3.1.1 The data deluge: the data explosion fueled by Internet of Things

The evolution of ICT (Information and Communication Technology) can be seen as composed of different periods, alternating between the dominance of centralized and then decentralized approaches. In the 1950s, the mainframes open the era of large central facilities providing batch processing, general-purpose computing facilities for large enterprises. Minicomputers enlarged the market to medium-scale companies, but they were also generally centralized in "computing centers" due to maintenance and the specialized staff they required. It is with the emergence of PCs, initially introduced in the early 1980s, that computing became available to a larger audience and became decentralized, helped with the growth of Internet in the late 1990s. The use of the Internet was booming, increasing by two orders of magnitude between 1995 and 2015, reaching 45% of the world population nowadays, and nearly 88% of the North America population [1]. The Internet, with all the services it offers, lead again to the building of large computing infrastructures to host all the various servers that provide services to Internet users. Even if PCs are far more powerful than the terminals of the 1980s, there was (and still is) a tendency to move storage and computation to the large data center infrastructures, or to the "cloud." Netbooks that emerged in 2007 are representative of the idea that the user's terminal could have reduced computing power and storage—"all is in the cloud." At the same time, the introduction of the iPhone by Apple triggered the era of smartphones. As, at that time, they were far lower performing than PCs, they strengthened the usage of the "cloud," and larger data centers were built. Mobile access overtook Internet desktop access in 2014 in the United States, and it is now the major interface to the web for most users [2].

But we are now entering a new era, where the Internet is not only for humans, but also allows machines to talk to machines: the "Internet of Things (IoT)." Beside the "Human Internet," where humans are on one side and are users or providers of information from the cloud, now machines are interconnected and can directly communicate with the cloud. Gartner predicts that 25 billion connected

Optical Interconnects for Data Centers. DOI: http://dx.doi.org/10.1016/B978-0-08-100512-5.00003-6

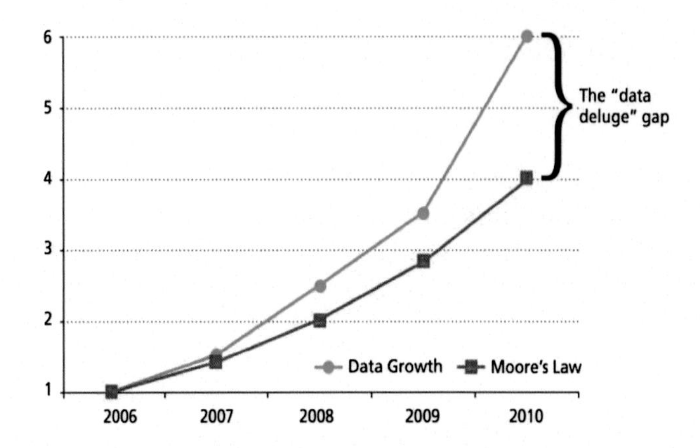

Figure 3.1 Data growth vs. Moore's Law trends. Data "deluge" means that we are heading towards a world where we will have more data available than we can process.
Source: From Duranton M, De Bosschere K, Cohen A, Maebe J, Munk H. HiPEAC vision 2015. https://www.hipeac.net/publications/vision/; 2015.

"Things" will be in use in 2020, and there are already 4.9 billion in 2015 [3] with a large market potential. Together with the data created or requested by humans, this will cause an exponential growth of the digital world. This explosion is called the "Data Deluge". The term "Data Deluge" was coined in 2003 [4] in the context of scientific data management to describe the massive growth in data volume generated in research, which was rapidly dwarfing all the data previously collected. Since then, the ability to generate vast quantities of data has outpaced the infrastructure and support tools.

In 2015 the world generated over 8.5 ZB (10^{21} bytes) of new data, and in 2020 more than 44 ZB [5] will be created, which is roughly is equal to 62 times the amount of all the grains of sand on all the beaches on earth. Global Internet traffic in 2019 will be equivalent to 66 times the volume of the entire global Internet in 2005 [6]. It is clear that optical interconnects, will be the main communication medium for long haul, regional, and even metro links (Fig. 3.1).

If we compare this growth with Moore's law [7] (transistor density doubling every 2 years), it is clear that data show a higher exponential growth than computation capacity, and this unprecedented trend is forcing us to reevaluate how we work with data in computer systems, and will lead to even bigger data centers to store and process data. But this will certainly drive more decentralized data processing and storage: instead of sending "raw" data to the cloud, the data will be processed, analyzed, and compressed as early as possible to extract only the relevant information that will be transmitted over long distances. The advent of federated clouds, mesh-distributed environments, increasing storage and broadband capacities available to home users, "*edge* [8] *and fog*" *computing*, metro and regional delivery networks, and "microservers" that will process and cache data at all levels of the communication chain; all these will contribute to avoid overloading the main backbones and central data centers.

As a result, the massive unified data stores and processing resources that are currently in large data centers might again become fragmented over countless systems, from Cyber Physical Systems (sensor data) to private people's personal NAS/SAN devices and federation of a variety of clouds.

3.1.2 IoT and cyber physical systems: the new challenges for data centers

However, data centers, even with the distributed computing facilities offered by IoT systems and Cyber Physical systems, will remain the main locations of data transmission, processing and storage. Cisco predicts [10]: "*by 2019, 1.8 ZB will be traditional traffic, while 8.6 ZB will be cloud data center traffic. Traffic between data centers and end users will reach 1.9 ZB annually (18% of total data center traffic), while 0.9 ZB (9%) will come from traffic between data centers and 7.6 ZB (73%) will be coming within the data center.*" It is therefore of the utmost importance to have very high bandwidth, low latency, and affordable solutions to cover this need for communication within data centers. To allow accessing and using such an amount of information in the most effective way at the data center level, and to enable the continued growth of data transfer and storage over the Internet, low cost ($<$1$/Gbps), high speed ($>$100 Gbps per link), and efficient ($\sim$1 pJ per transmitted bit of data) transmission solutions are needed for [0–2 km] interconnects at very high volumes ($\sim$10 million units for four channel-modules, accounting for 3.3 billion $ in 2020, [11]). Optical interconnects are key elements to ensure this communication throughput at the right cost and size.

The major capital expenditure of data centers is switches, servers, memory, and interconnect between these units. As the switch components become larger and increase in Input/Output bandwidth, the optical transceivers need to sit closer to the switch, and be positioned on the switch board, limiting the nonefficient electrical path between the two. This location requirement adds new constraints on the optical transceiver packaging, as mid-board transceivers are preferred to conventional standalone modules. As we will see later, new solutions for optical interconnect, already well-established to interconnect data centers and racks inside data centers, will be a key ingredient to decrease power consumption at the board and event at the component level.

A rule of thumb is that above 100 Gbps/m photonic solutions are more efficient. Below that limit, photonics might still be cost-effective, or have more power efficiency or density. Due to the high bandwidth increase even at chip level, this demonstrates the trend of photonics becoming the prominent interconnect technology, even at board (and even chip) level.

3.2 What are data centers?

3.2.1 A brief taxonomy of data centers

There are several ways to classify data centers. Of course, size is one. Due the ever increasing need to store and process data, their physical size is also growing. Larger data centers raise the need for extended transmission reaching higher than 500 m with higher than 25 Gbps bandwidth. Data center architectures are also flattening, requesting more links. But there is still an upper limit in their power consumption: a common estimate is that it should not be above 20 MW (mainly due to cost reasons).

But another way to classify data centers is their quality of service or their reliability. The Uptime Institute created a classification of data centers in four tiers [12]: Tier 1 is composed of a single path for power and cooling distribution, without redundant components offering 99.671% availability. Tier 2, is still composed of a single path for power and cooling distribution but with redundant components, and allows 99.741% availability. Adding multiple active power and cooling distribution with redundant components defines Tier 3, which provides 99.982% availability. Finally, fault tolerant data centers with higher than 99.995% availability are part of Tier 4. We see that reliability of components is very important for data centers, and photonics devices should also offer a very high reliability (and possibilities for redundant solutions), maintainability is also very important: changing a photonic component should be easy and should not imply shutdown of a large part of the system for a long period of time. Hot-plug is a must.

As stated above, there are plenty of ways to characterize data centers. Another one is to consider whether the company that is using them owns them or if the data center resources are outsourced. The last category is growing for economical reasons: owning a data center and keeping it up-to-date requires a large investment and is not the core business of most data center users. "Hiring" data center resources is the most flexible solution, allowing dynamic increase or decrease of the resources according to the evolution of needs. And you can adapt the nature of the data center to a particular need. Data center resource providers have *purpose-built* data centers. For example, there are shared data (Amazon, Rackspace, etc.), and data centers specialized to "cache" data (Akamai, Limelight, etc.). Some are more specialized for storage data, others for providing computing resources to a large number of users, either through applications (e.g., Google, Facebook, etc.) or using resource virtualization. Virtualization allows creation inside a real machine (e.g., a processor node) of one or several "virtual" machines that seem to the end-user to be a complete physical machine, even if in practice the physical resources (CPU, memory, IO, storage) are shared. This allows "consolidating" loads, therefore always being able to use the maximum physical resources. There are also specialized data centers, e.g., the ones used by Telecom companies, or for high frequency trading. High Performance Computing (HPC) centers can belong to a specialized category: instead of providing resources for a large number of not so demanding applications but for a large number of users, they provide resources for few applications that take a large part of the machine. By oversimplification, we can say that a cloud data server provides a computing core to a million applications (therefore for a million users), while an HPC machine provides a million cores to only one application. Anyway, all providers try to attract customers with an offer adapted to the users' needs.

All the different specialized data centers have different structures, but they share the same basic elements: compute nodes, memory, interconnect, and storage. Only the ratio and organization may vary. This also impacts on the communication and therefore on the photonic devices requirements: total bandwidth, bandwidth per channel, number of channels, latency, etc. Data centers are generally built with some kind of hierarchy; they are composed of a rack of servers (blade server). There are less than 20 servers per rack. Depending on the application, the number of servers can be up to 100,000 (Google centers).

3.2.2 Energy is a key challenge for data centers

The progress of computer technology is fueled by the so-called Moore's law [7], which describes the exponential increase of the number of transistors that can be integrated on a silicon area over time. It is expected that this transistor density increase will continue for the next few years, thanks to innovations like vertical transistors (FinFET), new materials, or 3D monolithic [13], etc. Even before the end of Moore's law, we are already facing the end of Dennard's scaling [14]: in old technology nodes, going to the next node not only allowed double the number of transistors, but also nearly double the running frequency while reducing the voltage, therefore having a nearly constant energy density. Unfortunately, at the beginning of the new millennium, the technology nodes were not so friendly: even if the transistor density continued to double, the running frequency hardly improved and the energy density began to increase with the number of devices, both due to leakage and active power. The first consequence was the limitation of performance of single thread processors, leading to the development of multi, then many cores from around 2005. The more drastic constraint comes from the increase of energy per unit of surface. For computing systems, the limit of what can be afforded has been reached, and therefore the main challenge now is to reduce energy consumption. This is obvious for mobile systems, to increase battery life, but also for servers because of the cost of ownership (electricity cost) and the limit of the energy than can be evacuated by the active area of silicon.

The high electrical consumption of hardware generates heat, and therefore increases the cost of the cooling devices (and their energy consumption) and the overall cost of the complete data center.

Electric consumption in large data centers may be up to several tens of MW, and affects a large part of the cost of a data center (its TCO—Total Cost of Ownership). Indeed, operating costs must be added to the cost of construction, which is—if we do not include the servers themselves—roughly proportional to the power consumption, because of the cost of power distribution and cooling systems. The buildings themselves must be sized according to the heat removal. Overall, it remains the depreciation cost of the servers themselves (renewed every 2–3 years, much more frequently than the rest of the facility, scheduled to last at least 10 years) which is not directly proportional to the energy consumption. Since about 2007 [15], the cost of energy and the cost of the data center power and cooling infrastructure became as important as the cost of the IT equipment, and drives the choice of the global infrastructure.

Data centers worldwide are estimated to have consumed 268 TWh in 2012 [16], which is about 1.4% of the worldwide electricity consumption at that time and is a 52 TWh increase since 2007. Projections shows a large increase in the share of data center power consumption in worldwide electricity consumption by 2020.

In the particular domain of HPC, if we continue using the current technology, the power (only for the CPU) for an Exaflop computer will be equivalent to the power use of the complete Bay Area (San Francisco and San Jose [17]), and the cost of the electricity bill of the complete machine will be of the order of 1 M€/day!

According to Matt [18], the average power usage for a typical dual-processor 450 W 2U server is indicated in Table 3.1.

The losses in AC/DC and DC/DC are due to the inefficiency of the power conversion processes, and are roughly proportional to the power consumed by the active parts, similarly for the fans. Large data centers receive electrical power at voltages in the order of 100 KV which are generally converted into low DC voltage supplied to a battery system, and then converted into intermediate voltage (~380 V) to be routed to the server racks where the voltage is further reduced to the one required by the components (~1 V). Primary power systems are doubled and a diesel generator coupled thereto. In case of power failure, the batteries provide power to the data center during the time of starting up the generator.

As shown in Table 3.1, the power of the active electronic parts accounts for 41% (plus 16% for the disks) of the total energy of this server blade.

It is difficult to assess the energy effectively consumed in the active transistors and by the leakage current, but with the current technology nodes (28 nm and beyond), the energy dissipated by communication is far higher than that consumed by the active transistors used for computing. nVidia shows [17] and [40] that a 64-bit double precision operation takes approximately 20 pJ, while an on-chip access of 256-bits in a 8 kB SRAM consumes 50 pJ, 10 mm of 256-bits bus 256 pJ, 40 mn of 64-bit on-chip bus 1 nj, and an efficient off-chip link 500 pJ and DRAM read/write 16 nJ. The relative cost between communication and computation grows with each new technology node, and the wire delay in ps per mm

Table 3.1 Average power usage for a 450 W 2U server rack

	AC/DC losses	DC/DC losses	Fans	Drives	PCI cards	Processors	Memory	Chip set
Power (W)	131	32	32	72	41	86	27	32
Percentage	29%	7%	7%	16%	9%	19%	6%	7%

is not improving. It is therefore clear than transmitting bit on wires inside dies, and even worse between dies, makes a major contribution to the power dissipation of the active electronics in a server. If optical interconnects can offer the same bandwidth, latency, and cost at lower energy dissipation than the current electrical interconnect, it will allow decreasing global dissipation of the system, and will have a big impact on the TCO.

We have seen that it is important to reduce the consumption of data centers, and especially since the trend is towards a continuous increase of the share of the power consumption in the overall cost, because consumption of servers increases with their performance (about 12% per year), but their purchase pricc should remain stable. As the cost of the servers is also an important fraction of the total cost of a data center, important for their short life compared to the rest of the facilities, it is important that the gains in power consumption should be made without significantly increasing the cost of servers.

One way to decrease the overall consumption is to make it more proportional to the performance: a data center is generally used at full power only for a very small part of the time.

In recent years, a lot of progress has been made on the proportionality of the consumption of CPUs versus their performance. These gains are not obtained today by DVFS (Dynamic Voltage and Frequency Scaling—adapting the voltage and the frequency of the CPU to the effective load), due to the low (below 1 V) supply voltage of modern technology that did not allow a large variation (except some technologies like FDSOI [19]), but by completely powering off some cores and reallocating their workload on other cores of the same die. It is not possible to completely remove power from a server, because new requests can reach them every few ms, and even if there is no need for the CPU, you may need data from disks, or to retain data in DRAM for subsequent faster access.

3.2.3 From data centers to racks to blades to chips: various architectures

There are various architectures for data servers, depending on their specialization. A trend announced by Intel is the extreme modularity of servers; the idea is to provide boards on which we can mount mezzanine boards to configure the server according to the exact needs of the user. It is called "rack disaggregation" [20]: all the resources that currently exist in a rack, compute, storage, networking, power distribution, are put into discrete boards. Then they can be grouped together, or distributed on the rack and changed independently. The backplane has an optical interconnection providing an Ethernet or PCIe protocol. Reference [20] shows the importance of photonic interconnect for this architecture: "*Intel's photonic rack architecture, and the underlying Intel silicon photonics technologies, will be used for interconnecting the various computing resources within the rack. We expect these innovations to be a key enabler of rack disaggregation*" said Raejeanne Skillern (Intel's director of marketing for cloud computing).

3.3 Data communication requirements

For data center owners, the cost of the interconnection network (mainly switches) is considered high [21]. Fortunately, for most data center—or for "cloud" data centers—the required bandwidth is well below the needs of HPC. In addition, the latency of the network is hidden by the disks, and is therefore not critical for today's "big data" applications. However, this could change if the cost of SSD decreased sufficiently to compete with hard drives. Network latency would become dominant, leading to an increase of network capacity to benefit from the performance of the SSD. The problem of the cost of the interconnection network could become a major issue, if not solved by cheap photonic solutions.

Table 3.2 shows the orders of magnitude of latency access to data according to wether they are in DRAM, SSD, hard drive, locally in the same rack server, or further.

It should be noted in this table that accessing the data of another server in the same rack adds approximately 300 μs of time access because of the network crossing time, and that accessing the data of another server in another rack adds yet an additional 200 μs. These interconnect delays have a low impact on disk access, intrinsically very long, but make access to data in the DRAM of another server almost as long as those for the Flash memories of this other server (Fig. 3.2).

However, the announcement of new Non Volatile Memories (NVMs) like the 3D Xpoint by Micron and Intel [22] will change the rules of the game: these new memories will have a drastic impact on the current memory hierarchy and will require a rethink of the architecture of servers, because they will bring large amounts of memory near the processor, memory faster than Flash memories (Intel says its 3D Xpoint is 1000 time faster than NAND) and as compact as DRAM (even more, it is claimed 10 times denser than DRAM). As they keep their content even without power, they could replace magnetic hard disks if the cost per bit is acceptable. Instead of having SRAM, DRAM, Flash, and HDD, the memory hierarchy could be reduced to SRAM (for fastest access and ultra low latency) and NVM. This will also have a drastic impact on the communication infrastructure that could not "benefit" from the latency of current disks; therefore, the interconnect will need to further improve its bandwidth but keeping a very low latency. Theses are the challenges that optical interconnect will have to cope with to be paired with these new NVMs. An example of a novel architecture for servers is the announcement by HP of "The Machine" [23]. This architecture will "*use electrons for computation,*

Table 3.2 Storage hierarchy in a data server

	Cluster (30 racks)	Local rack (80 servers)	One server
DRAM	30 TB, 500 μs, 10 MB/s	1 TB, 300 μs, 100 MB/s	16 GB, 100 ns, 20 GB/s
Disk	4.80 PB, 12 ms, 10 MB/s	160 TB, 11 ms, 100 MB/s	2 TB, 10 ms, 200 MB/s
Flash	600 TB, 600 μs, 10 MB/s	20 TB, 400 μs, 100 MB/s	128 GB, 100 μs, 1 GB/s

Source: From Barroso LA, Clidaras J, Hölzle U. The datacenter as a computer: an introduction to the design of warehouse-scale machines. 2nd ed. San Rafael (CA): Morgan & Claypool; 2013.

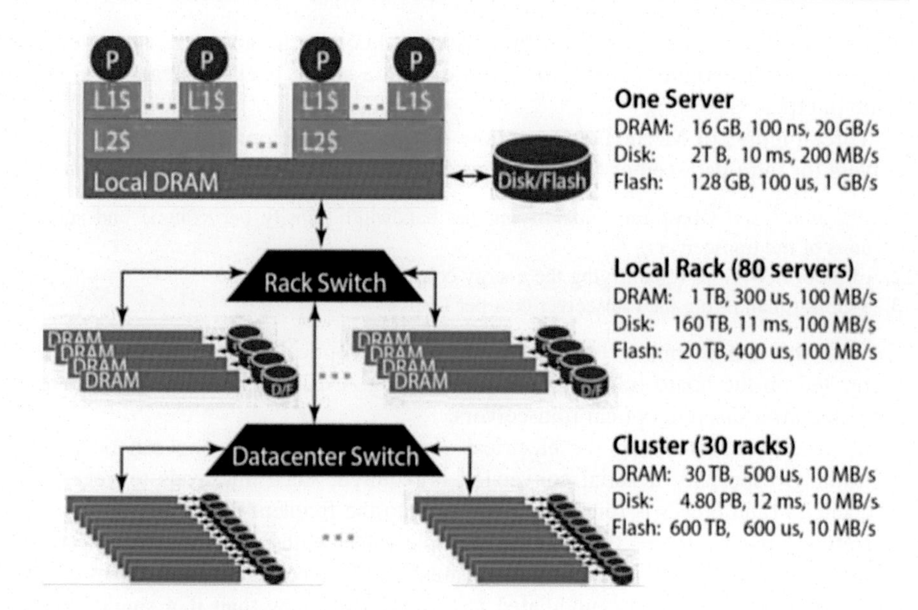

Figure 3.2 Storage hierarchy in a data center.
Source: From Barroso LA, Clidaras J, Hölzle U. The datacenter as a computer: an introduction to the design of warehouse-scale machines. 2nd ed. San Rafael (CA): Morgan & Claypool; 2013.

photons for communication, and ions for storage [39]." Photonic interconnects (and NVMs or Memristors) are key elements for this new architecture proposal.

Providing the microprocessors, switches, or FPGAs with Terabit per second (Tbps) optical transmission capabilities has emerged as a new requirement in exascale cloud data centers and HPC. The combination of Silicon Photonics and 3D interconnecting technologies stands as a unique solution for fulfilling this need while addressing the key factors of success that are: more Gbps per cm^2, less pJ per byte, and less $ per Gbps.

3.4 Optical interconnect: a solution to energy and bandwidth requirements?

Since the end of the last century, optical interconnections have supported the tremendous growth of worldwide data traffic, by providing fiber optics based technologies for long haul, metro, and access networks.

Current optical interconnects are generally used at the systems integration level or beyond, such as optical fibers connecting Internet devices across countries or oceans, and rack connections in data centers. A scaled-down version of this concept, called silicon photonics, promises lower communication energy cost, higher bandwidth, and low manufacturing cost in existing silicon foundries.

The technology is compatible with existing silicon technology. It can even be scaled down for communication between dies in a single package, leading to photonic interposers.

Three Figures of Merit (FoM) are now driving the development of optical transceivers for intra-data-center-transmissions and HPC systems:

1. Gbps/mm^3 and Gbps/mm^2 (quantifying the bandwidth density per volume and surface units of the transceivers),
2. pJ/bit or mW/Gbps (quantifying the energy consumption of the links),
3. $/Gbps (quantifying the transceiver cost per Gbps).

With the aim of improving the three FoM, the optical transceiver needs to be integrated on the board (closer to the transmitting ASIC). Silicon interposers are a next step after discrete optical transceivers.

In general, transforming the electrical signal into an optical one can be power consuming, especially if serial conversion is required, since this leads to very high frequencies (and power is roughly proportional to the frequency for CMOS devices). Therefore, optical systems that support multiple wavelengths may be preferable from an energy point of view: it allows the transmission of several streams of signal on the same medium, each stream modulated at a lower frequency than if a single stream with the same bandwidth would have been used. WDM (Wavelength Division Multiplexing), that allows sending multiple wavelengths over the same fiber, scales communication speeds linearly: adding an extra channel of 25 Gbps adds just 25 Gbps bandwidth to a single fiber. Switching off unused channels can save energy if not all the bandwidth is required at a particular moment in time.

3.4.1 Optical interconnect between data centers

The current data server infrastructure is almost completely based on optical interconnect. Where in the past fiber was mostly used for communication between data centers, it has now replaced copper wiring inside data centers.

Communication between data centers is based on single-mode fiber technology. 100 Gbps per wavelength through the use of advanced modulation techniques, and the use of WDM enable to achieve >3 Tbps. However, this technology is limited by several factors:

- Speed: dispersion in the fibers causes bit smearing. This can be solved through optical dedispersion, or through signal processing at the fiber ends, but that needs high-speed processing, and therefore energy.
- Power: high intensities from high-power lasers cause nonlinear signal distortion effects in fibers.

These limitations come from the combination of the medium (quartz glass) and the wavelengths used, and cannot be easily solved. For now, 100 Gbps technology per channel is used [24], 400 Gbps is in the labs. Further developments in WDM, advanced modulation techniques, and enhanced protocols will further increase the bandwidths.

Table 3.3 **Cost of different interconnect solutions**

Technology	Cost of energy	Power consumption for 10 Gbps link	Power per Tbps
Copper	3500$ per year	10 W	1 kW/Tbps
Traditional optical interconnect	700$ per year	2 W (10 Km)	200 W/Tbps
VCSEL or Silicon Photonics (see below for the definition)	70$ per year	0.2 W	20 W/Tbps

The following table (Table 3.3) gives an example of cost and energy for different solutions:

3.4.2 Optical interconnect between racks

Communication between systems, and between racks within systems, is based on both single-mode and multimode fiber technology. Multimode fiber technology is cheaper (relaxed mechanical precision on the optoelectronic and inter optic interfaces), but limited in distance (on the order of 100 m) and speed (more dispersion than single-mode fibers, 10 Gbps is common per channel).

With the latest advances, single-mode and multimode are becoming competitive. However, once built, the communication backbone of a data center is more or less fixed, locking the owner into a particular technology for several years. For example, Google uses multimode technology, while Facebook uses single-mode technology in new data centers: "*We are committed to making the transition from multi-mode fiber to single-mode duplex because it will be less expensive. Silicon photonics must be manufactured to scale. … the cost gets down to where we want it, about $1 per 1 Gbit/s.*" [25].

3.4.3 Optical interconnect between blades and boards

Optical communication technology is also being developed for creating connections within systems, between components. Such components that directly connect the processing devices to optics are coming to market already, a trend that is referred to as "moving the (optical) connector from the edge of the PCB to the middle." This evolution also makes optoelectronic chip interconnects possible. HPC profits the most from this development; for other applications, the cost compared to copper interconnection is the main factor for acceptance.

The laser diode is an important part of the transmitter, and producing them in a controlled, reproducible, and cost-effective way is important. Most modern transmitters are using VCSEL (Vertical Cavity Surface-Emitting Laser) where the beam is emitted perpendicularly to the top surface. The fabrication process of VCSEL

Table 3.4 **On-board VCSEL-based optical modules**

Device	Key features	Electrical connector	Optical connector	Reference
Avago MicroPOD	12 × 10 Gbps	MegArray	Prizm connector	[27]
Finisar BOSA	12 × 25 Gbps	Electrical interposer	MT ferrule connector	[28]
SAMTEC FireFLY	12 × 14 Gbps	UEC5/UCC8	MMF Pigtail	[26]

allows testing at several stages, leading to a more controlled yield and quality. The final step is a test on-wafer, before mounting, allowing reduction in the overall cost. Furthermore, due to their structure, they offer high coupling efficiency with optical fibers.

Optical transceivers using VCSEL transmitters and photo-receivers are commercially available (Table 3.4). They use Surface Mount Technology (SMT) or equivalent techniques, so that the package can be placed close to the host chip on the motherboard. Typically, they provide Electrical/Optical and Optical/Electrical conversion. Depending on the connecting system, they can be directly mounted on the PCB using BGA balls, or electrically connected to the board through a high-speed electrical connector. These devices have a typical footprint less than 20 × 20 mm, and FoM of ~40 Gbps/cm^2, ~20 mW/Gbps, <100 fJ/bit dissipated energy, and ~2 $/Gbps [26].

The transmission is made through Multi-Mode Fiber (MMF) enabling low-cost fiber attachment techniques, but limiting the transmission reach (<100 m at 25 Gbps with On-Off Keying modulation). In addition, MMF does not support WDM.

However, the current optical transceiver solutions for intra data center applications suffer from:

- Limited performances: VCSELs-based transceivers can only provide 25 Gbps per fiber up to 100 m of Multi-Mode Fiber (MMF) at a cost of ~2$/Gbps, for about 20 mW/Gbps;
- Cost that does not meet the investments capability of the web-players [25]: these are for the traditional WDM-transceivers (Externally Modulated Lasers—EML, and Directly Modulated Lasers—DML, from InP technologies), which can provide 4 × 25 Gbps but at higher cost [11].
- Packaging represents a high cost share. This is especially true for single-mode optics because today nearly all DML and EML in the market require a hermetic package, and single-mode optical alignment is expensive.

For reference, electrical links for System-In-Package transceivers are at about 25 GB/s over 5 m for 6 mW/Gbps and 0.2$/Gbps. However, at data rates near 25 Gbps, many signaling and interconnect issues are appearing; signals traveling across PCBs lose most of their integrity in only a few centimeters, requiring special PCB material at very high cost. This is a further handicap for electrical interconnect at very high data rate.

The current performances and limitations of optical interconnect solutions are summarized on Fig. 3.3.

Optical integration has the promise to significantly reduce cost while meeting the needed performances. This is especially true when a large fraction of the functions in a transceiver can be integrated on the same platform with high yield. In the past years, a lot of attention has been paid to silicon photonic optical passive and active devices integrated using CMOS compatible processes, with mass-production capabilities. Silicon photonics transceivers are now entering the market in between VCSELs, and EML/DML [29].

Silicon photonics stands as the unique technology, with the potential to achieve the aggressive FoM of >100 Gbp/cm^2, ~2 mW/Gbps, and <1$/Gbps, with >1 km-reach capabilities. The technology is developed to be fully compatible with the existing CMOS infrastructure, and now runs on research and production foundries [30,31]: ST Microelectronics and IBM have led the way and introduced silicon photonic lines in their fabs. Silicon-photonic-based modules typically use a Photonic Integrated Circuit (PIC) comprising silicon waveguides, silicon modulators, and Germanium-on-Silicon waveguide photodiodes. Silicon-photonic WDM components achieve state-of-the-art performances [32], while having reduced footprints (divided by a factor of 100 and 10,000 respectively as compared with Indium Phosphide and Silica-on-Silicon-based devices). The integration of the PIC and its electronic driving and reading ICs (EIC) can be monolithic [33] or hybrid [34]. This leads to significant reduction in the power consumption as the PIC-to-EIC interconnects are shortened [35], avoiding the need for matching resistors, which are responsible for a large fraction of power consumption of the drivers. Optical modules are

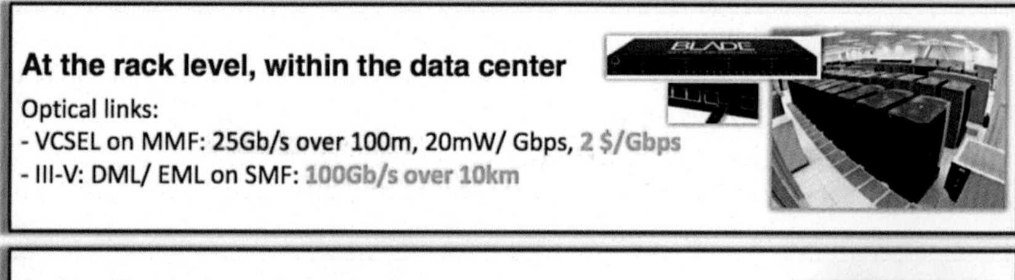

At the System-In-Package level

Electrical links:
- 25 Gbps over 5m, 6mW/Gbps
- 0.2 $/Gbps

Optical links:
- VCSELs, closer to the ASIC (Surface Mountable engines)
- BW densities

At the chip level
- Memory access is a bottleneck
- Access more memories, deported further, while keeping low latencies

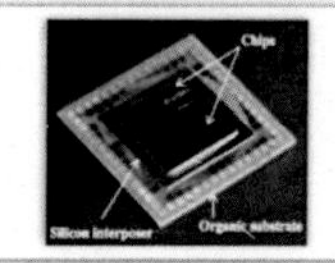

Figure 3.3 Current performances and limitations of photonic interconnects within data centers.

now available [36], in QSFP form factor, providing 100 Gbps. Eight parallel fibers are used, carrying each 25 Gbps (4 × 25 Gbps Transmit and 4 × 25 Gbps Receive) over up to 2 km of SMF. The EIC/PIC assembly is wire bonded to the PCB.

3.4.4 Optical interconnect between chips and dies

Due to the cost, bandwidth, and energy requirements, the application of optoelectronics communication is also moving at chip level. Designing high performance chips is becoming more and more expansive because of the exploding complexity and validation time (see Fig. 3.4 and [37]).

Even technology scaling is facing severe difficulties: cost, variability, power density, and timing uncertainties. In order to pursue technology and design integration, More Than Moore technologies, such as 3D integration, are becoming increasingly attractive, and they are enablers to bring silicon photonics at the chip level. With 3D technologies, and Through Silicon Vias (TSV), it is possible to stack various dies together. The 3D technologies are opening a full range of new application possibilities, by integrating more devices from potentially different technologies (CMOS, Flash, DRAM, Photonics, etc.). Full 3D can be envisaged by stacking vertically dies on top of each other, or Interposers—also called 2.5D—by stacking horizontally dies (called chiplets) onto silicon substrate (Fig. 3.5).

The advantages of 3D technologies are numerous:

- Shorter communication distance between dies, thus reducing communication load and then reducing communication power consumption;
- Possibility of stacking dies from various heterogeneous technologies, like stacking memory on top of logic, Flash, or photonic, in order to benefit from the best technology where it best fits;
- Improve system yield and cost by partitioning the system in a divide and conquer approach: multiple similar dies are fabricated, tested, and sorted before the final 3D assembly, instead of fabricating ultra large dies with very reduced yield—exemplified by the first commercial FPGA interposer-based from Xilinx.

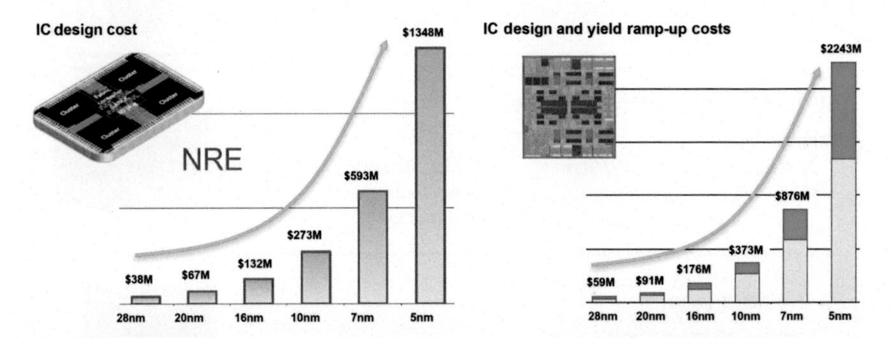

Figure 3.4 Exponential cost of the design of chips in small technology nodes.
Source: From Mourey, B. What chipmakers will need to address growing complexity, cost of IC design and yield ramps, http://electroiq.com/blog/2015/06/what-chipmakers-will-need-to-address-growing-complexity-cost-of-ic-design-and-yield-ramps/; 2015.

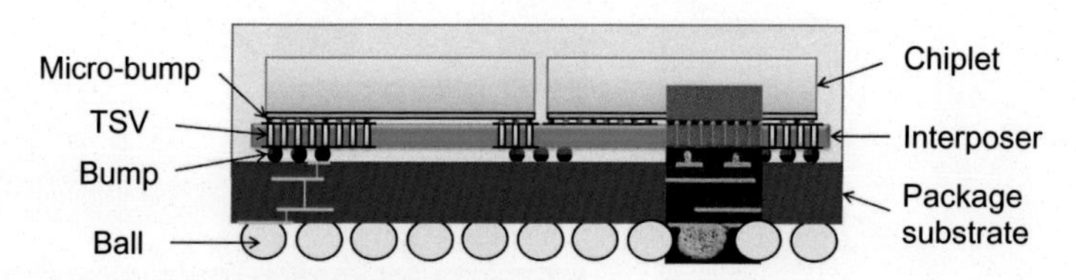

Figure 3.5 2.5D integration with silicon interposer.

Table 3.5 Roadmap for interposer technologies

	Today	2016–18	2020
Technology	Metallic interposer	Active interposer	Photonic interposer
# chiplet per interposer	1–4 chiplets	5–9 chiplets	>10 chiplets
On-chip bandwidth	≤ 250 Gbps	≤ 2 Tbps	> 4 Tbps (>2×)
Number of cores	≤ 16	≤ 36	> 72 (>2×)
Power for on-chip communication	≈ 1 W	≈ 20 W	≈ 20 W (≈1×)

All these advantages will facilitate the design of high performance processors: faster and shorter communication, allowing denser computation and reduced cache sizes; use of the appropriate technology to efficiently partition the many-core architecture; stack DRAM on top of core arrays, as with WideIO, or use 3D DRAM (like with Hybrid Memory Cube); implement power management and system Ios, e.g., using photonics within an active interposer; partition the many-core in regular 3D stackable tiles in order to improve system cost, with mask reuse, and overall yield reduction [38].

Using 2.5D technology enables the use of a silicon photonics interposer, where electrical links are replaced by optical links which achieve increased information bandwidth, with the goal of achieving energy per transmitted bit lower than ~1 pJ. Silicon photonics allows a denser integration of current processor and memory devices, while maintaining a tight energy budget. They also overcome one large part (up to 80%) of the component cost of standard photonic solutions: the cost of the packaging.

Table 3.5 gives some characteristics of the various interposer technologies and some forecasts.

Photonics can be integrated in a two-step approach:

First, a chiplet making the interface between electrical signals and photonics can be used instead of a pure electrical interconnect chiplet (off-chip photonics approach, see Fig. 3.6),
Then, when the technology is mature, a full photonic interposer can be used to interconnect the chiplets together in a single package (in-package photonics) (see Fig. 3.7).

For example, building on previous photonics devices for high-speed communications using silicon-compatible technology (Figs. 3.8 and 3.9), the CEA's Hubeo+ project is building a demonstrator to interconnect microprocessors and memories through a photonic interposer.

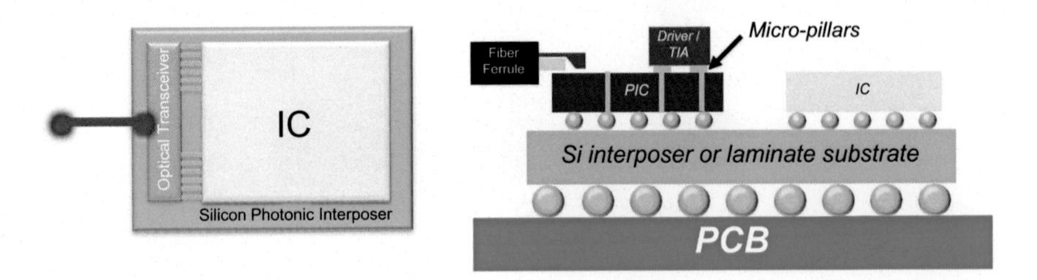

Figure 3.6 Off-chip photonics.

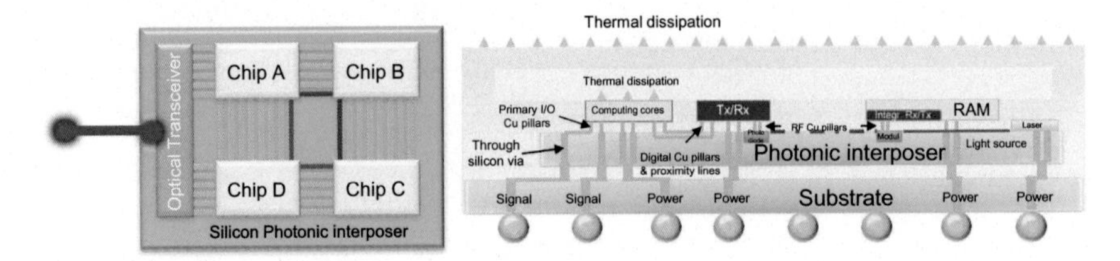

Figure 3.7 In-package photonics.

Figure 3.8 Flip chip assembly with micro-bumps for high speed electrical interconnects.

Fig. 3.10 shows the system architecture and an example of the 3D packaging technology required to implement a Silicon photonic interposer.

3.5 Conclusion

Over the years, as the requirements for higher bandwidth constantly increases, photonic interconnects are acknowledged to be an efficient solution for shorter distances. Today, the cross-over point with electrical interconnects is a [bandwidth×distance] product of ~100 Gbps/m, as illustrated in Fig. 3.11. With its capability to achieve

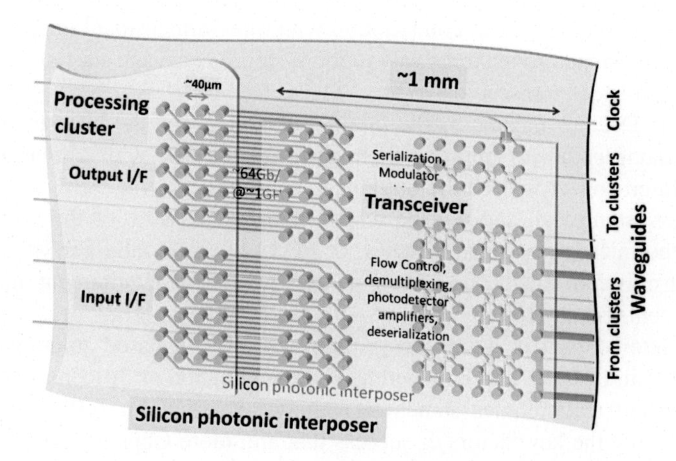

Figure 3.9 3D-stacked die with RF copper-pillar assembly.

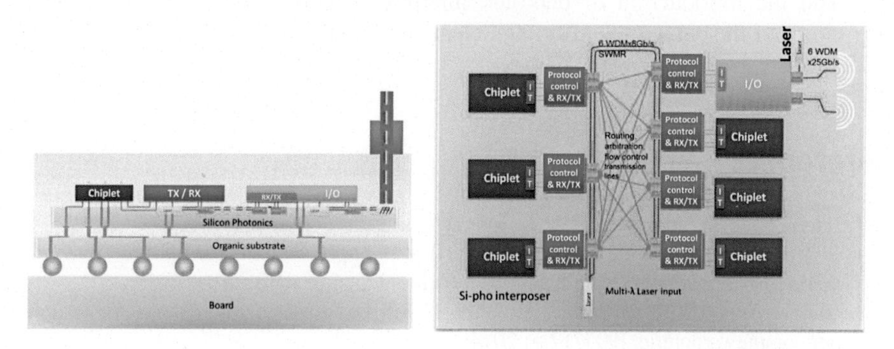

Figure 3.10 Cross-section of assembly view showing the 3D integration on a silicon photonic interposer and the system architecture view showing the multipoint optical NoC between the processing devices (chiplets).

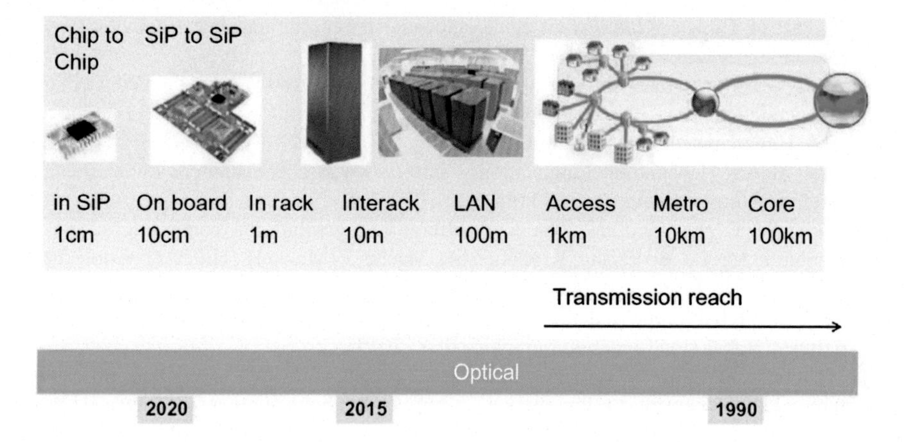

Figure 3.11 Evolution of the use of photonics interconnects over the years.

low cost, Silicon photonics, in combination with three-dimensional integration technologies, is envisioned to make its way to the system in package starting in 2020, for providing high-speed transceivers to big ASICs.

The "Data Deluge" created by the Internet of Things and the Internet of Humans will force data centers to improve their performance, but with a constant power envelope. Interconnect will be the bottleneck as it will consume 80% of the power, 40% of the system cost, and will be responsible for 50−90% of the performance: providing the microprocessors, switches, or FPGAs with Terabit per second (Tbps) optical transmission capabilities has emerged as a new requirement in exascale cloud data centers and HPC.

Copper interconnect is power consuming and VCSEL-based solutions are currently limited in distance and bandwidth. The combination of Silicon Photonics and 3D interconnect technologies stands as a unique solution for fulfilling the need, while addressing the key factors of success that are: more Gbps per cm^2, less pJ per byte, and less $ per Gbps. Photon will replace electrons for transporting information, and the introduction of photonic interposers will bring optical interconnect ever closer to the compute and storage elements of data servers.

References

[1] Internet World Stats, <http://www.internetworldstats.com/stats.htm>.
[2] Bosomworth D. Mobile marketing statistics, <http://www.smartinsights.com/mobile-marketing/mobile-marketing-analytics/mobile-marketing-statistics/>; 2015.
[3] Gartner says 4.9 billion connected "things" will be in use in 2015, <http://www.gartner.com/newsroom/id/2905717>; 2014.
[4] Hey T, Trefethen A. The data deluge: an e-science perspective. In: Berman F, Fox G, Hey T, editors. Grid computing: making the global infrastructure a reality. Chichester: John Wiley & Sons, Ltd; 2003.
[5] IDC. The digital universe of opportunities: rich data and the increasing value of the Internet of Things, <http://www.emc.com/leadership/digital-universe/2014iview/executive-summary.htm>; 2014.
[6] CISCO. The zettabyte era—trends and analysis, <http://www.cisco.com/c/en/us/solutions/collateral/service-provider/visual-networking-index-vni/VNI_Hyperconnectivity_WP.html>; 2015.
[7] Intel. Moore's law and intel innovation, <http://www.intel.com/content/www/us/en/history/museum-gordon-moore-law.html>.
[8] wikipedia. Edge computing, <https://en.wikipedia.org/wiki/Edge_computing>.
[9] Duranton M, De Bosschere K, Cohen A, Maebe J, Munk H. HiPEAC vision 2015, <https://www.hipeac.net/publications/vision/>; 2015.
[10] CISCO. Global Cloud Index (GCI), <http://www.cisco.com/c/dam/assets/sol/sp/cloud_index/global-cloud-index-infographic.pdf>; 2015.
[11] Lightcounting. growing investments of web 2.0 companies into networking infrastructure adds excitement to the market, <http://www.lightcounting.com/News_022615.cfm>; 2015.
[12] UptimeInstitute. Tier classification system, <https://uptimeinstitute.com/tiers>.

[13] Johnson RC. CoolCube 3D goes monolithic, <http://www.eetimes.com/document.asp?doc_id=1326213>; 2015.

[14] Dennard RH, Gaensslen F, Yu H-N, Rideout L, Bassous E, LeBlanc A. Design of ion-implanted MOSFET's with very small physical dimensions. IEEE J Solid State Circuits SC−9 1974;(5):256−68.

[15] Belady, CL Data centers in the data center, power and cooling costs more than the it equipment it supports, <http://www.electronics-cooling.com/2007/02/in-the-data-center-power-and-cooling-costs-more-than-the-it-equipment-it-supports/>; 2007.

[16] Internet-science.eu. D. 8.1. Overview of ICT energy consumption, <http://www.internet-science.eu/sites/eins/files/biblio/EINS_D8%201_final.pdf>.

[17] Lanfear T. GPU computing and the future of HPC, <http://orap.irisa.fr/wp-content/uploads/2015/04/NvidiaTimothyLanfear.pdf>; 2015.

[18] Matt. Average power use per server, <http://www.vertatique.com/average-power-use-server>; 2015.

[19] Beigne E, Valentian A, Miro-Panades I, et al. A 460 MHz at 397 mV, 2.6 GHz at 1.3 V, 32 bits VLIW DSP embedding F MAX tracking.. IEEE J Solid-State Circuits 2015;50(1):125−36.

[20] Skillern, R. Meet the future of data center rack technologies, <http://www.datacenterknowledge.com/archives/2013/02/20/meet-the-future-of-data-center-rack-technologies/ > ; 2013.

[21] Barroso LA, Clidaras J, Hölzle U The datacenter as a computer: an introduction to the design of warehouse-scale machines. 2nd ed. San Rafael (CA): Morgan & Claypool; 2013.

[22] Intel and Micron Produce Breakthrough Memory Technology, <http://newsroom.intel.com/community/intel_newsroom/blog/2015/07/28/intel-and-micron-produce-breakthrough-memory-technology>; 2015.

[23] Labs HP. The machine, <http://www.labs.hpe.com/research/systems-research/themachine/>; 2014.

[24] FINISAR. Optical transceivers, <https://www.finisar.com/optical-transceivers>.

[25] O'Shea D. Optical feels the facebook effect, <http://www.lightreading.com/optical/100g/optical-feels-the-facebook-effect/a/d-id/714866>; 2015.

[26] Verdiell JM. Advances in onboard optical interconnects: a new generation of miniature optical engines. In: DesignCon; 2013.

[27] Vaughan D, Hannah R, Fields M. Applications for embedded optic modules in data communications. In: White Paper; 2011.

[28] Finisar. Optical engines: 25G BOA (Board-Mount Optical Assembly), <https://www.finisar.com/optical-engines/fbotd25fl2c00>.

[29] Luxtera. FOR IMMEDIATE RELEASE Luxtera Debuts 1310nm 100G - PSM4 QSFP 28 Module and Silicon Photonics Chipset at OFC 2015, <http://www.luxtera.com/luxtera/LuxteraOFC2015PressRelease2232015.pdf>; 2015.

[30] Fowler D, Baudot C, Fedeli JM, Caire B, Virot L, Leliepvre A, et al.. Complete Si-Photonics device-library on 300 mm wafers. OFC; 2014.

[31] Boeuf F. Recent progress on silicon photonics R&D and manufacturing on 300 mm wafer platform. Semicon Europa; 2014.

[32] Sciancalepore C, Lycett RJ, Dallery JA, Pauliac S, Hassan K, Harduin J, et al. Gallagher, Sylvie Menezo, Badhise Ben Bakir. Low-Crosstalk fabrication-insensitive echelle grating demultiplexers on silicon-on-insulator. IEEE PTL 2015;27(5).

[33] Narasimha A., Abdalla S., Bradbury C, et al. An ultra low power CMOS photonics technology platform for H/S optoelectronic transceivers at less than $1 per Gbps. In: OMV4, OFC; 2010.

[34] Denoyer G, Chen A, Park B, Zhou Y, Santipo A, Russo R. Hybrid silicon photonic circuits and transceiver for 56 Gb/s NRZ 2.2 km Transmission over single mode fiber. In: ECOC; 2015.

[35] Saeedi S, Menezo S, Emami A. A 25Gbps 3D-integrated CMOS/silicon photonic optical receiver with -15 dBm sensitivity and 0.17 pJ/bit energy efficiency. In: Proc. IEEE OIC; 2015.

[36] Luxtera. Luxtera debuts 1310 nm 100G-PSM4 QSFP28 module and silicon photonics chipset at OFC 2015, <http://www.businesswire.com/news/home/20150323005646/en/Luxtera-Debuts-1310nm-100G-PSM4-QSFP28-Module-Silicon#.VZf7pvlIeUk>; 2015.

[37] Mourey, B. What chipmakers will need to address growing complexity, cost of IC design and yield ramps, <http://electroiq.com/blog/2015/06/what-chipmakers-will-need-to-address-growing-complexity-cost-of-ic-design-and-yield-ramps/>; 2015.

[38] Yvan, T. In (NoCS), Eighth IEEE/ACM international symposium on networks-on-chip, ed., (NoCS), Eighth IEEE/ACM international symposium on networks-on-chip. 2014.

[39] <http://h30507.www3.hp.com/t5/Innovation-HP-Labs/Electrons-for-compute-photons-for-communication-ions-forstorage/ba-p/115067>.

[40] Sekar, D. The Dally-nVIDIA-Stanford prescription for exascale computing, <http://www.monolithic3d.com/blog/the-dally-nvidia-stanford-prescription-for-exascale-computing>.

Part II

Materials and Components

Indium phosphide (InP) for optical interconnects

4

M. Lebby[1,2,3,4], *S. Ristic*[5], *N. Calabretta*[6] *and R. Stabile*[6]
[1]OneChip Photonics Inc., Ottawa, ON, Canada, [2]Glyndwr University, Wales, United Kingdom, [3]Lightwave Logic Inc., Longmont, CO, United States, [4]University of Southern California, Los Angeles, CA, United States, [5]McGill University, Montreal, QC, Canada, [6]Eindhoven University of Technology, Eindhoven, The Netherlands

4.1 Introduction

InP is one of the few semiconductors that can provide both active and passive optical devices. InP has found widespread use in telecommunications and other applications, mainly for the production of active devices because it is a direct bandgap semiconductor. Epitaxial InGaAsP and InGaAlAs layers can be grown on InP and uniquely enable lasers that emit light in the 1550 and 1310 nm wavelength windows, where the optical fiber exhibits small attenuation and dispersion, respectively. Other high-performance active devices such as photodetectors, electro-absorption and phase modulators, switches, attenuators, semiconductor optical amplifiers, and wavelength converters have been developed over the years as well. In addition, InP epitaxial layers can be designed to provide optical waveguides that are transparent in 1550 and 1310 nm wavelength windows, and a variety of high-performance passive devices have been developed, including: various types of optical waveguides, directional couplers, multimode interferometer couplers, arrayed waveguide gratings, Echelle gratings, Bragg gratings, ring filters, polarization converters, and others.

Unlike InP, Si does have native oxide. Also, silicon dioxide can be deposited on silicon as a several-micrometer-thick, low-stress, high-quality film that has low optical loss. It can provide hermetic encapsulation to silicon, electrical isolation, and importantly, high-refractive-index-contrast optical waveguides that has led to a miniaturization of silicon passive optical circuits. In addition, InP wafers have a higher level of fragility and suffer from higher defect levels in larger format sizes, which is one of the major reasons why InP formats are typically 100 mm or smaller. Unlike silicon wafers that come in sizes as large as 450 mm [1], the largest commercially viable InP wafers are 100 mm. Larger sized 150 mm InP wafers were available a decade ago, but had limited commercial interest [2] and were subsequently pulled out of the market. It is expected that there will be renewed commercial interest with 150 mm InP wafers in the near future [3]. Compared to InP, silicon has a major disadvantage in that it is an indirect bandgap semiconductor and inherently cannot provide usable optical gain and lasing. Regardless, silicon is

Optical Interconnects for Data Centers. DOI: http://dx.doi.org/10.1016/B978-0-08-100512-5.00004-8

cost-competitive for some applications due to the large size of it's wafers and compact size of it's passive circuits. In this manuscript, we argue that this cost-competitiveness relative to the incumbent InP only applies for very-large-volume applications, and it is questionable that it applies to the data center optical interconnect markets. Unlike silicon, InP offers monolithic integration of active and passive components, which can simplify packaging of optical circuits used in data centers, significantly reducing the form factor and the cost. We further argue that InP offers high-performance components for data center interconnects whose monolithic integration can produce electro-optic circuits that lead to significant energy savings, and, in turn, operational cost savings. Moreover, it is already expected that 150 mm InP wafers will be more widely commercialized in the near future, which will further increase the cost competitiveness of InP relative to Si for data center applications.

We continue by presenting the most popular InP monolithic integration approaches in Section 4.2. In Section 4.3, we then present the state-of-the-art in InP photonic integrated circuits (PICs) for data center applications, and also provide a comparison to their Si counterparts. We end the chapter with Section 4.4, where the future trends for InP technology are discussed.

4.2 InP photonic integration platforms

Regardless of the application, photonic integration has long been recognized for its importance in decreasing optical circuit footprint, decreasing packaging cost, and reducing the inter-element optical coupling losses. Naturally, these benefits of photonic integration extend to data center interconnects applications. There are several insightful reviews on the history and state-of-the-art of InP monolithic PICs [4–6], and here we summarize the most popular integration approaches, as illustrated in Fig. 4.1.

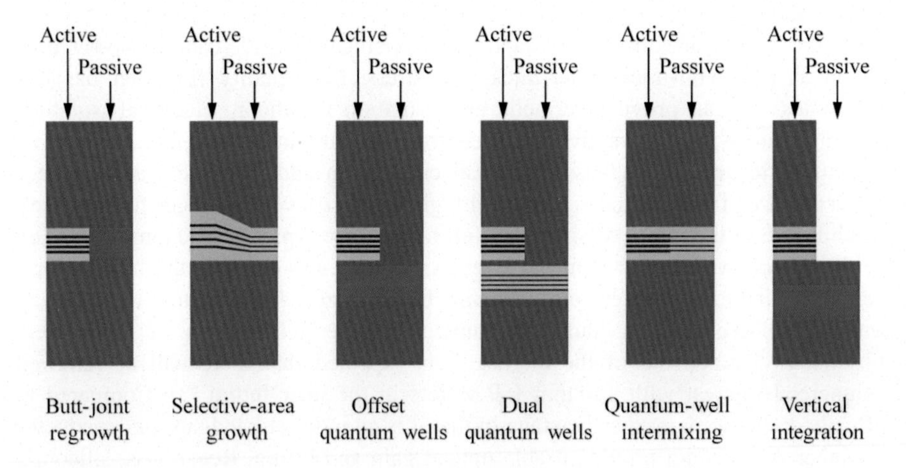

Figure 4.1 Typical InP monolithic integration approaches.

Butt-joint regrowth is considered to be the most flexible integration platform because it allows integration of epitaxial layers of different bandgaps with minimal restrictions in terms of the design and functionality of the integrated elements. A dielectric hard mask is deposited and opened over the patterned area where the regrowth will take place, while protecting the rest of the wafer. Three or more bandgaps of active and passive elements are usually integrated using this approach. This approach offers small optical coupling losses between the elements, but it requires good control of the growth process in order to manage the yield, as well as the optical reflections at the interface.

Similar to butt-joint regrowth, in the *selective area growth* approach, a patterned dielectric mask is used to define layers of different bandgaps. In the selective area growth approach, however, the area of the dielectric mask surrounding the elements to be grown is designed so that the larger the mask area, the smaller the precursor consumption over the mask area, and, in turn, the higher the growth rate of the element surrounded by the mask area and the thicker the corresponding layers. For quantum-well layers, the increased thickness implies electronic transitions of lower energies, i.e., lower effective bandgap. Selective area growth offers low reflections and optical coupling losses between the elements. Similar to butt-joint regrowth, it requires good control of the growth process. Another drawback of this approach is that it inherently produces thickness variations on the wafer surface that can complicate further lithography and processing steps.

The epiwafer for an *offset quantum-well* integration platform contains an active, quantum-well layer stack on top of a passive waveguide core layer. In this integration platform, the active layer stack is patterned and wet-etched everywhere except in the parts of the wafer intended to provide gain. Following the etching that defines active and passive areas, p-doped cladding is regrown and shared by both active and passive elements. Hydrogen can be diffused or implanted in the p-doped cladding for the passive elements to reduce the intervalence-band optical absorption. Laser gratings can be easily patterned before the p-cladding regrowth, because the active layer stack is thin and easy to planarize. A unique aspect of this integration platform is that the optical mode stays localized in the passive waveguide core and only evanescently couples to the active layers. Consequently, one of the main drawbacks of this integration platform is small optical confinement in the active layers. Optical loss and reflection at the active/passive interfaces are small, and the interface is angled relative to the waveguide axis in order to further reduce optical reflections. The bandgap of the passive core can be engineered to be relatively close to the wavelength of operation, with some optical absorption penalty, so that the Franz-Keldysh effect can be used for intensity or phase modulation. The most appealing aspect of the platform is the simplicity of the regrowth. The regrowth is done over a very small topography, and it is a regrowth of a simple, binary InP, which can be easily outsourced and performed by external vendors with high yield, without a need for special growth optimization.

A *dual quantum-well* integration platform in most aspects is similar to the offset quantum-well platform, except that the bottom layers are not passive. Rather, both top and bottom layers are active, quantum-well layers, where the top layers may form a laser gain medium and the bottom layers may form an efficient modulator. Here, the presence of the bottom quantum-wells will induce a degree of optical loss, i.e., intra-cavity absorption, and potentially some nonlinear interaction if it is not properly designed. Obviously, the bottom active layers are not suitable for long, passive routing due to a relatively large absorption in the quantum-well, but this technique can be combined with quantum-well intermixing in order to increase the bandgap in the passive sections.

In *quantum-well intermixing*, multiple bandgaps are formed by selective implantation and subsequent high-temperature annealing of a single stack of quantum-wells, where the size of the bandgap is proportional to the annealing time. Optical losses and reflections at the interfaces are very low. Intermixing can be done before the growth of the p cladding in order to avoid diffusion of zinc, which is typically used as a p dopant in active InP devices. Although a drawback of this approach is that a regrowth is likely required, this is a simple, easy-to-outsource regrowth of binary InP, similar to the offset quantum-well and the dual quantum-well integration approaches.

The last, but not the least, of the popular integration platforms is the *vertical integration platform*. Here, optical waveguide cores of different bandgaps are stacked on top of each other, separated by shared lower-index cladding material, typically InP. The optical mode guided by a waveguide core is squeezed into an underlying waveguide core by the use of lateral waveguide tapers. This adiabatic coupling requires interaction lengths of 10–100 second of micrometers in order to avoid optical radiation losses, and the PICs integrated in this platform typically have larger footprints compared to the other integration platforms. The thicknesses of the InP claddings that separate different cores, as well as the thicknesses of the waveguide cores, have to be carefully optimized in order to minimize undesirable interaction between the vertically stacked devices, while enabling efficient transfer of light between them. As a consequence, the optical confinement factor in each device can be restricted. The unique and important aspect of this technology, however, is that it does not require regrowth. The fabrication of the PICs in this approach can be done using well-established, high-yield processes developed for consumer electronics at a number of foundries that are not required to have regrowth capabilities. Due to the fact that outsourcing to external foundries and technology standardization is an important and increasingly popular approach to commercialization, vertical integration is a promising low-cost platform that is competitive for data center interconnects applications.

4.3 State-of-the-art in InP photonic integrated circuits (PICs) for data centers

InP PICs are an important technology for both transceiver and switching data center applications. In this section, the emphasis is placed on the state-of-the-art solutions that contain a few (Section 4.3.1) to hundreds of components (Section 4.3.2) on a single chip. In addition, the state-of-the-art in InP PICs is compared to their Si photonic counterparts.

4.3.1 Transceiver InP PICs by vertical integration

The vertical integration platform introduced in the previous section has been developed both in academia [7,8] and industry [9]. More recently, OneChip Photonics Inc., in Ottawa, Canada, has demonstrated vertically integrated transceiver PICs for data center interconnects [10–12] after having developed the integration technology to address large-volume, low-cost, fiber-to-the-home markets [13]. This particular regrowth-free flavor of vertical integration by OneChip Photonics is termed Multi-Guide Vertical Integration (MGVI) platform.

4.3.1.1 Transmitter PICs

OneChip Photonics Inc. developed both transmitter and receiver PICs using the MGVI platform, as a technology demonstration for 100GBASE-PSM4 and 100GBASE-LR4 standards. Fig. 4.2 illustrates the epitaxial layers of a 100G Ethernet (100GE) transmitter [10], including a distributed feedback laser (DFBL), an electro-absorption modulator (EAM), InGaAsP passive waveguide (PWG), and a diluted-waveguide spot-size converter (SSC) matched to a cleaved single-mode fiber. Both the DFBL and the EAM are based on InGaAlAs multiquantum-well (MQW) stacks. The DFBL in MGVI is an undercut, inverted-mesa structure, where the P cladding is at the bottom and the InP N cladding is at the top. The undercut is implemented in an aluminum quaternary (Al-Q) aperture layer, surrounded by InP etch-stop layers. This laser arrangement relies on a good contact resistance of N-doped InP that allows for the N contact metal to be made narrow and moved to

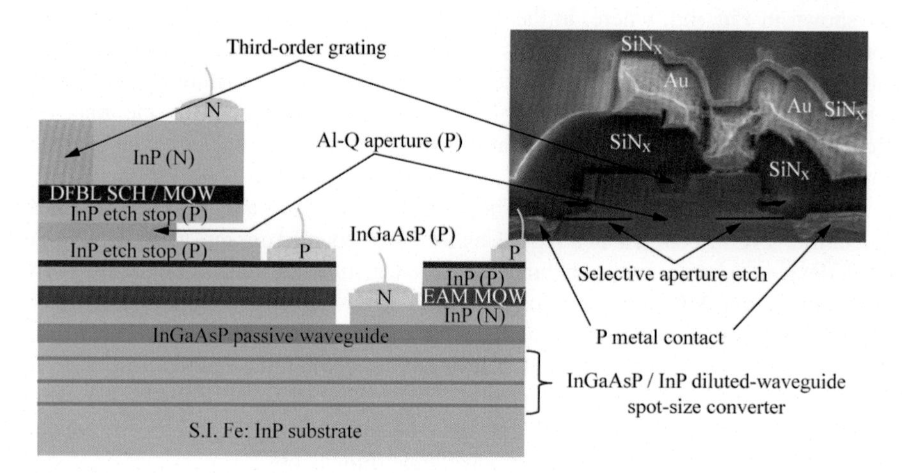

Figure 4.2 MGVI epitaxial layers for 100GE transmitter and a scanning-electron-microscope image of laser cross-section.

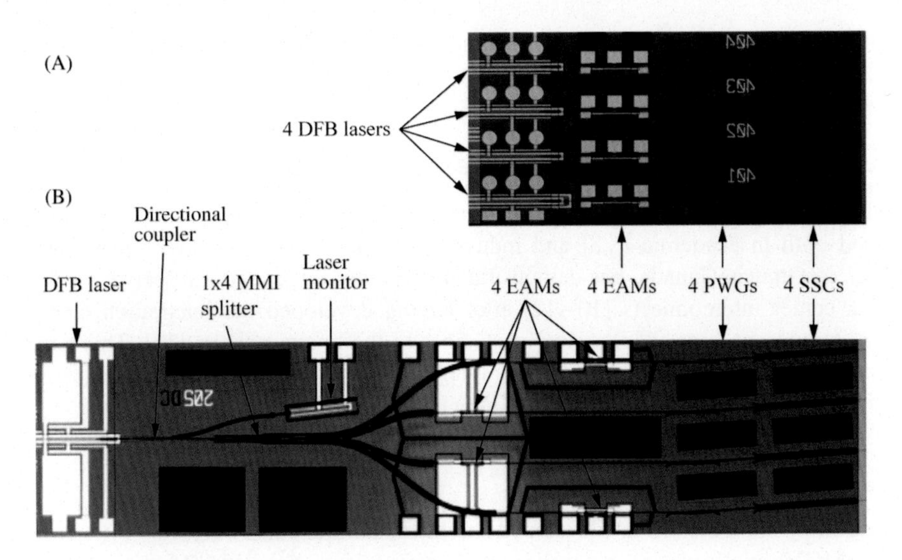

Figure 4.3 (A) Transmitter array of 4 EMLs (2.1 mm × 1 mm), and (B) PSM4 transmitter (4.1 mm × 1 mm).

the side of the wide mesa in order to minimize the optical absorption due to the metal, even for very thin top N cladding. A thin top cladding, in turn, allows for the grating to be etched from the surface, eliminating the need for regrowth. Another uncommon aspect of the MGVI laser, besides being regrowth-free, is that it uses a third-order grating fabricated by i-line projection lithography. Both of these aspects of the MGVI platform make it easy to outsource to a number of foundries, which significantly lowers the cost of the technology.

The micrographs for 100GBASE-LR4 and 100GBASE-PSM4 transmitter PICs, are shown in Fig. 4.3, where, in the first fabrication run, the 100GBASE-LR4 chip consists of an array of four electro-absorption-modulated lasers (EMLs) without a monolithically integrated multiplexer. The addition of a multiplexer to the 100GBASE-LR4 PIC, such as the Echelle grating developed for the matching receiver PIC shown below, is straightforward as the transmitter epitaxial structure contains the necessary PWG layer.

The EML that is at the heart of both 100GBASE-LR4 and 100GBASE-PSM4 transmitter PICs is shown in Fig. 4.4 as an enlarged micrograph view of a singulated device, as well as a schematic that better illustrates the topography corresponding to the MGVI integration platform. Fig. 4.4 also illustrates the use of lateral tapers as a means of adiabatically transferring light between vertically stacked waveguides. In the MGVI platform, the smaller the bandgap of a waveguide, the closer to the top the waveguide is, not only because of the higher corresponding intervalance absorption, but also because of the higher corresponding refractive index that requires smaller waveguide dimensions.

Figs. 4.5 and 4.6 show the room temperature DC as well as high-frequency results for the EML device [10]. As shown in Fig. 4.5, the DFB laser provides a

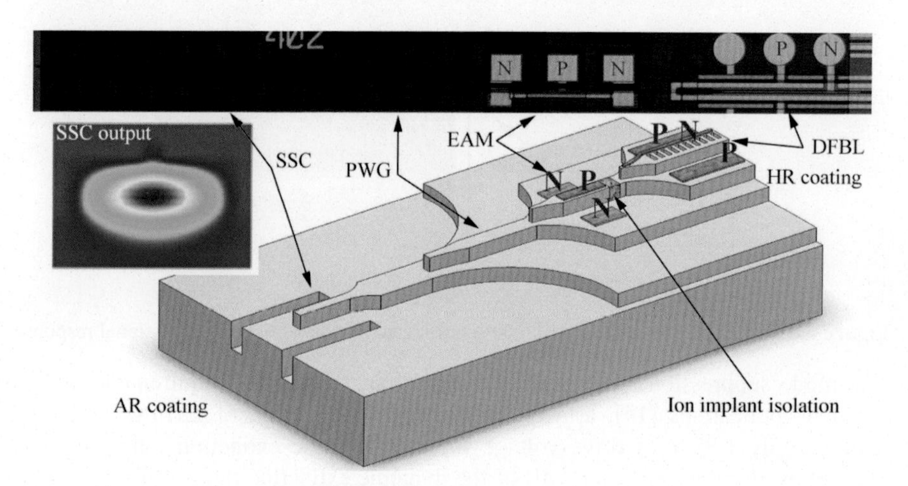

Figure 4.4 Electro-absorption modulated laser (EML) transmitter in the MGVI platform. Total chip length is 2.1 mm.

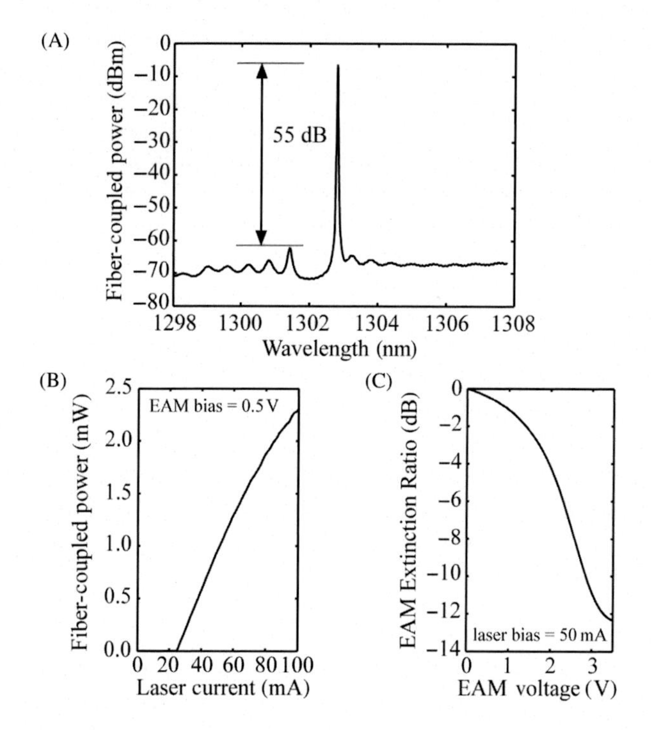

Figure 4.5 Room-temperature, continuous-wave response of the transmitter EML PIC: (A) DFB laser spectrum, (B) PIC output light versus DFB laser bias current, and (C) EAM DC extinction ratio.

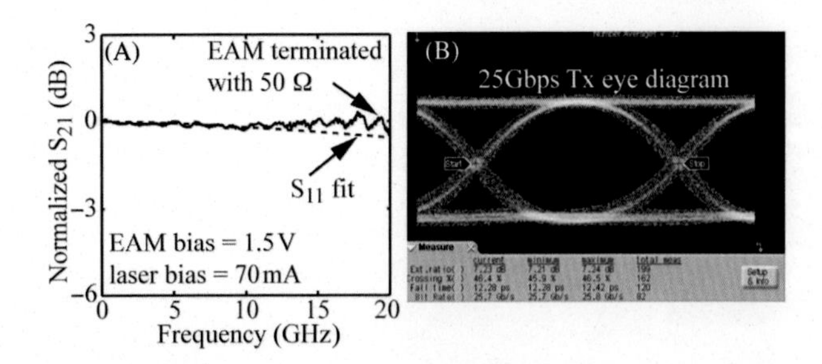

Figure 4.6 AC responses of the EAM: (A) small-signal response, and (B) large-signal response.

side mode suppression ratio of 55 dB, which far exceeds the requirements for the targeted applications [14], as well as the threshold current of 22 mA. The EAM requires only 1.85 V of drive voltage for a 6 dB static extinction ratio, which in turn, allows for the necessary 4 dB of the dynamic extinction ratio. With the help of the monolithically integrated SSC, the PIC outputs around 2.3 mW of power into a cleaved single-mode fiber, when the EAM is biased in its ON state, i.e., at 0.5 V. The MGVI SSC designed to match a single-mode fiber typically has around 1.1 dB of insertion loss, where 0.8 dB is due to the mode mismatch loss and 0.3 dB is due to the propagation loss. The SSC also has a great 1 dB alignment tolerance of ± 2 μm, and can easily be designed to be polarization independent.

There are two straightforward ways to further improve the fiber-coupled power of the EML PIC. First, the DFB laser used in this demonstration is the same design developed by OneChip Photonics for directly modulated applications [13]. Making the laser longer and redesigning its grating and epitaxial layers for continuous wave operation will significantly improve the output power. Second, the EAM insertion loss of 4 dB can be reduced to 2 dB by minimizing the optical absorption due to the P metal electrode. As 100GBASE-LR4 is a temperature-controlled operation, no temperature penalty is expected for the output optical power, and less than 2 dB of power penalty arises from a monolithic integration of EMLs operating at four different wavelengths. A similar maximum power penalty of 2 dB arises for the single-wavelength operation of the MGVI 100GBASE-PSM4 transmitter PIC as the temperature is scanned over the full range for this uncooled application.

Fig. 4.6 shows small-signal and large-signal responses for the EAMs terminated with 50 Ω. Both off-chip and monolithically integrated, on-chip, 50 Ω resistors have been successfully used in terminating the EAMs. Due to the bandwidth limitation of the lightwave network analyzer (20 GHz), the small-signal bandwidth that far exceeds 20 GHz had to be estimated by extracting small-signal parameters using an S_{11} fitting procedure, which predicts a bandwidth of 50 GHz. The measured EAM small-signal response and the predicted response obtained from the S_{11} fitting procedure show excellent agreement up to 20 GHz, as seen in Fig. 4.5A. The large predicted bandwidth of 50 GHz is further verified by the large-signal response (Fig. 4.6B), where the EAM does not cause any eye closure at 25 Gbps. The rise and fall times measured at the output of the EAM are practically those of the modulator driver electronics, i.e., around 12 ps.

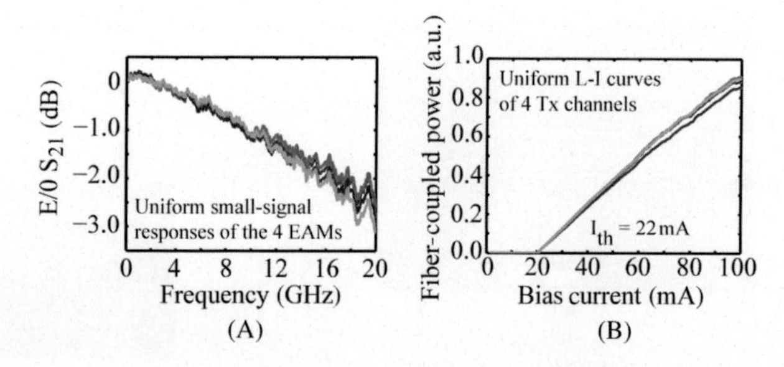

Figure 4.7 100GBASE-PSM4 Transmitter: (A) small-signal responses of the 4 EAMs, and (B) fiber-coupled power versus laser bias current for the four channels.

The peak fiber-coupled power for the 100GBASE-PSM4 PIC is −7 dBm per channel (the EAM biased in the ON state). The DFB, EAM, and SSC designs are the same as those for the 100GBASE-LR4 PIC, and the fiber-coupled power can be improved the same way; redesigning the DFB laser for the continuous-wave operation and reducing the EAM P metal optical loss. The 1×4 optical power splitting in the fully integrated 100GBASE-PSM4 PIC is achieved using two stages of 1×2 MMIs, where the total insertion loss of the cascade is less than 1 dB, and the total channel imbalance is less than 0.5 dB.

Fig. 4.7 shows normalized DC and small-signal AC responses of the four channels, demonstrating an excellent uniformity among the channels. The small-signal bandwidth is the same as that for the 100GBASE-LR4 PIC shown in Fig. 4.6A, except that here it is measured without the necessary 50 Ω termination and thus appears smaller. In both 100GBASE-LR4 and 100GBASE-PSM4 PICs, the small-signal curves are smooth and without oscillations at the DFB laser's relaxation oscillation frequency of 4 GHz, which means that the monolithic chips cause no significant parasitic optical reflections that can affect the stability of the DFB laser.

4.3.1.2 Receiver PICs

Fig. 4.8 shows micrographs of fully integrated 100GBASE-PSM4 and 100GBASE-LR4 receiver PICs. The 100GBASE-PSM4 receiver is an array of four identical high-speed photodetectors (PDs) with polarization-independent SSCs. The 100GBASE-LR4 PIC has a single polarization-independent SSC at the input (surrounded by two auxiliary SSCs in the figure), an Echelle grating de-multiplexer (DeMux), and four high-speed PDs. The Echelle grating is implemented in the PWG layer very similar to that in the transmitter PICs so that the Echelle grating can be used as a multiplexer in the transmitter PIC with minimal modifications.

Fig. 4.9A shows a normalized wavelength dependent responsivity of the 100GBASE-LR4 receiver PIC. In addition, Fig. 4.9B shows the same for a receiver designed for coarse wavelength division multiplexed (CWDM) 10 Gbps applications, such as 40GBASE. The two receivers have the same layout and almost identical footprint.

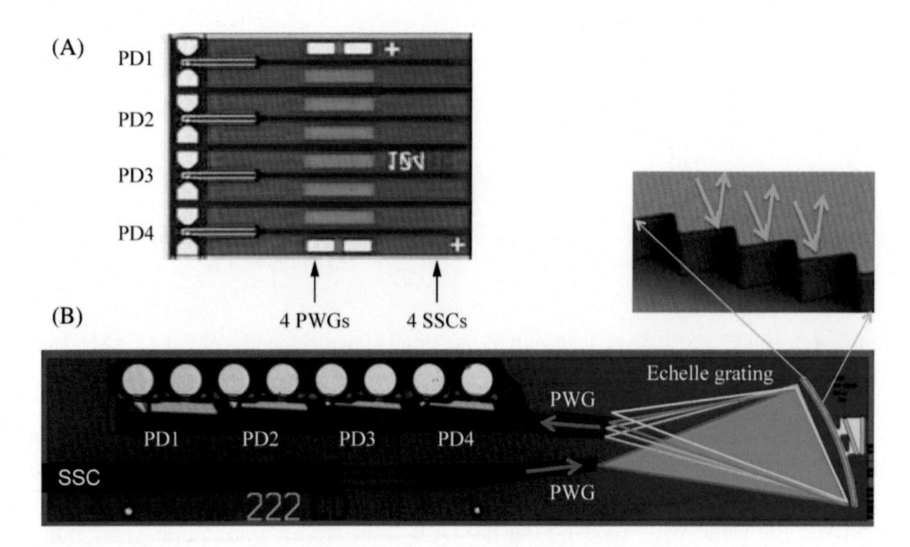

Figure 4.8 (A) PSM4 receiver PIC (1.4 mm × 1 mm), and (B) LR-4 receiver PIC (0.45 mm × 2.5 mm). A CWDM receiver PIC, not shown here, is almost identical in layout and footprint (0.45 mm × 2.2 mm) to the LR-4 receiver PIC.

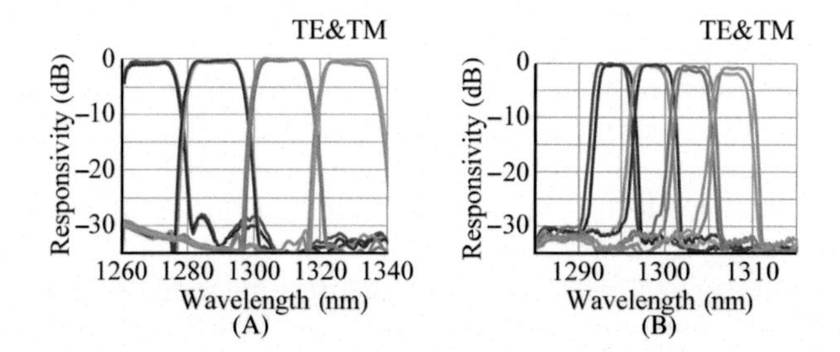

Figure 4.9 Normalized wavelength dependent responsivities of: (A) LR-4 type of receiver, and (B) CWDM receiver.

The measured responsivities of the four channels vary between 0.25 and 0.3 A/W for 100GBASE-LR4 PIC and between 0.32 and 0.39 A/W for the CWDM PIC. With simple modifications in both design and fabrication, it is predicted that these numbers will exceed 0.35 and 0.4 A/W, respectively. The 1 dB passbands for the two receivers are quite uniform and vary in the ranges between 2.9 − 3.0 nm and 13.0 − 13.6 nm, respectively. These results, as well as the polarization-dependent loss, polarization-dependent wavelength, and adjacent/nonadjacent crosstalk are summarized in Table 4.1 below. By measuring a series of intentionally varied test structures, not fully integrated as receiver PICs, an Echelle grating design with polarization dependent wavelength of close to 0.1 nm has been achieved and will be implemented in a fully integrated receiver, which otherwise would have been measured to have a value of around 0.6 nm. With the improved polarization-dependent

Table 4.1 LWDM and CWDM receiver results for four DeMUX channels

	Fiber-coupled responsivity (A/W)	1 dB Passband (nm)	Polarization dependent wavelength (nm)	Polarization dependent loss (dB)	Crosstalk (dB)	
					Adj.	Nonadj.
LWDM	0.25–0.30	2.9–3.0	0.6	0.5–0.9	−18	−32
CWDM	0.32–0.39	13.0–13.6	0.57–0.68	0.35–0.55	−30	−30

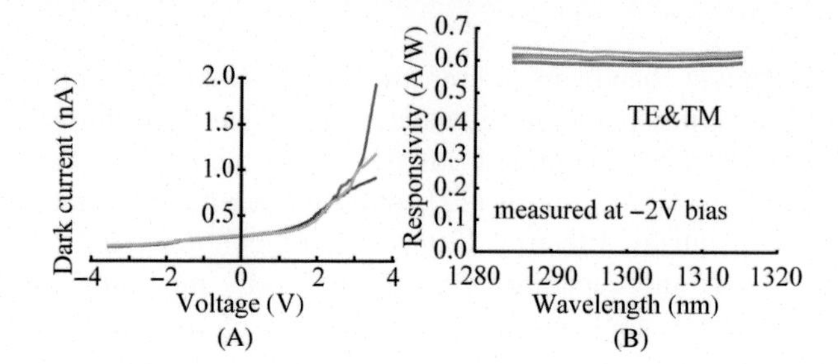

Figure 4.10 100GBASE-PSM4 receivers: (A) dark current, and (B) 16 (overlapping) responsivity versus wavelength responses for devices selected from multiple wafers.

wavelength, the adjacent crosstalk of the 100GBASE-LR4 receiver is expected to improve from −18 dB to be closer to −25 dB.

The basic DC results for the 100GBASE-PSM4 receivers are shown in Fig. 4.10. Fig. 4.10A shows voltage-dependent PD dark currents from three different wafers, where the dark currents are below a very low value of 0.5 nA for −2V, which is the expected operating bias voltage. Fig. 4.10B shows 16 responsivity vs. wavelength measurements, for both TE and TM polarizations, selected from various parts of multiple wafers. The 100GBASE-PSM4 receiver PICs show excellent uniformity of their responses. A minor design optimization will improve the responsivity value from around 0.6 A/W to the desirable level of 0.75 A/W.

Lastly, Fig. 4.11 shows the small-signal and large-signal responses for the the100GBASE-PSM4 receivers, which are also similar to those of the 100GBASE-LR4 receivers. Fig. 4.11A illustrates the uniform small-signal bandwidths for the four channels, measured to be 17.5 GHz. As shown in Fig. 4.11B, these devices provide excellent eye opening when operated at 25 Gbps. The MGVI EML transmitter is used as the source in this large-signal experiment.

4.3.2 Transceivers: other integration approaches

OneChip Photonics is one of a few companies that have demonstrated fully-integrated transmitter or receiver PICs in InP for data center applications. Another notable demonstration is NTT's 4 × 25 Gbps transmitter [15]. In both

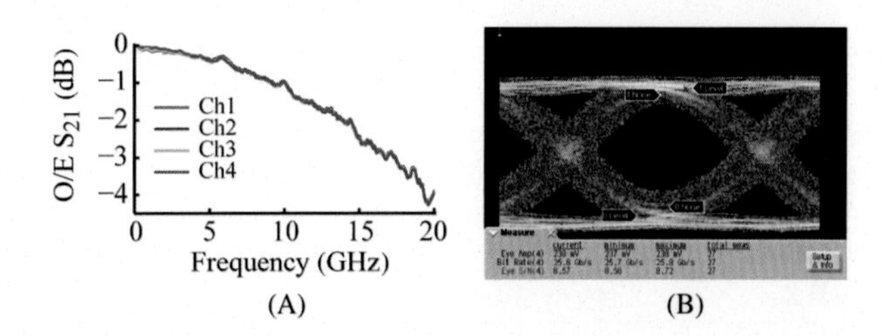

Figure 4.11 100GBASE-PSM4 receivers: (A) the small-signal responses of the four channels, and (B) the large-signal response of a single channel, where the MGVI EML was used as the source.

cases, three different bandgaps were integrated: laser quantum-wells, EAM quantum-wells, and passive routing waveguide, where Onechip's 100GBASE-PSM4 transmitter also has integrated SSC layers. Component makers and institutions, such as Oclaro [4,16], Finisar [17], Avago, Lumentum (former JDSU) [18], Hitachi [19], Mitsubishi Electric [20], and Fraunhofer HHI [4] have the integration platforms and capabilities to produce fully monolithically integrated InP PICs for data center applications, but, to the best of our knowledge, they have demonstrated only partial circuits, such as lasers monolithically integrated with EAM or Mach-Zehnder modulators. There are several reasons why fully-integrated monolithic InP PICs have not yet reached their full potential in the data center transceiver market. They are discussed in Section 4.4, together with the predicted future of InP monolithic integration.

4.3.3 InP PICs for optical switching

Photonic switches can enable simultaneous wavelength, time, and space switching operation. They have the potential to route transparently massive bandwidth optical signals, independently of signal wavelength, number of multiplexed wavelength channels, and data format. These features can be fully exploited for telecom and datacom network applications. The highest connectivity circuits have been based on low loss and high connectivity 3D MEMS switching engines. Connectivity can already extend to several hundreds of connections [21], but there are concerns about the scaling properties in terms of equipment cost, modularity, and size. Furthermore, the slow switching speeds do not support packet-based traffic, which increasingly permeates the network. Particularly in datacom, it is increasingly important to realize scalable optical switching engines that are capable of fast nanosecond-timescale reconfiguration.

Photonic integrated circuits offer the exciting opportunity to create single chip switching solutions for high capacity data routing by using different approaches and technologies. Interferometers offer the potential for low-current, and therefore low-power operation. Mach-Zehnder interferometers have been configured as 2×2 elements in the larger monolithic, multistage 16×16 and 32×32 switch matrices on LiNbO3 [22–24]. More recently 4×4 and 8×8 switch matrices have been fabricated with Silicon-on-Insulator technology [25,26]. Circuit level challenges have focused on improving the modest switch extinction ratios and relaxing the precision requirements for voltage control. Although fast switching operation in the order of few nanoseconds is feasible, the high accumulated losses in these materials prevent cascadability and thus scaling to higher photonic switch radix.

InP semiconductor optical amplifier technology can not only provide nanosecond switching time operation, but also optical amplification to compensate for the on-chip losses. Moreover, the optical gain profile can be properly tailored to cover a larger optical spectrum ranging from 850 to 2000 nm.

Until recently, InP space switch architectures have focused on broadcast and select optical switch fabrics [27,28] or cross-point architecture [29], and have been limited to a connectivity of 4×4. Research also highlights the 1×100 switches based on the use of phase arrays [30]. Systems integrators keep requiring higher connectivity, and photonic integration technology has now evolved to the point where a number of academic and industry research groups are making complex integrated circuits with hundreds of components. For packet-compliant optical switching circuits, this now means tens of connections for each switch fabric. Work with Wang et al. led to one of the first lossless 16×16 optical switch matrices [31]. This first design was implemented on an all-active epitaxy with amplifying splitters and shuffle networks, ensuring excellent loss compensation. However, the widespread use of amplifying waveguides required high operating current and is expected to compromise the dynamic range. Subsequently, an active-passive rearrangeably nonblocking integrated 16×16 SOA-based switch has been fabricated, characterized, and evaluated in Stabile et al. [32] to allow further enhancements in terms of optical power handling and noise performance. This switch is designed implementing a hybrid-Tree-Benes architecture, where the central five stages are collapsed to one stage by using a 4×4 broadcast and select switch architecture. The input and output switching stages are implemented with 2×2 broadcast and select switching elements. Fig. 4.12 shows a photograph of the switch circuit. The complete circuit includes a total number of 480 components within an area of 4.0 mm $\times$ 13.2 mm. The three cascaded SOA gate stages are implemented across

Figure 4.12 Composite image of the first 16×16 active-passive optoelectronic switch. Waveguides are visible as black lines against the gray background. Gold electrodes for the SOA-gates are visible at the six columns of active islands as light boxes [32].

six columns of 30 μm wide and 500 μm long quantum-well active islands on a selective area epitaxial regrown InP/InGaAsP wafer. The circuit is fabricated with an eight-mask process. The switch is demonstrated with a control interface for reconfigurable routing and power equalization between multiple connections. The 10 Gb/second data is routed with 28.3 dB/0.1 nm OSNR and $<10^{-9}$ bit error rate.

The 2×2 energy efficient elements from UCAM make use of Mach-Zehnder interferometric switches to provide low-loss high-speed routing together with short semiconductor optical amplifiers, and have already been studied to emulate switches up to 128×128 port count, with excellent results for energy efficiency and power penalty [33]. Other architectures have reported demonstration on Data Vortex and cross-bar switches based on ring resonators [34,35], with low power consumption [36].

The reported photonic switches are based on space switching operation. Wavelength routing switching is an alternative technology to implement photonic switching architectures. Notable examples include the MOTOR device that is an 8×8 monolithically integrated in InP optical switch based on tunable wavelength converters [37]. The DOS project is also employing tunable wavelength converters and AWGR [38]. The IRIS project demonstrated a multistage architecture based on a combination of space switches and wavelength routing switches that utilizes fast wavelength converters and 40×40 AWG [39]. These architectures are limited by the quality of the wavelength converted signals, the poor dynamic range operation, and high power consumption despite the circuit compactness.

Combining space and wavelength switching circuits, photonic switches can ultimately implement circuits with wavelength, space, and time switching operation [40,41]. An example of a photonic wavelength, space, and time cross-connect switch is illustrated in Fig. 4.13 [41]. The nonblocking optical cross-connect has 4 inputs, and each input can process and forward 4 different wavelengths to the 16 output ports. The modular cross-connect processes the WDM inputs in parallel by the respective optical modules, and forwards the individual wavelength channels to any output port according to the switching control signals [42]. Each optical module consists of a 1:N splitter to broadcast the WDM channels to the N wavelength selective switches (WSS). The WSS consists of two AWGs and M SOA-based optical gates. The broadband operation of the SOA enables the selection of any wavelength in the C-band. Moreover, the amplification provided by the SOA compensates the losses introduced by the two AWGs. Fig. 4.13 shows the photonic integration of four optical modules for wavelength, space, and time switching of four WDM inputs in parallel. The chip has been realized in a multiproject wafer (MPW) and its dimension is 6 mm $\times$ 4 mm. The chip includes a total of 112 elements (4 SOA preamplifiers (DC bias), 64 SOAs as gates (RF bias), 32 AWGs, and 12 1×2 MMI couplers). Experimental results show a crosstalk <-30dB, lossless and error-free operation with <1 dB penalty for 10 Gbps WDM signals.

Thus, recent advances in the complexity and functionality of integrated optical switch fabrics offer considerable promise for next generation networking. Lossless and nanosecond order operation, high energy efficiency and good extinction ratio, good OSNR and full connectivity, wavelength, space, and time switching operation

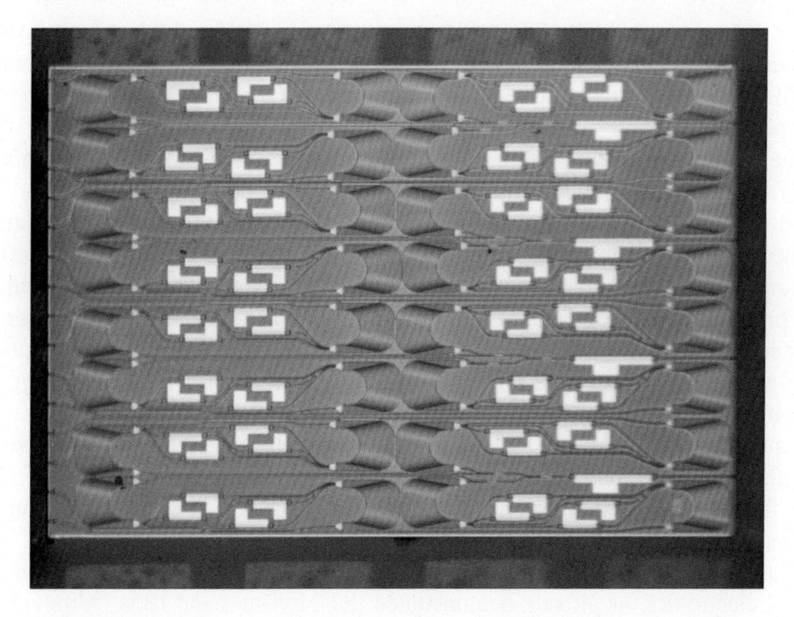

Figure 4.13 Monolithically integrated active-passive wavelength, space, and time cross-connect switch [41].

have been demonstrated for the highest state-of-the-art level of connectivity, though on different chips. Moreover, there are still some issues like polarization independency, power leveling, and capacity to be addressed in order to validate these fast switches in high performance systems, but sustained technology improvements and pressures on cost, size, optical power budgets, and efficiency of packaged parts and modules continue to drive integration.

4.3.4 Comparison of InP and Si photonics

Si photonics has been identified as the main competitor to the incumbent InP photonics for data center applications. Here, the InP solutions are compared to their Si counterparts in terms of optical performance, costs, and scalability.

A simple comparison of InP to Si transceiver solutions is obscured by the presence of multiple integration platforms and ways to fabricate PICs, in both material systems. Not all of the various InP integration platforms provide the same degree of PIC cost-effectiveness, footprint, and performance. Similarly, Si PICs can be implemented in micrometric silicon-on-insulator (SOI) platforms with a 3 μm-thick Si waveguide layer, such as the one pursued by Kotura (now Mellanox), or Si wire SOI platforms, where the Si waveguide layer thickness is around 200–300 nm (typically 220 nm), such as the platforms developed by IBM or Luxtera. Micrometric PICs tend to have larger footprints because of larger minimum banding radii compared to their Si wire counterparts. However, they offer lower propagation loss, lower polarization dependence, and lower reflections at the interfaces. For example, a PIC such as the 100GBASE-PSM4 receiver shown in Fig. 4.8A, which does not require waveguide bending but does require polarization-independent operation, may be more easily implemented in a micrometric SOI platform compared to the Si wire platform.

The comparison of InP and Si transceiver technologies also depends on the available fabrication tools and foundry infrastructure, which play an important role in the ultimate cost and performance of the PICs. E-beam lithography, for example, has been a popular choice for long-haul telecommunication InP devices for defining high-resolution features, such as laser gratings. However, due to the cost and volume pressures presented by data center transceiver applications, e-beam lithography loses some of its competitiveness. A competitive edge maybe obtained by replacing the e-beam lithography with other alternatives, such as DUV stepper lithography, without significantly sacrificing the performance of the devices. Similarly, in Si photonics, the 90 nm node foundries can achieve better performing components than their 130 nm node counterparts. For example, the 90 nm technology can be used to define the sharp tips of the inverted, buried SSCs for edge-coupling, whereas the 130 nm DUV steppers cannot do this task as easily.

Si photonics is often praised for its capacity for monolithic integration with CMOS electronics, as already demonstrated, e.g., by Intel and IBM. Although lagging behind in this regard, InP PIC can also be integrated with, for example, HBT electronics. The superior high-frequency performance of the HBT electronics becomes more important as the channel speeds increase to 50 GHz and beyond. This may prove to be an important obstacle to Si squeezing out InP in the data center space, the same way CMOS electronics did not fully win over GaAs electronics in wireless applications.

Even though a fair comparison of InP and Si technologies needs to address all these aspects, it is hard to dispute the cost advantage of Si photonics due to higher quality of Si wafers, with lower defect densities, smaller footprints of SOI PICs, and larger wafer sizes, especially before a more mature commercialization of 150 mm InP wafers takes place. However, two important insights are often overlooked when the cost advantage of Si photonics is argued.

First, Si photonics technology, similar to the CMOS technology, is cost-effective only for very large volumes, and it is not clear that the data center markets can be considered to fall into that category in the foreseeable future. In part, Si photonics requires large markets to be competitive, because it requires larger upfront development expenses, compared to InP, even though fabrication price per wafer may be lower. Also, as argued in Figure 5 of a recent roadmap [43], smaller wafers may be more economical to fabricate in a foundry than larger ones, if the market is not very large. In the figure, it is stated that a 150 mm (or 6-inch) wafer is more competitive than a 100 mm (or 4-inch) wafer only when the PIC market exceeds 20 million mm^2 per year. Reviewing Figs. 4.3 and 4.8 of this chapter, it can be estimated that 5 mm^2 is a reasonable transceiver PIC chip area size (for both LR-4 and PSM4, and other similar standards), while keeping in mind that the area can be further minimized. This means that 100 mm wafers may be more economical than 150 mm ones for a foundry that provides four million ports per year, or less.

Similarly, it makes economical sense for a foundry to upgrade its 150 mm line to a 200 mm line only if the annual sales reach 32 million ports. For reference, it is predicted that the total annual market for data center optical ports will reach 50 million by 2019 [44]. This means that the whole market can be served with only two InP foundries with 150 mm lines, and it means that the InP technology is well positioned to serve the data center transceiver markets. The demands by data centers to increase performance and reduce cost means that companies designing PICs need to think carefully about the economics of fabrication—as per the silicon industry 40 years ago. As the data center markets grow, the argument to increase InP wafers to 150 mm becomes stronger, and this will be the single most important means of bringing the fabrication cost below the desired $1/mm^2. There is also the opportunity that the business will drive even larger InP wafers to 200 mm, but that is probably beyond year 2020.

Second, what is neglected in the comparison of InP to Si photonics is the fact that InP can provide integrated components that have a superior performance relative to their Si counterparts. One way this superior performance could be leveraged is as energy savings. If a Si PIC has an insertion loss or optical power penalty that is around 2 dB larger relative to an InP counterpart, this would translate into significantly larger electricity consumption and, in turn, it would represent a lost deployment opportunity of additional servers, following the reasoning presented in Ref. [45]. The same is true for the modulator and laser energy efficiency.

The Quantum-Confined Stark Effect (QCSE), exploited for absorption and phase modulation in InP quantum-well modulators, is a much stronger effect compared to the plasma dispersion effect in bulk Si, which is the most popular modulation mechanism in Si photonics. The OneChip Photonics 50 GHz EAM requires less than 2 V of drive voltages, whereas the typical Mach-Zehnder Si modulator of comparable speed requires around 6 V of drive voltage. Because the drive voltage, insertion loss, and extinction ratio of modulators are interdependent quantities, if the Si modulator was designed to have the same low drive voltage as the InP modulator, its insertion loss would be significantly higher. The performance difference between InP and Si modulators is only expected to be more pronounced as channel speeds extend beyond 50 Gbps.

While it is possible to monolithically integrate lasers in InP with minimal optical power loss penalty, laser integration in Si photonics can easily incur 1 dB of loss, or more. In the wafer bonding approach, where an InP membrane is transferred onto the SOI wafer, the thermal impedance of the underlying buried-oxide layer is detrimental for high-temperature, i.e., uncooled operation of the laser.

The Si photonics receivers typically use Ge PDs that can have dark currents in the range of 15–20 nA. OneChip Photonics InP integrated PDs have dark currents less than 0.5 nA. An additional 1 dB, or more, of optical power penalty can be expected from Si photonics receivers based on Si wire waveguides, as they are inherently polarization sensitive and require either use of a polarization diversification scheme or polarization splitting grating couplers.

In conclusion to the comparison between InP and Si photonics, it can be stated that Si photonics has not yet proved to dominate the data center markets as it had been expected to. The shortfall can be at least in part assigned to the suboptimal performance of the Si photonic components, and perhaps inflated expectations of the data center markets in the near future. One important battle where Si photonics has been winning so far, and which may propel Si photonics to be first-to-market, is the use of external foundries and standardized device libraries. OneChip Photonics has been a pioneer and a rare example of a company doing the same with InP photonics. InP component providers typically use in-house foundries and specialized processes.

4.4 Future trends

Here, we present several future trends that can be expected for InP photonics.

4.4.1 Optoelectronic integrated circuits (OEICs)

Integration of optics and electronics is an important future trend that will reduce the overall cost and footprint of transceivers, as well as power consumption. The basic benefits of OEICs for transmitters and receivers have been well understood and generally accepted. Although optics/electronics integration is still in its infancy (see Fig. 4.14 showing a University of Southern California 10-year technology roadmap for OEICs), some of the notable efforts are IBM's CMOS integration with Si photonic wire technology [46], and the COBRA InP Membrane on Silicon technology and BiCMOS/InP photonics integration [43].

4.4.2 Heterogeneous integration

Photonic integration is recognized for becoming increasingly important in lowering the cost and the footprint of data center optics. However, the timely entrance of fully monolithic InP transceivers for data center markets has been determined by the following factors.

Until recently, the industry tolerated a large CFP-type package form factor. As the market pressure for smaller packages increases, such as the QSFP and Micro-QSFP-type of packages, in order to increase the I/O port density, the photonic circuits have to be made commensurately smaller, which means that an increased degree of monolithic integration is inevitable.

Most of the transceiver cost is presently associated with the packaging. Therefore, the monolithic solutions that decrease the cost of packaging to meet the latest expectations for transceiver costs of \$1/Gbps (at 400 Gbps) or less, as explained in Section 4.4.3, are becoming increasingly important.

Certain types of present and future transceiver standards and solutions may provide additional technological challenges to full monolithic integration. For example, monolithic integration of uncooled CWDM transmitters requires integration of more than a single set of laser quantum-wells in order to cover the wide optical

Roadmap/year	2015	2016	2017	2018	2019	2020	2021
Modules/TxRx	40 Gb/s		100 Gb/s		400 Gb/s	Technology cost barrier	*100 Gb/s*
Data rate density	10 Tb/s/1U	25 Tb/s/1U	50Tb/s/1U		100 Tb/s/1U	Technology cost barrier	*400 Tb/s/1U*
Form factor	CFP4	QSFP		*DSFP*	Technology cost barrier	*SFP*	*microSFP?*
Typical Link length	< 10 km	< 10 km	< 2 km	< 2 km	< 2 km	< 2 km	< 2 km
Industry wish ($/Gb/s at 400Gb/s) for 2km links	5		2	Technology cost barrier	*1*		*0.5*
Supplier plan ($/Gb/s at $400Gb/s) for 2km links		2		*1*		*0.5*	
Typical link length	10-100 m	10-100 m	10-100 m	5-50 m	5-50 m	1-25 m	1-25 m
Industry wish ($/Gb/s at 400Gb/s) for 10-100m links	1		0.5	Technology cost barrier	*0.25*		*0.05*
Supplier plan ($/Gb/s at $400Gb/s) for 10-100m links		1		*0.25*		*0.15*	
S/C chip line rate	10-25 Gb/s		50 Gb/s	Technology cost barrier	*100 Gb/s*		*>1000 Gb/s*
Level of integration	Disc driver/TIA	Integrated Driver/TIA	OEIC Integrated driver/TIA 50 Gb/s		Technology cost barrier	*OEIC 100 Gb/s w/other integrated functions*	
Energy/bit (pJ/b)	20		10	*4*	Technology cost barrier	*2*	*<1*
Modulation formal	NRZ	NRZ/PAM4	NRZ/PAM4/8	Technology cost barrier	*NRZ/PAM4-16*	*Coherent?*	
Die cost for PIC/OEIC (2.5mmx2.5mm) ($)	2000		1000	Technology cost barrier	*500*		*250*
TSV processes	3D baseline PIC	3D 25 GHz PIC	3D 50 GHz PIC	*3D 50 GHz OEIC*	Technology cost barrier	*3D 100 GHz OEIC*	
Chip interconnect	SiP PIC 3D stacking 10-25 GHz		TSV + 3D stack for SiP PIC 25 GHz	Technology cost barrier	*TSV + 3D stack for 50 GHz OEIC (SiP or III-V)*		
AC/DC wafer test	10-25 G		50 G		*100 G*		*>100 G*
Packaging platform	NH 25G PIC OSA	NH 25 G OEIC (SiP or III-V) OSA	NH 50G OEIC OSA (SiP or III-V)		Technology cost barrier	*NH 100 G OEIC OSA (SiP or III-V)*	
Generic packaging design (G-PIC, G-OEIC)	10 G PIC	25G PIC & OEIC	50 G OEIC	*Multichannel 50 G OEIC*	Technology cost barrier	*Multichannel 100 G OEIC*	
Packaging performance/real estate	10 Gb/mm^2		40 Gb/mm^2 (including driver/TIA)		*100 Gb/mm^2 (including OEIC)*	Technology cost barrier	*160 Gb/mm^2 (including OEIC)*
FC bumping design	25 GHz		50 GHz (micro bump)		Technology cost barrier	*100 GHz (micro bump)*	
Packaging performance per channel	25 GHz/Ch		50 GHz/Ch		Technology cost barrier	*100 GHz/Ch*	
Packaging channels per real estate	100/mm^2	200/mm^2	500/mm^2		*1000/mm^2*	Technology cost barrier	*10,000/mm^2*
Fiber technology	300 m 10 G (<1D array)		100 m 25 G (1D/MCF)	*50 m 50 G (1D/MCF); 2km MCF*	Technology cost barrier	*Up to 10 km MCF*	
Connectors and spot size converters for coupling	1xN and NxN <1dB	Beam expanded 1xN and NxN <0.75dB		Beam expanded 1xN and NxN <0.5dB		*Beam expanded 1xN and NxN <0.25dB*	

Figure 4.14 Photonic interconnect market-pull technology roadmap.

bandwidth for the full temperature range. This would have to be done in addition to integrating the modulator and the passive waveguide layers. In the future, using quantum-dot lasers that have wider optical bandwidths may alleviate this challenge, although, so far, they have not been widely used in data center applications.

Full monolithic integration requires very good control of the processing yield. Monolithic integration implies a multiplication of the yield of the subsequent processing steps, rather than an addition, which would be the case for the yield of two independently processed, nonintegrated components. Because the processing yield is ultimately an issue of cost, it may not always be economically feasible to follow the path of full integration. This may especially be true when a product with new processing steps is introduced.

Similar to the processing yield, the single-mode yield of DFB lasers needs to be addressed for a fully integrated transmitter to be economically feasible. The grating and cavity designs that address the single-mode yield issue need to be considered. This additional effort is worthwhile because, compared to the DBR laser, the DFB laser can have a smaller footprint, better spectral purity, and may be simpler to control.

Considering these points about monolithic integration, and considering that partially integrated transceiver components, such as an EAM-modulated laser PIC (EMLs), seem to be more readily available compared to their fully integrated counterparts, the early penetration of the InP into the data center market will likely involve integration of these PICs with Si photonics in a heterogeneous fashion. For example, an EML can be fabricated independently from the rest of the Si photonic circuits, and then, in a backend fashion, flip-chip bonded to the Si PIC, where the optical coupling between the two chips can be done using butt coupling [47] or surface-grating coupling [48]. Compared to the hybrid integration approaches adopted by, for example, Scorpios or Aurrion, flip-chip bonding integration does not require investment in customized co-processing of InP and Si. P-down flip-chip bonding of lasers in this fashion, when done to the Si substrate with the buried-oxide removed, is an excellent way to extract the laser's heat and improve its uncooled operation. This integration approach also allows flexibility in choosing how to divide the components between the InP and Si wafers. For example, a transceiver provider may choose the modulator to be in InP, if that component is superior to its Si counterpart, and may choose the multiplexer to be a low-loss Si Echelle grating that may be provided as a standardized component library of superior performance. This integration approach may also be a reasonable compromise for scaling of the I/O of InP PIC switches, where banks of SOA can be flip-chip bonded onto Si photonic passive circuits, so that both gain of the InP chip and compactness of the Si chip are leveraged.

4.4.3 Inexpensive and innovative integrated photonics

InP photonics is under cost pressure coming from Si photonics, as well as customers who want to see data center transceiver costs fall down to $1/Gb per second at 400 Gb/s. Heterogeneous integration and optics/electronics integration will lower the cost of InP solutions, but still, it is argued that inexpensive integration platforms will grow in importance. Regrowth-free vertical integration, easy to outsource to an external foundry, is presented as such a platform that has merit in potentially meeting these metrics, although other technologies such as high-speed electro-optical polymers, as those developed by Lightwave Logic, may also suffice [49,50]. High-speed electro-optical polymer technology, in particular, is expected to have a strong impact as the material can achieve scalability in cost and volume through low-cost fabrication techniques, such as spin-coating of the organic optical modulator and its integration with Si photonics as the active component of the device [51]. A market survey (see Fig. 4.14) of data center interconnects and assembly and packaging was completed in the first half of 2015. It includes input from major social media, Internet, telecommunications, data-communications, and computer/server companies, as well as smaller private optical components suppliers. The survey identifies important technology cost-barriers between current capabilities and cost points that will significantly stimulate photonics manufacturing demand by opening new markets. Thus, an important goal for optical suppliers of PICs in demonstration and development is to address these needs. Metrics have been established in photonic interconnects, packaging, test, communications, etc., spanning 7 years for the industry.

The technology cost-barriers on the roadmap in Fig. 4.14, colored purple, are identified by comparing the desires of industrial consumers of photonics technologies with current technological forecasts. In general, industry forecasts for key metrics such as $/Gb per second are 5 − 10 times higher than industry wishes. For example, if industry wishes to purchase an optical link for a data center using single mode fiber, at a reach of 2 km, in the timeframe of 2020, fiber optic transceivers will need to adhere to a cost-performance metric of $1/Gb per second at 400 Gb/second. In other words, the link needs to cost $400 (i.e., $200 each end). For shorter links on the order of 10−100 m, the cost-performance metric increases to $0.25/Gb per second at 400 Gb/second (i.e., $100 link where each end is $50). Industry market forecasts do not see this wish being fulfilled; thereby creating a gap. This gap is an opportunity for PIC suppliers to innovate new designs for their customers. PIC suppliers need to close the cost-performance gap for photonic interconnect customers by accelerating the timeline so that customers can see PIC products at competitive pricing before they expect them.

Integrated photonics projects must also address the key concern of "scalability," which, in the eyes of the data center industry, is being able to scale technology at the correct cost levels to achieve their technical metrics. It has been recognized throughout the stakeholder community that even though technologies may be deemed scalable in volume, the art of innovative integrated photonics is being able to show accelerating scaling of volume and cost structures before consumer-level volumes are created.

The advances in technology to close the gap between industry wishes and delivered technology for scalable product development will drive the scope of the projects that will be run at PIC suppliers. Key projects will address the advancement of simple PICs to an optoelectronic-integrated circuit (OEIC) that fully integrates both photonic and electronic components onto a semiconductor chip or stacked chip. Initial designs, with the support of modeling and simulation modules, will integrate electronic functions such as laser drivers, transimpedance amplifiers (TIAs), and digital support circuitry for more efficient data interfaces. Further work is expected to drive levels of integration higher, to allow for more efficient signaling, and ASIC functionality. The impact of fully integrating components will allow smaller footprints, and lower power consumption, together with increased line rates.

References

[1] Haddadin M, Radloff S. 450 mm SEMI physical interface standards: architecture and efficiency. In: Advanced semiconductor manufacturing conference (ASMC), 26th annual SEMI; 2015.

[2] Warner K, Oakley D, Donnelly J, Keast C, Shaver D. Layer transfer of FDSOI CMOS to 150 mm InP substrates for mixed-material integration. In: Indium phosphide and related materials conference proceedings, international conference; 2006.

[3] Van Thourhout D, Roelkens G. Heterogeneously integrated SOI compound semiconductor photonics. In: Conference: ECOC-35th European conference on optical communication. 09/20/2009–09/24/2009 at Vienna; 2009.

[4] Smit M. Generic InP-based integration technology, today and tomorrow. Proceedings of advanced photonics congress. Colorado: Colorado Springs; 2012, IM2A. 1.

[5] Van der Tol JJGM, Oei YS, Khalique U, Nötzel R, Smit MK. InP-based photonic circuits: comparison of monolithic integration techniques. Prog Quantum Electron 2010; 34:135–72.

[6] Coldren LA, Nicholes SC, Johansson L, Ristic S, Guzzon RS, Norberg EJ, et al. High performance InP-based photonic ICs—a tutorial. J Lightwave Technol 2011;29(4): 554–70.

[7] Menon VM, Xia F, Forrest SR. Photonic integration using asymmetric twin-waveguide (ATG) technology: part II—devices. IEEE J Sel Top Quantum Electron 2005;11(1): 30–42.

[8] Xia F, Menon VM, Forrest SR. Photonic integration using asymmetric twin-waveguide (ATG) technology: part I—concepts and theory. IEEE J Sel Top Quantum Electron 2005; 11(1):17–29.

[9] Gokhale MR, Studenkov PV, Ueng-McHale J, Thompson J, Yao J, Van Saders J. Proceeding of optical fiber communications conference, Atlanta (GA), PD42-1; 2003.

[10] Ristic S, Pimenov K, Tolstikhin V, Logvin, Y. 25 Gbps re-growth-free externally modulated laser for Multi-Guide Vertical Integration in InP. In: Proceeding of OSA conference on integrated photonics research, silicon and nano-photonics (IPR), Rio Grande, IW5A.6; 2013.

[11] Ristic S, Florjanczyk M, Lebby, M. Optoelectronic integrated circuits for 100G Ethernet and coherent networks based on Multi-Guide Vertical Integration platform. In: Proceeding of optical fiber communications conference, San Francisco (CA), Tu3H.6; 2014.

[12] Tolstikhin V, Wu F, Logvin Y, Ristic S, Tang Y, Pimenov K, et al. 100Gb/s receiver photonic integrated circuits in InP for applications in fiber-optics interconnects. In: Proceeding of OSA conference on integrated photonics research, silicon and nano-photonics (IPR), Rio Grande, IW5A.4; 2013.

[13] Watson C, Tolstikhin V, Pimenov K, Wu F, Logvin Y. On-chip emitter for regrowth-free Multi-Guide Vertical Integration in InP. In: Proc. IEEE photonics conference (PHO), Arlington (WA), TuQ3; 2011.

[14] IEEE 802.3 Ethernet Working Group. IEEE 802.3 ETHERNET. [Online]. Available from: http://www.ieee802.org/3/; 2010 [accessed 05.09.15].

[15] Kanazawa S, Fujisawa T, Ohki A, Ishii H, Nunoya N, Kawaguchi Y, et al. A compact EADFB laser array module for a future 100-Gb/s Ethernet transceiver. IEEE J Sel Top Quantum Electron 2011;17(5):1191–7.

[16] Oclaro. Insights—Oclaro. [Online]. Available from: http://www.oclaro.com/insights/; 2015 [accessed 07.09.15].

[17] Adams DM, Isaksson M, Wesstrom J-O, Erriksson U. Transmission performance of monolithically integrated Y-branch tunable laser with zero-chirp Mach-Zehnder modulator. Electron Lett 2007;43(9):522–4.

[18] Kozody P, et al. Thermal effects in monolithically integrated tunable laser transmitters. In: 20th annual IEEE semiconductor thermal measurement and management symposium, San Jose (CA); 2004, p. 177–183.

[19] Makino S, et al. Uncooled CWDM 25-Gbps EA/DFB lasers for cost-effective 100-Gbps Ethernet transceiver over 10-km SMF transmission. In: Proc. optical fiber communications conference, San Diego (CA), PDP21; 2008.

[20] Saito T, Yamatoya T, Morita Y, Ishimura E, Watatani C, Aoyagi T, et al. Clear eye opening 1.3 μm-25 / 43Gbps EML with novel tensile-strained asymmetric QW absorption layer. In: 35th European conference on optical communications, Vienna; 2009, p. 1–2.

[21] Madamopoulos N, Kaman V, Yuan S, Jerphagnon O, Helkey RJ, Bowers JE. Applications of large-scale optical 3D-MEMS switches in fiber-based broadband-access networks. Photon Netw Commun 2010;19:62–73.

[22] Murphy E, Murphy TO, Ambrose AF, Irvin RW, Lee BH, Peng P, et al. 16 × 16 non-blocking guided-wave optical switching system. J Lightwave Technol 1996;14:352–8.

[23] Okayama H, Kawahara M. Prototype 32 × 32 optical switch matrix. Electron Lett 1994;30:1128–9.

[24] Duthie PJ, Wale MJ. 16 × 16 single chip optical switch array in Lithium Niobate. Electron Lett 1991;27:1265–6.

[25] Lee B, et al. Monolithic silicon integration of scaled photonic switch fabrics, CMOS logic, and device driver circuits. J Lightwave Technol 2014;32:743.

[26] Das Mahapatra P, Stabile R, Rohit A, Williams KA. Optical crosspoint matrix using broadband resonant switches. J Sel Top Quantum Electron 2014;20(4):5900410.

[27] Stabile R, et al. Multipath routing in a fully scheduled integrated optical switch fabric. In: European conference on optical communications, Turin, We.8.A.6-1/3; 2010.

[28] Wang H, et al. Lossless multistage SOA switch fabric using high capacity monolithic 4 × 4 SOA circuits. In: Proc. optical fiber communications conference; 2009a.

[29] Varrazza R, Djordjevic IB, Yu S. Active vertical-coupler-based optical crosspoint switch matrix for optical packet-switching applications. J Lightwave Technol 2004;22 (9):2034–42.

[30] Soganci IM, Tanemura T, Zaitsu M, Takenaka M, Nakano Y. Monolithic 100-port photonic switch. In: Proc. European conference on optical communications, post-deadline paper PD1.5, 2010.

[31] Wang H, Wonfor A, Williams KA, Penty RV, White IH. Demonstration of a lossless monolithic 16 × 16QW SOA switch. In: Post-deadline paper, Proc. European conference on optical communications; 2009b.

[32] Stabile R, Albores-Mejia A, Williams KA. Monolithic active-passive 16 × 16 optoelectronic switch. Opt Lett 2012;37(22):4666–8.

[33] Cheng Q, Wonfor A, Wei JL, Penty RV, White IH. Demonstration of the feasibility of large-port-count optical switching using a hybrid Mach-Zehnder interferometer-semiconductor optical amplifier switch module in a recirculating loop. Opt Lett 2014;39:5244–7.

[34] Liboiron-Ladouceur O, et al. The data vortex optical packet switched interconnection network. J Lightwave Technol 2008;26:1777–89.

[35] Bergman K, et al. Design, demonstration and evaluation of an all optical processor memory-interconnection network for petaflop supercomputing. In: ACS interconnects workshop, http://lightwave.ee.columbia.edu/?s=research&p=highperformance_computing_systems#dv; 2010.
[36] Biberman A, Hendry G, Chan J, Wang H, Preston K, Sherwood-Droz N, et al. CMOS-compatible scalable photonic switch architecture using 3D-integrated deposited silicon materials for high-performance data center networks. In: Proc. optical fiber communications conference, Los Angeles (CA), OMM2; 2011.
[37] Nicholes SC, Mašanović ML, Jevremović B, Lively E, Coldren LA, Blumenthal DJ. The world's first InP 8 × 8 monolithic tunable optical router (MOTOR) operating at 40 Gbps line rate per port. In: Proc. OFC 2009 post deadline paper B1, San Diego (CA); 2009.
[38] Ye X, Yin Y, Mejia P, Proietti R, Akella V, Yoo SJB. DOS—a scalable optical switch for datacenters. In: ANCS, La Jolla (CA); 2010.
[39] Gripp J, Duelk M, Simsarian JE, Bhardwaj A, Bernasconi P, Laznicka O, et al. Optical switch fabrics for ultra-highcapacity IP routers. J Lightwave Technol 2003;21:2839–50.
[40] Stabile R, Rohit A, Williams KA. Monolithically integrated 8 × 8 space and wavelength selective cross-connect. J Lightwave Technol 2013;32(2):201–7.
[41] Calabretta N, Williams KA, Dorren H. Monolithically integrated WDM cross-connect switch for nanoseconds wavelength, space, and time switching. In: Proc. European conference on optical communications Valencia; 2015.
[42] Miao W, Luo J, Di Lucente S, Dorren H, Calabretta N. Novel flat datacenter network architecture based on scalable and flow-controlled optical switch system. Opt Express 2014;22:2465.
[43] JePPIX. JePPIXRoadmap2015 [Online]. Available from: http://www.jeppix.eu/document_store/JePPIXRoadmap2015.pdf; 2015 [accessed 08.09.15].
[44] Lightwave. Web 2.0 companies will buy a lot of optical modules says LightCounting—Lightwave [Online]. Available from: http://www.lightwaveonline.com/articles/2015/07/web-2-0-companies-will-buy-a-lot-of-optical-modules-says-lightcounting.html; 2015 [accessed 09.09.15].
[45] IEEE 802.3 Ethernet Working Group. 400GbE using Nyquist PAM4 for 2km and 10km PMD-rao_3bs_01a_0115.pdf [Online]. Available from: http://www.ieee802.org/3/bs/public/15_01/rao_3bs_01a_0115.pdf; 2015 [accessed 08.09.15].
[46] Asefa S, et al. A 90 nm CMOS integrated nano-photonics technology for 25 Gbps WDM optical communications applications. In: Electron devices meeting (IEDM), San Francisco (CA); 2012.
[47] Kurata Y, Hashizume Y, Aozasa S, Itoh M, Hashimoto T, Tanobe H, et al. Heterogeneously integrated PLC with low-loss spot-size converter and newly developed waveplate PBS for DC-DP-16QAM eeceiver. J Lightwave Technol 2015;33(6):1202–9.
[48] Song B, Contu P, Stagarescu C, Pinna S, Abolghasem P, Ristic S, et al. 3D integrated hybrid silicon laser. In: 41st European conference on optical communications, Valencia, We.2.5.5; 2015.
[49] Goetz FJ, Ashton A, Eaton DF, Arduengo AJ, Simmons HE, Runyon JW. Stable free radical chromophores, processes for preparing the same, nonlinear optic materials and uses thereof in nonlinear optical applications, US 20120267583 A1; 2012.
[50] Priye V, Mickelson AR. Multiple scale analysis of low index subwavelength slot waveguide electrooptic modulator. In: Proc. international conference on microwave and photonics (ICMAP), Dhanbad; 2013.
[51] Zelibor T. Powered by the speed of light. Lightwave logic investor presentation; 2015, p. 10–11.

Photonic crystal cavities for optical interconnects

5

Liam O'Faolain
SUPA, School of Physics and Astronomy, St Andrews, KY, Scotland

Over the past decades, the clock speed and wiring density in microprocessors has risen dramatically. As a result of this increase, the metal wires interconnecting components on the chip have become less and less efficient and, now, $50-80\%$ of a computer processor's power consumption occurs in these interconnects [1]. The limited capacity of electrical interconnects is a major problem for communications between racks of computers, communication between boards, and even at shorter distances, such as between chips.

Optics has important advantages over electrical signals for interconnections, e.g. the absence of resistive loss. The limit on data transmission in an optical waveguide can be as high as 100 Tbit/second, and data transfer at high bit rates is much more energy-efficient as the line does not need to be charged to the operating voltage. Consequently, this has triggered a significant deployment of optics in computing. There is an ever-increasing demand for bandwidth, and future generations of computers, servers, and data centers are predicted to require huge numbers of optical links. However, the current state-of the-art using millimeter scale devices is orders of magnitude away from required efficiencies, in terms of power, volume, and cost.

In general, the energy consumption of a photonic device is proportional to the product of the photon loss rate and the volume [2]. Compact optical resonators are therefore interesting as a means to reduce power consumption. Four broad flavors exist, silica based micro-toroids [3], silicon based micro-ring resonators [4], photonic crystal cavities [5], and plasmonic cavities [6]. While the micro-toroids show the lowest loss rate, and the plasmonic cavities the lowest volume, silicon-based photonic crystal cavities provide the best compromise between the two, making them very promising for energy efficient optical interconnects.

5.1 Photonic crystal background

The photonic crystal is an artificial material consisting of a periodic variation of the refractive index with the effect that the propagation of photons is modified in a manner similar to that of electrons propagating in a semiconductor. Light that scatters in each period may interfere destructively or constructively, and can make the crystal transparent or opaque. For certain geometries and sufficiently high refractive index contrasts, the propagation of light may be completely forbidden, giving what is known as the *photonic band gap* (*PBG*). At its simplest, the PBG provides an

Optical Interconnects for Data Centers. DOI: http://dx.doi.org/10.1016/B978-0-08-100512-5.00005-X

extremely effective mirror, unlike metals which always exhibit some absorption, and is the basic principle behind most photonic crystal devices, allowing light to be manipulated and confined on the wavelength scale. Indeed, photonic crystals provide the best compromise between confining light to small spatial volumes while maintaining long photonic lifetimes [7].

The photonic crystal is generally considered to have been invented by Yablonovitch [8] and John [9] in 1987. In 1996 Thomas F Krauss and Richard de la Rue demonstrated the first two-dimensional (2D) photonic crystal at telecommunications wavelengths, and showed how photonic crystals can be fabricated using the tools of the semiconductor industry. Since that development the field has expanded rapidly.

5.1.1 Theory

In a dielectric medium with refractive index $n(r)$, the propagation of light is governed by Maxwell's equations, which can be solved to determine the electric (**E**) and magnetic (**H**) fields. In a linear, isotropic, homogeneous (LIH) material, we can write Maxwell's equation for the **E**-field as follows:

$$\frac{c^2}{n^2}\nabla^2\mathbf{E}(\mathbf{r},t) - \frac{\partial^2\mathbf{E}(\mathbf{r},t)}{\partial t^2} = 0$$

The general solution for **E** as a function of t and **r** is a superposition of plane waves of the form by expressing the solution as harmonic modes:

$$\mathbf{E}(r,t) = E_0 e^{i(\mathbf{k}.\mathbf{r}-i\omega t)}$$

where ω is the angular frequency and **k** the wave vector. The relationship between ω and **k** is known as the dispersion relation and has linear dependence in a LIH material:

$$\frac{\omega}{k} = \frac{c}{n}$$

In a photonic crystal, $n(\mathbf{r})$ is a periodic function such that $n(\mathbf{r}) = n(\mathbf{r}+\mathbf{R})$, where **R** is the lattice vector. Using the Floquet − Bloch theorem, the solution can be written as:

$$\mathbf{E}(r,t) = E_0 u_k(\mathbf{r}) e^{\mathbf{i}(\mathbf{k}.\mathbf{r}-\omega t)},$$

where $u_k(\mathbf{r})$ is a function with the same periodicity as $n(\mathbf{r})$. In other words, the solution is a plane wave with a periodic envelope, and is usually described as a Bloch mode. A corollary of the Floquet − Bloch theorem is that modes with wave vectors that differ by $2\pi/\mathbf{R}$ are identical. The dispersion relation is strongly modified by the

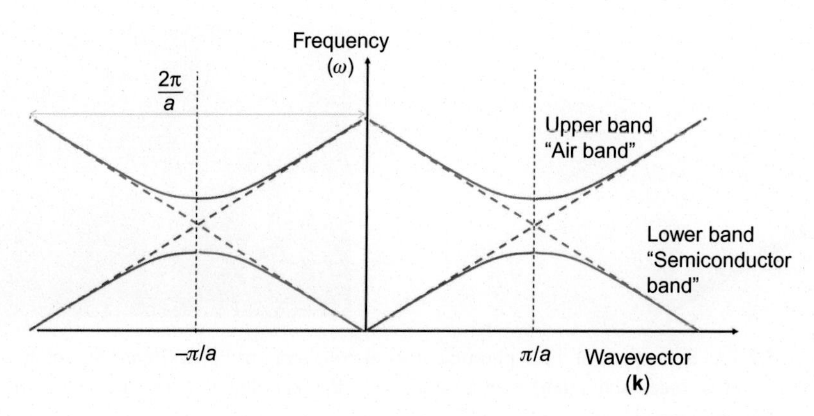

Figure 5.1 Schematic showing how bands arise for a one-dimensional photonic crystal (period a). The dispersion curve is periodic in **k** with a periodicity of $2\pi/a$. Due to reciprocity, the dispersion curve for a wave propagating in the negative direction is the same as that propagating in the positive direction.

periodicity, with the result that there is no longer a linear dependence between ω and **k**. The curvature of the dispersion curve and the repetition in **k** causes gaps to appear in the dispersion curve, giving rise to the formation of "bands."

Fig. 5.1 shows a schematic band diagram for a one-dimensional photonic crystal with lattice constant a. Band gaps arise at **k**-vectors equal to integer multiples of $2\pi/a$. As a consequence of the Floquet − Block theorem, all the necessary information is contained in the region $0 < \mathbf{k} < \pi/a$, which is known as the *reduced Brillouin zone*.

In 1887, Lord Rayleigh gave a proof that any material with a periodic modulation of the refractive index exhibited band-gaps [10], an important feature of photonic crystals. In two- and three-dimensions, the existence of an omni-directional band gap is non-trivial and careful design is required to realize a photonic crystal with an omni directional band gap (namely the realization of high refractive index contrast, see Joannopoulos et al. [11] for a full discussion). Nonetheless, partial band gaps are still of very significant interest, with most modern devices falling into this category. Photonic crystals of the type shown in Fig. 5.2 possess a 2D PBG in the $x - y$ plane, and use total internal reflection to confine light in the z direction. As TIR theoretically provides perfect reflection, for a region of ω and **k** values, operation very similar to that of a 3D photonic crystal may be realized. An important advantage of this configuration is its compatibility with semiconductor processing (to be discussed further in the following section).

As a result, this family of photonic crystals have the highest performance and most versatility. A wide range of designs and components can be created by rearranging or omitting air holes.

5.1.2 Photonic crystal cavities: the ultimate confinement of light

As photons have no rest mass or charge, fields or potential wells cannot be used to control and manipulate them. Light can only be manipulated using mirrors, in one shape or

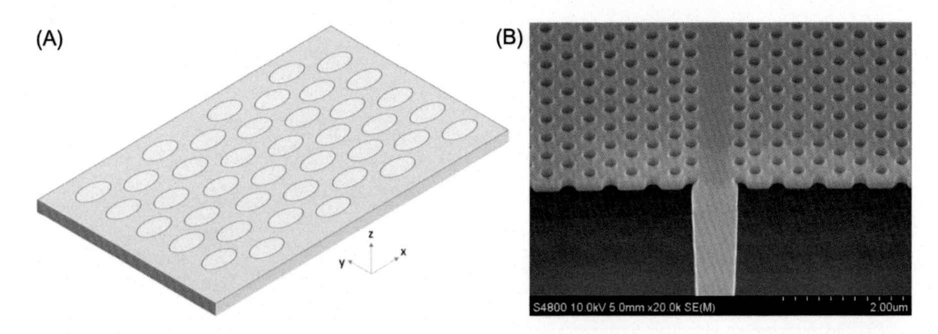

Figure 5.2 (A) Schematic of a 2D photonic crystal consisting of holes (*light blue*) etched in a high refractive index slab (slab) with air cladding. (B) Scanning electron microscope image of a silicon photonic crystal.

form. Confining light on the wavelength scale is particularly challenging: mirror reflectivities in excess of 99.9999% would be required to confine light in a 1-μm long Fabry − Perot cavity for 1 ns. Metal mirrors simply cannot provide such reflectivies.

A 2D photonic crystal can be designed to operate as a mirror by choosing its parameters such as a PBG (the frequency region in which there is no solution to the wave equation). The configuration of Fig. 5.2, consisting of a triangular lattice of holes etched in a refractive index slab, gives a 2D band gap centered on a wavelength of 1550 nm, when the period is 420 nm, hole radius is 120 nm, and a slab with a thickness of 220 nm with $n = 3.46$. Omitting one row for holes from the lattice creates a guided mode inside the band gap. Light is then confined by total internal reflection in the z-direction, and by the PBG in the $x - y$ plane giving what is known as a W1 photonic crystal waveguide.

Omitting a single hole from the lattice gives the simplest photonic crystal cavity, termed the H1 cavity. Light inside this region cannot propagate in the $x - y$ plane by virtue of the 2D PBG, and is confined very tightly to this region. Such a cavity was first demonstrated in 1999 by Painter et al. [12] The mode volume was extremely small: 0.31 $(\lambda/n)^3$, though out-of-plane losses were very severe due to weak confinement in the z direction. As a result, the photon lifetime, τ_0, was less than 0.2 picoseconds.

The quality factor is generally used to describe the quality of the optical cavity or resonator, and is defined as 2π times the energy stored in the cavity divided by the energy lost per optical cycle. In the case of the H1 cavity above, the Q-factor was 250 (where $Q = \omega\tau_0$, also approximately equal to $\lambda/\Delta\lambda$). At the time, it was believed that a full 3D PBG was required to increase the Q-factor; however, it was soon realized that, by careful design, total internal reflection could be used to prevent light escaping. Noda et al. made the key breakthrough in 2003 [13] and realized that by "gently confining" light to the cavity, the losses could be greatly reduced. Due to the small size of the mode of a photonic crystal cavity, it is made up of a large number of components with a wide range of **k**-vector values. The optical loss is then determined by the proportion of these which are not guided by total

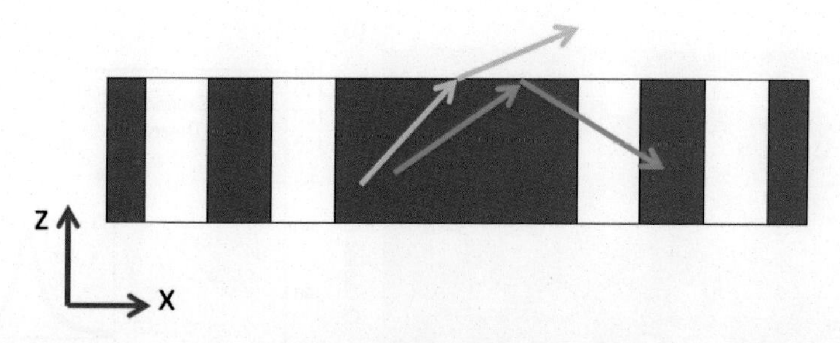

Figure 5.3 Cross-section of the photonic crystal cavity. The *green arrow* indicates a component of the mode that is confined by total internal reflection. The *yellow arrow* indicates a mode component whose angle of incidence on the surface is a greater angle and thus may leak out of the slab.

internal reflection, see Fig. 5.3. Noda's insight was that an abrupt change in the mode profile generated additional **k**-vector components (given by the Fourier transform of the mode profile) that could leak out of the slab.

By designing the photonic crystal cavity such that the mode profile had a Gaussian envelope, the spread of **k**-vector components was reduced, dramatically increasing the Q-factor. One approach to achieving this was to adjust the position and size of air holes at the extremities of the cavity, modifying the Bragg reflection condition. A phase mismatch between the reflections at different interfaces is introduced which weakens the magnitude of the Bragg reflection, allowing light to penetrate deeper into the photonic crystal and altering the mode profile with respect to the original design.

The L3 cavity is one of the most popular photonic crystal cavity designs and is realized by omitting three holes from the hexagonal lattice illustrated in Fig. 5.2, as can be seen in Fig. 5.4A (an extension of this design family are the L5, L7s,... corresponding to the omission of five, seven,... air holes). In the simple case, the Q-factors are very low, on the order of a few thousand. By reducing the radius of the marked holes in Fig. 5.4A, and by shifting them way from the cavity, a more Gaussian mode profile is attained with a consequent improvement in the Q-factor. The nominal radius, r, was given by $r/a = 0.3$, where a is the lattice constant. The optimum Q-factor is obtained for a radius reduction of $\Delta r/a = 0.06$ and a shift of $\Delta x/a = 0.19$.

Values for Q-factor as high as 110,000 have been realized [14] with this approach, see Fig. 5.4B. The difference between the simulated and measured values arises from fabrication imperfections (to be discussed in more detail in the following section).

The next evolution of the photonic crystal cavity was the double- and multi-heterostructure designs [15], which are based on a photonic crystal waveguide, in which the lattice constant is varied between the cavity and mirror regions, creating a potential well that confines light. In the L3 cavity, confinement is provided by the PBG created by the 2D lattice; in heterostructure cavities the mirror is created as a consequence of the mode gap effect; light is incident on a region in which no mode

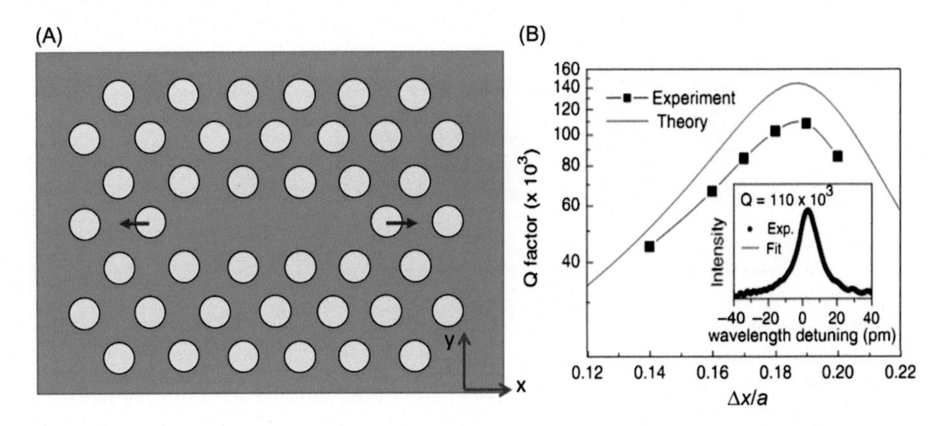

Figure 5.4 (A) Schematic showing how an optimized L3 can be realized. (B) Experimental and simulated results showing how the Q-factor may be increased by optimization of the cavity.
Source: Reprinted with permission from Galli M, Portalupi S, Belotti M, Andreani L, O'Faolain L, Krauss T. Light scattering and Fano resonances in high-Q photonic crystal nanocavities. Appl Phys Lett 2009;94:071101. Copyright © 2009. AIP Publishing LLC.

exists. The evanescent behavior of the mode is different in the two systems, with more control provided with the heterostructure approach. The double heterostructure quickly achieved record Q-values before being surpassed by the multiple-heterostructure, the current record holder with $Q = 9$ million [5].

Design optimization by means of genetic algorithms promises even more sophisticated designs. Recently, this approach has given L3 cavities with Q-factors of one million [16].

5.1.3 Fabrication

Two-dimensional photonic crystals have been realized in many different material systems; here, we focus on the silicon platform as it is suitable for low-cost mass production [17]. The fabrication of a silicon photonic crystal typically starts with a material known as Silicon-On-Insulator (SOI). The top silicon layer is 220 nm, chosen so that only a single mode is confined in the vertical direction. A 2 μm thick layer of silicon dioxide provides optical isolation from the silicon substrate. Traditionally, electron beam lithography (EBL) is used, as very high resolution patterning is possible. The throughput is generally low but, nonetheless, EBL remains the most popular approach to defining photonic crystals. The process works by applying a suitable resist to the SOI material (composed of polymer) and exposing it to a precisely controlled beam of high energy electrons (30–100 keV). The incident electrons break bonds in the polymer chain allowing these regions to be removed in a suitable chemical (known as a development) [18].

Reactive ion etching is used to transfer the pattern from the resist into the top silicon layer. The surface of the material is exposed to chemically reactive ions and

Figure 5.5 Scanning electron microscope image of a photonic crystal pattern etched in silicon.

the chemistry is chosen such that the resist experiences minimal erosion while the ions react with the silicon to form gaseous by-products [19]. Key concerns are the smoothness of the side walls and the verticality of the etched side walls, and careful optimization is required. A number of different processes and systems are available. The sulphur hexafluoride and carbon tetramethane gas combination gives particularly high quality etching, see Fig. 5.5.

The silicon dioxide is removed from underneath the photonic crystal using hydrofluoric acid, giving what is known as a membrane photonic crystal. The purpose of this step is to increase the refractive index contrast in the vertical direction, thus maximizing the range for which the guided mode is confined by total internal reflection.

The membrane photonic crystal is most popular with the research community, but lacks robustness and is difficult to integrate with other components such as heaters or electronic control circuitry. Retaining the silicon dioxide lower cladding gives mechanical stability, but breaks the vertical symmetry of the structure, allowing coupling between the TE-like and TM-like modes of the silicon slab, increasing propagation losses. Applying a layer of a spin-on-glass that fills the holes of the photonic crystal and hardens in after thermal treatment is the preferred solution. The reduced refractive index reduces the operating region making the design more challenging, but once the condition for TIR is met the device exhibits very low propagation losses [20].

5.2 Mass production

Very precise fabrication is required to realize high performance photonic crystals. Due to the high refractive index contrast and the tight confinement of light, small defects have serious consequences; the hole side wall angle must be vertical to within a few degrees [21]; the roughness should be at the nanometer level; and, in many of the current designs, the position of each hole must be controlled with nanometer precision. Deep Ultra Violet (DUV) Photolithography is a must for the mass production of silicon optical devices. However, the photomasks (also known as reticles) used typically have a 5–10 nm design grid, which gives the minimum position increment, and conflicts with the latter requirement. To circumvent this restriction it was necessary to develop a new design technique, known as dispersion adaption [22]. Here, the modifications to the photonic crystal lattice are compatible with this design grid. The design methodology is based on the gentle confinement method, and the lattice modifications required are determined by a combination of numerical simulations and analytical calculations. Despite using relatively large modifications of the lattice (10s of nanometers), the approach allows us to envelop the mode gently. Such large modifications also provide improved disorder stability compared to other photonic crystal cavity designs.

Similar to Kuramochi et al. [23], the dispersion adapted (DA) cavity is realized by locally modulating the width of a W1 photonic crystal waveguide. A graded potential well is created by adapting the W1 dispersion curve (hence the name) such that the confined state that results has a Gaussian shape. Such a shape is typically achieved via the construction of a cavity from a guiding central part surrounded by sections of "soft" mirrors, into which the field exponentially decays. By carefully adjusting the mirror regions, the overall mode shape will be very close to Gaussian. The dispersion curve of the waveguide and the corresponding cutoff frequencies as a function of hole-shift are calculated using a frequency-domain eigensolver [24]. An analytic fit is made to the dispersion curve and the cutoff frequencies for different hole-shifts and this is used to build the cavity [5]. A schematic of the DA cavity, with an L3 for comparison, is given in Fig. 5.6.

Using the advanced CMOS research environment of IMEC, DA cavities were fabricated. The process started with SOI wafers, from SOITEC, with a 220 nm device layer and 2 μm buried oxide. Using an ASML PAS5500/1100 step-and-scan system operating at 193 nm, the patterns were defined in resist and developed. The pattern was then transferred to the top silicon layer using an inductively coupled plasma low pressure and high density etch system with a HBr/O_2 chemistry [25]. After the photonic crystal etch, the buried oxide layer was removed using diluted hydrofluoric acid.

The resonant scattering technique [14] was used to characterize the photonic crystal cavities. Resonant scattering has the advantage that it gives the unloaded, or intrinsic, Q-factor of the cavity, unlike in-plane characterization techniques that use coupling waveguides that modify the behavior of the cavity. In this technique, a laser or broad band light source of fixed polarization is tightly focused on the device under test at normal incidence. The reflectance from a photonic crystal cavity is measured using a crossed-polarization setup. The cavity is oriented at 45 degree with respect to the axis of polarizer and analyzer, maximizing the coupling between the input and output signals. More details are provided in Portalupi et al. [26].

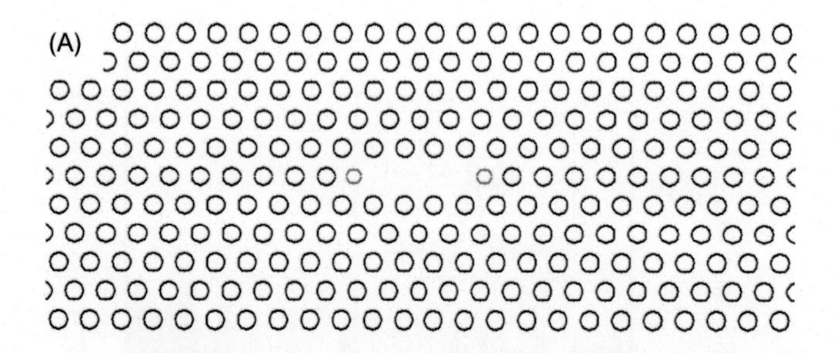

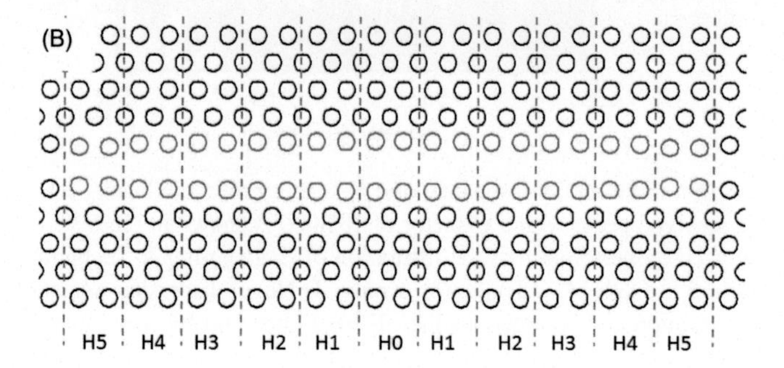

Figure 5.6 (A) The L3 cavity is realized by removing three holes at the center, and the Q-factor is optimized by shifting and shrinking the holes (shown in *green*) on each side of the cavity. (B) The DA cavity is realized by adjusting the position of the holes on each side of the W1 waveguide.
Source: Reproduced by permission of Welna K, Portalupi S, Galli M, O'Faolain L, Krauss T. Novel dispersion-adapted photonic crystal cavity with improved disorder stability. IEEE J Quantum Electron, 2012;48:1177–83.

A typical spectrum is shown in Fig. 5.7. The sharp peak close to the wavelength of 1553 nm is the first order cavity resonance. The inset gives a closer look at the spectrum around the resonance peak. The red dots give the measured RS signal, and the black line the Fano fit.

It is possible for interference to occur between light scattered from the background photonic crystal (continuum of states) and scattering due the cavity resonance (discrete state), known as the Fano effect [19], which gives rise to an asymmetric line shape rather than the more normal Lorentzian function. The line width of the resonance is measured to be $\sim$8.2 pm, corresponding to a Q-factor of 189,000. For comparison, L3 and L5 cavities were included in the same fabrication run, and the highest Q-factor measured was 27,500 and 85,500 respectively. By optimizing the etch conditions, further improvements in the Q-factors can be realized.

The tolerance of designs to fabrication imperfections is a key concern for mass production. In commercial environments high yields are crucial, and it is often the worst performing cavity in a given batch that limits the overall performance. To investigate the robustness of different photonic crystal cavity designs, varying levels of hole-diameter disorder were deliberately introduced into the designs of a set of

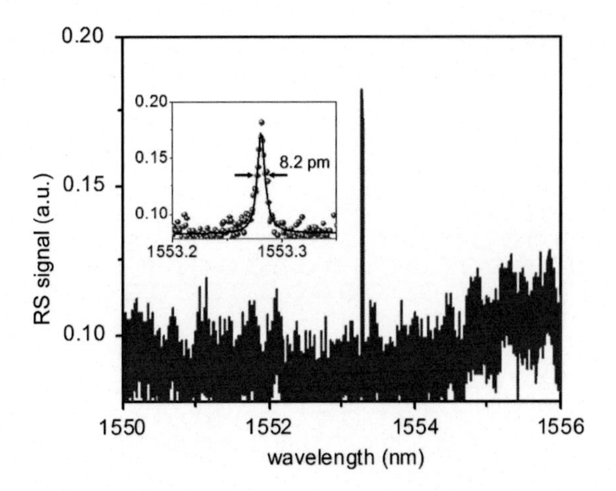

Figure 5.7 The first order cavity resonance of a DA cavity measured using resonant scattering. The resolved peak, with a Q-factor of 189,000 is shown in the inset.
Source: Reproduced by permission of the Institution of Engineering & Technology from Welna K, Portalupi S, Galli M, O'Faolain L, Krauss T. Novel dispersion-adapted photonic crystal cavity with improved disorder stability. IEEE J Quantum Electron, 2012;48:1177–83.

fabricated DA and heterostructure cavities. The DA cavity had a 400 nm lattice with a hole size of $r/a = 0.25$, while the heterostructure employs a center period of 420 nm, a mirror-period of 410 nm, and a hole size given by $r/a = 0.26$. The deliberate disorder was implemented as a pseudo-random Gaussian variation of the hole-radii, expaf; $\left[(r - r_{\text{mean}})^2/(2\sigma^2)\right]$, where σ is the standard deviation.

Similar to the effect observed for L3 cavities [20], the measured Q-factor drops monotonically with increasing disorder (Fig. 5.8). However, two differences between the DA and the HS cavities can be seen: firstly, the DA cavity has a higher Q-factor in the absence of deliberate disorder ($\sigma_{\text{delib}} = 0$) and secondly, the decay of the Q-factor is less steep for the DA cavity. The following formula shows how the different factors contribute to the experimentally measured Q-factor (Q_{cavity}):

$$\frac{1}{Q_{\text{cavity}}} = \frac{1}{Q_{\text{design}}} + \frac{1}{Q_{\text{disorder}}} \tag{5.1}$$

where Q_{design} is the simulated Q-factor (i.e., without any disorder), and Q_{disorder} is the disorder contribution, consisting of the fabrication induced intrinsic disorder (e.g., sidewall roughness) and the deliberately created disorder. Following [27], we have the following:

$$\frac{1}{Q_{\text{disorder}}} = A\left(\frac{\sigma_{\text{delib}} + \sigma_{\text{intrinsic}}}{a}\right)^2 \tag{5.2}$$

where $\sigma_{\text{intrinsic}}$ is the intrinsic disorder, arising from side wall roughness, σ_{delib} is the deliberately introduced disorder, a is the lattice constant, and A is a constant. In the

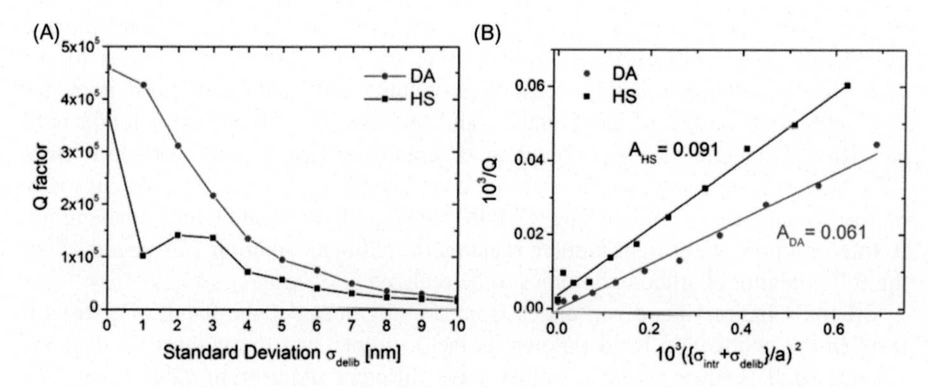

Figure 5.8 (A) Experimental Q-factor versus the standard deviation of the disorder (σ_{delib}) that was deliberately introduced for the dispersion adapted (DA) and heterostructure (HS) cavities. (B) Plot of the values in (A) against the relative quadratic disorder contribution of intrinsic and deliberate disorder. The A-parameter is determined by the slope of the straight line fit according. *Source*: Reproduced with permission from Welna K, Portalupi S, Galli M, O'Faolain L, Krauss T. Novel dispersion-adapted photonic crystal cavity with improved disorder stability. IEEE J Quantum Electron, 2012;48:1177−83.

absence of deliberate disorder, Eq. (5.2) gives what is usually referred to as the fabrication limited Q-factor, Q_{fab}.

It might be thought that the fabrication dependent Q_{fab} should be identical for both DA and HS designs, as the cavities have been fabricated on the same chip, and have therefore experienced identical fabrication conditions. However, using the design factors (calculated from 3D Finite Difference Time Domain simulations), of 2,700,000 and 5,500,000 for DA and HS, respectively, it is found that fabrication limited Q-factor, accounting for intrinsic disorder, only, is higher for the DA cavity, at 553,000, than for the HS cavity at 404,000. The difference is a consequence of the parameter A in Eq. (5.2), which is actually design dependent.

In White et al. [20], the value for σ_{intr} was estimated to be on the order of $0.5-1$ nm for the fabrication process used here. The disorder parameter A can now be determined from a straight line fit with $\sigma_{intr}=0.5$ nm, as shown in Fig. 5.8B. A is inversely proportional to the disorder limited Q-factor and thus, a lower value for the A parameter corresponds to a more robust design cavity; a lower value is therefore advantageous. This behavior is indeed shown in Fig. 5.8B where the A parameters for the DA and HS designs are 0.061 and 0.091, respectively. Notably, despite the higher design Q-factor, the HS cavity is inferior to the DA for all the disorder levels considered here. A key design principle is, therefore, to optimize designs in the knowledge that some disorder will always be present in practice, particularlyin the multi-step processes used in CMOS fabrication.

5.3 Light emission

It is well known that silicon is a very poor light emitter as a consequence of its indirect band gap. Nonetheless, the realization of a silicon-based CMOS compatible light source would be truly revolutionary. The absence of a cheap, efficient, and electrically driven silicon-based light emitter creates a significant barrier to the wider use of photonics in low- cost, high-volume applications, a major issue for key fields such as optical interconnects and bio-sensing. The current solution is to add III – V materials through bonding or epitaxy giving efficient on-chip lasers [28]; however, the use of costly material and complex processing cancels out many of the advantages of silicon and CMOS, making mass manufacture challenging. A true group-IV nano light emitter remains the ultimate solution and would allow the full potential of silicon photonics to be realized.

Silicon's indirect band gap arises from the alignment of the bands. The lowest level in the conduction band (known as the X point), and the highest level in the valence band (known as the Γ point), have different momentum values, meaning that the radiative recombination of an electron-hole pair requires some additional momentum, which can be provided by a phonon. However, the probability of such a three-particle interaction is low, making radiative recombination inefficient.

A considerable amount of research has been devoted to improving the luminescence of silicon; silicon nanocrystals have exhibited optically pumped gain [29] in the visible and near-infrared wavelength ranges; the incorporation of rare earth dopants has enabled optically pumped transparency from erbium-doped silicon nitride nano-cavities [30]; and highly-doped, strained germanium has resulted in a 270 μm long electrically pumped laser [31].

While these approaches have considerable potential, no single solution provides all the desired characteristics of a silicon light source, namely electrical pumping, operation at sub-band gap wavelengths, room temperature operation, small size, and narrow emission linewidths.

5.3.1 Photoluminescence

Optically active defects provide another approach to improving the luminescence of silicon. The interaction of free carriers with such defects aids in conserving momentum during recombination, creating luminescence lines in the emission spectrum, see Fig. 5.9, including the important telecommunication windows. Dislocation loops have been used to realize broad area electrically driven LEDs operating near the silicon band-edge [32] and "A-centers" have demonstrated stimulated emission at cryogenic temperatures [33]. These results, although the luminescence was at undesirable wavelengths and temperatures, highlight the usefulness of optically active defects. In 1991, Bestwick et al. [34] and Singh et al. [35] observed that exposing silicon to plasmas caused the formation of a range of emission lines and bands, and attributed them to optically active defects at the surface that resulted from the treatment. Such an approach has the advantages of providing emission in the correct temperature and wavelength ranges, and being suitable for incorporation with the process used in CMOS fabrication [36].

As outlined in Section 1, there have been huge strides in the fabrication of nanophotonic devices since the 1990s. In particular, the high Purcell factors available with photonic crystal nanocavities are extremely powerful and can be used to enhance the emission of light [37]. The environment surrounding an emitter influences its emission through the local density of states. By coupling

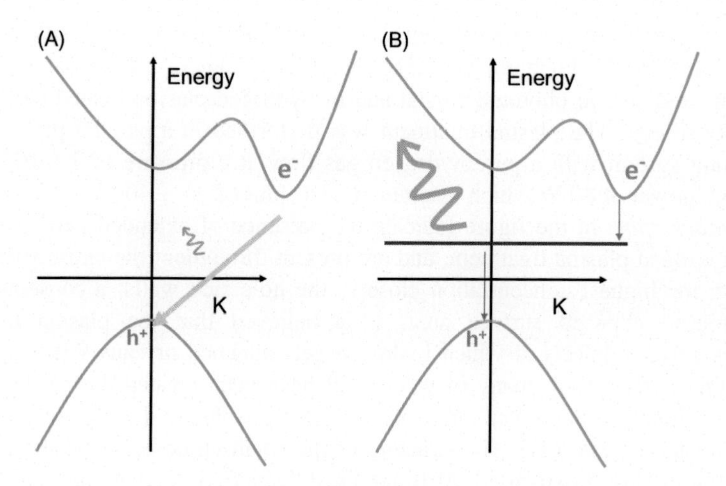

Figure 5.9 Schematic of the silicon band-diagram, showing: (A) That the momentum mismatchmakes the probability of photon emission small, and (B) The presence of a mid-gap point defect provides momentum conversation due to its broadness in momentum space, facilitating radiative recombination (*red horizontal line*).

the emitter with a resonator, it's spontaneous emission rate can be increased if there is a spectral and spatial overlap between the emitter and the mode of the resonator. The enhancement factor is known as the Purcell factor, F_p, [38] and is given by the following equation for a single emitter with dipole moment "d":

$$F_p = \frac{3Q_{\text{eff}}(\lambda_c/n)^3}{4\pi^2 V}\left|\frac{\vec{d}\cdot\vec{f}(\vec{r}_e)}{\vec{d}}\right|^2 \frac{\Delta\omega_c^2}{4(\omega_e-\omega_c)^2+\Delta\omega_c^2}$$

where $\frac{1}{Q_{\text{eff}}} = \frac{1}{Q_e} + \frac{1}{Q_c}$ is the effective quality factor, thus accounting for the different line width of the emitter and of the cavity (the subscripts c and e refer to cavity and emitter, respectively). V is the cavity mode volume, and n is the refractive index. The second and third terms refer to the degree of spatial and spectral overlaps between the emitter and cavity mode. In general, these factors are less than unity, but if perfect spatial and spectral overlaps are assumed the Purcell factor is given by the well-known expression [31]:

$$F_p = \frac{3Q_{\text{eff}}(\lambda_c/n)^3}{4\pi^2 V}$$

Thus, to achieve high enhancement, a cavity emitter system with high Q-factor and low mode volume is required. The L3 photonic crystal cavity is therefore an excellent platform for the enhancement of optically active defects. Its Q-factor is high, in excess of 100,000 [14], mode volume is relatively low, 0.76 $(\lambda/n)^3$, and reliability is very good (yields approaching 99% are typical even in a research environment).

Fig. 5.10 shows planar view and cross-sectional TEM images (carried out using a JEOL JEM 2010 TEM operating at an acceleration voltage of 200 kV) of a hydrogen plasma-treated silicon photonic crystal and of hydrogen plasma-treated Crochzalski [39] (CZ) silicon. The plasma treatment was performed in a parallel plate reactive ion etching system with a pure hydrogen gas flow at a pressure of 1×10^{-1} mbar, and an RF power of 40 W which resulted in a DC bias of ≈ -400 V.

The *black spots* in the figure indicate the presence of extended defects induced by the hydrogen plasma treatment, and are present throughout the entire silicon surface with the highest concentration close to the hole side walls; a consequence of the increased exposed surface area. It is believed that the plasma treatment generates surface defects at which hydrogen gets attached or causes it to attach to pre-existing defects [40], many of which will have been created during the etching process. It should be noted that luminescence has been observed in the literature from etching related defects [41]. The structure of the resulting defect population may be seen in Fig. 5.10B, a defocused off-Bragg cross-section view. A number of different types of defects may be observed at different penetration depths, in the 10 nm range below the surface, nanobubbles dominate, with sizes of a few nanometers. Their exact nature is unknown, but they most likely consist of agglomerates of vacancies. Deeper into the silicon layer, platelets [42] (*white arrows* in Fig. 5.10B)

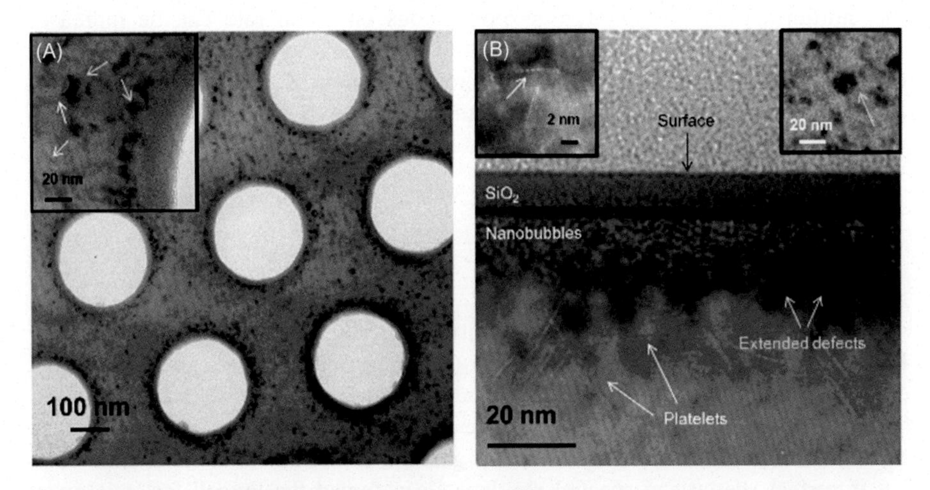

Figure 5.10 Transmission electron microscopy (TEM) images of a hydrogen plasma treated silicon photonic crystal. (A) Planar view TEM showing an increased defect concentration close to the hole sidewalls. The inset shows a close-up view of a single hole. (B) Cross-sectional (CS) view of a plasma-treated Crochzalski silicon sample, with white arrows indicating platelets and yellow arrows pointing to extended defects. The top left inset gives a high resolution image of the platelets, and in the top right inset the typical coffee bean shape of a dislocation loop is visible.
Source: Reproduced with permission from Shakoor A, Lo Savio R, Cardile P, Portalupi S, Gerace D, Welna K, et al. Room temperature all-silicon photonic crystal nanocavity light emitting diode at sub-bandgap wavelengths. Laser Photon Rev 2013;7:114–21.

predominate. Their mean diameter is 10–15 nm and they occupy the (100) plane crystal plane, parallel to the surface, and the {111} planes. Dark traces are located between the two nanobubble and platelet regions which exhibit the "coffee bean" shape typical of dislocation loops. An example is shown in the top right inset of Fig. 5.10B. Defects of this nature have been reported in the literature [36]; however, their luminescence is not yet fully understood.

Room temperature confocal photoluminescence measurements and the resonant scattering method were used to measure the PL and Q-values of the nanocavity modes. A CW laser diode was used to excite the modes of the nanocavity. The lasing wavelength was 640 nm, above the band gap energy of silicon, and was absorbed generating electron hole pairs. The spot was focused to a spot of 1 μm^2 centered on the nanocavity by means of a high numerical aperture microscope objective (NA = 0.8). The emitted light was collected with the same objective and coupled into a spectrometer.

Photoluminescence spectra for photonic crystal nanocavities with different lattice constants, along with a Crochzalski (CZ) silicon sample (for comparison purposes), are shown in Fig. 5.11. The photoluminescence signal from the CZ silicon is extremely weak and the plasma-treated photonic crystal nanocavities show an enhanced, relatively flat background signal with strong, sharp peaks at wavelengths corresponding to the resonances of the nanocavity modes. The background PL enhancement is a consequence of the incorporation of optically active defects

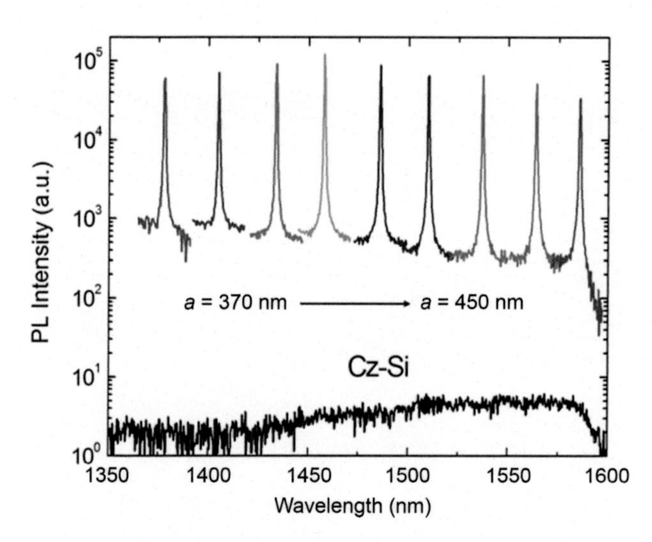

Figure 5.11 Photoluminescence of hydrogen-treated photonic crystal nanocavities, with lattice constants (a) ranging from 370 to 450 nm. The peaks correspond to the fundamental mode of each cavity. For comparison, the emission spectrum on Crochzalski silicon is included. The emission lines of the treated nanocavities are over four orders of magnitude stronger.
Source: Reproduced with permission from Shakoor A, Lo Savio R, Cardile P, Portalupi S, Gerace D, Welna K, et al. Room temperature all-silicon photonic crystal nanocavity light emitting diode at sub-bandgap wavelengths. Laser Photon Rev 2013;7:114–21.

into the silicon. The peaks occur due to the increased extraction efficiency of the photonic crystal cavity and Purcell enhancement. Overall, a 40,000 times enhancement of the PL signal at room temperature is observed relative to CZ silicon. The emission lines may be tuned throughout the 1300 − 1600 nm range, demonstrating the potential of these devices for Wavelength Division Multiplexing and related applications.

5.3.2 Electroluminescence

Photonic crystal fabrication is fully compatible with CMOS fabrication (see Section 5.1) allowing *pin* diodes to be integrated by means of multiple lithography and ion implantation steps, providing a monolithic source for the electron-hole pairs that recombine at optically active defects. Doped regions that extend into the photonic crystal are created, giving a low resistance path to guide the charge carriers and ensure that the maximum amount of electron-hole pairs recombine inside the optical mode. The doped "finger" structure can be seen in Fig. 5.12.

The n-type finger-like arm of the device was realized by doping the SOI with phosphorus using ion implantation to a level of 10^{19} P/cm^3. The p-type finger was doped with boron to a level of 10^{19} B/cm^3. The two regions were separated by a distance of 500 nm. This separating region has p-type doping of 10^{15} B/cm^3 (the background doping level of the as-bought SOI wafer). The alignment between the doped regions and the photonic crystal was carefully carried out using EBL. It is known that optical losses are increased in doped silicon; however, photonic crystal

Figure 5.12 Schematic of fabricated photonic crystal nanocavity and pin junction (the doped regions are superimposed on the SEM image of an actual device). The beam indicates the active region and the direction of emission.
Source: Reproduced with permission from Shakoor A, Lo Savio R, Cardile P, Portalupi S, Gerace D, Welna K, et al. Room temperature all-silicon photonic crystal nanocavity light emitting diode at sub-bandgap wavelengths. Laser Photon Rev 2013;7:114–21.

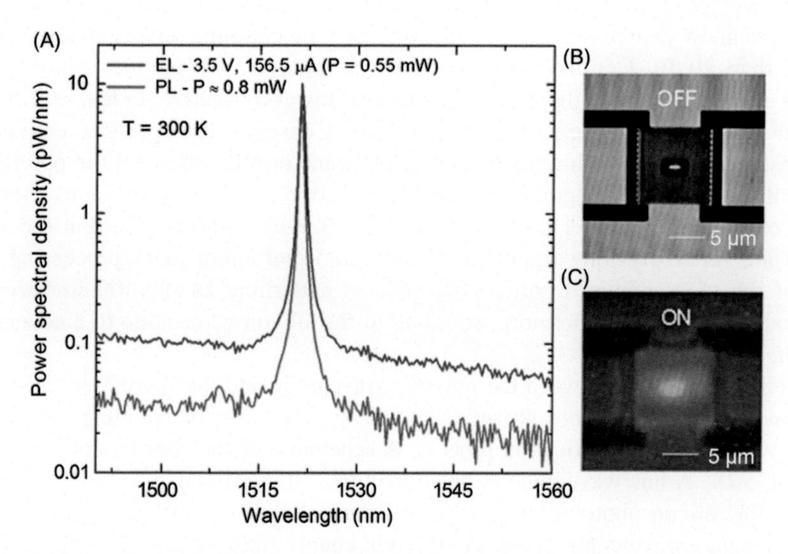

Figure 5.13 (A) Electroluminescence (EL) and Photoluminescence (PL) spectra from a plasma treated PhC nanocavity. (B) SEM image of the photonic crystal (top view), and (C) A filtered IR picture of the device showing clear electroluminescence. The electroluminescence is recorded at an applied voltage of 3.5 V, with a current of 156.5 μA, thus consuming electrical power of 0.55 mW.
Source: Reproduced with permission from Shakoor A, Lo Savio R, Cardile P, Portalupi S, Gerace D, Welna K, et al. Room temperature all-silicon photonic crystal nanocavity light emitting diode at sub-bandgap wavelengths. Laser Photon Rev 2013;7:114–21.

cavities with Q-factors in excess of 40,000 can be realized even for doping levels of $1 \times 10^{18}\ \mathrm{cm}^{-3}$.

When the junction is forward biased carriers are injected efficiently into the intrinsic region by virtue of the doped fingers giving rise to electroluminescence. A comparison between electrical and optical pumping is shown in Fig. 5.13 (made with the same apparatus as for Fig. 5.11).

The power spectral density (expressed in pW/nm) is a key measure of a device's performance for wavelength selective applications (WDM). The device shown in Fig. 5.13 achieves a value of approximately 10 pW/nm, the current record for silicon-based light emitters [36].

5.4 The fiber coupling problem and its solution

Interfacing photonics chips with optical fibers is a crucial challenge for any optical device or system. A single mode optical fiber typically has a core diameter of 8 μm, orders of magnitude larger than the dimensions of high refractive index contrast nanophotonic waveguides (100 – 500 nm). The large mismatch between the optical modes of each component results in coupling losses that are often more than 20 dB. Coupling between fibers to polymer waveguides can be much more efficient ($\sim < 1$ dB loss), due to the lower refractive index, ~ 1.5, and larger dimensions of the polymer waveguide. Consequently, spot size converters, consisting of a tapered interface region, that transform the mode in the polymer to that of the silicon waveguide is a popular solution. Such spot size converters are very effective, with fiber-to-fiber losses of ~ 3.5 dB reported [43], but the structures are relatively long (hundreds of microns) and entail extra processing. The tip of the taper required particularly precise patterning as sub-100 nm sizes are typically required. Furthermore, coupling to the silicon waveguide to a nanocavity results in additional loss.

Vertical coupling between the polymer waveguide and the silicon photonic crystal cavity is a much more attractive option, giving superior coupling efficiencies even with a simpler fabrication process. A schematic of the configuration is shown in Fig. 5.14. A bus waveguide is positioned on a low refractive index spacer layer above the silicon photonic crystal cavity, similar to that proposed in Qiu [44]. At the resonance wavelength of the cavity, light couples into the cavity from the waveguide mode.

The fabrication process is similar to that outlined earlier (see Section 5.1.3). After the etching of the photonic crystal into the silicon, a layer of silicon dioxide is put down, typically by means of a spin-on-glass (SOG) process. SOG has excellent planarization and void filling properties, and ensures that the holes of the photonic crystal are completely filled with oxide. The mechanical stability of oxide embedded photonic crystals is far superior to that of membraned photonic crystals, allowing for their integration into complex CMOS processes. The material index of the SOG is ~ 1.4 and is only slightly less than that of thermally grown silicon oxide. The asymmetry in the cladding indices is small and can limit the Q-factor introduced, but in optimized systems, Q-factors as high as one million have been reported [46]. A layer of silicon nitride is then deposited (typically 500 nm thick)

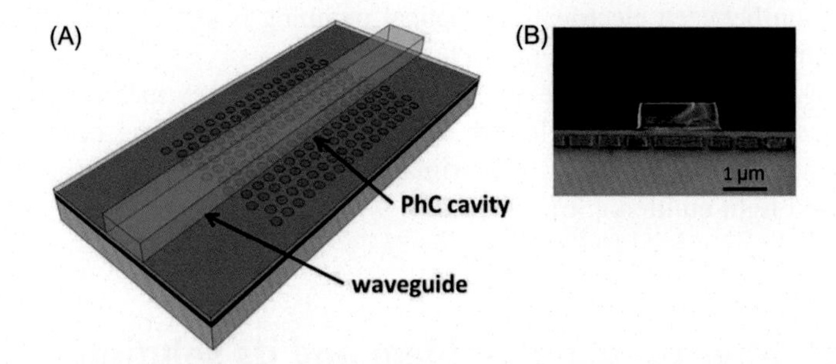

Figure 5.14 (A) A schematic representation of a bus waveguide is vertically coupled to a PhC cavity. (B) A Scanning electron microscopy image of an actual device.
Source: Reproduced with permission from Debnath K, Welna K, Ferrera M, Deasy K, Lidzey DG, O'Faolain L. Highly efficient optical filter based on vertically coupled photonic crystal cavity and bus waveguide. Opt Lett 2013;38:154–6 [45].

using Plasma Enhanced Chemical Vapor Deposition, and the waveguide defined. Silicon nitride is preferred for its CMOS compatibility, but a range of other materials, such as polymers, can also be used.

The reflection and transmission of the photonic crystal filter can be calculated using [47]:

$$T = \left(\frac{Q_{\text{total}}}{Q_{\text{cavity}}}\right)^2 \quad (5.3)$$

$$R = \left(\frac{Q_{\text{total}}}{Q_{\text{coupling}}}\right)^2 \quad (5.4)$$

Where Q_{cavity} is given by Eq. (5.1), Q_{coupling} describes the coupling rate between the waveguide and the cavity and:

$$\frac{1}{Q_{\text{total}}} = \frac{1}{Q_{\text{cavity}}} + \frac{1}{Q_{\text{coupling}}} \quad (5.5)$$

The system thus operates as a compact and efficient optical filter, and is ideal for use in integrated optical circuits for WDM applications.

Precise and accurate control of filter bandwidths and drop efficiencies is an important requirement for all optical filters. In the traditional laterally-coupled photonic crystal filters [48] the coupling (Q_{coupling}) is discrete as it is determined by the number of rows of holes between waveguides and cavities, whereas, in the vertically-coupled configuration, coupling may be smoothly varied by changing the spacer thickness. In the in-plane configuration, the line defect bus waveguide (typically created by removing a row of holes from the lattice) disrupts the symmetry of the photonic crystal lattice introducing k-components to the mode that lie inside the light cone (see Section 5.1.2), a significant disadvantage. The resulting optical loss has an impact on the value of Q_{cavity}, and 3D simulations show that placing an in-plane waveguide three rows away from an unmodified L3 cavity (i.e., no gentle confinement) reduces Q_{cavity} from 4982 to 2690_y which clearly has a detrimental effect on the extinction ratio of the optical filter, Eq. (5.3). With vertical coupling, there is a greater scope for optimization of the cavity design to find the optimum coupling condition.

In 2005, Min Qiu showed in a simulation that vertical coupling was possible between a silicon waveguide and a photonic crystal silicon cavity [14], but achieving efficient transfer of light between waveguides and cavities with different effective refractive indices is non-trivial. The coupling efficiency is determined by two factors, the spatial overlap between the evanescent modes of the cavity and the waveguide, and the **k**-vector match between the two modes. The spatial overlap is determined by the thickness of the barrier layer, but in conventional approaches, e.g., the directional coupler, a **k**-vector match can only be achieved by using similar

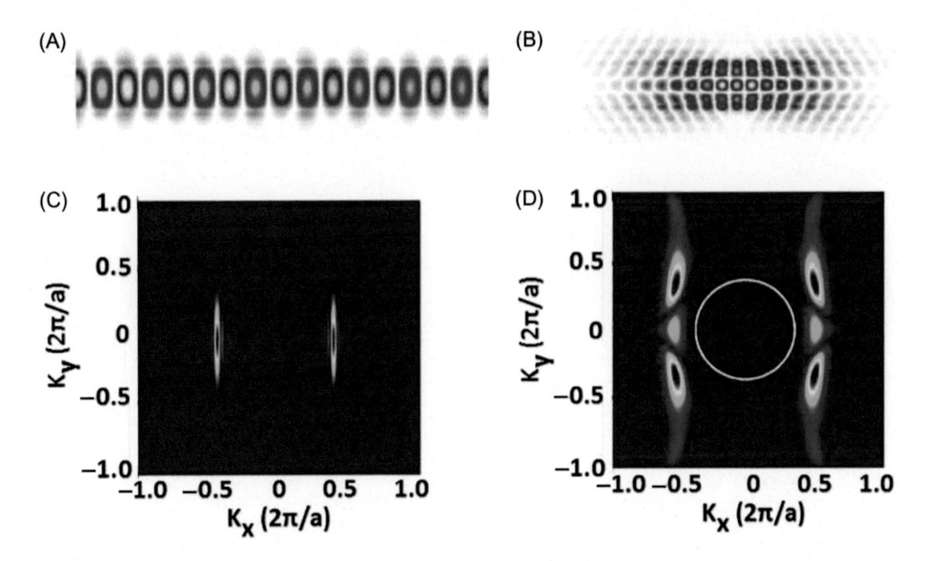

Figure 5.15 The electric field distribution in the waveguide (A) and cavity (B). The corresponding K-space distribution of the dominant field; (C) waveguide; and (D) cavity.

materials. The situation with a photonic crystal nanocavity is dramatically different, as due to their ultra-small mode volume the **k**-vector distribution is much richer, as shown in Fig. 5.15.

The waveguide shows a distribution that is centered on the **k**-vector value corresponding to the modal refractive index, Fig. 5.15C, with the width depending on the waveguide length. The distribution of the photonic crystal nanocavity has a complex distribution that is, while also centered on the **k**-vector corresponding to the effective index of the slab, much broader. Options for achieving **k**-vector matching are therefore provided, allowing for the possibility of efficient coupling between a low effective refractive-index waveguide mode and a silicon nanocavity. The advantages of the two material systems, namely low loss waveguiding and efficient fiber coupling of dielectrics, and the compactness and enhanced electro-optic optics of silicon nanophotonics, may thus be accessed.

Clearly, the effective refractive index (n_{eff}) of the waveguide will play a significant role in determining the **k**-vector matching, and this is illustrated in Fig. 5.16. The overlap integral between the **k**-vector distributions of the cavity and waveguide is numerically calculated as a function of the n_{eff}. Experimentally, a range of effective refractive indices can be obtained by varying the waveguide composition and dimensions. With a polymer, such as SU8 (material index of 1.56), values of n_{eff} in the range 1.47–1.49 may be achieved for dimensions on the order of $3 \times 2\ \mu\text{m}$. Efficient fiber coupling (~ 2 dB) is also obtained with such waveguides. The material refractive index of a dielectric such as silicon nitride is much higher, at $n \sim 1.88$, providing access to $n_{\text{eff}} = 1.55 - 1.6$. Silicon nitride has the significant advantages of enhanced CMOS compatibility due to its very high thermal stability

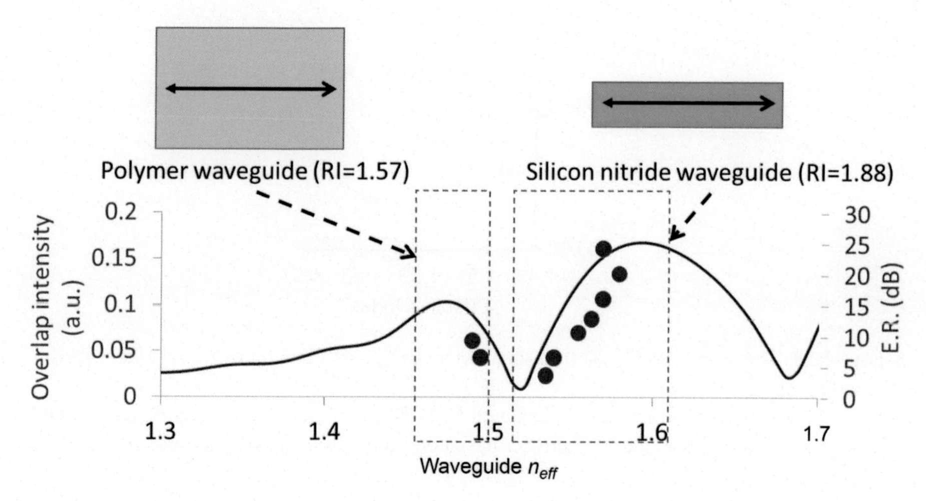

Figure 5.16 The effect of the waveguide effective refractive index on the extinction ratio of the photonic crystal filter. Both the composition and the dimension of the waveguide is altered so as to span a range of effective refractive index (n_{eff}).

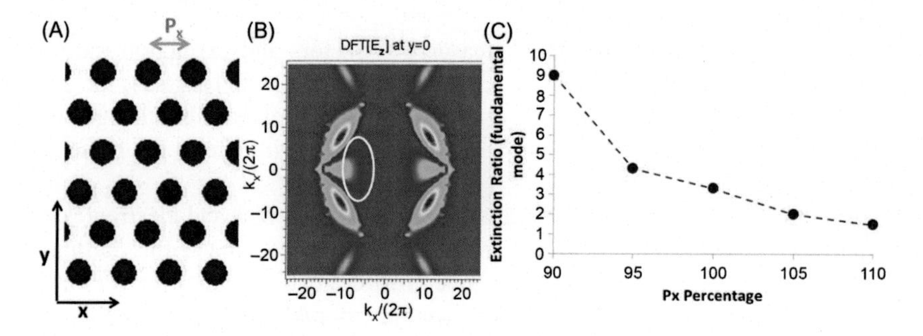

Figure 5.17 (A) the normal hexagonal photonic crystal lattice, showing the x-component of the lattice constant, *Px*; (B) The **k**-vector space of the PhC cavity with the region of interest circled; (C) Extinction ratio as a function of *Px* percentage: 100% corresponds to the normal value.

and its potential for high performance Photonic Integrated Circuits [49]. The *red circles* in Fig. 5.16 show the experimentally measured extinction ratios for a range of waveguides.

The photonic crystal nanocavity itself may also be optimized, Fig. 5.17, for more efficient coupling to the bus waveguide. Starting from the normal triangular lattice, the intensity of the **k**-vector distribution in specific regions can be altered by modifications of the lattice. For example, the lattice constant in the x direction (denoted as *Px*) may be "squeezed" or "stretched" with respect to its original value of 0.5a. The mode shape is indirectly altered due to changes in the band structure and the penetration of the mode into the lattice, directly due to the mode's interaction with particular holes, allowing the extinction ratio of the resulting optical filter

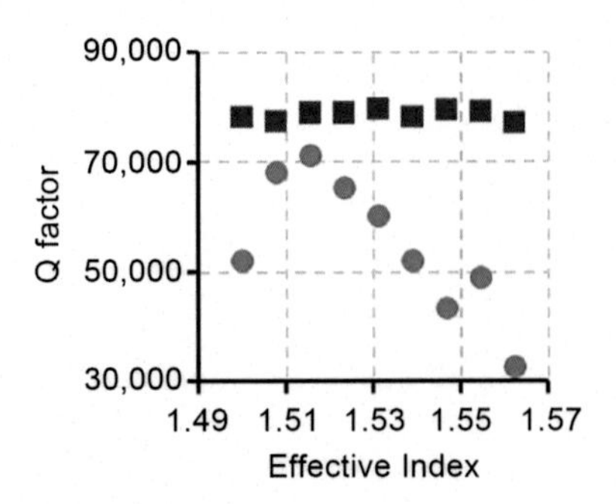

Figure 5.18 The values of Q_{cavity} (red squares) and Q_{total} as function of waveguide effective refractive index.
Source: Reproduced with permission Debnath K, Welna K, Ferrera M, Deasy K, Lidzey DG, O'Faolain L. Highly efficient optical filter based on vertically coupled photonic crystal cavity and bus waveguide. Opt Lett 2013;38:154–6.

to be varied for a given application. Enhanced flexibility in fine-tuning the coupling efficiency for different photonic crystal designs and waveguide compositions is thus provided.

By comparing the experimentally measured Q-factors and extinction ratios, the values of Q_{cavity} and $Q_{coupling}$ can be extracted, using Eqs. (5.4)–(5.5), the results are shown in Fig. 5.18. The value of Q_{cavity} is largely unaffected by changes in the coupling, highlighting the low perturbation of the nanocavity due to the presence of the bus waveguide.

The popular SOI is limited to lateral integration of photonics and electronics components [50]. A number of research groups have now recognized this failing and introduced multiple layers of deposited silicon to dramatically improve the power- and cost-efficiencies. Such 3D photonic stacking assemblies have the ability to be scaled up, increasing device density and functionality without an electronics layer, similar to that used in high volume and high density electronic circuit layouts used for making DRAM (Dynamic Random Access Memory) and microprocessors. The use of a vertically-integrated silicon nitride layer is even more attractive, as the propagation losses can be an order of magnitude lower than silicon [51], with high performance wavelength multiplexing/de-multiplexing also demonstrated. Vertical coupling to photonic crystals, while a more complex design challenge, is simpler to fabricate requiring less components than equivalent circuits that use spot-size converters [44].

5.5 Optical modulation

Electro-optic modulation is a key function in photonic integrated circuits for optical interconnects. A time-varying voltage, carrying information, must be converted into a time-varying optical intensity (supplied by a CW external source) for transmission across the optical link. The application of an electric field to a material can result in changes to the real and imaginary parts of the refractive index [52]. The Pockels effect, the Kerr effect, and the Franz – Keldysh effect are traditionally used in semiconductor materials to realize modulation based on either electro-refractive or electro-absorption modulation, however, these are weak in pure silicon at the telecommunications wavelengths (1.3 and 1.55 μm). With thermo-optic effects too slow for useful datacommunication, the plasma dispersion effect has become the preferred mechanism. When the concentration of free carriers in silicon changes, there is a corresponding change in the absorption coefficient leading to a modification of the refractive index (the Kramers – Kronig relation). This change in refractive index is usually transformed into an intensity modulation by means of a Mach – Zehnder Interferometer. The incident beam is divided into two paths with the refractive index of one varied and then recombined. Depending on the change in refractive index, the two beams will interfere constructively or destructively, thereby controlling the exiting light. The condition for destructive interference can be written as follows:

$$\frac{2\pi}{\lambda}\Delta nL = \pi \tag{5.6}$$

Where L is the length of the arms of the MZI and Δn is the change in refractive index between arms. Mach – Zehnder optical interferometers have been extremely popular, however, as the plasma dispersion effect is relatively weak (with the maximum change in refractive index, Δn, on the order of 1×10^{-3}) the device is typically a few millimeters long, see Eq. (5.6), reducing the integration density and increasing the power consumption during operation.

By realizing resonant structures, compact devices that give large changes in transmission even with such small changes in refractive index can be realized [53]. The resonance condition for a simple Fabry – erot resonator, see Fig. 5.19A, is given by:

$$nd = m\frac{\lambda}{2}$$

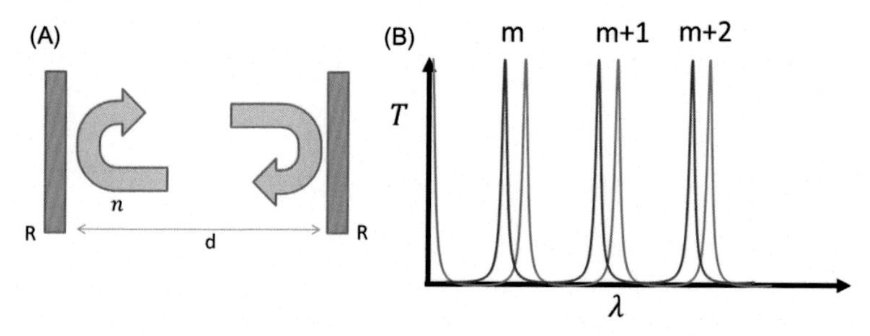

Figure 5.19 (A) Schematic of a Fabry – Perot cavity. (B) Example transmission for different values of the refractive index, n, in red and blue.

If the refractive index changes, then the resonance condition is satisfied at a different set of wavelengths (λ'):

$$(n + \Delta n)d = m\frac{\lambda'}{2}$$

The modulation depth or extinction ratio is a key characteristic of any modulator, and is defined as the ratio between the maximum ($I_{\max}$) and minimum transmission ($I_{\min}$), usually quoted in decibels, $10\log_{10}(I_{\max}/(I_{\min})$. Large modulation depths are important for long transmission distances, as are good bit error rates and high receiver sensitivity.

In a resonant modulator, the change in transmission as a function of the change in refractive index determines the extinction ratio of the modulator, which in turn depends on the resonance Q-factor, illustrated in Fig. 5.19B. The Q-factor, which is the ratio between the energy stored in the resonator and the energy lost per optical cycle, is given by:

$$Q = \frac{\lambda}{\Delta\lambda}$$

High Q-factors are advantageous for achieving high extinction ratios; however, the narrow linewidth can limit the bandwidth of the data channel. In general, Q-factor values of $10{,}000 - 20{,}000$ provide the best compromise [54].

Photonic crystal cavities are particularly attractive for optical modulation as they allow the strongest confinement of photons, thus minimizing the device capacitance, while providing high intrinsic Q-factors that enable high extinction ratio modulation. The low mode volume enables power consumption that can be one to two orders of magnitude lower than ring resonator based modulators [46,47] and the short cavity length provides a large free spectral range [55], ultimately limited by the $300 - 400$ nm wide PBG, thus potentially supporting almost an order of magnitude more WDM channels than the ring resonator approaches.

Notomi et al. of NTT Corp. pioneered the use of photonic crystal electro-optic modulators and demonstrated a system in which two photonic crystal waveguides are coupled together via a PhC cavity, giving rise to an appropriate filter response. Such an approach is limited by the challenge to couple light efficiently into the small optical mode of the silicon based waveguide. By using the vertical coupling technique of Section 5.4 a more powerful and practical system can be realized that has potential for terabit/second per mm^2 data communication with a single bus waveguide.

5.5.1 High speed (GHz) electro-optic modulation

Research on silicon waveguide-based optical modulators began in the mid-1980s [56]. Initially, modulation rates were generally slow ($\sim$megahertz data rates), as the devices operated on the basis of carrier injection, which was limited by the long carrier lifetime of silicon. The first silicon modulator that could operate at speeds above 1 GHz was demonstrated by Intel [57], and marked a milestone in the field.

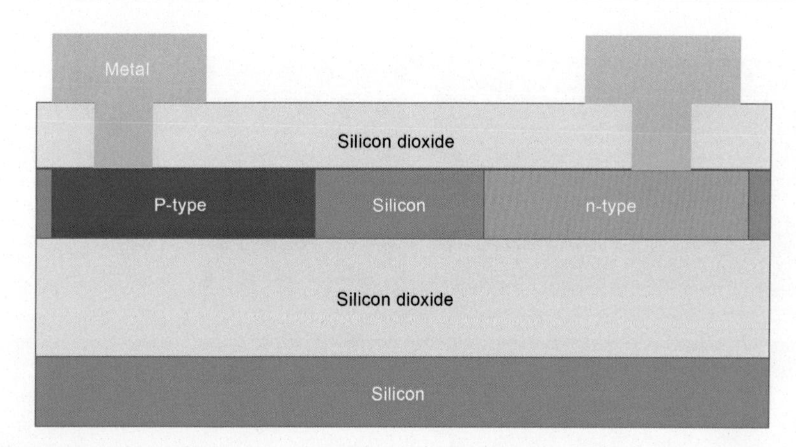

Figure 5.20 Schematic of pin junction for use in an optical modulator. The waveguide is typically situated in the center region.

The use of a *pn* junction (with an embedded waveguide) is one of the most popular approaches to modify the carrier density, see Fig. 5.20. When the junction is forward biased, electrons and holes are injected into the active region.

To create the *pn* junction, regions of silicon are selectively doped using photo- or electron beam-lithography and ion implantation of boron and phosphorous (for the p- and n- regions respectively.

It is well known that doping introduces optical loss via free carrier absorption, and it is important to design the device carefully so as to minimize the interaction of the mode with the lossy regions. Similar to that used earlier for electroluminescent devices, Section 1, the photonic crystal modulator is realized with "fingers" extending into the photonic crystal regions. In Cardile et al. [58], *Q*-factors of 40,000 were measured in silicon with a uniform doping of 10^{18} cm^{-3}. The use of higher doping densities in smaller areas therefore are not detrimental to the device operation.

Resonant modulators have the advantage that they modulate light only at particular wavelengths (the resonant wavelengths of the photonic crystal cavity in this case) allowing wavelengths far from the resonance to pass almost unaffected, as noted by Xu et al. [59]. Multiple photonic crystal modulators with different resonant wavelengths may therefore be cascaded along a single bus waveguide, and independently modulate different wavelengths, see Fig. 5.21. Precise control over the resonance wavelength of each photonic crystal cavity is a prerequisite. One approach is to change the lattice constant of the photonic crystal, however, a 1 nm change in period will change the resonance wavelength by approximately 3 nm, an insufficient degree of control.

A superior approach is to change the position of selected holes in the photonic crystal. Fig. 5.22 shows the effect of changing the position of the four holes marked in Fig. 5.21. The level of positioning accuracy of modern lithography can satisfactorily give the precision required, potentially allowing cascading of large numbers of devices. The Lorentzian response of each photonic crystal cavity filter [41]

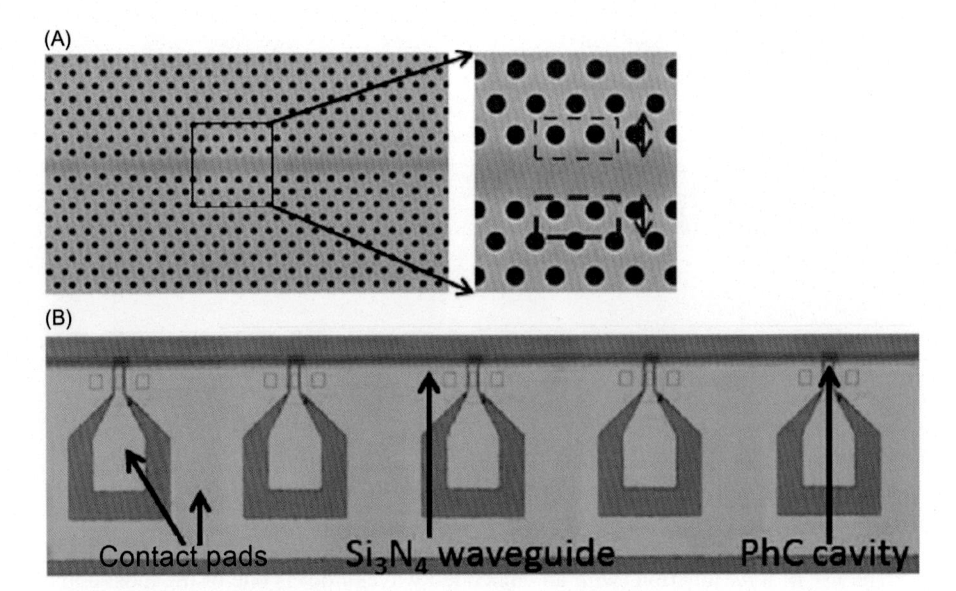

Figure 5.21 (A) SEM image of Photonic Crystal cavity. (B) Image of five cascaded photonic crystal modulators.
Source: Reproduced with permission from Debnath K, O'Faolain L, Gardes FY, Steffan AG, Reed GT, Krauss TF. Cascaded modulator architecture for WDM applications. Opt Express 2012;20:27420–8.

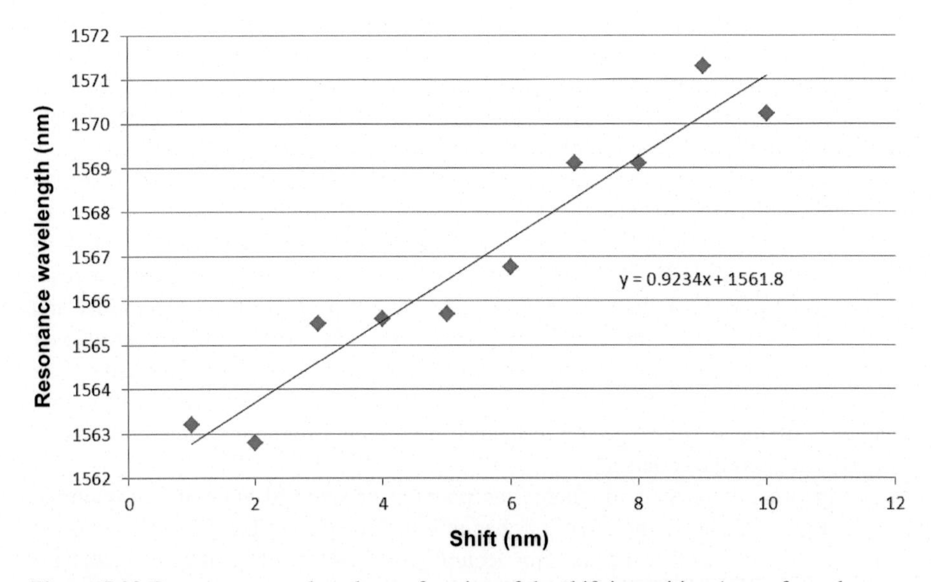

Figure 5.22 Resonance wavelength as a function of the shift in position (away from the waveguide) of the four holes marked in Fig. 5.17A.

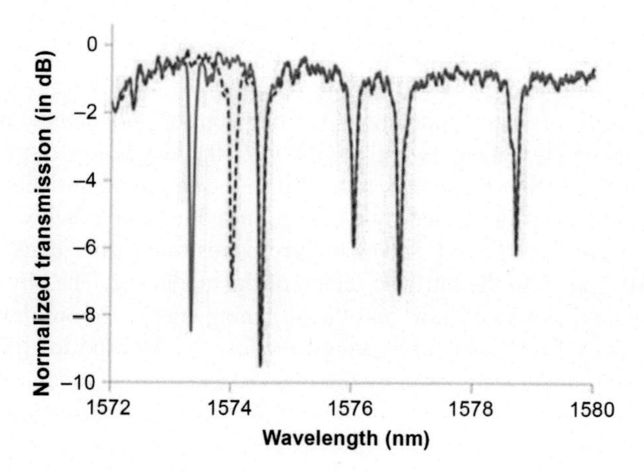

Figure 5.23 Transmission for five cascaded photonic crystal cavities (blue). Each cavity may be electro-optically tuned individually (red).

and the crosstalk requirements of the particular application, place a limit on the spacing between wavelength channels.

Vertically-coupled photonic crystals are completely compatible with the processes required to create *pn* junctions. The ion implantation is generally carried out first, then the photonic crystal is etched and the spacer layer put down. Finally, the contacts are deposited and the bus waveguide defined [60]. To provide optimum performance, the waveguide-cavity coupling is optimized so the total Q-factor (Q_{total}) of the cavity is around 15,000 (with Q_{cavity} typically on the order of 70,000). The ultimate limit on modulation speed is given by the photon life time in the cavity, which is approximately 12 ps at this Q-factor, corresponding to a modulation limit of tens of GHz. The extinction ratio depends on the coupling coefficient, see Eq. (5.3), and was designed to be 10 dB or above to reach the target bit error rates.

The transmission spectrum of a typical system is shown in Fig. 5.23, with each of the five transmission dips corresponding to one of the five cavities. Each cavity can be controlled independently allowing low crosstalk modulation of each channel. Due to the effects of disorder [20], the spacing between the channels is not even. Integrated micro heaters can be used to correct the resonance wavelengths of each device. Alternatively, trimming techniques such as laser assisted local oxidation can be employed [61].

The optical modulation properties of the devices were tested by applying a high-speed electrical data stream that forward biased the junction. The driving signal had a peak-to-peak voltage of 700 mV, and open eye diagrams were obtained at 0.5 and 1 Gbit/second. As forward biasing was employed, the response time of the modulator was limited by the recombination time of the injected carriers, which was approximately 1 ns in these devices, imposing a limit of about 1 Gbit/second.

Power consumption is a key metric for optical modulators. In this measurement, the AC component of the power consumption was calculated as $P_{ac} = V_{rms}^2/R$, where $R \sim 100\ \mathrm{k}\Omega$, giving a value of 0.625 μW, or ~0.6 fJ/bit at 1 Gbit/second modulation. The DC bias consumed 38 μW (or 38 fJ/bit for 1 Gbit/second operation). The switching energy is the energy required to make a transition from the OFF to ON state, a more fundamental indication of the device performance. Considering an input voltage swing of 700 mV, the extinction ratio was 3.5 dB at 1 Gbit/second, which corresponds to a shift in the resonance wavelength shift of 20 pm. In turn, this shift is caused by a change in the refractive index of the silicon device layer of $\Delta n = 4.48 \times 10^{-5}$, which corresponds to a carrier density change of $\Delta n = 6.7 \times 10^{15}\ \mathrm{cm}^{-3}$ in the intrinsic region of the modulator. The physical volume of the intrinsic region was 2.2 $\mu\mathrm{m}^3$ and the switching energy for our device is therefore approximately 1.6 fJ, which is amongst the lowest ever reported [62].

To avoid the limitation imposed by the free carrier lifetime, modulators based on carrier depletion were proposed in 2005 [63], with 50 Gbit/second modulation speeds demonstrated [64]. When a reverse bias is applied to a *pn* junction, the depletion region expands, thus modifying the carrier density and the refractive index. In general, relatively low levels of doping are used, on the order of $5 \times 10^{17}\ \mathrm{cm}^{-3}$, to ensure that a wide, ~200 nm, depletion region is formed that shows a good spatial overlap with the optical mode. To provide resistive contacts, heavily doped regions ($\sim 1 \times 10^{19}\ \mathrm{cm}^{-3}$) are added, see Fig. 5.24.

Depletion is potentially a very fast effect, rise and fall times of 7 ps are predicted [56]. The frequency of response is then given by the RC time constant of the circuit, making the optimization of the highly doped regions and the contact layout very important.

Fig. 5.25A shows the dependence of the resonance wavelength of a vertically-coupled photonic crystal cavity for different applied voltages (reverse bias). Carriers are extracted, resulting in a positive change in the refractive index red shifting the resonance.

Carrier depletion is a relatively weak effect and responsivities of 20 pm/V are typical with the simple junction design of Fig. 5.24. The use of interdigitated *pn*

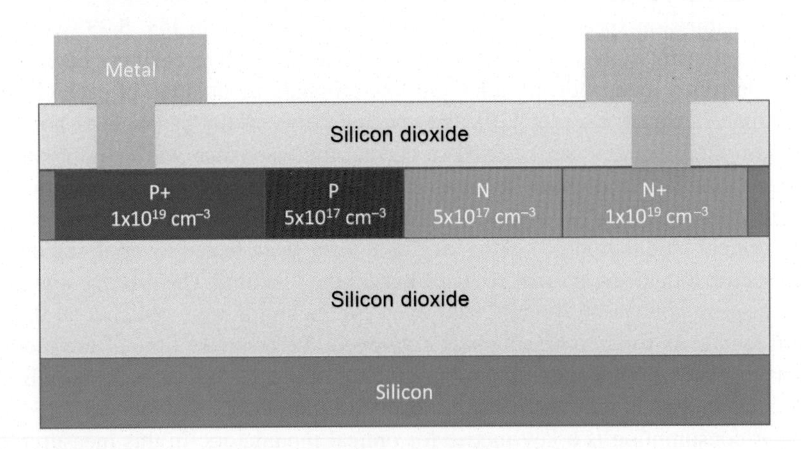

Figure 5.24 Schematic of a carrier depletion modulator (cross-section).

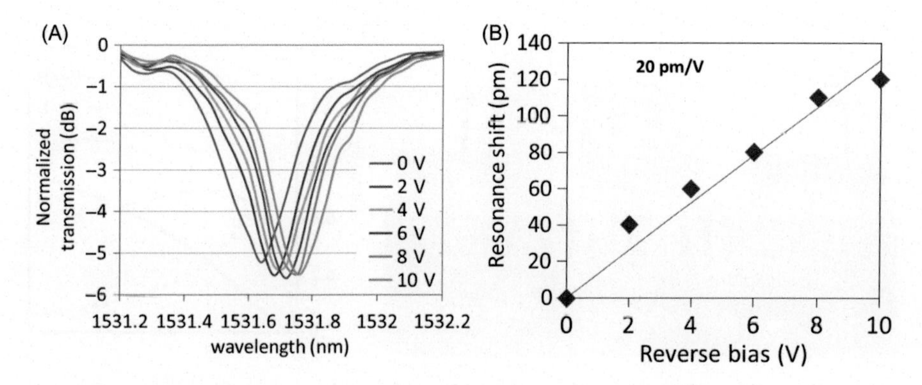

Figure 5.25 The response of a vertically-coupled photonic crystal cavity with reverse biased pn junction. (A) Transmission spectra at different applied voltages. (B) Resonance shift as a function of applied voltage.

junctions [65] can provide a better overlap with the optical mode and increase in the response.

5.6 Photo-detection

The conversion of an optical signal back into an electrical signal is a further essential functionality required in all photonic integrated circuits. In the telecoms window (centered on 1550 nm), the photon energy is less than the band gap energy of silicon, meaning the material is transparent, which is helpful for the realization of low loss waveguides and high Q resonators but problematic for the creation of photo-detectors. Hybrid solutions, in which absorbing materials are attached to the chip are a possible solution; however, a monolithic process would be preferred. Much work has gone into the development of germanium based photo-detectors via direct growth onto the silicon substrate, but CMOS compatibility is compromised. High performance devices have been realized, but achieving high quality germanium growth is challenging, as the introduction of large numbers of lattice defects must be avoided, which would result in high dark currents (i.e., a current when the detector is not illuminated). Furthermore, the growth of thick (~micron scale), high purity germanium layers is a highly specialized process that limits the number of foundries that could be used for mass production. In 2006, it was discovered by Knights et al. that by deliberately introducing defects into the silicon lattice, thereby creating deep-levels within the band gap, photons with sub-band gap energies could be absorbed to generate a photocurrent [66].

Defects can be introduced using standard ion implantation, making the approach highly CMOS compatible. To extract the photo generated carriers, a *pn* junction, very similar to that outlined in Fig. 5.20, is used. With this approach, Geis et al. have realized a photodiode with a responsivity of 0.8 A/W when biased at −20 V, which was based on a silicon rib waveguide, highlighting the potential of defect

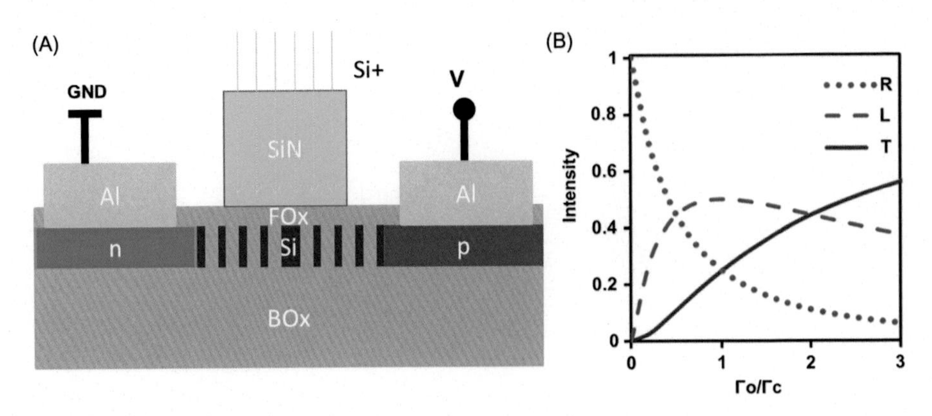

Figure 5.26 (A) Cross-section of a PhC cavity, with integrated pin diode, vertically-coupled to a silicon nitride waveguide. Silicon ion implantation is used to create defects in the silicon resulting in photodetector operation. (B) Analytic plot of transmittance (T), reflectance (R), and optical loss (L) against coupling.
Source: Reprinted with permission from Debnath K, Gardes FY, Knights AP, Reed GT, Krauss TF, O'Faolain L. Dielectric waveguide vertically coupled to all-silicon photodiodes operating at telecommunication wavelengths. Appl Phys Lett 2013;102:171106. Copyright 2013, AIP Publishing LLC.

mediated absorption [67]. Nonetheless, the weak absorption coefficient is a major disadvantage, which forces long devices to be used.

As with optical modulators, resonant enhancement is a powerful means to reduce the size of such photodetectors [68], with the added advantage in that the wavelength selectivity required for de-multiplexing WDM channels is automatically provided. Photonic crystal cavities provide further length reductions and free spectral range enhancements with respect to ring resonators [61]. Fig. 5.26A shows a schematic of a photonic crystal photodetector vertically coupled to a silicon nitride bus waveguide [69]. Defects in the silicon are created by the implantation of silicon ions at 6 MeV, introducing lattice damage without modifying the chemistry of the material. As for modulators, the vertically-coupled configuration provides the advantages of the SiN platform and in some ways, this device can be considered as a dielectric waveguide with detection capability. The possibility of facilitating coupling between such dissimilar materials is unique to photonic crystals, see Section 5.4.

At the resonance wavelength, light is coupled into the cavity mode. Some of the coupled light is lost due to scattering from defects, some couples back to the waveguide mode contributing to the reflection and transmission components, and of the remainder some is absorbed by the defects in the cavity giving rise to a photocurrent. There is an optimum coupling rate at which the absorption is maximized cavity, which can be controlled via the thickness of the spacer layer and the k-space overlap. Fig. 5.26 plots transmittance (T), reflectance (R), and optical loss (L) against the $Q_{coupling}/Q_{cavity}$ ratio. Optical loss is calculated as $L = 1 - R - T$ and comprises both scattering and absorption. If absorption is much greater than

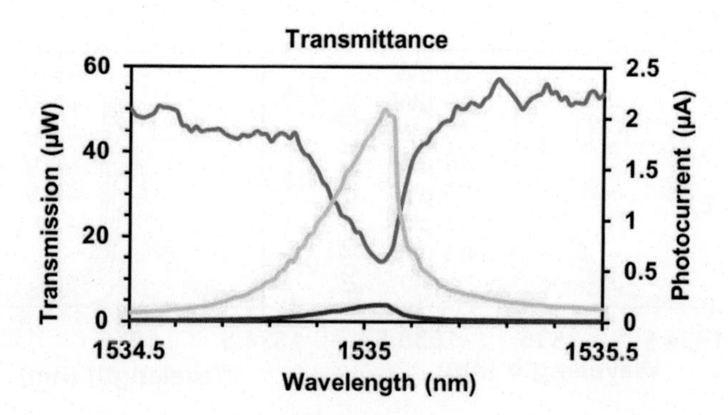

Figure 5.27 The optical spectra in the silicon nitride waveguide immediately after the PhC cavity is shown in blue. The photocurrent at applied biases of 0 and −10 V and photocurrent at 0 V are shown in red yellow respectively.
Source: Reprinted with permission from Debnath K, Gardes FY, Knights AP, Reed GT, Krauss TF, O'Faolain L. Dielectric waveguide vertically coupled to all-silicon photodiodes operating at telecommunicationwavelengths. Appl Phys Lett 2013;102:171106.

scattering, then the photocurrent maximum is reached when $Q_{coupling}$ and Q_{cavity} are equal, similar to the critical coupling condition for micro-ring resonators. Notably, a quantum efficiency of 50% is possible for all values of the absorptance if the above conditions are met.

The photocurrent spectra when the device is unbiased and with reverse bias of 10 V is shown in Fig. 5.27. The devices were tested by coupling light from a tunable laser into the silicon nitride waveguide, and the photocurrent was measuring using a pico-ammeter. The reverse bias has the effect of increasing the extraction of the photogenerated carriers from the active region and improving the efficiency from 7 to 93 mA/W. The wavelength sensitivity is also clearly evident. In this particular device, the Q_{cavity} value was estimated to be 35,000 before implantation, corresponding to an equivalent scattering loss of approximately 17 dB/cm. The defect implantation was estimated to give absorption equivalent to 25 dB/cm. By improving the Q_{cavity} value or by increasing the implantation dose, substantial improvements in efficiency can be expected.

An asymmetry is evident in the spectral response in Fig. 5.27, which is due to thermal effects due to carrier recombination. A fraction of the photogenerated carriers will recombine releasing their energy as heat, redshifting the resonance wavelength of the photonic crystal dynamically as the lasing wavelength is tuned.

Fig. 5.28A shows the photocurrent spectra measured over a range of coupled powers. As the incident power is increased the asymmetry very quickly becomes apparent. Fig. 5.28B shows the effects of the reverse bias that, with increasing bias, extracts an increasing fraction of the generated carriers before they can recombine, reducing the heating effect. A small blue shift is apparent for reverse biases between 0 and 2 V. At higher values, ohmic heating becomes important and a redshift occurs. Thus, by adjusting the power and reverse bias, a wide range of operating regimes are possible.

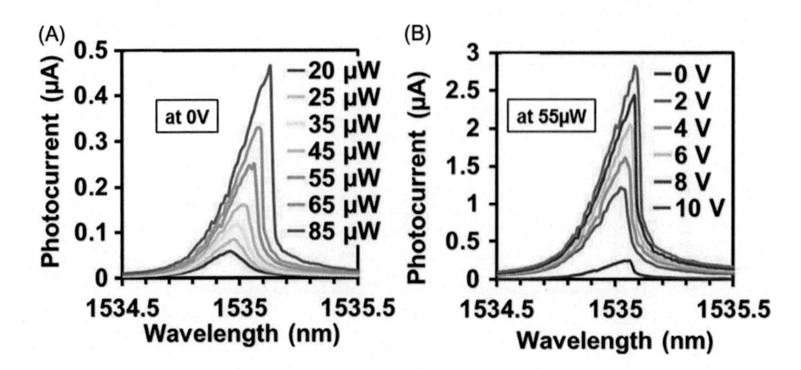

Figure 5.28 (A) Photocurrent spectra measured at different input optical powers at 0 V reverse bias. (B) Measured photocurrent spectra for different applied reverse bias, when the input optical power is held constant at 55 μW.
Source: Reprinted with permission from Debnath K, Gardes FY, Knights AP, Reed GT, Krauss TF, O'Faolain L. Dielectric waveguide vertically coupled to all-silicon photodiodes operating at telecommunicationwavelengths. Appl Phys Lett 2013;102:171106.

Such photo-detectors may be cascaded easily using the same lithographic tuning technique as shown in Fig. 5.22, providing an optical receiver that does not use Arrayed Waveguide gratings or Echelle gratings, supplying a much more compact solution. Additionally, as the losses incurred in the use of spot size converters, grating couplers, AWGs, or Echelle gratings, significant reductions in the optical power budget are also possible, which brings the effective responsivity (photocurrent relative to power from the optical fiber) of a photonic crystal photodiode very close to that of III − V/germanium demultiplexed photodetection systems.

5.7 Outlook

The advantages of photonic crystal cavities for optical interconnects are contained in a single number: the interference order of the resonator. The smaller this number, the larger the free spectral range and the smaller the resonator length, and thus the capacitance. A photonic crystal resonator typically operates on the fundamental cavity mode, and thus is the logical extreme of this principle.

Due to their large Free Spectral Range, large arrays of single mode photonic crystal cavities may be fabricated, each with a different cavity resonance. The width of the PBG is on the order of 300 − 400 nm, and is the ultimate limit on the available bandwidth. This scalability, combined with the low switching energies, small footprint, and low insertion losses, makes our architecture very attractive for WDM applications. Even with low modulation speeds reported here, such a FSR could support 1000 channels providing for terabit/second data transmission with a single bus waveguide, and providing a very attractive platform for the integration on electro-optical printed circuit boards.

References

[1] Miller DAB. Device requirements for optical interconnects to silicon chips. Proc IEEE 2009;97:1166–85.

[2] Notomi M, Nozaki K, Shinya A, Matsuo S, Kuramochi E. Toward fJ/bit optical communication in a chip. Opt Commun 2014;314:3–17.

[3] Schliesser A, Del'Haye P, Nooshi N, Vahala KJ, Kippenberg TJ. Radiation pressure cooling of a micromechanical oscillator using dynamical backaction. Phys Rev Lett 2006;97:243905.

[4] Chen L, Preston K, Manipatruni S, Lipson M. Integrated GHz silicon photonic interconnect with micrometer-scale modulators and detectors. Opt Express 2009;17:15248–56.

[5] Sekoguchi H, Takahashi Y, Asano T, Noda S. Photonic crystal nanocavity with a Q-factor of ~9 million. Opt Express 2014;22:916–24.

[6] Miyazaki HT, Kurokawa Y. Squeezing visible light waves into a 3-nm-thick and 55-nm-long plasmon cavity. Phys Rev Lett 2006;96:097401.

[7] Notomi M. Manipulating light with strongly modulated photonic crystals. Rep Prog Phys 2010;73:096501.

[8] Yablonovitch E. Inhibited spontaneous emission in solid-state physics and electronics. Phys Rev Lett 1987;58:2059–62.

[9] John S. Strong localization of photons in certain disordered dielectric superlattices. Phys Rev Lett 1987;58:2486–9.

[10] Strutt J. (Lord Rayeleigh) "On the maintenance of vibrations by forces of double frequency, and on the propagation of waves through a medium endowed with a periodic structure. Philos Mag 1887;24:145–59.

[11] Joannopoulos JD, Johnson SG, Winn JN, Meade RD. Molding the flow of light, 2nd ed.; 2008. Princeton University Press.

[12] Painter O, Lee RK, Scherer A, Yariv A, O'Brien JD, Dapkus PD, et al. Two-dimensional photonic band-gap defect mode laser. Science 1999;284:1819–21.

[13] Akahane Y, Asano T, Song B, Noda S. High-Q photonic nanocavity in a two-dimensional photonic crystal. Nature 2003;425:944–7.

[14] Galli M, Portalupi S, Belotti M, Andreani L, O'Faolain L, Krauss T. Light scattering and Fano resonances in high-Q photonic crystal nanocavities. Appl Phys Lett 2009;94:071101.

[15] Song B-S, Noda S, Asano T, Akahane Y. Ultra-high-Q photonic double-heterostructure nanocavity. Nat Mater 2005;4:207–10.

[16] Lai Y, Pirotta S, Urbinati G, Gerace D, Minkov M, Savona V, et al. Genetically designed L3 photonic crystal nanocavities with measured quality factor exceeding one million. Appl Phys Lett 2014;104:241101.

[17] Matsuo S, Shinya A, Kakitsuka T, Nozaki K, Segawa T, Sato T, et al. High-speed ultracompact buried heterostructure photonic-crystal laser with 13 fJ of energy consumed per bit transmitted. Nat Photon 2010;4:648–54.

[18] Rai-Choudhury P. Handbook of microlithography, micromachining, and microfabrication: microlithography; 1997 [chapter 2].

[19] Jansen H, de Boer M, Legtenberg R, Elwenspoek M. The black silicon method: a universal method for determining the parameter setting of a fluorine-based reactive ion etcher in deep silicon trench etching with profile control. J Micromech Microeng 1995;5:115.

[20] White TP, O'Faolain L, Li J, Andreani LC, Krauss TF. Silica-embedded silicon photonic crystal waveguides. Opt Express 2008;16:17076−81.

[21] Tanaka Y, Asano T, Akahane Y, Song B-S, Noda S. Theoretical investigation of a two-dimensional photonic crystal slab with truncated cone air holes. Appl Phys Lett 2003;82:1661.

[22] Welna K, Portalupi S, Galli M, O'Faolain L, Krauss T. Novel dispersion-adapted photonic crystal cavity with improved disorder stability. IEEE J Quantum Electron, 2012;48:1177−83.

[23] Kuramochi E, Notomi M, Mitsugi S, Shinya A, Tanabe T, Watanabe T. Ultrahigh-Q photonic crystal nanocavities realized by the local width modulation of a line defect. Appl Phys Lett 2006;88:041112.

[24] MIT Photonic Bands, see http://ab-initio.mit.edu/wiki/index.php/MIT_Photonic_Bands.

[25] Selvaraja S, Jaenen P, Bogaerts W, Van Thourhout D, Dumon P, Baets R. Fabrication of photonic wire and crystal circuits in silicon-on-insulator using 193-nm optical lithography. J Lightwave Technol 2009;27:4076−83.

[26] Portalupi S, Galli M, Belotti M, Andreani L, Krauss T, O'Faolain L. Deliberate versus intrinsic disorder in photonic crystal nanocavities investigated by resonant light scattering. Phys Rev B 2011;84:045423.

[27] Gerace D, Andreani LC. Effects of disorder on propagation losses and cavity Q-factors in photonic crystal slabs. Photon Nanostruct Fundam Appl 2005;3:120−8.

[28] Tanabe K, Watanabe K, Arakawa Y. III-V/Si hybrid photonic devices by direct fusion bonding. Sci Rep 2012;2:349.

[29] Pavesi L, Negro LD, Mazzoleni C, Franzo G, Priolo F. Optical gain in silicon nanocrystals. Nature 2000;408 440−444

[30] Gong Y, Makarova M, Yerci S, Li R, Stevens M, Baek B, et al. Observation of transparency of erbium-doped silicon nitride in photonic crystal nanobeam cavities. Opt Express 2010;18:13863−73.

[31] Camacho-Aguilera RE, Cai Y, Patel N, Bessette JT, Romagnoli M, Kimerling LC, et al. An electrically pumped germanium laser. Opt Express 2012;20:11316−20.

[32] Ng W, Lourenco M, Gwilliam R, Ledain S, Shao G, Homewood K. An efficient room-temperature silicon-based light-emitting diode. Nature 2001;410:192−4.

[33] Cloutier S, Kossyrev P, Xu J. Optical gain and stimulated emission in periodic nanopatterned crystalline silicon. Nat Mater 2005;4:887−91.

[34] Henry A, Monemar B, Lindström JL, Bestwick TD, Oehrlein GS. Photoluminescence characterization of plasma exposed silicon surfaces. J Appl Phys 1991;70:5597−603.

[35] Singh M, Weber I, Konuma M. Evidence for intrinsic point defect generation during hydrogen-plasma treatment of silicon. Phys B 1991;170:218−22.

[36] Shakoor A, Lo Savio R, Cardile P, Portalupi S, Gerace D, Welna K, et al. Room temperature all-silicon photonic crystal nanocavity light emitting diode at sub-bandgap wavelengths. Laser Photon Rev 2013;7:114−21.

[37] Noda S, Fujita S, Asano T. Spontaneous-emission control by photonic crystals and nanocavities. Nat Photon 2007;1:449.

[38] Purcell EM. Phys Rev 1961;69:681.

[39] The Czochralski process is a method of growing high purity single crystals of semiconductors (e.g. silicon) The most important application may be the growth of large cylindrical ingots, or boules, of single crystal silicon which are cut into discs or wafers and used in the electronics industry.

[40] Watanabe MO, Taguchi M, Kanzaki K, Zohta Y. DLTS Study of RIE-Induced Deep Levels in Si Using p + n Diode Arrays. Jpn J Appl Phys 1983;22:281−6.

[41] Savio RL, Portalupi SL, Gerace D, Shakoor A, Krauss TF, O'Faolain L, et al. Room-temperature emission at telecom wavelengths from silicon photonic crystal nanocavities. Appl Phys Lett 2011;98:201106.

[42] Ghica C, Nistor LC, Vizireanu S, Dinescu GJ. Annealing of hydrogen-induced defects in RF-plasma-treated Si wafers: ex situ and in situ transmission electron microscopy studies. Phys D Appl Phys 2011;44:295401.

[43] McNab S, Moll N, Vlasov Y. Ultra-low loss photonic integrated circuit with membrane-type photonic crystal waveguides. Opt Express 2003;11:2927–39.

[44] Qiu M. Vertically coupled photonic crystal optical filters. Opt Lett 2005;30:1476–8.

[45] Debnath K, Welna K, Ferrera M, Deasy K, Lidzey DG, O'Faolain L. Highly efficient optical filter based on vertically coupled photonic crystal cavity and bus waveguide. Opt Lett 2013;38:154–6.

[46] Song B-S, Jeon S-W, Noda S. Symmetrically glass-clad photonic crystal nanocavities with ultrahigh quality factors. Opt Lett 2011;36:91–3.

[47] Xu Y, Li Y, Lee RK, Yariv A. Scattering-theory analysis of waveguide-resonator coupling. Phys Rev E 2000;62:7389–404.

[48] Notomi M, Shinya A, Mitsugi S, Kuramochi E, Ryu H-Y. Waveguides, resonators and their coupled elements in photonic crystal slabs. Opt Express 2004;12:1551–61.

[49] Morichetti F, Melloni A, Martinelli M, Heideman R, Leinse A, Geuzebroek D, et al. Box-shaped dielectric waveguides: a new concept in integrated optics? J Lightwave Technol 2007;25:2579–89.

[50] Masaud TMB, Tarazona A, Jaberansary E, Chen X, Reed GT, Mashanovich GZ, et al. "Hot-wire polysilicon waveguides with low deposition temperature. Opt Lett 2013;38:4030–2.

[51] Biberman A, Preston K, Hendry G, Sherwood-Droz N, Chan J, Levy JS, et al. Photonic network-on-chip architectures using multilayer deposited silicon materials for high-performance chip multiprocessors. J Emerg Technol Comput Syst 2011;7:7.

[52] Reed GT, Mashanovich G, Gardes FY, Thomson DJ. Silicon optical modulators. Nat Photon 2010;4:518–26.

[53] Xu Q, Schmidt B, Pradhan S, Lipson M. Micrometre-scale silicon electro-optic modulator. Nature 2005;435:325–7.

[54] Buckwalter JF, Zheng X, Li G, Raj K, Krishnamoorthy AV. A monolithic 25-Gb/s transceiver with photonic ring modulators and Ge detectors in a 130-nm CMOS SOI process. IEEE J Solid-State Circuits 2012;47:1309–22.

[55] For a simple Fabry-Perot cavity, the free spectral range is given by $\lambda^2/2nl$, and l is on the order of 1–10 μm for a PhC cavity. However, It should be noted that PhC cavities have a complex mode shape that the equations given in this section.

[56] See Ref. [49] for a full review.

[57] Liu A, Jones R, Liao L, Samara-Rubio D, Rubin D, Cohen O, et al. A high speed silicon optical modulator based on a metal-oxide-semiconductor capacitor. Nature 2004;427:615–18.

[58] Cardile P, Franzò G, Savio RL, Galli M, Krauss TF, Priolo F, et al. Electrical conduction and optical properties of doped silicon-on-insulator photonic crystals. Appl Phys Lett 2011;98:203506.

[59] Xu Q, Schmidt B, Shakya J, Lipson M. Cascaded silicon micro-ring modulators for WDM optical interconnection. Opt Express 2006;14:9430–5.

[60] Debnath K, O'Faolain L, Gardes FY, Steffan AG, Reed GT, Krauss TF. Cascaded modulator architecture for WDM applications. Opt Express 2012;20:27420–8.

[61] Lee HS, Kiravittaya S, Kumar S, Plumhof JD, Balet L, Li LH, et al. Local tuning of photonic crystal nanocavity modes by laser-assisted oxidation. Appl Phys Lett 2009;95:191109.

[62] This analysis follow the approach outlined in: Chen L, Preston K, Manipatruni S, Lipson M. Integrated GHz silicon photonic interconnect with micrometer-scale modulators and detectors. Opt Express 2009;15248–56.

[63] Gardes FY, Reed GT, Emerson N, Png C. A sub-micron depletion-type photonic modulator in Silicon On Insulator. Opt Express 2005;13:8845–54.

[64] Thomson D, Gardes F, Fedeli J-M, Zlatanovic S, Hu Y, Kuo B, et al. 50-Gb/s Silicon Optical Modulator. IEEE Photon Technol Lett 2012;24:234–6.

[65] Ziebell M, Marris-Morini D, Rasigade G, Crozat P, Fedeli J-M, Grosse P, et al. Ten Gbit/s ring resonator silicon modulator based on interdigitated PN junctions. Opt Express 2011;19:14690–5.

[66] Bradley JDB, Jessop PE, Knights AP. Silicon waveguide-integrated optical power monitor with enhanced sensitivity at 1550 nm. Appl Phys Lett 2005;86:241103.

[67] Geis MW, Spector SJ, Grein ME, Schulein RT, Yoon JU, Lennon DM. CMOS-Compatible All-Si High-Speed Waveguide Photodiodes With High Responsivity in Near-Infrared Communication Band. IEEE Photon Technol Lett 2007;19:152.

[68] Logan DF, Velha P, Sorel M, Rue RMDL, Knights AP, Jessop PE. Defect-enhanced silicon-on-insulator waveguide resonant photodetector with high sensitivity at 1.55 um. IEEE Photon Technol Lett 2010;22:1530–2.

[69] Debnath K, Gardes FY, Knights AP, Reed GT, Krauss TF, O'Faolain L. Dielectric waveguide vertically coupled to all-silicon photodiodes operating at telecommunication wavelengths. Appl Phys Lett 2013;102:171106.

Types and performance of high performing multi-mode polymer waveguides for optical interconnects

6

M. Singh[1] *and K. Weidner*[2]
[1]Sumitomo Bakelite, Tokyo, Japan, [2]Dow Corning, Auburn, MI, United States

6.1 Introduction

As discussed in previous chapters, a strategy to manage the increased requirements for interconnects of higher data bandwidths, higher density components, and lower power consumption is to bring optics into racks and closer to microprocessors. Implementing this strategy over past years has involved VCSEL opto-electric transceivers, typically mounted on motherboards and connected with fiber optic cables across the backplane of racks. This approach has been effective, but as the number of devices continues to increase, with next generation systems and interconnect densities developing, routing the mass number of fiber interconnects has become cumbersome and has limited the ability to achieve close proximity to the microprocessor. Polymer waveguides (PWG) offer a viable solution to this interconnect challenge with the potential to provide an even higher density of interconnects in a very manageable form factor that is ultimately lower in cost and complexity. Albeit the optical losses are inherently higher for polymer waveguides than fibers, current state-of-the-art materials have reasonably low losses of ≤ 0.05 dB/cm that are appropriate for the lengths required to route interconnects in motherboards and backplanes.

An ultimate form factor for the realization of polymer waveguides is to integrate them directly into printed circuit boards (PCB). State-of-the-art materials have demonstrated the ability to accomplish this feat. System integrators consider it high risk to venture into this new and unfamiliar PWG interconnect technology, given the high cost and level of integration of advanced PCB, so adoption is cautious and slow. An alternate form factor for PWG is with flyovers or flex cable assemblies that connect externally to the PCB; this involves minimal to no integration with the rest of the system, and therefore is much less risky than fully-integrated PCB. There are current products today using fibers routed on flex panels that accomplish this task, such as the Molex flex plane, or the TE Light Ray, so adoption of PWG versions can be relatively quick. PWGs offer a much simpler fabrication, and ultimately a lower cost than fiber routed assemblies.

The channel density possible with PWG technologies has the potential to be much greater than that of optical fiber or copper interconnects. For one, a PWG

Optical Interconnects for Data Centers. DOI: http://dx.doi.org/10.1016/B978-0-08-100512-5.00006-1

link is able to directly cross another PWG link in the same plane. With respect to copper electrical interconnects that have to be routed on several layers, plus additional ground layers to minimize electromagnetic interference, a single PWG layer can eliminate several electrical layers. In addition to direct crossings, PWG technology can be fabricated into a variety of optical devices such as splitters and couplers. The ability to fabricate this variety of optical functions in a single layer enables a PWG panel to provide much more functionality in a unit area compared to fibers. Fabrication techniques of polymer waveguides vary, depending on the exact material system being used, but all methods are able to provide the variety of optical functionality; a list of some of these approaches is shown in Fig. 6.1A.

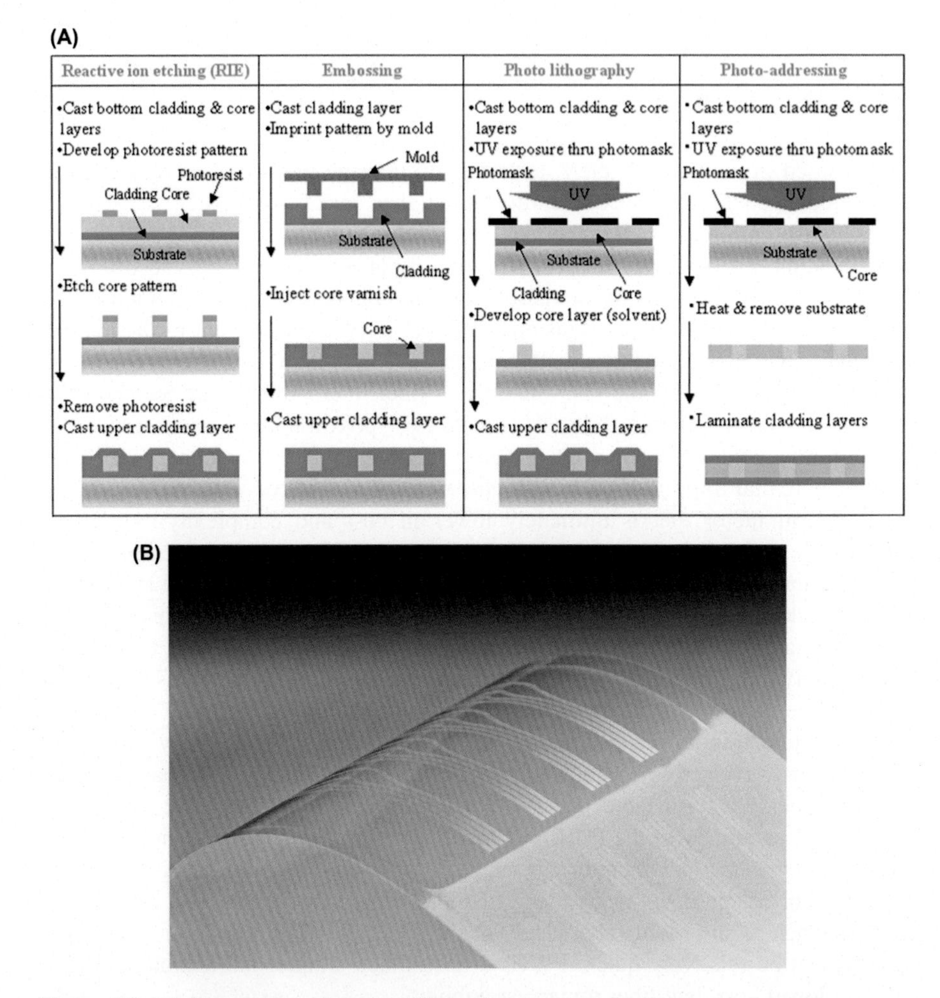

Figure 6.1 (A) Comparison of various fabrication processes for polymer waveguides. (B) Polymer waveguide after fabrication.

PWGs have been presented as a simple, low-cost option to meet the increasing interconnect challenge, but there are significant demands on the material system to produce these PWGs. Performance evaluation is not yet standardized for PWG technology. The International Electrotechnical Commission (IEC) has established a special committee, TC-86 JWG-9, to set these standards. Many factors are involved in measuring the propagation and insertion loss for polymer waveguide such as: optical source, fiber size, coupling technique, alignment method, etc. The IEC committee is developing the standards for PWG. First, the optical absorption has to be a minimum at the operating wavelength, which is typically 850 nm for the majority of VCSEL opto-electrical transceivers; loss values <0.05 dB/cm are reasonable to expect for polymers. These loss values have to remain low through fabrication, and through standard reliability requirements as defined by the Telecordia—GR-468 and JPCA —PE02 − 05-01S specifications typical for high performance computing applications. Further, there are many other requirements that may vary depending on the specific applications, such as flexibility, chemical inertness, toughness, etc. PWG research started in the 1970s for application in optical interconnects. Many Japanese, American, and European companies focused on PWG with different materials, fabrication processes, and target applications. Various materials were used to fabricate polymer waveguide such as: polynorbornene, silicone, acrylate, epoxy, siloxane, polysilane, polyimide, etc. The authors of this chapter will explain how these performance requirements can be met, with a focus on polynorbornene and silicone material systems, and their readiness for adoption as a viable interconnect technology for high performance computing.

6.2 Polynorbornene

Polynorbornene is one of the cyclic olefin polymers with high optical transparency at 850 nm and high glass transition temperature at 250°C or higher, which makes it a good material candidate for optical waveguides [1]. Fig. 6.2 shows the general structure of polynorbornene and functional groups. These functional groups, X and Y, can be used to synthesize the optical, mechanical, and other properties of polynorbornene.

Polynorbornene-based materials have the capability of simple, quick, and mass production compatible fabrication processes. Waveguide fabrication with polynorbornene does not need a wet development process for core patterning, which leads to smooth sidewalls for the waveguide core. Details of the fabrication process with polynorbornene material will be explained later in this chapter.

Polynorbornene materials show low propagation loss at 850 nm wavelength. The propagation loss for 12ch array waveguide is 0.03 dB/cm measured by the cut-back method. The propagation loss in this range is acceptable for short-reach optical interconnect from chip-to-chip or board-to-board.

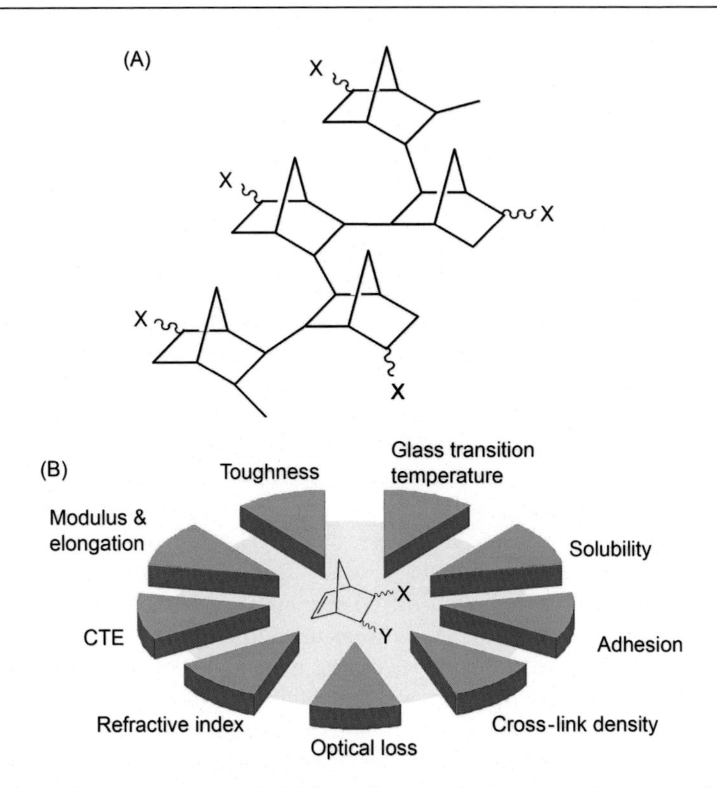

Figure 6.2 (A) General structure of addition of homo-polymerized norbornene; and (B) Schematic of tuning polymer properties via functional groups X and Y.

6.2.1 Polynorbornene waveguide fabrication

The polynorbornene waveguide fabrication process is based on the photo-addressing method. Details of the photo-addressing method are shown in Fig. 6.1. In photo-addressing, first the bottom clad and core layers are formed by film-casting methods such as spin coating. After this, photo patterning is performed by photo-masking, followed by heating to form the core structures. Then the top layer of clad is cast as a film. The polymer layers include a monomer and photo-initiator for curing. The curing reaction after UV exposure under a high-temperature atmosphere turns the UV-exposed area into a cladding with a low refractive index ($n = 1.53$) due to selective cross-linking reactions at the UV-exposed area. The area protected by the photo-mask has a higher refractive index ($n = 1.55$), forming the waveguide cores. The photo-addressing method has several advantages: being a simple process, having no chemical development, and having smooth sidewall roughness of core features.

Polynorbornene materials have the capability to be fabricated with a graded refractive index for the clad to core. Graded index polymer waveguides have been fabricated and evaluated by different groups, and are benefical due to their low crosstalk, and low coupling loss with optical fiber. Ishigure Lab at Keio University in Japan is a leading research lab in the field of graded index polymer waveguide research where they use the Mosquito method to fabricate graded index polymer waveguide [2,3]. Graded index waveguides can be fabricated with polynorbornene material using the same fabrication method as described above, with some additional steps [4]. Fig. 6.3 shows the refractive index profile and core shape of a graded index polynorbornene polymer waveguide.

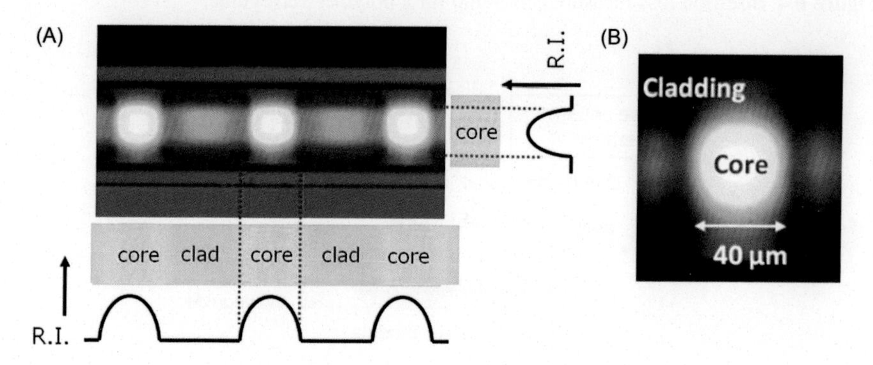

Figure 6.3 (A) Refractive index profile of graded index polymer waveguide. (B) Core of graded index polymer waveguide.

6.2.2 Polynorbornene waveguide performance

The cut-back method is commonly used to measure propagation loss of polymer waveguides. Fig. 6.4 shows the setup for insertion loss measurement of a polymer waveguide using 850 nm VCSEL, PD, and GI 50 μm multi-mode fiber, to make the test results similar to practical applications. We measured insertion loss data of polymer waveguides with various lengths. The polynorbornene materials demonstrate optical losses of <0.04 dB/cm under these test conditions.

The main application for polymer waveguides is targeted for high performance computing (HPC), servers, and optical modules such as QSFP and AOC. In all of these applications, performance of the polymer waveguides at high data rates such as 10/25/100 Gbps is critical. High bandwidth data transmission capability of the polymer waveguides has been demonstrated in various research works [5,6].

Fig. 6.5 shows the setup for high bandwidth data transmission tests of polymer waveguides with a VCSEL-based optical engine. High-bandwidth data transmission using an optical interconnection was evaluated under Non-Return-to-Zero (NRZ), a pseudo-random bit sequence (PRBS) of 2^{23-1} with the O/E modules. A differential signal was generated by a pulse pattern generator and transformed to an optical signal by a VCSEL. The optical signal was reflected at micro-mirrors, and guided into a 120 mm long waveguide. After that, the optical signal was transformed to an electrical signal by a PD. The output signal was detected in a sampling oscilloscope with a coaxial cable. The VD was operated at 16 Gbps data rates per channel (Fig. 6.6).

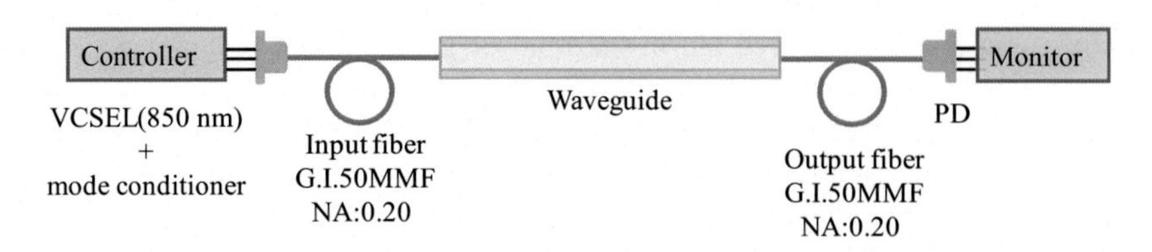

Figure 6.4 Insertion loss measurement setup for a polymer waveguide.

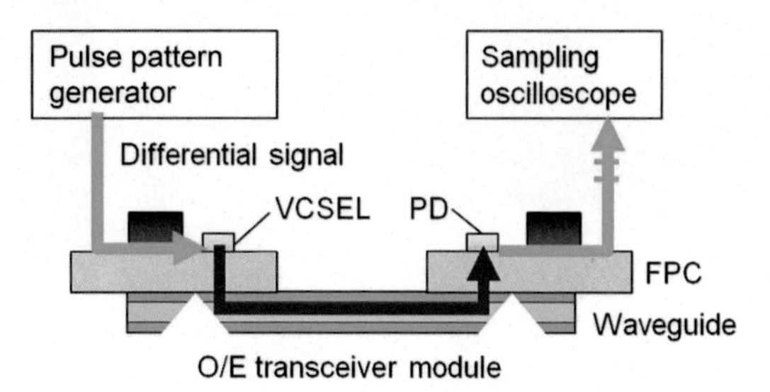

Figure 6.5 Evaluation setup for high-bandwidth data transmission test for O/E transceiver module.

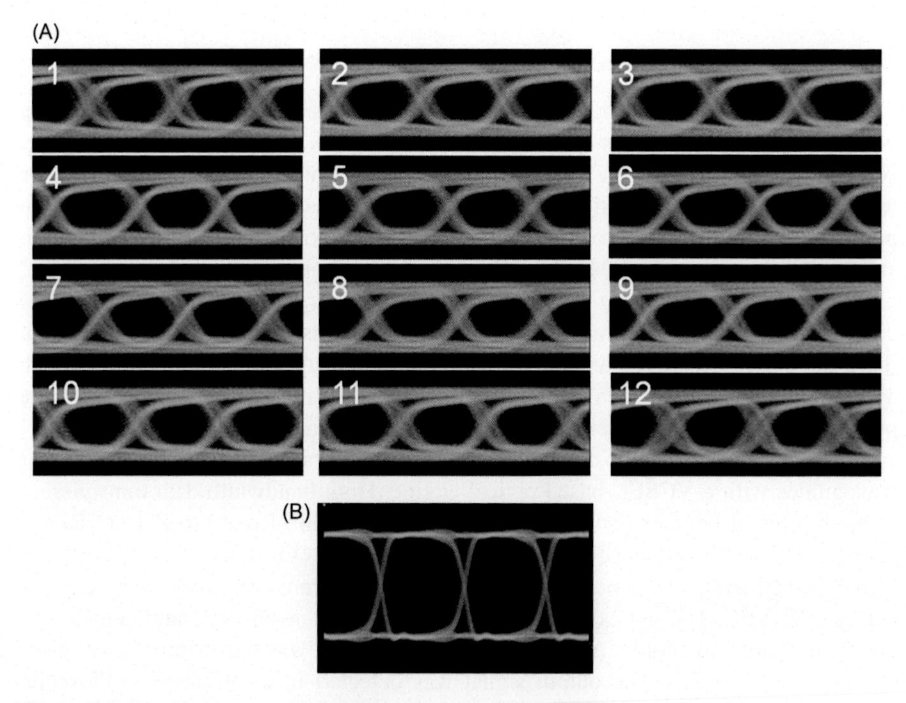

Figure 6.6 Eye diagrams at (A) 16 Gbps through O/E module interconnection. (B) Reference without device under test.

Table 6.1 Reliability data for polymer waveguide fabricated with polynorbornene

Reliability test items	Condition	Result
High temperature test	1000 h at 125°C[a]	Stable[b]
Low temp. test	500 h at −40°C[a]	Stable[b]
High temp. and humidity test	85 ± 2°C, 85 ± 5% (RH), 1000 h[a]	Stable[b]
High temp. and humidity test	(A). 75°C, 85 ~ 95% RH, for 8 h[a](B). 80 ~ 100% RH (C). −40°C, for 8 h 5 cycles	Stable[b]
Thermal shock test	Temp. Time[a] (1) −40 ± 3°C 30 min (2) 85 ± 3°C 30 min 500 cycles	Stable[b]

[a]Telcordia—GR-468, JPCA—PE02—05-01S.
[b]Within the range of measurement repeatability.

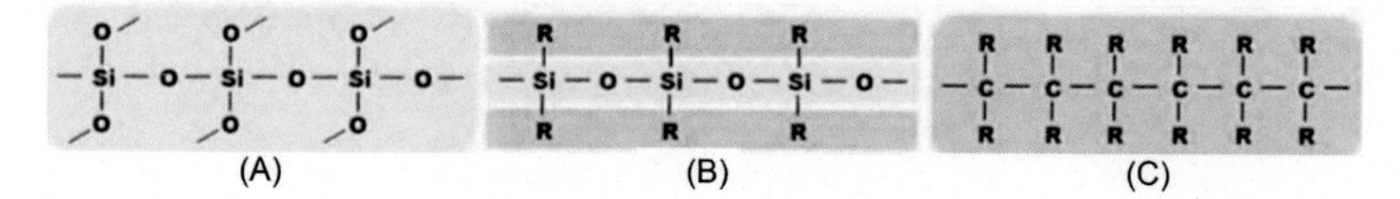

Figure 6.7 Representation of (A) glass; (B) silicone polymers; and (C) organic polymers.

Concerns with the reliability of polymer waveguides is a challenging issue to be utilized in optical interconnect systems for servers, switches, etc., in data centers. Different materials have been tested for reliability of polymer waveguides. The table below shows the reliability of polymer waveguides fabricated with polynorbornene-based materials (Table 6.1).

6.3 Silicones

Silicones are a polymer type characterized by a silicon–oxygen bond that can provide glass-like properties, making silicones typically more inert, thermally and optically stable, and optically clear, than most organic polymers. But unlike glass, the other bonding sites to the silicon not bonded to oxygen can consist of other components ranging from single elements to more complex functional groups. This added functionality allows silicones to behave much more like organics, and enables the silicone polymers to be tailored to fit their purpose and attain the desired curing, refractive index, optical, mechanical, and rheological properties (Fig. 6.7).

Silicone materials can be characterized further by just how many oxygen bonds attach to the silicon. This allows a broad range of materials, ranging from long chained high molecular weight polymers to short chained molecular species, to be used to produce materials with brittle, highly cross-linked networks, or flexible structure after cure. This provides an expansive toolbox of options for formulating

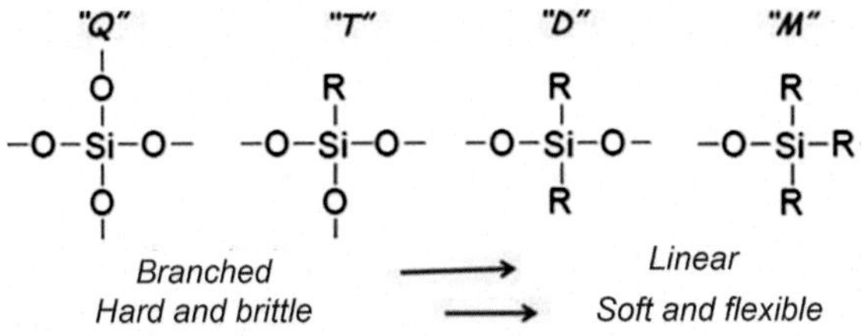

"Q" "T" "D" "M"

Branched → Linear

Hard and brittle → Soft and flexible

Figure 6.8 Variants of silicone chemistry.

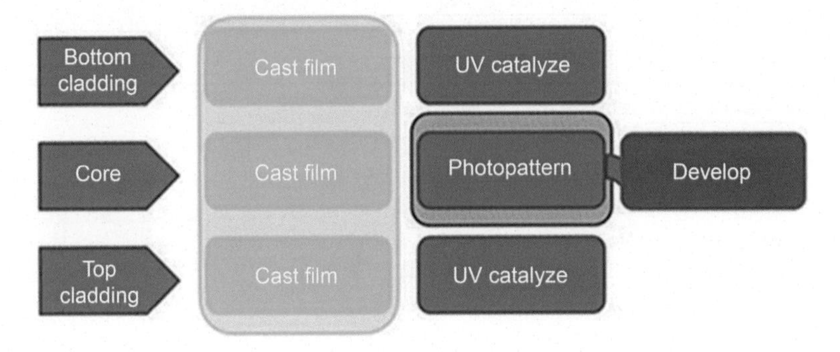

Figure 6.9 Process flow for fabricating PWG from liquid solutions.

the silicone polymer waveguide materials, and an ability to tune the final materials properties. Examples of this "tunability" can be seen by the past efforts to transition early rigid material systems into a flexible material system [7]. Further tuning of a material system was reported to improve mechanical properties and enable PCB integration, as reported by R. John et al. [8] (Fig. 6.8).

A base process flow for the fabrication of a full PWG build is shown in Fig. 6.9; there may be variants on this process flow, depending on the exact material system used, if bakes or other cure techniques are required, and if other patterning structures are included.

Spin coating is one of the simplest film casting techniques that can achieve very precise and uniform film thickness; it is widely available in most lab environments, and is an easy method to rely on for R&D. However, this technique is not very applicable for PWG fabrication of larger panels to use in motherboards and backplanes; substrate type is limited to rigid materials, and waste of excessive PWG materials becomes significant as the panel size increases. Casting films by doctor blade have been demonstrated as a viable alternative for PWG fabrication meeting a tolerance of $<2\%$ variation [7]. This method requires a coating "blade" with a defined gap over the substrate that pushes a bead of liquid solution in front of it at a defined rate, leaving a film of the solution behind it. A meniscus is formed in the trailing edge, and the final film thickness is self-metered by the gap, it will be some fraction of the gap height defined by this meniscus and any non-solid content that may be in the solution. This method, as used for PWG fabrication to date, is simple and flexible,

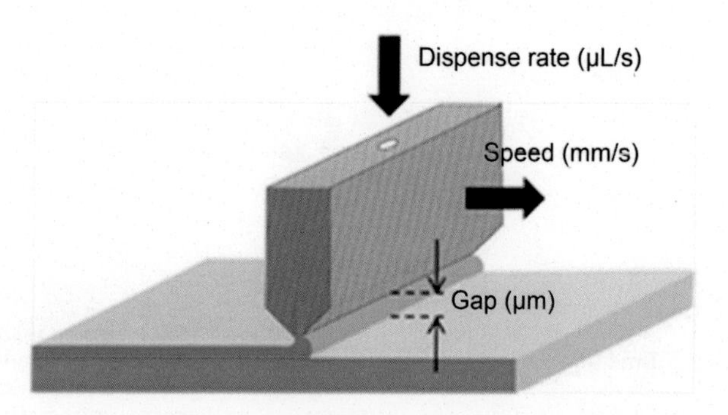

Figure 6.10 Diagram for casting films by slot die coating.

but it is also fairly manual and dependent on the operator, so it is most practical for early prototype work and less so for large-scale manufacturing. Slot die coating, on the other hand, is well aligned with large-scale manufacturing. Slot die coating is a mature technology developed mostly for the display industry in past years, so well optimized equipment and methods are available on the market. Material is pumped from a reservoir through a precision die with manifold channels to evenly spread the material across the coating width. The final thickness of the film is metered by the exact volume pumped through the die and the rate at which the die moves over the substrate. Retaining the material in an enclosed reservoir and dispensing only what is required is ideal for PWG fabrication since it retains the high cleanliness and clarity of the product, as well as minimizing waste. The contained system, along with the pump rate metered deposition, remove much of the operator dependence and is more feasible for large-scale production environments. Both of these deposition methods can coat films at rates greater than 5 mm/seconds, which will coat most panel sizes under consideration in less than 2 minutes (Fig. 6.10).

The waveguide structures are formed by photolithography; to be more specific, a "negative resist" photolithography approach. Negative resist lithography is when regions of the material exposed to UV energy are cured, and the remaining uncured portions are removed with a developing solution. For this, a photo-activated catalyst is included in the formulation that initiates the cure mechanism required for the specific polymer system. The specific wavelength of absorption may vary depending on the photo catalyst, but is commonly in the near-UV region around 355–375 nm. Mercury arc lamps commonly used for lithography are broad spectrum and cover this wavelength range well. Given the intensity of collimated mercury arc lamps, and the required dose of most photopolymers, curing of the entire pattern can be performed in less than 2 minutes. Masked photolithography is a mature technology used in many other industries and viable production equipment is available. Previous work has also shown that single wavelength lasers can be used to selectively cure PWG materials and when combined with a precision XY stage can "write" the PWG patterns [7]. Laser direct imaging is very effective for low-cost tooling to manufacture prototypes, but may not be as efficient as masked lithography for large-scale production. To complete the fabrication of PWG core structures,

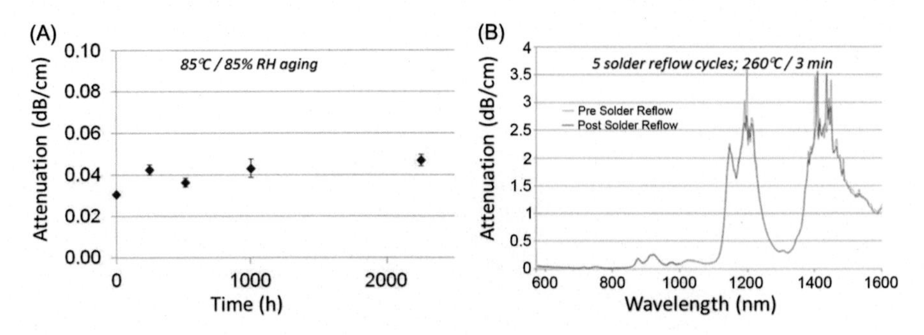

Figure 6.11 Silicone PWG reliability: (A) >2000 h aged at 85°C and 85% RH. (B) After five solder reflow cycles of 260°C for 3 min each.

the uncured portions of the film are removed by a developing solution. The top clad is processed much like the other layers, but variations in the process parameters can induce a gradient refractive index in the core, or not [8]. Each process step can be accomplished in 2 minutes or less, so looking forward to a large-scale production line and a full in line process a task time of 2 minutes should be realizable for the production of PWG panels.

Silicone polymer waveguides have a stable optical loss of <0.05 dB/cm. The materials have been shown to be reliable with minimal increases in optical loss after 2000 hours of aging at 85°C and 85% RH. These materials have also shown the ability to survive high temperature treatments, such as ones used for solder reflow depicted in Fig. 6.11B, Fig. 6.5 cycles to 260°C for 3 minutes each demonstrated no increase in loss. In addition to thermal stability, silicones are also rather inert to many chemicals, and have been shown to integrate into PCBs, surviving via drilling, through hole plating, and solder shock testing [9].

The research team at the Center for Advanced Photonics and Electronics (CAPE) at the University of Cambridge, UK, has performed extensive studies to understand the data rate limits of multi-mode polymer waveguides, and has demonstrated error free data rates of 40 Gb/seconds over a 1 m PWG link [10]. By launching pico-second pulse conditions and measuring the pulse broadening due to dispersion, this team was able to predict 100 Gb/seconds data rates are achievable with the gradient index polymer waveguides [11]. These results, along with further work to demonstrate 56 Gb/seconds PAM-4 data transmission [12], help support the claims that multi-mode polymer waveguides will meet the data rate requirements for the next few generations of high performance computing.

6.4 Connectors and coupling

Coupling of light into polymer waveguides is challenging in both in-plane or out-of-plane cases. PMT connectors used for in-plane coupling, while turning mirrors are used for out-of-plane coupling.

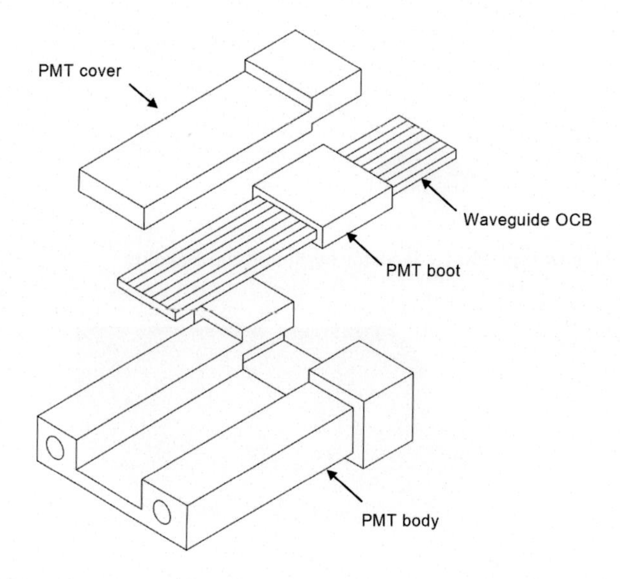

Figure 6.12 PMT connector details.
Source: Reproduced from IEC JWG9 document.

6.4.1 In-plane coupling

Connectors for polymer waveguides have been researched and developed by a few companies. MT is the standard connector for 12 channel ribbon fibers. A similar and compatible connector for polymer waveguide is named the PMT connector. PMT connectors are compatible with any MT connector. The main difference between MT and PMT connectors is in the internal design. The MT connector has grooves to accommodate the fibers, while PMT connectors do not have any such structure. The details of the PMT connector are shown in Fig. 6.12. The PMT connector consists of three main parts, as shown in the figure below:

1. Main body
2. Cover
3. Boot

The outer housing is similar to an MPO connector, and a PMT connector can fit into an MPO housing. Sometimes a clip is used to attach two connectors if the MPO housing is not attached.

The above-mentioned connector is a traditional PMT connector which has a higher assembling cost, due to having three parts and some associated assembling issues. A new kind of PMT connector is proposed by Sumitomo Bakelite to improve the assembly process. The proposed PMT connector is shown below in Fig. 6.13. The new kind of PMT connector has only one part, so no assembling process is needed. The polymer waveguide can be inserted, aligned, and fixed with optical glue in a single process.

The ability to pattern clad materials in some of the PWG material systems, such as silicones, enables other versions of in-plane connectors to simply passively align

Figure 6.13 The new type PMT connector by Sumitomo Bakelite.

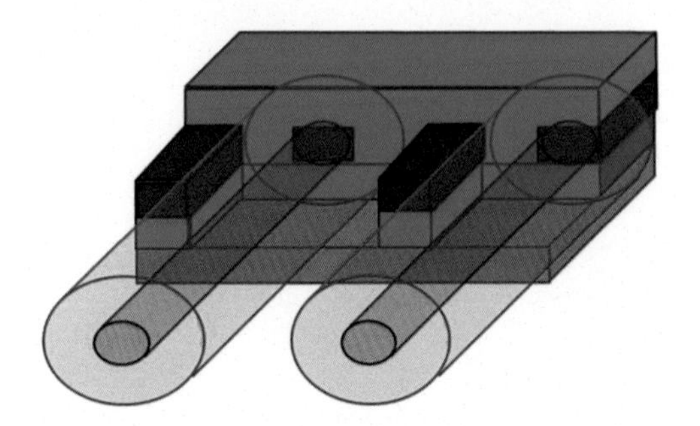

Figure 6.14 Direct fiber attachment with passive alignment structures.

to the core structures, with features registered in the core pattern for precise alignment. Fig. 6.14 is an example of this passive alignment; in this case the fibers can be directly attached to the polymer waveguides.

6.4.2 Out-of-plane coupling

Out-of-plane coupling capability of polymer waveguides is critical to enable its use with optical engines. Fabrication of a turning mirror on polymer waveguides is one approach to solve this. A discrete mirror can also be used, but this approach is more complicated due to an increased number of parts, and associated handling and alignment. Mirrors can be fabricated on polymer waveguides using various methods such as:

- Laser ablation or deicing (total internal reflection mirror);
- Embossing (total internal reflection mirror);
- Metal mirror (gold or silver metal).

Amongst these, the total internal reflection (TIR) mirror is easy to fabricate and is low loss. Gold polished mirrors also have low loss, but the fabrication process is not capable of mass production, due to the complicated fabrication process. Embossing is a mass production capable method, but repeatability is one key issue.

Total internal reflection mirrors with 45 degree angle have been fabricated on polynorbornene waveguides with laser ablation, as shown in Fig. 6.15A and B. The loss for each TIR mirror is 0.5 dB/mirror.

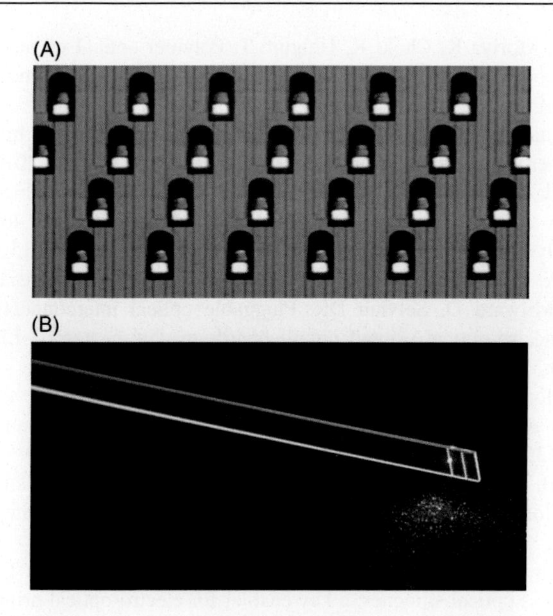

Figure 6.15 (A) 45-degree mirrors on polymer waveguide. (B) Polymer waveguide with 45-degree mirrors.

6.5 Conclusions

Multi-mode polymer waveguides have been well developed in the past few years, and are ready to be implemented into the next generation of high performance computing systems, data centers, and sensors. The fabrication process for polymer waveguides with various materials systems have been well defined, with mature technologies that can quickly ramp to large-scale production. A variety of methods to connect data signals in and out of the polymer waveguides have been developed and are available. High reliability has been demonstrated over the past few years, increasing confidence that these materials can withstand the lifecycle of a high performance system. Further, data rate studies have shown multi-mode polymer waveguides will perform at target data rates over the next several generations of systems, so this technology will be applicable for many years to come.

References

[1] Mori T, Takahama K, Fujiwara M, Watanabe K, Owari H. Optical and electrical hybrid flexible printed circuit boards with unique photo-defined polymer waveguide layers. SPIE photonics west; 2010.

[2] Hsu H-H, Hirobe Y, Ishigure T. Fabrication and inter-channel crosstalk analysis of polymer optical waveguides with W-shaped index profile for high-density optical interconnections. Opt Express 2011;Vol. 19(No. 15):14018–30.

[3] Kinoshita R, Moriya K, Choki K, Ishigure T. Polymer optical waveguides with GI and W-shaped cores for high-bandwidth-density on-board interconnects. J Lightwave Technol IEEE/OSA 2013;31/24:4004–15.
[4] Singh M, Kitazoe K, Moriya K, Horimoto A. High reliability and high density graded index polymer waveguides for optical interconnect. Opt Commun 2016;362:33–5.
[5] Ito Y, Terada S, Singh MK, Arai S, Choki K. Demonstration of high-bandwidth data transmission above 240 Gbps for optoelectronic module with low-loss and low-crosstalk polynorbornene waveguides.In: Proc. 62 ECTC; 2012, p. 1526–31.
[6] Pitwon RCA, Wang K, Graham-Jones J, Papakonstantinou I, Baghsiahi H, Offrein BJ, Dangel R, Milward D, Selviah DR. Pluggable optical interconnect technologies for polymeric electro-optical printed circuit boards in data centers. J Lightwave Technol 2012;30(21):3316–29.
[7] Dangel R, Horst F, Jubin D, Meier N, Weiss J, Offrein B, et al. Development of versatile polymer waveguide flex technology for use in optical interconnects. J Lightwave Technol 2013;V31(N24):3915.
[8] Swatowski B, Amb C, Hyer M, John R, Weidner W. Graded index silicone waveguides for high performance computing. In: IEEE optical interconnect conference proceeding; 2014. p 133.
[9] John R, Amb C, Swatowski B, Weidner W, Halter M, Lambrecht T, et al. Thermally stable, low loss optical siliocnes: a key enabler for electro-optical printed circuit boards. J Lightwave Technol 2015;V33(N4):814.
[10] Bamiedakis N, Chen J, Westbergh P, Gustavsson J, Larsson A, Penty R, et al. 40 Gb/s data transmission over a 1-m-long multimode polymer spiral waveguide for board-level optical interconnects. J Lightwave Technol 2015;V33(N4):882.
[11] Chen J, Bamiedakis N, Vasil'ev P, Edwards T, Brown C, Penty R, et al. Graded-index polymer multimode waveguides for 100 Gb/s board-level data transmission. In European conference on optical communications ECOC; 2015.
[12] Bamiedakis N, Wei J, Chen J, Westbergh P, Larsson A, Penty R, et al. 56 Gb/s PAM-4 data transmission over a 1 m long multimode polymer interconnect. In: Lasers and electro-optics CLEO; 2015.

Design and fabrication of multimode polymer waveguides for optical interconnects

7

T. Ishigure[1] and M. Immonen[2]
[1]Keio University, Yokohama, Japan, [2]Meadville Aspocomp International Limited, TTM Technologies, Inc., Salo, Finland

7.1 Introduction

The research and development history of polymer optical waveguides is quite long. In IBM Journals of Research and Development published in 1977 [1], we can find that relating terms such as "polymeric thin films," and "integrated optical techniques" already exist. Over the last 40 years, polymer optical waveguides have been a prospective passive optical component, although a wide variety of applications of polymer waveguides as an active device have been proposed. The major applications as a passive component has been optical links on board, and a concept for chip-to-chip optical interconnects already existed even in the 1980s, although optical fiber links had not been fully deployed in the telecommunication networks. As is developed for optical fibers, the appropriate numbers of propagating mode in polymer optical waveguides has been discussed for a long time. From the integration tolerance point of view, multimode waveguide achieved by a large core size (~50 μm) should be promising, while in order to apply the polymer waveguides to optical fiber links composed of single-mode fibers, single-mode waveguides have been highly anticipated. So, both multimode and single-mode polymer waveguides have been developed almost equivalently. However, since the late 1990s, multimode fiber (MMF) links combined with GaAs-based vertical cavity surface emitting lasers (VCSELs) have drawn much attention for computer-coms. Originally, modal dispersion of MMFs were a big technical issue for updating the bit-rate of MMF links over 10 Gbps, but the development of a new MMF with low dispersion currently allows even 25 Gbps optical links combined with a specific launch condition [2,3]. These MMF links are widely deployed in data center networks to realize rack-to-rack optical interconnects. Then, the optical links are gradually penetrating into on-board areas, so the polymer optical waveguides are expected to be integrated on board [4,5], and connected with MMFs. Thus, in this chapter, multimode polymer waveguides are focused on, and the design and fabrication for low-loss, low-crosstalk, and high bandwidth density multimode polymer optical waveguides are discussed.

Optical Interconnects for Data Centers. DOI: http://dx.doi.org/10.1016/B978-0-08-100512-5.00007-3

7.2 Structure of multimode polymer optical waveguide

7.2.1 Step-index core

In multimode optical waveguides, multiple modes of lightwave propagate. The number of modes depends on the core size, the refractive index difference between the core and cladding, and the wavelength of light. In the case of optical fiber, the core diameter of MMFs currently deployed in high-speed data link is standardized to be 50 μm [6], so the core size of multimode polymer waveguide has been designed to have the same core size as multimode optical fibers. However, since lithography has been widely applied to fabricate polymer waveguides, the core shape of polymer waveguides is, in general, square or rectangular. Hence, a square-shaped core with a size of 50 × 50 μm has been almost the standard size over the last couple of decades. When multiple cores are aligned in parallel to obtain a waveguide array, the interchannel pitch in polymer waveguides has been 250 μm, which could also be designed to have a good match with an MMF bundle (MMF ribbon with a standard pitch of 250 μm). The cross-sectional structure of widely deployed multimode polymer waveguides is shown in Fig. 7.1A. Almost all the polymer waveguides previously reported have been of step-index (SI) type, in which the refractive index is uniform in the whole core. Here, as mentioned above, the refractive index difference between the core and cladding, which determines the numerical aperture (NA) of the waveguide directly affects the number of propagating modes. However, the NA of SI polymer optical waveguides previously reported have been of a wide variety: the NA is simply determined by the materials composing the core and cladding. In order to confine the light highly into the core, the NA of SI polymer optical waveguides previously reported has been as high as 0.3.

Meanwhile, over the last decades, the application of polymer optical waveguides to on-board optical interconnects has been drawing much more attention. Although similar applications were reported even in the 1980s to the 1990s, such as optical printed circuit boards (OPCBs) integrated with polymer optical waveguides, they were mainly supposed to be utilized for point-to-point links [7]: the waveguides just connected a light source and a photo detector. Hence, just the insertion loss of the waveguides had been the critical parameter. However, current technical trends are in the extension of board-to-board optical wiring realized with MMF links to the vicinity of LSI mounted on board. Therefore, polymer optical waveguides integrated on PCBs are anticipated to be connected with MMFs at the board edge, so the structural differences between polymer waveguides and MMFs such as core shape, NA, and refractive index profile become a crucial issue. When the output light from a 50 × 50 μm square-shaped core waveguide is coupled to a 50 μmø GI MMF, the light exiting from the periphery region that is not overlapped with the GI-MMF core causes coupling loss. Hence, the current trends in polymer waveguide structure are in a smaller sized core: 35 × 35 μm and less [4]. In response to the small core size, the interchannel pitch is reduced to 62.5 μm.

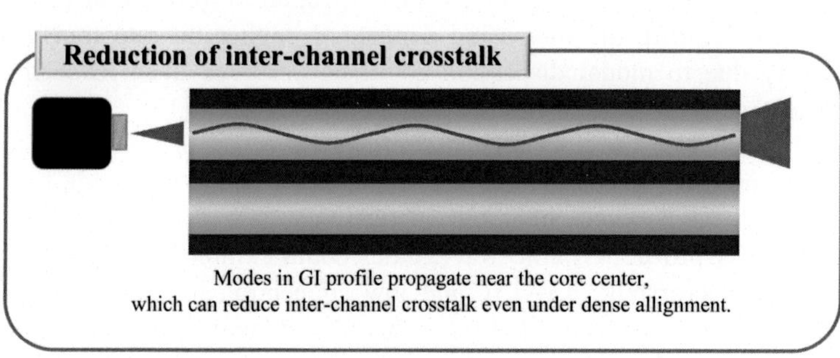

Figure 7.1 Structure of multimode polymer waveguides.

7.2.2 Graded-index core

We have focused on the OPCB applications with polymer optical waveguides as illustrated in Fig. 7.2 [8]. A pair of PCBs are connected with MMF ribbon(s) that correspond to just the board-to-board link, while multimode polymer waveguides are integrated on both PCBs, one of which is coupled to a light source and the other to a photodiode. At the edges of both boards, the MMF and waveguides are connected. Hence, the polymer waveguides should have the same structure as the MMF to be connected. Then, the two cores can completely overlap, and the mode fields between the MMF and waveguides coincide, resulting ideally in no

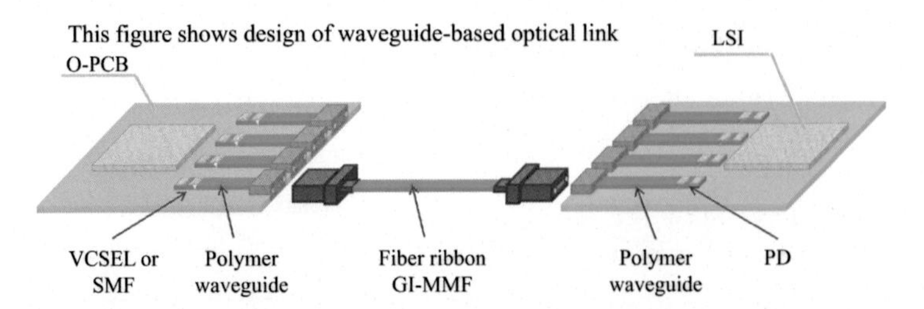

Figure 7.2 Multimode polymer waveguide based link.
Source: From R. Kinoshita, K. Moriya, K. Choki, and T. Ishigure, "Polymer Optical Waveguides With GI and W-Shaped Cores for High-Bandwidth-Density On-Board Interconnects,"J. Lightw. Technol.,31 (24), pp. 4004-4015 (2013).

connection loss to be observed at the connection points. From this technical point of view, we proposed the introduction of the GI structure, as shown in Fig. 7.1B, in polymer waveguides in 2007 [9]. In general, the highest technical advantage of the GI core over the SI counterpart could be regarded as low modal dispersion. We have already experimentally demonstrated the high-speed data transmission capabilities of GI core polymer waveguides [10]. For on-board applications, the link distance should be 1 m at longest, and generally tens of centimeters long or even shorter waveguides are expected to be used on board [11]. For such an optical link, modal dispersion in the SI core waveguide has been regarded as negligible, so there has been no need to form GI cores in polymer waveguides from the dispersion point of view. However, with the increasing demand in higher bit rate transmission, a power penalty due to modal dispersion could be observed even in such a short-reach optical link in the case of SI-core polymer waveguides [12].

We have focused on the tight optical confinement effect of GI core as the other benefit for forming GI cores in polymer waveguides. Since the mode-field is not extended in the whole core area, but is confined near the core center with a Gaussian intensity profile, GI core waveguides could exhibit low propagation loss, low inter-channel crosstalk [13–15], and low connection loss with light sources/photodiodes via 45° mirrors [16]. Hence, we also demonstrated that the GI core allowed very narrow interchannel pitch maintaining low crosstalk [17].

7.3 Fabrication method

As previously mentioned, photolithography has been the major fabrication method, so how to fabricate GI cores has been another technical issue. Over the last couple of years, several prospective fabrication techniques have been proposed, and we mainly focus on the following three methods: soft lithography [18], photo-addressing [8], and the Mosquito method [19]. The first two methods are similar to photolithography, because a UV exposure through photomasks is required to form the waveguide patterns. Therefore, the waveguides obtained have GI square cores. Soft lithography is a well-known method for fabricating SI core waveguides, and the photo-addressing method was developed at Sumitomo Bakelite

Co., Ltd. Meanwhile, the Mosquito method is a unique photomask-free fabrication method. By means of the Mosquito method, circular GI cores can be formed. In the next section, procedures of the three methods are described in more detail.

7.3.1 Softlithography method [18]

For the initial trial of the soft lithography method, we utilized a UV curable and cross-linkable acrylate polymer, TPIR-202, for the waveguide fabrication. The material is supplied by Tokyo Ohka Kogyo Co., Ltd. There are four steps in the fabrication of a waveguide: (1) preparation of an elastomeric stamp with a relief pattern; (2) formation of under-cladding by pattern transfer; (3) GI core formation; and (4) over-clad fabrication. These procedures are schematically represented in Fig. 7.3. First, we prepare an elastomeric stamp with poly dimethyl siloxane (PDMS) by replicating the relief pattern on a master mold (a rigid photo-resist on a substrate). Here, a negative photo-resist, TMMF S-2000 supplied by Tokyo Ohka Kogyo Co., Ltd., is used. Subsequently, mixture of a PDMS prepolymer and a catalyst for cross-linking is cast against the master mold, followed by curing to obtain a stamp. The stamp is removed from the master mold. In previous reports on soft lithography for SI core polymer waveguides, it was found that the propagation loss of waveguides was significantly sensitive to the surface roughness of the master mold, which corresponds to the surface roughness of the core-cladding boundary. Hence, much attention has been focused on how to obtain a smooth sidewall of the master mold, stamp, and finally the core-cladding boundary.

Meanwhile, as we have emphasized in our previous publications [9,14], the propagation loss of waveguides with GI cores is less sensitive to the core-cladding imperfection. Therefore, we pay less attention to the surface smoothness of the stamp in terms of the propagation loss.

After the PDMS stamp with a relief pattern is obtained, the TPIR-202 monomer is cast on a substrate, and the PDMS stamp is pressed over the TPIR-202 monomer layer against the substrate to project the relief pattern on the monomer layer, as shown in Fig. 7.3A. Then, the TPIR-202 monomer is cured under UV exposure

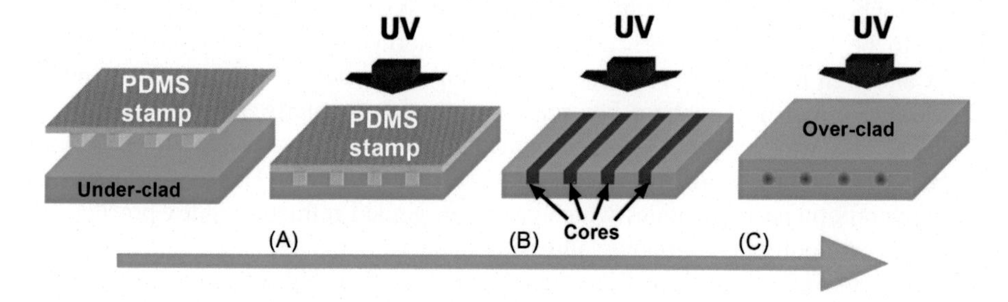

Figure 7.3 Schematic diagram of the fabrication process of GI-core polymer optical waveguides.
Source: From Ishigure T, Nitta Y. Polymer optical waveguide with multiple graded-index cores for on-board interconnects fabricated using soft-lithography. Opt Express 2010;18 (13):14191–201.

using a mask aligner (LA 310k, Nanometric Technology Inc.). Here, curing of the TPIR-202 monomer is completed within a few seconds exposure time with a power of 80 mW/cm^2 at 365 nm wavelength. After the curing process, the PDMS stamp is peeled off, and then the grooves for the cores are formed on the under-cladding. Next, the grooves are filled with the monomer for the core by coating it on the under-cladding. Then, the excess amount of the core monomer is wiped off, and also cured under UV exposure, as shown in Fig. 7.3B. Here, we use the TPIR-202 monomer and a dopant with a refractive index higher than that of the TPIR-202 polymer. The refractive index of the TPIR-202 polymer is 1.548 at a wavelength of 589 nm (D-line), and we use diphenyl sulfide (DPS) for the dopant with a sufficiently high refractive index, ($n_d = 1.633$). We have already confirmed that DPS has a good miscibility as a dopant with poly methyl methacrylate (PMMA) [9,20], a typical acrylate polymer. Near-parabolic refractive index profiles are formed during the core polymerization process. In the curing step of the under-cladding, we do not cure the monomer completely by adjusting the UV exposure time. Even if the under-cladding is less than half cured, the cladding layer maintains the relief pattern on it, because the TPIR-202 has a cross-linking structure. The TPIR-202 monomer and DPS molecules filled in the grooves are able to diffuse into the cladding layer. Finally, a concentration distribution of DPS is formed in the core region, which corresponds to the graded refractive index profile. The core layer is not cured completely, since we also need to form a gradual refractive index distribution at the boundary between the core and the over-cladding.

In the final step, the over-cladding is formed using the TPIR-202 monomer, as shown in Fig. 7.3C, and whole the waveguide is completely cured under sufficient UV exposure. On the other hand, the soft lithography is capable of fabricating both GI- and SI-core waveguides with the same material system. The way to control the index profile is quite simple: just by adjusting the curing condition of the under-cladding and core region. Hence, we fabricate waveguides with SI-cores using the same TPIR-202 and DPS system.

7.3.2 Photo-addressing method [8]

For the photo-addressing method, a specific polymer material, polynorbornene, needs to be used. Polynorbornene is a cyclic olefin resin with a high refractive index and high temperature resistance. Since polynorbornene is an amorphous (non-crystalline) polymer, it realizes low-scattering loss. Particularly, low optical loss is achieved at a wavelength of 850 nm, which is an emission wavelength of GaAs-based VCSELs widely utilized in very-short-reach MMF links. Hence, we focus on the polynorbornene waveguides which can form graded refractive index profiles in the cores by modifying the photo-addressing method.

The photo-addressing method is schematically shown in Fig. 7.4. Fig. 7.4A shows the original fabrication technique of polynorbornene-based SI-square core polymer waveguides. First, the precursor for the cores is coated on a substrate using the doctor-blade technique. The coated layer is pre-baked, followed by UV exposure through a photo-mask. The coated layer includes a monomer and photo-initiator. The curing reaction after the UV exposure under a high temperature atmosphere turns the UV-exposed

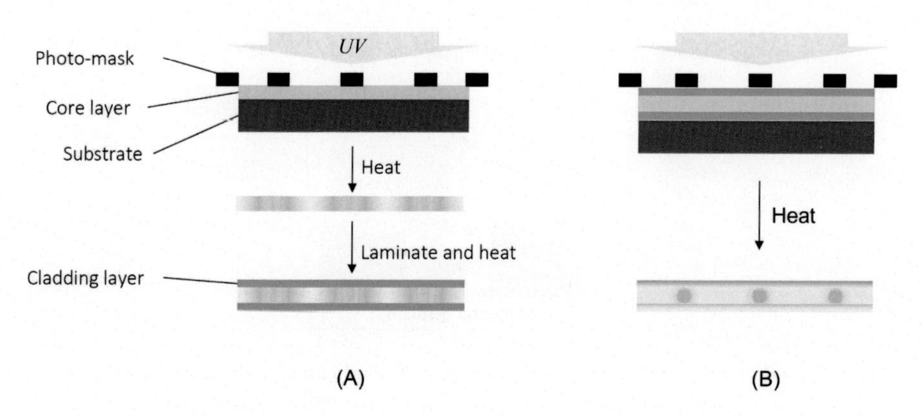

Figure 7.4 Fabrication technique of the photo-addressing method for (A) SI-core and (B) GI-core waveguide).
Source: From Kinoshita R, Moriya K, Choki K, Ishigure T. Polymer optical waveguides with GI and W-shaped cores for high-bandwidth-density on-board interconnects. J Lightw Technol 2013;31(24):4004–15.

area into a cladding with a low refractive index ($n_d = 1.53$) due to selective cross-linking reactions at the UV-exposed area. On the other hand, the area protected by the photo-mask has a higher refractive index ($n_d = 1.55$), forming the waveguide cores. Next, the patterned film is laminated between two cladding films with a uniform refractive index of 1.52 to obtain an SI-square core waveguide, as shown in Fig. 7.4A. Here, by adjusting the curing condition, a concentration distribution of the high-index polymer is formed in the horizontal direction due to the monomer diffusion, which corresponds to the refractive index profile in the horizontal direction [21]. Hence, the index profile in the horizontal direction can be varied from SI to parabolic (GI). In contrast, the index profile in the vertical direction is fixed to be an SI profile.

Meanwhile, a fabrication method of GI-cores (in both horizontal and vertical directions) is shown in Fig. 7.4B. The precursors for core and claddings are prepared and coated onto a substrate one by one. Various kinds of refractive index modifiers can be incorporated in the precursor prior to use. The coated layers are pre-baked and exposed to UV light through a photo-mask to obtain a waveguide, as well. Here, the index modifiers added to the precursor contribute to the formation of an index profile in the vertical direction. Thus, in this method, the fabrication process is much simpler than the method shown in Fig. 7.4A.

7.3.3 The Mosquito method [19]

For the Mosquito method, a micro-dispenser system (Musashi Engineering Co., Ltd., ML-808FXcom) is used, which consists of a pressure-adjustable digital dispenser and a robot (SHOT MASTER 300 DS) to draw desired dots and lines on a substrate with highly precise positioning functions. In the following section, we introduce a very simple fabrication method for GI-core polymer waveguides: the Mosquito method. The procedure of the Mosquito method is

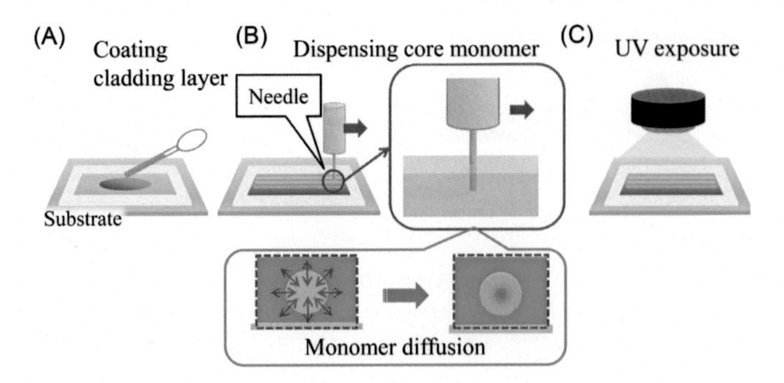

Figure 7.5 GI circular-core waveguide fabrication method named "Mosquito method". *Source:* From Soma K, Ishigure T. Fabrication of a graded-index circular-core polymer parallel optical waveguide using a microdispenser for a high-density optical printed circuit board. IEEE J Sel Top Quant Electron 2013;19(2):3600310.

shown in Fig. 7.5. As the waveguide material we use UV curable silicone resins (FX-W712 for core: monomer viscosity is 12,000 cPs, FX-W713; for cladding: monomer viscosity is 10,000 cPs supplied by ADEKA Co.) [22].

1. First, the viscous monomer for the cladding is coated on a substrate.
2. Next, the viscous monomer for the core is dispensed into the cladding layer by inserting the tip of the syringe needle into the cladding layer, and then the needle is scanned horizontally as shown in the inset.
3. Finally, both core and cladding are cured under UV exposure followed by post baking at 100°C. The desktop robot (SHOT MASTER 300 DS) for the dispenser is used for scanning the needle to fabricate cores.

As shown in the inset of Fig. 7.5B, the core and cladding monomers slightly diffuse into each other to form a concentration distribution before curing under the UV exposure. Since the two monomers are three-dimensionally cross-linking, and form a copolymer, the concentration distribution of the two monomers is fixed, which leads to the refractive index profile. Therefore, the index profiles are quite stable at higher temperatures compared to the dopant-based GI-core polymer waveguides we previously reported [14,18].

Uniformity and controllability of the core diameter dispensed by the needles are very important issues. Therefore, we focus on the relationship between the dispensed core diameter and dispensing conditions such as the scanning velocity of the needle, the dispensing pressure of the core monomer, and the inner diameter of the needle (100, 130, 150, and 190 μm).

The core diameter could be well controlled, and we found that the variation of the core diameter was approximately less than 5%. In addition, we experimentally confirmed that the reproducibility of the core diameter increases with decreasing inner diameter of the needle, and with increasing scan velocity. It was found that forming a smaller core requires lower dispensing pressure and higher speed needle-scan. Finally, we found that the core diameter could be as small as one-third to one-fifth of the inner diameter of the needle under the appropriate dispensing conditions. It should be emphasized that when we use needles with less than 150 μm inner diameter, a desirable core diameter (smaller than 50 μm) was successfully formed. In addition, higher needle-scan velocity contributes to form the core to be close to completely circular. Therefore, from a mass-production point of view, we can say the Mosquito method is a promising fabrication technique.

7.4 Characterization

7.4.1 Waveguide structure and refractive index profile

A core-cladding structure is confirmed by observing the cross-sections of the fabricated waveguides using a digital optical microscope. Fig. 7.6 shows examples of cross-sectional views of the waveguides fabricated using soft lithography, the photo-addressing, and the Mosquito methods. We confirm that multiple cores are aligned in parallel in all the waveguides.

In the soft lithography method, when we completely cure the under-cladding layer, the dopant, DPS in the core monomer barely diffuses into the cladding layer, and thus, near SI-cores are obtained, as shown in Fig. 7.6A. Compared to the cross-sections of the GI cores shown in Fig. 7.6B, we can visually confirm a clear boundary between the core and cladding due to the abrupt change of the refractive index at the boundary. From the interference fringe pattern measurement described below, we can find a slight gradation of refractive index at the boundary even in the SI-cores, but a quantitative analysis of the index profiles shows that the waveguides should be approximated as SI.

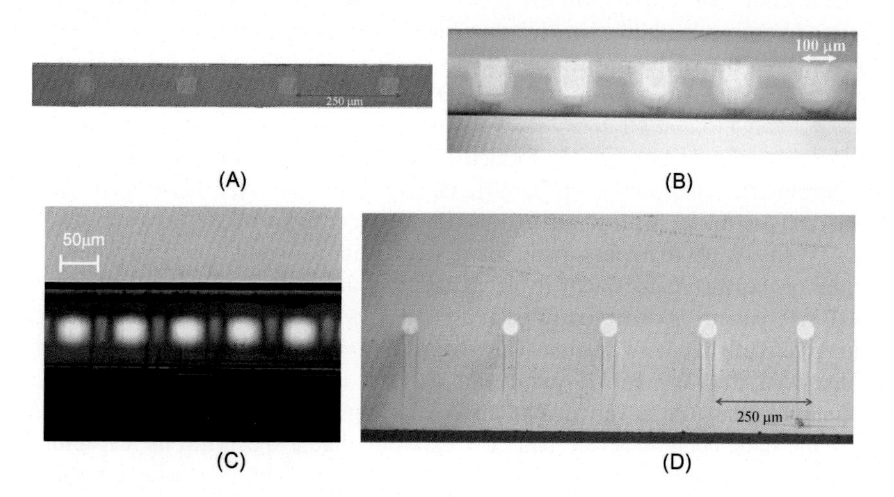

Figure 7.6 (A) Cross-section of 11-ch. polymer parallel optical waveguide with 40 × 40 μm SI-square-cores and with a 250-μm pitch (B) cross-section of 12-ch. polymer parallel optical waveguide with GI-square-cores with a 250-μm pitch (C) Cross-section of 5-ch. polymer parallel optical waveguide with 40 × 40 m GI-square-cores and with a 125-m pitch (D) Cross-section of 12-ch. polymer parallel optical waveguide with 40-μmø GI-circular-cores and with a 250-μm pitch.

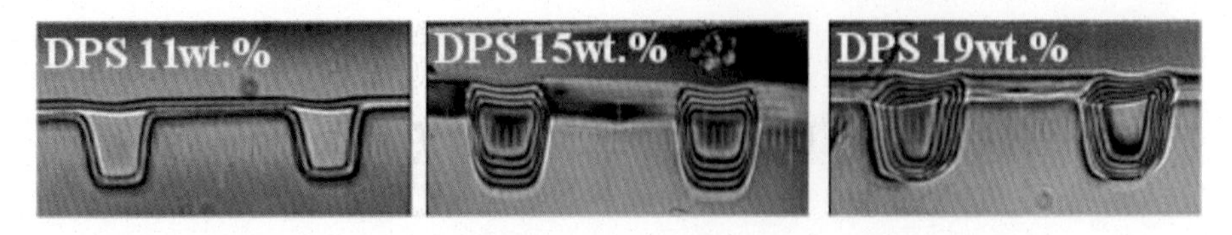

Figure 7.7 Interference fringe pattern observed on a cross-section of slab sample.

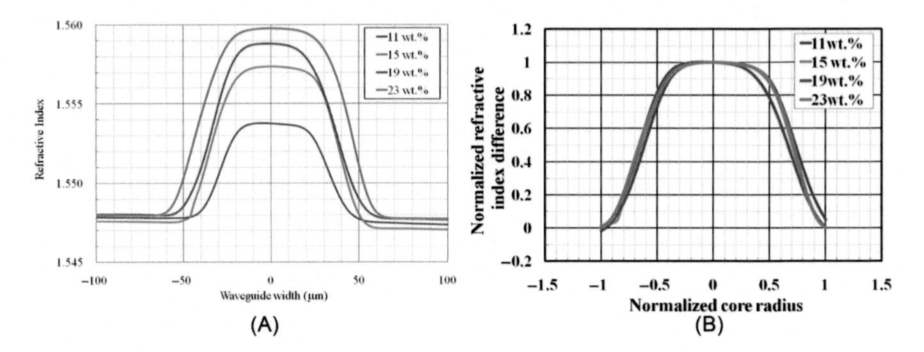

Figure 7.8 (A): Refractive index profile in one core of fabricated polymer waveguides with different dopant concentration (B) and the normalized.
Source: From Ishigure T, Nitta Y. Polymer optical waveguide with multiple graded-index cores for on-board interconnects fabricated using soft-lithography. Opt Express 2010;18(13): 14191–201.

Meanwhile, the cross-sectional photos of the waveguides fabricated using the photo-addressing method also shows a blah core-cladding boundary, as shown in Fig. 7.6C, so the shape of the cores looks close to a circle, while the original core shape was square. As shown in Fig. 7.6D, the Mosquito method allows formation of completely circular GI cores.

The refractive index profiles formed in the core regions are measured using an interferometric microscope (Mizojiri Optics Co.). A very thin slab sample (100–200 μm thick) is prepared, and the interferometric slab method is applied [23]. Fig. 7.7 shows photographs of the interference fringe pattern observed on a cross-section of the waveguides actually fabricated. As shown in Fig. 7.7, a contour map-like fringe pattern is observed in each core area, which indicates the existence of a gradual refractive index variation. The waveguides shown in Fig. 7.7A and B are fabricated using the same PDMS stamp with a 50 μm line width, while the dopant concentration in the core is varied. The under-cladding layer of all the samples is cured under the same conditions (the same UV exposure time), to confirm the capability of index profile formation through this fabrication process. With increasing dopant concentration, the number of dark fringes also increases, as shown in Fig. 7.7A and B. This indicates the large difference of the refractive index between the core and cladding in the waveguide with a high dopant concentration. From the interference fringe patterns shown in Fig. 7.7A and B, the refractive index profiles are calculated in the horizontal direction, and the results are shown in Fig. 7.8A and B. Here, we focus only on one

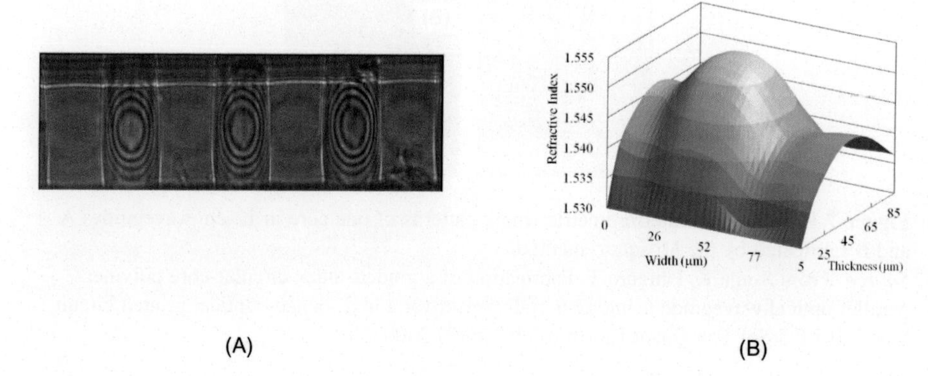

Figure 7.9 Interference fringe pattern observed in the waveguides fabricated using the Mostuito method. (A) measured interferometric fringe patterns (B) calculated index profile. *Source:* From Kinoshita R, Moriya K, Choki K, Ishigure T. Polymer optical waveguides with GI and W-shaped cores for high-bandwidth-density on-board interconnects. J Lightw Technol 2013;31(24):4004–15.

core, but we find almost the same index profiles are formed in the other cores. As estimated from Fig. 7.7, the index difference between the core center and cladding strongly depends on the dopant concentration, and even at a concentration of 23 wt.%, the NA of the waveguide is calculated to be as high as 0.191 from Fig. 7.8A. Therefore, the dopant concentration should be at least 25 wt.% to obtain an NA of 0.2 and beyond. In the case of polymer materials with no cross-linking structure like PMMA, a dopant concentration as high as 25 wt.% may cause deterioration in the thermal stability and mechanical strength. Therefore, TPIR-202 should be used with other monomers for the core that can copolymerize with TPIR-202 to stabilize the index profile.

Meanwhile, the index profiles formed in the waveguide cores fabricated using the photo-addressing method were also evaluated. Fig. 7.9 shows the results of the refractive index profile measurement using the interference microscope. As observed in Fig. 7.9A, clear concentric interference fringe patterns are observed in all the cores. It is noted that the fringe patterns are concentric circular shapes, which means a centrosymmetric near-parabolic refractive index profile is formed in the core region although the outer core shape is square. Indeed, a near-parabolic refractive index profile is confirmed by calculating the profile from the fringe pattern, as shown in Fig. 7.9B. The profiles obtained in the horizontal and vertical directions in Fig. 7.9B are fitted to a well-known power-law equation [23] expressed by Eq. (7.1), and the best-fit index exponent g in both directions is 2.5 on average.

$$n(r) = n_1\left[1 - 2\Delta\left(\frac{r}{a}\right)^g\right]^{1/2} \quad (0 \le r \le a) \tag{7.1}$$

$$\Delta = \frac{n_1^2 - n_2^2}{2n_1^2} \tag{7.2}$$

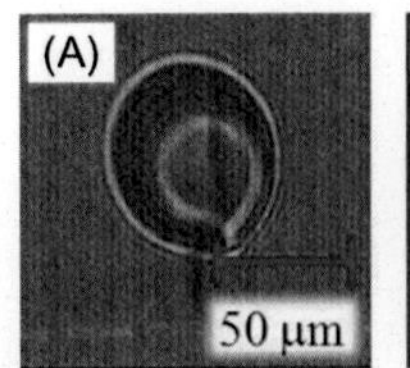

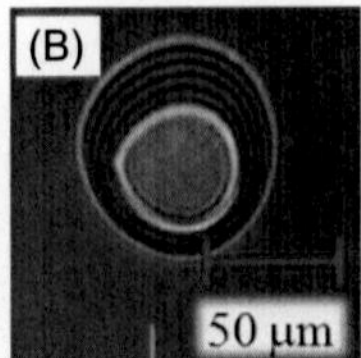

Figure 7.10 Measured interferometric fringe patterns of one core in 12-ch. waveguides A and B fabricated by the Mosquito method.
Source: From Soma K, Ishigure T. Fabrication of a graded-index circular-core polymer parallel optical waveguide using a microdispenser for a high-density optical printed circuit board. IEEE J Sel Top Quant Electron 2013;19(2):3600310.

where: r is the distance from the core center to the measured point; n_1 and n_2 are the refractive indices at the core center ($r = 0$) and the cladding, respectively; a is the half width of the core for the square core; and g is the index exponent. In Fig. 7.9, it is noteworthy that the index profile in the horizontal direction (in the width axis) has a trench outside the parabolic distribution, making it resemble a W-shaped profile. This index trench could be formed due to the monomer diffusion in the photo-addressing method.

We have already confirmed that W-shape profiles decrease the inter-channel crosstalk in parallel waveguides even compared to GI-core one [24,25]. Hence, automatic index trench formation is preferable from the inter-channel crosstalk point of view.

From the index profile and far field pattern (FFP) measurements, the NA of the GI polymer waveguides fabricated using the photo-addressing method is calculated to be between 0.19 and 0.23. The NA of the GI-core waveguide is lower than the NA of the conventional SI-core waveguides (~0.3). However, since polymer waveguide links connected with silica-based MMFs are the primary goal, the NA of the waveguides should be the same as that of MMF (0.21) for a low-connection loss.

The refractive index profile of the waveguides fabricated utilizing the Mosquito method is measured in the same way. Measured results on two waveguides with different core diameters are shown in Fig. 7.10 (Waveguide A: small core; Waveguide B: large core). Concentric interference fringes are observed in the circular core regions, although slight concentricity deviation is observed. The refractive index profile calculated from the fringe pattern shown in Fig. 7.11A and B shows the 3D index profile data on Waveguide A. From the data shown in Fig. 7.11B an almost symmetric index profile is observed, so the concentricity deviation observed in the fringe pattern exhibits little influence on the profile.

It is obvious that a graded refractive index distribution is formed approximately in the areas of 60 and 90 μm diameters in waveguides A and B, respectively, as shown in Fig. 7.11A. The core size of Waveguide A we estimate from Fig. 7.10 is slightly larger than the size (40 μm) we measure in the photo of cross-sections shown in Fig. 7.6D. This is probably due to the diffraction effect of the core edge in the image of the interference microscope, which is more obvious in smaller cores. Here, the refractive indices of the core (FX-W712) and cladding (FX-W713) polymers themselves are 1.526 and 1.511, respectively. Therefore, the ideal NA of the waveguides composed of these two polymers is calculated to be 0.213. On the other hand, the maximum

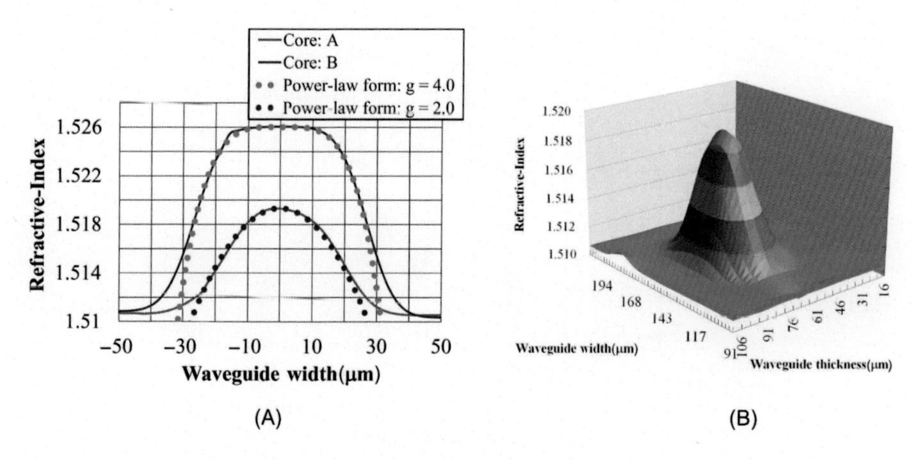

Figure 7.11 (A) 2D refractive index profiles and (B) 3D refractive index profile of a core in 12-ch. waveguide A fabricated by the Mosquito method.
Source: From Soma K, Ishigure T. Fabrication of a graded-index circular-core polymer parallel optical waveguide using a microdispenser for a high-density optical printed circuit board. IEEE J Sel Top Quant Electron 2013;19(2):3600310.

Table 7.1 The numerical apertures of waveguides A and B obtained from FFP measurement

NA	
Waveguide A 0.15–0.17	Waveguide B 0.22–0.23

refractive indices at the center of Waveguide A and B in Fig. 7.11A are 1.519 and 1.526, respectively. Consequently, the calculated NA of Waveguide A (with smaller cores) from the index profile data is 0.167. It is predictable that a small core tends to show a lower NA because the core and cladding monomers diffuse sufficiently before the curing process (Fig. 7.5B). The NAs of the two waveguides are experimentally evaluated using the FFP measurement system (Precise Gage, FFP 1005). The results of the two waveguides are summarized in Table 7.1. In the case of Waveguide B, the NA calculated from the angular distribution (angular width at 5% intensity) is 0.22 to 0.23, which is almost the same as the ideal value (0.213). On the other hand, the NA obtained from the FFP results for Waveguide A is 0.15 to 0.17, which is almost the same as the NA obtained from the refractive index profile.

7.4.2 Propagation loss

The propagation loss of the fabricated GI core waveguides is measured using the cut-back method. In this measurement, a halogen-tungsten lamp (white light source) was used, and a 1 m long, 50 μmø core GI-MMF worked as a launching probe to couple the light into a channel of the waveguide. Meanwhile, a 1 m long, 100 μmø

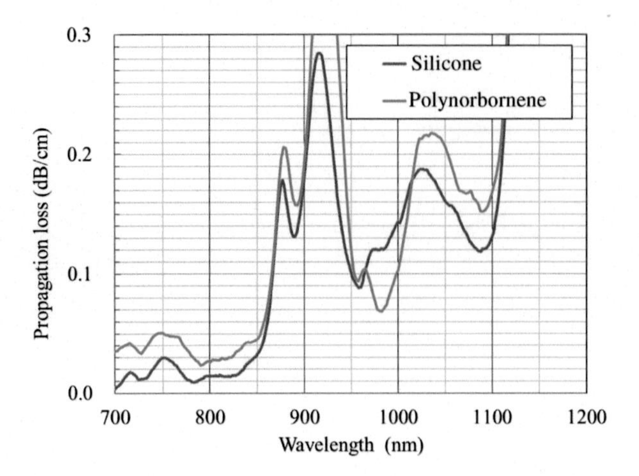

Figure 7.12 Loss spectrum of polynorbornene- and silicone-based waveguides.

core SI-MMF probe whose NA (0.29) is higher than the NA of the waveguide was used to collect all the output light from the channel of the waveguide and guide it to an optical spectrum analyzer.

For the waveguides fabricated using the photo-addressing method, we used a standard dicing saw for cutting the waveguide, and a 20 cm long waveguide was evaluated, while a 15 cm long waveguide fabricated using the Mosquito method was used, which should be long enough to accurately detect the optical power variation before and after dicing the waveguide.

The obtained propagation loss spectra for both waveguides are shown in Fig. 7.12. From the spectrum for the polynorbornene-based waveguide, the propagation loss at the 850 nm wavelength is found to be 0.0457 dB/cm, which is almost the same as or slightly less than the loss of SI-core polynorbornene-based waveguides [26]. For comparison, the loss spectrum of a GI-circular core silicone-based waveguide fabricated using the Mosquito method [19] is also shown in Fig. 7.12. The propagation loss at 850 nm is almost the same, while it is very interesting that the loss in a wavelength range from 950 to 1110 nm is slightly lower compared to the silicone-based waveguide. This difference is attributed to the absorption loss dependence of C−H bonds on the chemical structure.

In the case of the silicone-based spectrum, the propagation loss at 850 nm is 0.033 dB/cm, so it is almost the same or a rather lower loss than that (0.040 dB/cm) of the SI square-core waveguide fabricated by photolithography [4], or even that of the polynorobornene-based waveguide. Although the GI-core waveguides are not fabricated in a clean-room, such a low loss is achieved: one of the reasons for the low loss is the minimal effect of side-wall roughness in GI-cores.

In both spectra, there is a sharp absorption peak at 875 nm wavelength, and the left shoulder of the peak could influence the loss at 850 nm. It is already reported that this absorption peak is attributed to the fourth overtone of the carbon − hydrogen stretching vibration involved in aromatic rings [27]. Therefore, reducing the number of aromatic groups from the core polymer molecules can lead to lower the propagation loss at 850 nm.

7.4.3 Connection loss with MMF

In this section, we focus on the connection between multimode waveguides and GI-MMF, and then demonstrate how GI-core waveguides are advantageous for integrating them on PCBs. In the waveguide link model illustrated in Fig. 7.2: in the test setup, first the light was coupled to the waveguide via a 1 m SMF probe. We also used a 50 μmø MMF with an SI-circular core as a detection probe, substituting for a 50 μmø photodiode. For the waveguide to be evaluated, polynorbornene-based GI-core waveguides were utilized while conventional SI-core and GI-circular core waveguides with the same core size as the polynorbornene-based one were also evaluated for comparison. The output power from the waveguides and fibers were detected using an optical power-meter (ANDO AQ-2140 and AQ-2741) which has an active area sufficiently larger than the core sizes of the waveguides and fibers.

The measured connection losses for the light coupling from the SMF probe to the waveguide tested, and from the waveguide tested to the SI-core MMF are summarized in Table 7.2. The connection losses shown here are the averages for 11 or 12 channels in the tested waveguides. Then, the total connection losses including four connection points in the twin waveguide links are calculated as 1.77 dB for the GI square-core waveguide link and 2.97 dB for the GI circular-core waveguide link, while it increases to 3.56 dB for the SI-core waveguide.

7.4.4 Inter-channel crosstalk

The tight optical confinement effect of the GI core leads to the potential for low inter-channel crosstalk in GI-core polymer waveguides even under a very narrow pitch [28]. Actually, we simulated how we can decrease the inter-core pitch while maintaining an inter-channel crosstalk lower than −20 dB for both GI and SI cores. Then, we showed that the minimum pitch of GI-core waveguides could be smaller than that of SI-core waveguides.

Here, conventional SI-core polymer waveguides generally have had a pitch of 250 μm, while in order to realize higher-density wirings, recently, it tends to decrease to 62.5 μm or less [4]. One of the concerns in reducing the pitch is an increase in inter-channel crosstalk. Therefore, we experimentally compare the crosstalk of the GI square-core

Table 7.2 Connection loss at Connection point

	SMF-WG	WG-MMF
GI square (70-mm long)	0.41 dB	0.44 dB
GI circular (50-mm long)	0.77 dB	0.46 dB
SI square (50-mm long)	0.73 dB	0.73 dB

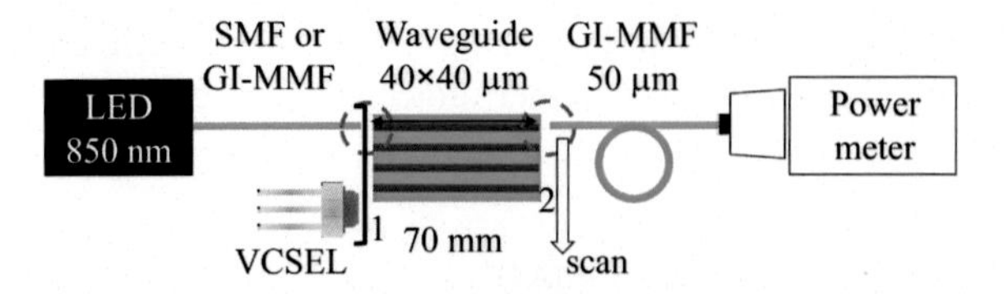

Figure 7.13 Experimental set up for measuring crosstalk.
Source: From Kinoshita R, Moriya K, Choki K, Ishigure T. Polymer optical waveguides with GI and W-shaped cores for high-bandwidth-density on-board interconnects. J Lightw Technol 2013;31(24):4004–15.

waveguides with different pitches: 62.5, 125, and 250 μm (the core size is 40 × 40 μm), which are excited via a 50 μmø GI-MMF probe, while direct coupling with the VCSEL is also adopted for comparison. The inter-channel crosstalk is measured using the setup schematically shown in Fig. 7.13. First, we measure the output power from the excited core after transmitting a 7 cm long waveguide. The tested waveguides have 11 channels aligned in parallel, so that the edge core is excited via a launching probe or via direct coupling with a VCSEL chip under 2 and 6 mA bias currents. Next, as shown in Fig. 7.13, the detection probe (GI-MMF) butt-coupled to the output facet of the waveguide is scanned horizontally to detect the output power from the other cores, as well as the output from the cladding area. Fig. 7.14A and B are the measured results. From Fig. 7.14, the GI-core waveguides show sufficiently low (less than −25 dB) inter-channel crosstalk except for the one with narrowest pitch (62.5 μm) even if the optical power is coupled to the high-order modes butt-coupled to a highly-biased VCSEL chip (over-filled launch). In addition, it is found from Fig. 7.14A that the inter-channel crosstalk depends on the launching condition. When the waveguide is directly connected to the VCSEL chip, although the light emitting area (spot size) is almost the same, the different launching NA depending on the bias current influences the crosstalk. Meanwhile in the case of a 50 μmø GI-MMF launch probe, although the spot size is wider than that of the VCSEL chip, the launch NA is about 0.2, which is almost the same as that of 2 mA biased VCSEL. Hence, the crosstalk values when coupled to the 2 mA biased VCSEL are almost the same as that excited via the 50 μmø GI-MMF launch probe. However, even in the worst case in Fig. 7.14A, a crosstalk value of less than −20 dB is maintained.

In Fig. 7.14B, the crosstalk values to the nearest channel are found to be −21.34, −26.05, and −27.18 dB for 62.5−, 125−, and 250 μm pitch GI-core waveguides, respectively. These crosstalk values in a range from −20 to −28 dB may not be sufficiently low. One reason for the relatively high crosstalk could be the spot size of the launch beam from the 50 μmø GI-MMF for 40 × 40 μm core of the waveguide: the light uncoupled to the waveguide core from the GI-MMF probe remains in the cladding as the cladding modes to increase the inter-channel crosstalk.

Meanwhile, the measured results of the interchannel crosstalk in the waveguide fabricated using the Mosquito method are shown in Fig. 7.15A. From Fig. 7.15A, the crosstalk values to the nearest channel are found to be −27, −31, and −39 dB for 62.5, 125, and 250 μm pitch GI-circular-core waveguides, respectively, which are lower than the values of the GI-square-core waveguides shown in Fig. 7.14. The

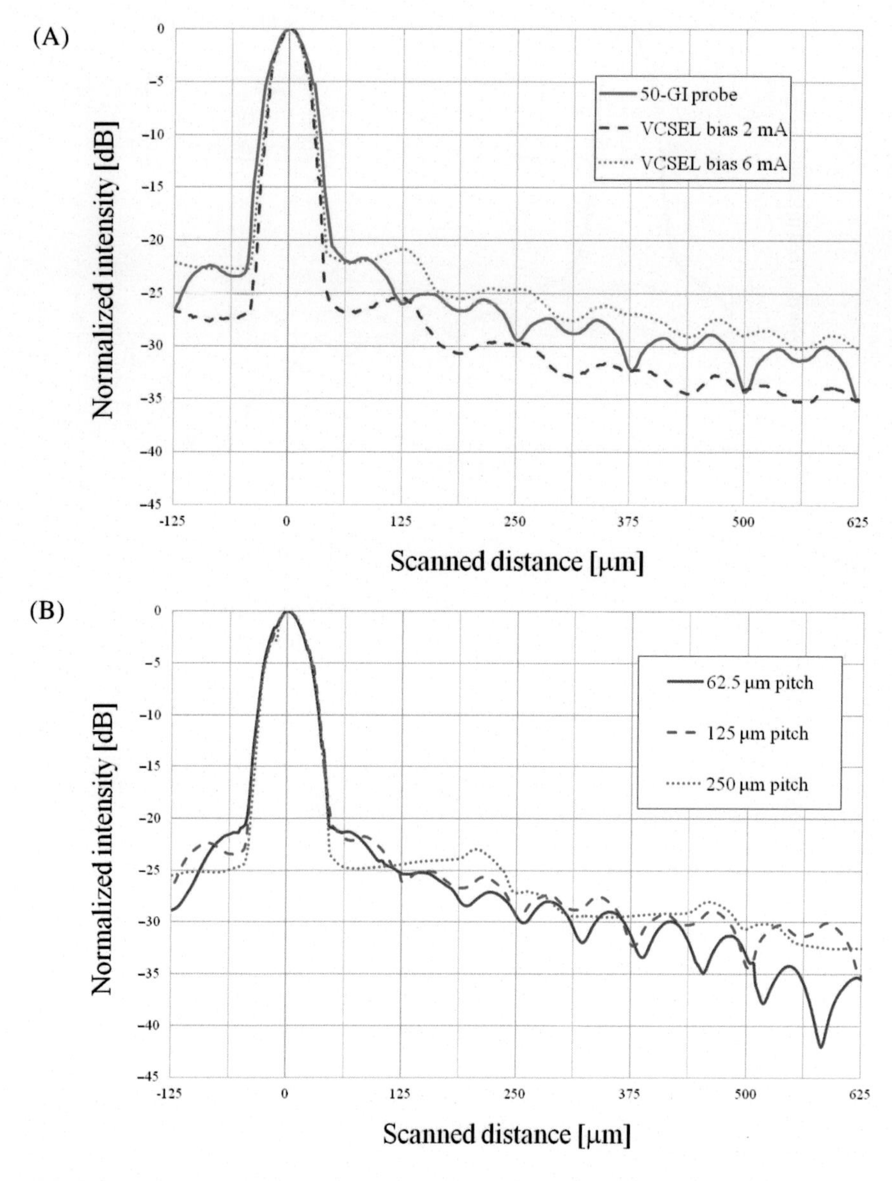

Figure 7.14 (A) Inter-channel crosstalk measurement results of the GI-core waveguide with a 125-μm pitch excited via a 50 -μmø MMF and VCSEL under bias currents of 2 and 6 mA. (B) Inter-channel crosstalk measurement results of the GI-core waveguide with 250-, 125- and 62.5-μm pitch excited via a 50 -μmø GI-MMF.
Source: From Kinoshita R, Moriya K, Choki K, Ishigure T. Polymer optical waveguides with GI and W-shaped cores for high-bandwidth-density on-board interconnects. J Lightw Technol 2013;31(24):4004–15.

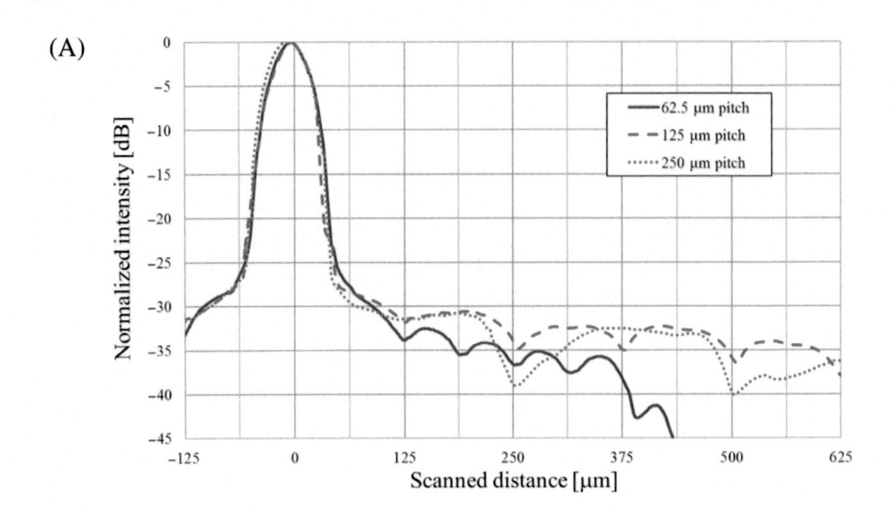

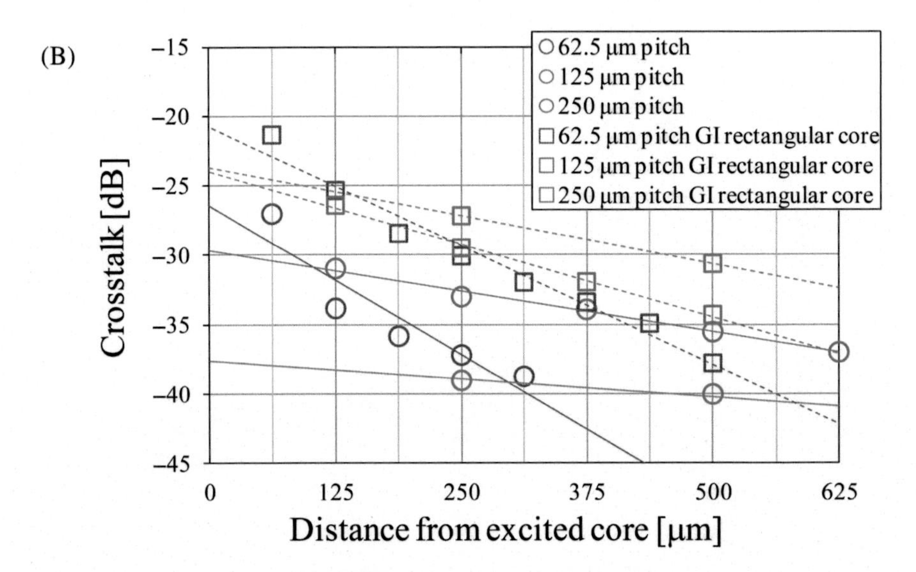

Figure 7.15 (A) Interchannel crosstalk in GI-core waveguides with 250-, 125- and 62.5-μm pitches excited via a 50-μmø GI-MMF. (B) Relationships between the interchannel crosstalk value and the inter-core distance.
Source: From Kinoshita R, Suganuma D, Ishigure T. Accurate interchannel pitch control in gradedindex circular-core polymer parallel optical waveguide using the Mosquito method. Opt Express 2014;22(7):8426–37.

Mosquito method allows us to obtain GI-circular core waveguides with sufficiently low interchannel crosstalk: the crosstalk from the nearest core is as low as −30 dB except for the narrowest pitch (62.5 μm) sample. Even in the case of the 62.5 μm pitch sample, the crosstalk is lower than −25 dB.

In actual optical links, the accumulated crosstalk from all the cores could deteriorate the signal integrity, because parallel signals propagate through all the channels in the waveguides simultaneously. Hence, we focus on the relationship between the crosstalk value and the interchannel distance. In Fig. 7.15B, the results of the fabricated waveguides using the Mosquito method are plotted with the previous results of GI-square-core waveguides. In the case of the GI-*circular*-cores (circle plots), the crosstalk values also exponentially decrease with increasing the interchannel distance: from the best-fit lines (solid lines), the slope values are found to be −0.005, −0.012, and −0.043 dB/μm for the 250, 125, and 62.5 μm pitch waveguides, respectively. Meanwhile, the GI-*square*-core waveguides fabricated by the photo-addressing method with 250, 125, and 62.5 μm pitches show the crosstalk slope values of −0.014, −0.022, and −0.034 dB/μm, shown in Fig. 7.15B as broken lines. The crosstalk slope is steeper with decreasing the pitch, which is the same tendency as those of GI-square-core waveguides. Therefore, the influence from the distant cores on the crosstalk would decrease while the crosstalk to the nearest channel would be high in the narrow-pitch waveguides, and we find in Fig. 7.15B that forming GI-circular-cores could decrease the dominant crosstalk from the nearest cores.

The 250 and 125 μm pitch square-core waveguides show steeper crosstalk slopes compared to the circular core waveguides with the same pitches. Contrastingly, the crosstalk in the 62.5 μm pitch circular-core waveguide rapidly decreases with the intercore distance, compared to the square-core samples. What we would like to emphasize here is that the circular-core waveguides always show lower crosstalk than the square-core waveguides, independent of pitch. This implies that the nearest-core crosstalk values are low. The lower crosstalk in the GI-circular cores than the square counterparts could be attributed to the thick cladding: low optical power confined in the thick cladding could decrease the crosstalk. We already discussed the optical power in the cladding and showed that optical power removal from the cladding is effective for decreasing the interchannel crosstalk [8,28]. Therefore, the effect of the cladding thickness on the crosstalk should be investigated in more detail.

7.4.5 Multichannel operation

In the previous sections, the fabricated waveguides are evaluated by launching one of the multiple cores. However, in actual optical links, parallel multiple cores need to be operated simultaneously. Hence, the optical properties of the waveguides should be discussed under the conditions of all channels operating simultaneously. In this section, we demonstrate a high-speed parallel signal transmission (11.3 Gbps × 12-Ch.) through a GI-core parallel optical waveguide for the first time, where we use parallel-optical transceivers employing 1060 nm VCSELs developed at Furukawa Electric Co., Ltd. [29]. A schematic representation of the experimental setup is shown in Fig. 7.16 along with a photo. One end of a 50 μmø GI-MMF ribbon with a 250 μm pitch is coupled to the transmitter, while the other end is connected via an MT connector to another MMF ribbon with a 250 μm/125 μm pitch conversion structure. Then, the 5 cm long polymer waveguide is butt-coupled to the MMF on the side of 125 μm pitch (with an MT connector), as shown in Fig. 7.16. After propagating through the waveguide, the output lights from the cores are connected to another 125 μm pitch GI-MMF ribbon to guide to the receiver array to measure the eye diagrams.

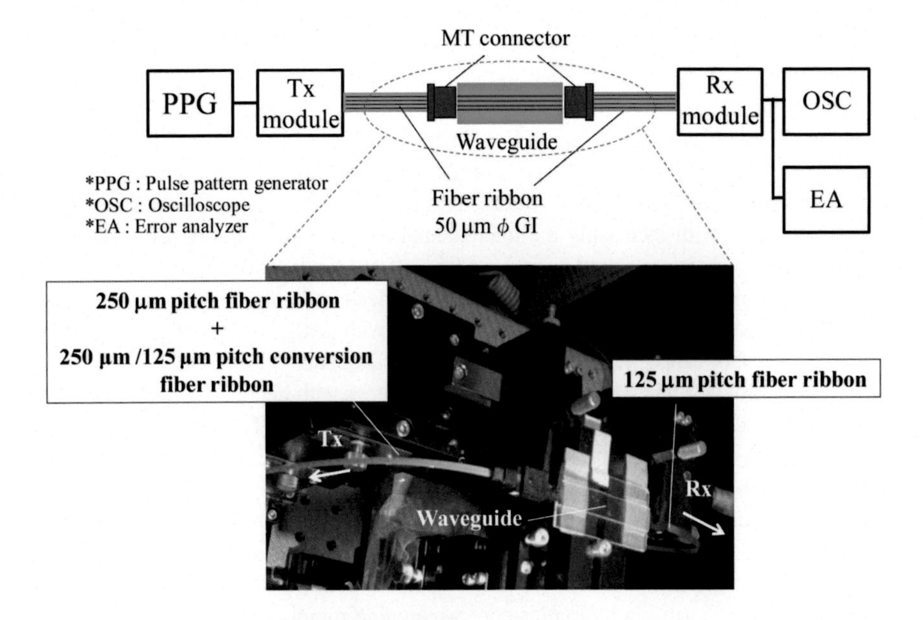

Figure 7.16 Experimental setup for multiple-core operation.
Source: From Kinoshita R, Suganuma D, Ishigure T. Accurate interchannel pitch control in gradedindex circular-core polymer parallel optical waveguide using the Mosquito method. Opt Express 2014;22(7):8426–37.

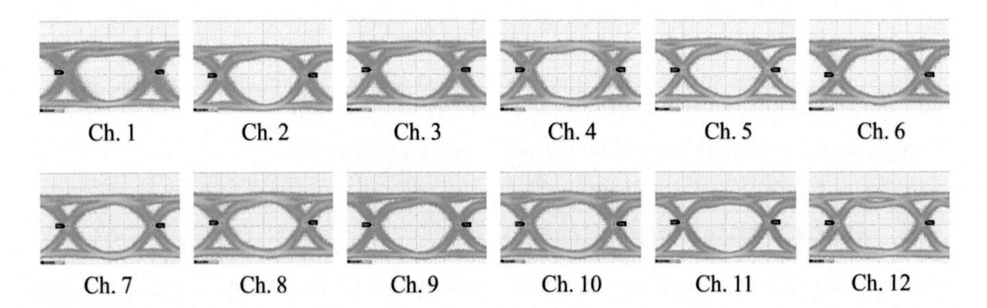

Figure 7.17 Eye diagrams for 11.3 Gbps parallel-optical signals after transmitting a 5-cm long waveguide with a 125-μm pitch.
Source: From Kinoshita R, Suganuma D, Ishigure T. Accurate interchannel pitch control in gradedindex circular-core polymer parallel optical waveguide using the Mosquito method. Opt Express 2014;22(7):8426–37.

The propagation loss of the same silicone based GI-core waveguide at 1060 nm was previously measured in [19] as 0.15 dB/cm, which is slightly higher than the propagation loss at 850 nm wavelength (0.033 dB/cm) due to higher vibrational absorption loss of carbon – hydrogen bonding. However, as the total propagation loss for a 5 cm long waveguide is calculated to be 0.75 dB (0.15 dB/cm $\times$ 5 cm) at 1060 nm, the high absorption loss at 1060 nm shows a small influence on the averaged insertion loss of 2–4 dB for the 12 channels. Since the pitch of the waveguide is precisely controlled to be 125 μm, an insertion loss is as low as 2–4 dB in all the channels.

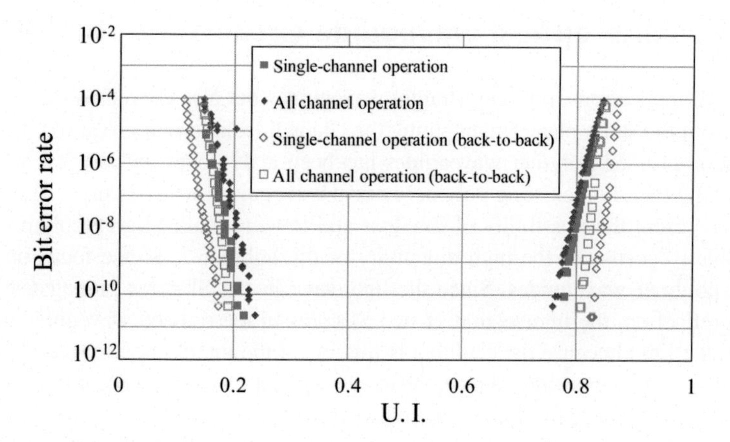

Figure 7.18 Bit error rate curves for 11.3 Gbps transmission when only Ch. 6 is operated and 12 channels are operated.
Source: From Kinoshita R, Suganuma D, Ishigure T. Accurate interchannel pitch control in gradedindex circular-core polymer parallel optical waveguide using the Mosquito method. Opt Express 2014;22(7):8426–37.

Fig. 7.17 shows the measured optical eye diagrams operated at after an 11.3-Gbps 2^{31-1} pseudo-random binary sequence (PRBS) bit-stream for each channel. In terms of a high-speed serial optical transmission, we demonstrated a 25 Gbps optical link utilizing just one core in a GI polymer parallel optical waveguide in a previous work [30]. Meanwhile, the results in Fig. 7.17 are the first demonstration of multichannel parallel-optical transmission in such a narrow-pitch GI-core waveguide. Since the pitch of the waveguide fabricated using the Mosquito method is accurately controlled to 125 μm, the insertion losses are low enough in all the channels, and thus, all eye diagrams observed in Fig. 7.17 are clearly open.

Next, we measure bit error rate (BER) bathtub curves using the setup in Fig. 7.18. Here, we focus on Chapter 6, Types and performance of high performing multimode polymer waveguides for optical interconnects, since it shows the lowest insertion loss, and in order to check the influence of interchannel crosstalk on BER, we compare the BER bathtub curve when all channels are operated to those of back-to-back, and when only Chapter 6, Types and performance of high performing multimode polymer waveguides for optical interconnects, is operated. Measured BER bathtub curves are shown in Fig. 7.18. From Fig. 7.18, the jitter margins are 0.45 U.I. or more, even with all channels operating. Meanwhile, the measured random jitter increment under all-channel operation compared to single-channel operation is almost 0.05 U.I. A slight jitter variation is also observed compared to two back-to-back patterns, as shown in Fig. 7.18. However, those jitter variations are caused mainly by the electric crosstalk, because the back-to-back curve under all channel operation shows larger jitter variation than under single-channel operation. Therefore, it is confirmed that the signal deterioration due to optical crosstalk is negligible even under a pitch as narrow as 125 μm.

7.5 Polymer optical waveguide circuit for optical PCB

For OPCB applications, not only straight waveguides but also curved and even crossing structures in the same plane (maintaining the channel isolation) are required. Indeed, the research on crossed polymer waveguides has been active over the last few years; they are normally based on existing polymer optical waveguides with SI cores [31,32]. These reports represent the possibility of low-loss and low-crosstalk SI-core polymer crossed waveguides, because of the high directivity of the lightwave. So we focus on GI-core crossed polymer waveguides. Since the lightwave in the SI cores propagates via total internal reflection, the intersection of two SI cores in the crossed waveguide no longer confines the light because the cladding is missing at the intersection, resulting in optical leakage loss at the crossings. Hence, the excess optical loss at the core intersection areas could still be a concern in SI-core crossed waveguides.

On the other hand, GI-core polymer waveguides are capable of confining the lightwave near the core center more tightly based on the refraction of light due to the index distribution; the GI-core waveguides exhibit low propagation loss and low interchannel crosstalk. Therefore, crossed waveguides comprised of GI cores potentially have lower loss than the SI counterparts. First, the light leakage loss of crossed waveguides with both SI and GI cores are simulated using a ray-trace simulation. Then, we show that a drastic reduction of the leakage loss is achieved in the GI-core crossed waveguide that is less than one-tenth of the leakage loss in the SI-core crossed waveguide. Next, we confirm the validity of the calculated results. We fabricate the GI-core polymer crossed waveguides (multiple crossings) using the photo-address method [8]. In this method, we use a polynorbornene resin whose refractive index could be controlled by intensity variations of UV exposure during the curing process.

It is difficult for the Mosquito method to fabricate crossed core structures on one plane. Since a needle has to scan on the crossed trajectory while dispensing a core monomer into a cladding monomer, the prior core dispensed is disturbed by the next needle scan for the crossing formation. Meanwhile, the photo-addressing method is a promising technique to realize a crossed waveguide.

Fig. 7.19 is an overview of a perpendicularly crossed GI-core waveguide fabricated using the photo-addressing method [34]. It is obvious that the crossing structure is

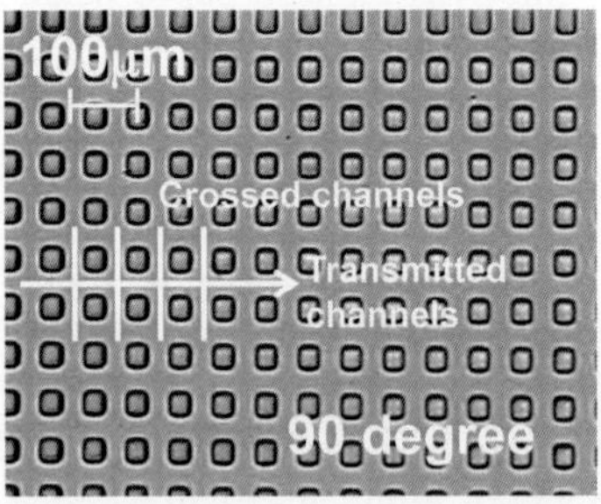

Figure 7.19 Overview of crossed GI-core polymer waveguide.
Source: From Ishigure T., Shitanda K., Kudo T., Takayama S., Mori T., Moriya K., and Choki K. Low-loss design and fabrication of multimode polymer optical waveguide circuit with crossings for high-density optical PCB, In: Proc. 63rd Electron. Compon. Technol. Conf.; 2013 pp.297–304.

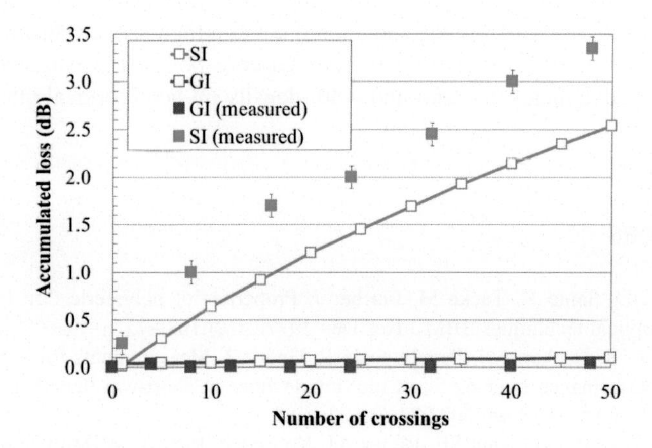

Figure 7.20 Comparison of accumulated loss in crossed polymer waveguides: comparison between calculated and measured results.

formed without any explicit defects. The losses of the crossed waveguides are experimentally evaluated by launching a core via a 50 μmø GI-MMF probe.

Fig. 7.20 shows the measured loss of the crossed waveguides that is plotted over the calculated loss using the ray trace method [33]. Here, the measured results of the SI-core waveguide shown in Fig. 7.20 are quoted from previous literature [32]. Since the loss values are manually read from a figure in [32], error bars are added to the plots in Fig. 7.20. Meanwhile, in terms of the calculated results of the SI-core waveguide in Fig. 7.20, the assumptions for the simulation, such as the waveguide structure and the launching conditions, do not necessarily agree completely with those for the waveguides in the literature. Hence, a disagreement is observed between the measured and calculated results, but the same tendency is observed: with increasing crossing number, the accumulated loss also increases but finally saturates. The measured losses in the SI-core crossed waveguide are noticeably higher than the calculated results.

On the other hand, the measured losses of the GI-core crossed waveguide are extremely low, and the simulated results show a very good agreement. The capability of GI-core waveguides to realize a low-loss polymer waveguide circuit including crossing structures is both experimentally and theoretically verified.

7.6 Summary

The design and fabrication for low-loss, low-crosstalk, and high bandwidth density multimode polymer optical waveguides are reviewed. The main function of past polymer optical waveguides had been just connecting a light source and a detector in such a short distance as several centimeters. Hence, the desired structure for polymer waveguides had been quite simple: large core with high NA to couple the light sufficiently from the light source. Then, the output light from the core had been received with a large-size detector. However, since MMF links are widely deployed in current short-reach networks, multimode polymer waveguides should have a high connectivity with the MMF links, and should cover a high-speed transmission. From these technical points of view, we believe GI-core polymer waveguides pave the way for great advantages in high-bandwidth-density on-board optical interconnects.

References

[1] Swalen JD, Santo R, Tacke M, Fischer J. Properties of polymeric thin films by integrated optical techniques. IBM J Res Dev 1977;21(2):168–75.

[2] Schlager JB, Hackert MJ, Pepeljugoski P, Gwinn J. Measurements for enhanced bandwidth performance over 62.5-μm multimode fiber in short-wavelength local area networks. J Lightw Technol 2003;21(5):1276–85.

[3] Pepeljugoski P, Golowich SE, Ritger AJ, Kolesar P, Risteski A. Modeling and simulation of next-generation multimode fiber links. J Lightw Technol 2003;21(5):1242–55.

[4] Dangel R, Berger C, Beyeler R, Dellmann L, Gmür M, Hamelin R, et al. Polymer-waveguide-based board-level optical interconnect technology for datacom applications. IEEE Trans Adv Pac 2008;31(4):759–67.

[5] Doany FE, Schow CL, Lee BG, Budd RA, Baks CW, Tsang CK, et al. Terabit/s-class optical PCB links incorporating 360-Gb/s bidirectional 850 nm parallel optical transceivers. J Lightw Technol 2012;30(4):560–71.

[6] Pepeljugoski P, Hackert MJ, Abbott JS, Swanson SE, Golowich SE, Ritger AJ, et al. Development of system specification for laser-optimized 50-μm multimode fiber for multigigabit short-wavelength LANs. J Lightw Technol 2003;21(5):1256–75.

[7] Kane CF, Krchnavek RR. Processing and characterization of benzocyclobutene optical waveguides. IEEE Trans Comp Pac Manufact Technol B 1995;18(3):565–71.

[8] Kinoshita R, Moriya K, Choki K, Ishigure T. Polymer optical waveguides with GI and W-shaped cores for high-bandwidth-density on-board interconnects. J Lightw Technol 2013;31(24):4004–15.

[9] Ishigure T, Takeyoshi Y. Polymer waveguide with 4-channel graded-index circular cores for parallel optical interconnects. Opt Express 2007;15(9):5843–50.

[10] Kosugi T, Ishigure T. Polymer parallel optical waveguide with graded-index rectangular cores and its dispersion analysis. Opt Express 2009;17(18):15959–68.

[11] Wang X, Wang L, Jiang W, Chen RT. Hard-molded 51 cm long waveguide array with a 150 GHz bandwidth for board-level optical interconnects. Opt Lett 2007;32(6):677–9.

[12] Yakabe S, Ishigure T, Nakagawa S. Link power budget advantage in GI-core polymer optical waveguide link for optical printed circuit boards. Proc. SPIE 2012; 8267:82670J.

[13] Takeyoshi Y, Ishigure T. Densely Aligned multichannel polymer waveguide with low inter-channel crosstalk. IEEE Photon Technol Lett 2007;19(19):1430–2.

[14] Takeyoshi Y, Ishigure T. High-density 2 X 4 channel polymer optical waveguide with graded-index circular cores. J Lightw Technol 2009;27(14):2852–61.

[15] Hsu H-H, Ishigure T. High-density alignment of graded-index core polymer optical waveguide and its crosstalk analysis with ray tracing method. Opt Express 2010;18(13): 13368–78.

[16] Morimoto Y, Ishigure T. Low-loss light coupling with graded-index core polymer optical waveguides via 45-degree mirrors. Opt Express 2016;24(4):3550–61.

[17] Kinoshita R, Suganuma D, Ishigure T. Accurate interchannel pitch control in graded-index circular-core polymer parallel optical waveguide using the Mosquito method. Opt Express 2014;22(7):8426–37.
[18] Ishigure T, Nitta Y. Polymer optical waveguide with multiple graded-index cores for on-board interconnects fabricated using soft-lithography. Opt Express 2010;18(13): 14191–201.
[19] Soma K, Ishigure T. Fabrication of a graded-index circular-core polymer parallel optical waveguide using a microdispenser for a high-density optical printed circuit board. IEEE J Sel Top Quant Electron 2013;19(2):3600310.
[20] Ishigure T, Horibe A, Nihei E, Koike Y. High-bandwidth, high-numerical aperture graded-index polymer optical fiber. J Lightw Technol 1995;13(8):1686–91.
[21] Mori T, Moriya K, Kitazoe K, Takayama S, Terada S, Fujiwara M, et al. Polymer optical waveguide having unique refractive index profiles for ultra high-density interconnection. In: Proc. Opt. Fib. Comm. Conf. Expo. Nation. Fib. Opt. Eng. Conf., 2012, OTu1.
[22] Hara K, Ishikawa Y, Shoji Y. Preparation and properties of novel silicone-based flexible optical waveguide. In: Proc. of SPIE, Boston; 2006, vol. 6376, pp. 63760K-1–63760K-10.
[23] Marcuse D. Principles of optical fiber measurements. New York, NY: Academic; 1981.
[24] Hirobe Y, Ishigure T. Four-channel polymer optical waveguide with W-shaped index profile cores and its low inter-channel crosstalk property. In: Proc. in Annu. IEEE-LEOS Meeting; 2008, pp. 443–444.
[25] Hsu H-H, Hirobe Y, Ishigure T. Fabrication and inter-channel crosstalk analysis of polymer optical waveguides with W-shaped index profile for high-density optical interconnections. Opt Express 2011;19(15):14018–30.
[26] Fujiwara M, Shirato T, Owari H, Watanabe K, Matsuyama M, Takahama K, et al. Novel opto-electro printed circuit board with polynorbornene optical waveguide. In: Proc. Int. Conf. Sol. St. Dev. Mat.; 2006, pp. 840–841.
[27] Tanaka A, Sawada H, Wakatsuki N. New plastic optical fiber using polycarbonate core. Fujitsu Sci. Tech. J 1987;23(3):166–76.
[28] Ishigure T, Ishiguro R, Uno H, Hsu H-H. Maximum channel density in multimode optical waveguides for parallel interconnections. In: Proc. 61st Electron. Compon. Technol. Conf.; 2011, pp. 1847–1851.
[29] Nasu H. Short-reach optical interconnects employing high density parallel-optical modules. J Sel Top Quantum Electron 2010;16:1337–46.
[30] Yakabe S, Ishigure T, Nakagawa S. Link power budget advantage of GI-core polymer optical waveguide for waveguide-based optical link. In: Technical Digest of Opt. Fiber Commun. Conf., OM2J.3, Los Angeles (CA), 4–9 Mar 2012.
[31] Bamiedakis N, Beals IV J, Penty RV, White IH, DeGroot Jr. JV, Clapp TV. Cost-effective multimode polymer waveguides for high-speed on-board optical interconnects. IEEE J Quant Electron 2009;45(4):415–24.
[32] Sakamoto T, Tsuda H, Hikita M, Kagawa T, Tateno K, Amano C. Optical interconnection using VCSELs and polymeric waveguide circuits. J. Lightw Technol. 2000;18:1487–92.
[33] Ishigure T, Shitanda K, Oizmi Y. Index-profile design for low-loss crossed multimode waveguide for optical printed circuit board. Opt Express 2015;23(17):22262–73.
[34] Ishigure T., Shitanda K., Kudo T., Takayama S., Mori T., Moriya K., and Choki K. Low-loss design and fabrication of multimode polymer optical waveguide circuit with crossings for high-density optical PCB, In: Proc. 63rd Electron. Compon. Technol. Conf.; 2013 pp.297–304.

Silicon photonics for multi-mode transmission

K. Kurata, Y. Suzuki, M. Tokushima and K. Takemura
Photonics Electronics Technology Research Association (PETRA), Tsukuba, Japan

8.1 Expectations for optical interconnection

Due to the rapid growth in the information society, there have been continuous demands for wider bandwidth in communication systems and higher performance in IT/NW equipment. Meanwhile, energy efficiency is another big concern for a sustainable society. To address these contradictory demands, optical technologies can contribute to a wider range of industry. Optical interconnection has attracted much attention in energy efficient equipment and systems, where the electrical I/O bandwidth of data communication faces its limit, called the I/O bottleneck [1]. Optical interconnection technology has been expected to solve this problem even for shorter distances. Attractive points for optical transmission are high-speed, high-density, and low power consumption performance which are summarized below.

8.1.1 Wide bandwidth over long distances

Optical interconnections have been employed in large systems, such as high performance computing and storage systems, where the performance of electrical interconnection is limited by the bandwidth and distance. As the data rate rises beyond 10 Gbps, low-cost MMF links, typically in the range of 100 m, have been widely employed in such systems. Demands for wider bandwidth, such as 100–400 gigabit Ethernet, due to the advances in higher performance system are continuously increasing.

8.1.2 High density

As the CMOS technologies keep up with Moore's law, the electronic interconnect barrier in terms of off-chip bandwidth will be the limiting factor in performance. To address the bottleneck, on-board electrical-optical convertors and optical PCBs will be integrated with the CMOS LSI.

8.1.3 EMI immunity

Fiber optic media is not subject to electromagnetic interference, radio frequency interference, or voltage surges, and so provides greater transmission reliability.

Optical Interconnects for Data Centers. DOI: http://dx.doi.org/10.1016/B978-0-08-100512-5.00008-5

Factory automations and robotics subjected to EMI have been the applications of optical interconnection. Optical interconnection in medical equipment such as real-time monitoring and diagnostics will be promising applications.

8.1.4 Reduced size and weight of cabling

Compared to copper, optical fiber is much smaller in diameter and lighter in weight. By replacing bulky copper cables in a computing rack, optical fiber cabling allows for ease of installation and improved cooling efficiency by producing more space for airflow. Another promising application of optical interconnection is data communication in aircraft and automobiles, where growing demands for transmission speeds must be addressed in relation to weight restrictions. EMI immunity also helps to reduce weight by eliminating the need for shielding. To realize optical interconnection, industrial problems such as lowest cost and mass production technology becomes the final problem. Silicon photonics technology is a strong answer to solve this matter. High-speed over 25 Gbps and small-size Si-photonics transmitter and receiver devices have also been reported [2]. Photonics integrated circuits are produced by a precise silicon wafer fabrication process. It is highly suitable for very small optical transceivers and for mass production. To establish silicon photonic module techniques is now required. Generally, optical coupling adjustment is one particular difficulty in manufacturing optical wiring systems. This difficulty forces us to use expensive parts and low production assembly procedures. Usually, a single mode transmission line is applied in an optical transceiver using silicon photonics technology. In this case, less than 1 μm alignment accuracy is required between the optical transceiver and the optical transmission line. It is considered that to get low cost and high productivity using commercialized equipment, applying multi-mode transmission lines is desirable. In this section, the authors propose a high density multi-mode wiring system with a chip scale silicon photonics transceiver. After a review of concepts and discussions of a grand design, the design of a chip scale optical transceiver named Optical I/O core produced using silicon photonics is introduced. Experimental results with connected multi-mode fiber are discussed. Finally, its applications and future prospects are introduced.

8.2 Multi-mode wiring for silicon photonics technology

8.2.1 Basic concept

Fig. 8.1 shows the basic concept of Optical I/O core compared with a conventional optical transceiver. The conventional optical transceiver consists of integrating discrete components, such as TOSA, ROSA, and ICs, on a printed circuit board. Though the optical/electrical interconnections are fixed in it and cannot be changed by users, it can be applied on only one type of transceiver. On the other hand, the authors propose that all optical/electrical functions, such as LDs, PDs, ICs, and so on, are integrated onto a new form factor of chip-scale silicon photonics substrate. By attaching customized

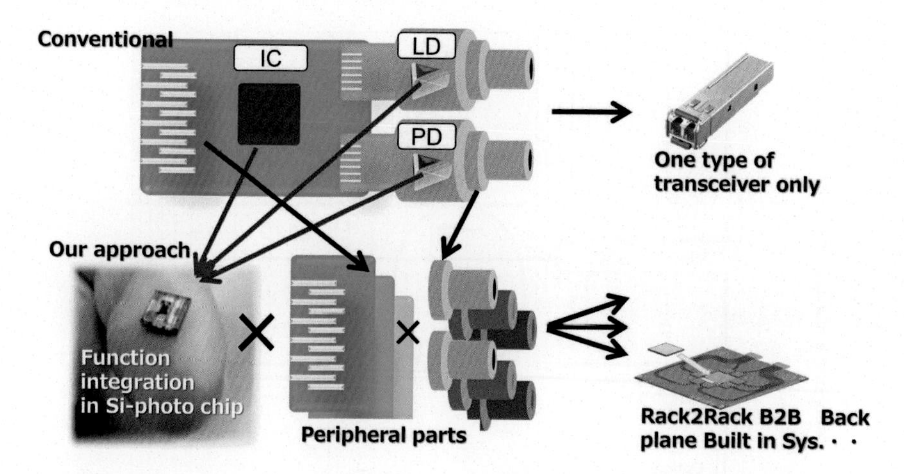

Figure 8.1 Basic concept of Optical I/O core as compared with a conventional optical transceiver.

optical/electricals to the new form factor, various kinds of customized optical transceivers can be formed for not only conventional AOCs, but also USBs, HDMIs, FPGAs, and so on. This new form factor is named Optical I/O core.

The Optical I/O core can be configured for various optical interconnection systems by combining peripheral components and equipment. Furthermore, by standardizing the optical/electrical interconnection of the Optical I/O core, electrical design makers without familiar optical components can easily adapt the optical interconnection into chips, boards, equipment, and so on. In the next section, the selection of an optical transmission line including wavelength with applicable transmission length and coupling tolerance will be discussed, based on the above-mentioned concept of the optical I/O core.

8.2.2 Multi-mode optical transmission line with a wavelength of 1.3 μm [3]

The authors propose a multi-mode optical interconnection with a wavelength of 1.3 μm for the optical I/O core. Low cost and easy connection is most important for accelerating deployment in several short reach interconnections of around less than 300 m, such as consumer or industrial applications that require higher speed, noise reduction, and low weight. Although silicon photonics is almost operated in single mode, high precision alignment to a single mode fiber is a serious challenge to realizing lower cost interconnection. Applying a multi-mode transmission configuration is considered suitable for relaxing the optical coupling requirement. Generally, 1/10 accuracy of core size is requested for a stable optical connection. Precise alignment forces us to use special machines, such as an active alignment machine, and to apply higher cost precision parts. On the other hand, the mounting accuracy of commercialized flip-chip bonding machines has been advanced roughly from 2 to

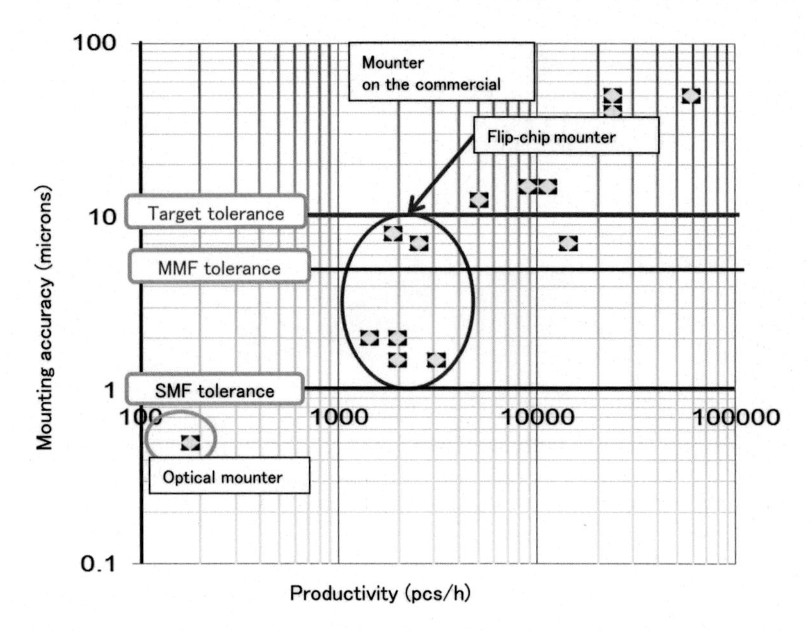

Figure 8.2 Relation of mounting accuracy and productivity in commercialized mounting equipment.

10 μm. Fig. 8.2 shows the relation of mounting accuracy and productivity in commercialized mounting equipment. The required optical coupling tolerance is also indicated. In the case of applying the usual multi-mode optical coupling, a highly-precise flip-chip bonding machine can be applicable. Because 10 μm optical coupling tolerance, which is comparable to multi-mode fiber (MMF) one is required, this leads us to use the usual flip-chip bonding machine. The applicable wavelength is longer than 1 μm, and more specifically, 1.3 and 1.5 μm, which are typically used for silicon photonics due to the higher transparency of silicon. Currently, transmission at 1.3 and 1.5 μm exhibits low propagation loss [4] in silicon photonic devices. The wavelength of 1.3 μm is more suitable for short reach interconnection due to the zero chromatic dispersion wavelength of optical fiber. On the other hand, as shown in Table 8.1, optical fibers tailored to wavelengths of 0.85 and 1.3 μm are already established as industrial standards. For example, in the current 100G-Ether standard, a wavelength of 0.85 μm covers the distance from 1 m to less than 100 m, and a wavelength of 1.3 μm with SMF of 4λ-WDM covers distances over 500 m. However, neither of them covers the distance from 100 to 500 m because of the limited transmission length for a wavelength of 0.85 μm and the wall of low cost for 1.3 μm WDM. A 1.3 μm MMF is a promising medium for solving this problem. For example, experimental results of 25 Gbps and 820 m transmission were reported by using graded index (GI) MMF optimized to 1.3 μm [5]. Furthermore, from a viewpoint of eye safety specification, the permitted launch power into an optical fiber for the 1.3 μm wavelength is larger than that for the 0.85 μm wavelength. This means achieving a wide dynamic range over 25 Gbps

Table 8.1 Current standards and applicable transmission medium for each standardized wavelength

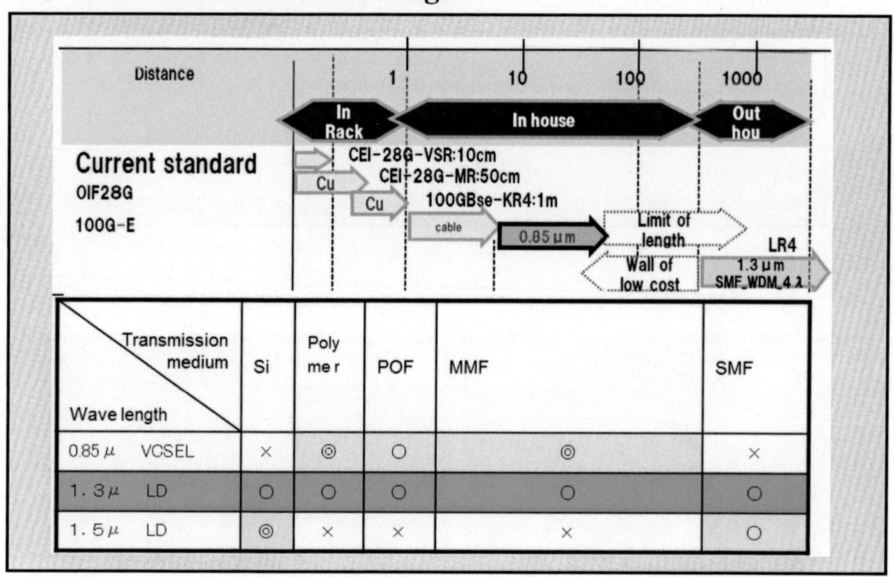

Transmission medium / Wave length		Si	Poly me r	POF	MMF	SMF
0.85 μ	VCSEL	×	◎	○	◎	×
1. 3 μ	LD	○	○	○	○	○
1. 5 μ	LD	◎	×	×	×	○

transmission in the future. Additionally, polymer waveguides and plastic optical fibers are also applicable in short-reach interconnection, as shown in Table 8.1. Therefore, the authors applied multi-mode optical transmission with a 1.3 μm wavelength for silicon photonics devices.

8.3 Chip-scale silicon photonics transceiver "optical I/O core" [6]

8.3.1 Design concept

As optical I/O cores can be used for solving the I/O bottlenecks of LSI chips by mounting around them, this transceiver size should be a similar size to the LSI chips. Because of heat suppression on a small chip, the power consumption should be low for driving silicon optical modulators and photodiodes at high-speed over 25 Gbps. Therefore, the authors employed a stacking structure of 28 nm CMOS drivers and TIAs chips, and a silicon photonics integrated chip to achieve small chip size. Furthermore, for easy mounting on the same board as the LSI chip, both optical and electrical interfaces of the optical I/O core are fabricated on the same flat surface, and it can be easily connected not only electrically, but also optically on the same printed circuit board with a polymer optical waveguide as LSI chip by flip-chip bonding. In consideration of these applications, the target specifications of the optical I/O cores were determined as listed in Table 8.2. The target data rate per channel ranges

Table 8.2 **Target specifications of the optical I/O cores**

Data rate	25–28 Gbps/ch
Size	5 mm × 5 mm
Number of channels	4 or 8 or 12
IC power consumption	5 mW/Gbps
Power-supply voltage	3.3 V/0.9 V (TX) 3.3 V/0.9 V (RX)
Wavelength	1.3 μm
Transmission medium	MMF

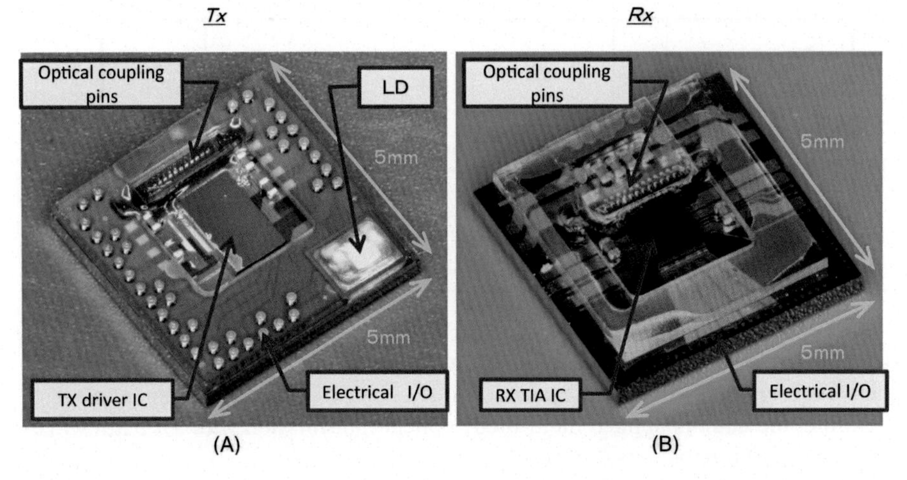

Figures 8.3 (A,B) Photographs of the TX (A), and RX (B) 4-channel × 25 Gbps parallel optical I/O cores.

from 25 to 28 Gbps. Both the TX and RX are 5 × 5 mm in size, and have 4-, 8-, or 12-channel I/Os. 1-V CMOS circuits (driver and transimpedance amplifier, TIA) were developed and used for the TX and RX ICs to achieve low-power consumption operation. The target power consumption per data rate of ICs was designed to be 5 mW/Gbps. The driver and TIA are basically driven at 0.9 V. A supply voltage of 3.3 V is applied to control the modulator bias and photodiode (PD) bias. The wavelength of the light sources is 1.3 μm, and the transmission medium is MMF.

8.3.2 Optical I/O core and its constituent elements

Photographs of the TX and RX 4-channel × 25 Gbps parallel optical I/O cores are shown in Fig. 8.3A and B, and their functional blocks are shown in Fig. 8.4A and B. The size of each Tx and Rx is 5 × 5 mm^2. The optical I/O core basically consists of a silicon-photonics platform, optical coupling pins as a newly designed optical coupling structure, a cover glass, a CMOS IC (driver or TIA), and decoupling capacitors. MZM modulators, grating couplers, spot size converters, LD

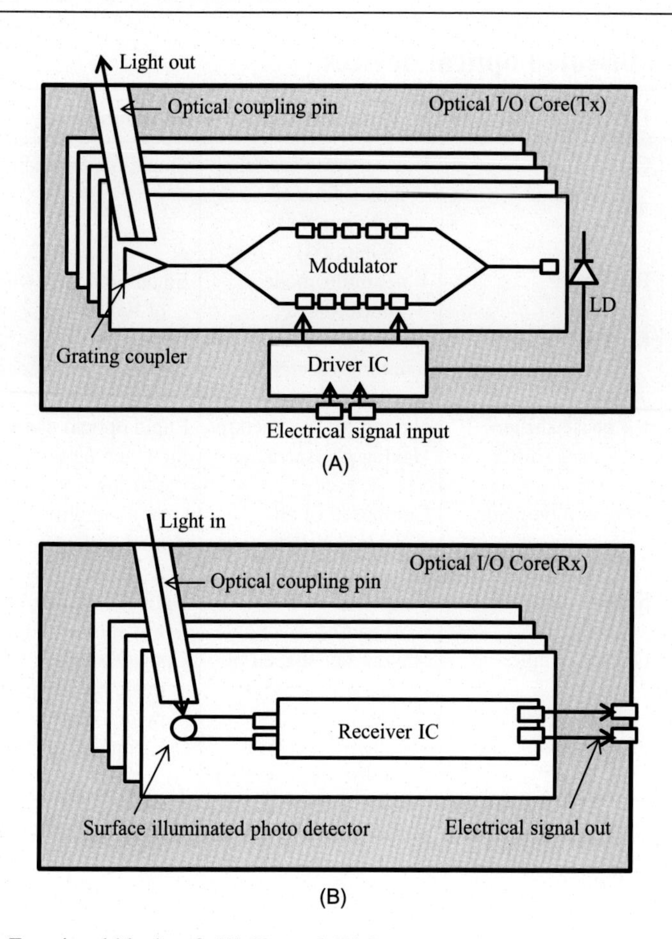

Figure 8.4 Functional blocks of: (A) Tx, and (B) Rx.

pedestal, and electrical wiring/pads are integrated on the SOI substrate in the Tx chip. Surface illuminated photo detectors and electrical wiring/pads are fabricated on the SOI substrate in the Rx chip. The driver and receiver chip are mounted on the chip by using the flip-chip mounting technique. In addition to these components, 1.3 μm Fabry – Perot laser diodes (FP-LDs) are mounted in the TX as a light source using a passive alignment technique. Newly designed optical coupling structures named optical coupling pins are the vertical waveguide.

8.3.3 Fundamental photonics devices

The dimensions of the waveguide are determined to allow only single mode guiding even for a short one. Only the optical output and input designs are different from the usual single mode silicon photonics chip. The Si photonics chip was fabricated from a silicon-on-insulator (SOI) wafer. The top silicon layer and the silicon

Table 8.3 **Fabricated optical devices**

Core type	Optical device	Structural features	Function
TX	Channel waveguide	Having cross-sectional dimensions of 340 nm (W) × 180 nm (H)	Single mode guiding of light
	Branch	1 × 2 multi-mode interferometer	Equal optical power splitting of light beam
	Coupler	2 × 2 multi-mode interferometer	Optical power re-splitting of light beams depending on their phases
	RF phase shifter	Having MOS structure	Rapid optical phase shifting
	DC phase shifter	Having *p-i-n* structure	Slow and large optical phase shifting
	Optical modulator	Composed of all devices listed above	Optical amplitude modulation for E-O conversion
	Spot-size convert (SSC)	Simple inverse taper in width	Optical coupling between waveguide and laser diode
	Grating coupler (GC)	Having fan-shaped Si slab with arc-type grating	Optical coupling to multi-mode optical pin or multi-mode fiber
RX	Photo detector (PD)	Having ~2 μm thick Ge for vertical input	O-E conversion

dioxide layer of the wafer beneath it were used as a core layer and an under cladding layer, respectively, for optical guiding. The top silicon layer was defined into several optical devices, as listed in Table 8.3. The cross-sectional dimensions of the waveguide were determined to allow a single mode guiding at a wavelength of 1310 nm. The photonics chip of the TX included spot-size converters (SSCs) for coupling to LDs, 1 × 2 multi-mode interferometers (MMIs) as optical power dividers, optical amplitude modulators, and grating couplers (GCs) for vertical output of optical signals.

The optical amplitude modulator is a key device which determines the operating bandwidth of the TX. Fig. 8.5A illustrates the optical modulator in the TX core that was demonstrated, and which had a symmetrical Mach-Zehnder interferometer (MZI) configuration. The optical modulator included RF and DC phase shifters. The RF phase shifters had a thin oxide film between the *n*- and *p*-doped regions, as illustrated in Fig. 8.5B. The oxide film allows electrical carriers to accumulate on both sides under a forward bias, and thereby causes efficient refractive index change through the carrier plasm effect. Therefore, the same kinds of RF phase shifters had the advantage of a large $V_\pi L$, contributing to the small chip size of the TX [7]. The RF phase shifter consisted of four segments, each of which was driven by being phase-matched with the propagating light. To make the MZI modulator operate at around a

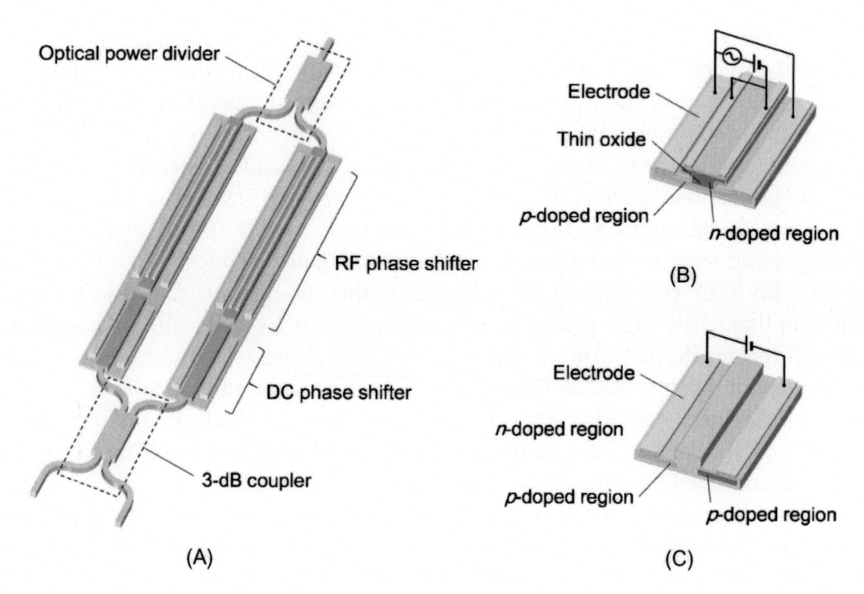

Figure 8.5 (A) Mach-Zehnder interferometer-type optical amplitude modulator including: (B) RF phase shifters, and (C) DC phase shifters in series.

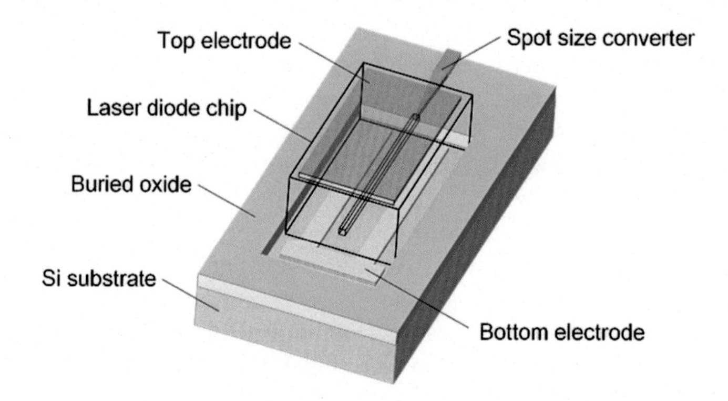

Figure 8.6 Laser diode (LD) chip fitted in an oxide trench directly on top of a Si base substrate. LD is coupled to an inverse-taper spot size converter.

through efficiency of -3 dB, direct-current (DC) phase shifters were added. The DC phase shifters had a *p-i-n* diode structure, and were operated under a forward bias condition. The effective index of the DC phase shifters is able to be shifted through the carrier plasma effect and, in addition, it is efficiently enhanced by the thermo-optic effect in the intrinsic silicon region. The *p-i-n* diode structure was not adopted for the RF phase shifter, since the thermo-optic effect is slow.

The way of providing a light source to the TX chip is notable in achieving its compact size. As shown in Fig. 8.6, a bare LD chip was fitted with its epi-side down in a deep oxide trench directly on top of the base Si substrate, which is advantageous to exhaust heat of the LD efficiently through the photonics chip. Besides, the overall height of the TX chip can be smaller than when a packaged LD module is mounted. This is preferable to make the top surface of the TX chip flat, and enable us to flip-chip bond the TX chip on a plastic circuit board. The LD was edge-coupled to an SSC facing a vertical wall of the oxide trench [8]. The SSC was a simple reverse taper of Si waveguide. Since the spot size of the LD used was about 3 μm in diameter, the thicknesses of the buried oxide (BOX) and the over-cladding oxide were set to be larger than 2 μm, to suppress optical loss of the SSC. The LD and SSC were aligned to each other within an error of ± 0.1 μm for efficient coupling. This high precision was accomplished by controlling the etching depth of the trench, and using a chip mounter with a mark alignment system. No optical isolator was inserted in front of the LD, which reduced chip size of the TX, and assembly cost. Instead, other optical components and their circuits in the TX photonics chip were carefully designed to suppress back reflection to the LD sufficiently, since it could put the LD at risk of unstable lasing [9].

The grating coupler (GC) as an output device is another feature of the photonics chip of the TX. As long as single polarization is used for the optical output, the GC can be comparable or better in coupling efficiency for coupling to an optical fiber with a compact size, while otherwise, a chip-end SSC may be better [10]. The GC had a fan shape to widen the output beam size efficiently in a short length, as shown in Fig. 8.7. Since the GC was to couple to an optical pin, which was a multi-mode waveguide, there was no need to exactly match the beam profile of the GC to that of the optical pin in order to achieve high coupling efficiency. The size and tilt angle of the light beam emitted from the GC was fitted only to those of the optical pins, which are about 35 and 8 μm, respectively, by adjusting the pitch, filling factor, and depth of its grooves. The upward directionality or the upward emitting efficiency was almost exactly the coupling efficiency to the optical pin. The thicknesses of the grating core, which is the top Si layer of the SOI wafer, and the BOX beneath it are crucial to the upward directionality, and were 180 nm and 3 μm, respectively. The core thickness was determined as a result of integration with the optical amplitude modulator, and was not optimized for the GC, while the BOX thickness was. As a result, the upward directionality of the GC in the TX demonstrated was about -4 dB. It was numerically demonstrated that the upward

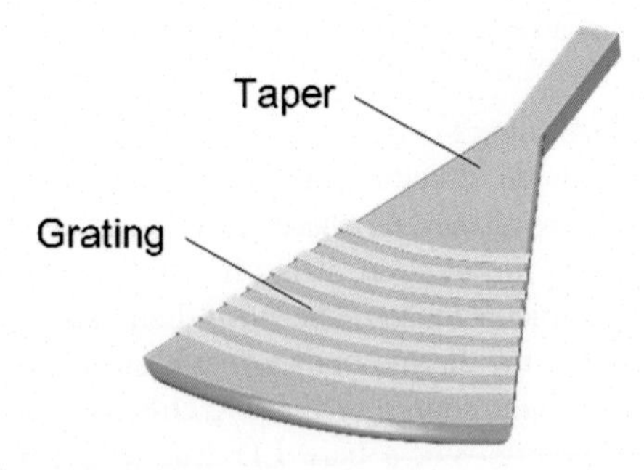

Figure 8.7 Optical grating coupler.

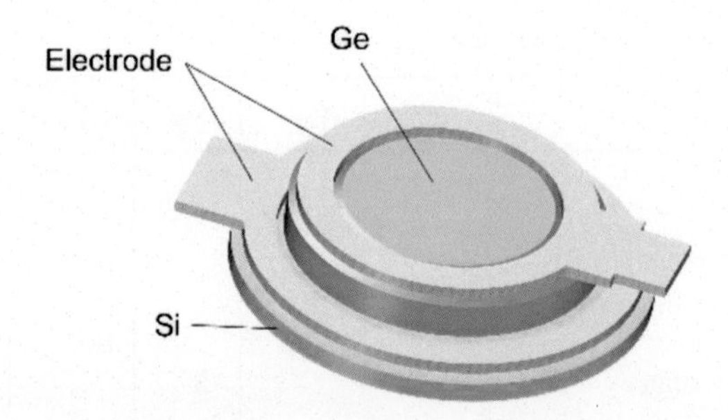

Figure 8.8 Surface illuminating Ge photo detector.

directionality could reach about −0.58 dB by increasing the core thickness to 200 nm for a wavelength of 1310 nm, keeping the groove depth shallow, and adding anti-reflection coating (ARC) layers on top [11,12]. Fabrication tolerance can also be improved by such a structure [13].

In contrast to the TX photonics chip, the RX photonics chip had only surface illuminating Ge *p-i-n* photo detectors (PDs) on it as a fundamental optical component. Since light output from the TX chip was not multiplexed in wavelength, it was directly received with a PD at the RX chip. The electrical current generated at the PD was transferred to a transimpedance amplifier (TIA) that was flip-chip bonded on the RX photonics chip. On top of the PD an optical pin was formed and coupled to an MMF through it. The PD itself on the RX photonics chip was formed by implanting boron ions to the top Si layer of an SOI wafer, epitaxially growing Ge on top, implanting phosphorous ions to the very shallow range of the Ge surface, and forming electrodes on the Si and Ge. Fig. 8.8 shows an illustration of the PD. The SOI wafer was used not for optical guiding, but for efficient reflection of input light at the back of the Ge in order to improve the sensitivity of the PD. By appropriately designing the SOI layer structures, the back-side reflection can be maximized. The thickness and diameter of the Ge were about 2 and 25 μm, respectively. The thickness was determined between the tradeoff of maximizing optical absorption by the Ge and suppressing carrier-drift times in the vertical direction. The diameter of the PD could be made smaller for a smaller RC delay, but it was finally determined to be larger than the bottom diameter of the optical pin to receive the whole light power input via the optical pin. An ARC of Si_3N_4/SiO_2 multiple layers was applied on the Ge to minimize optical reflection loss. As a result, the sensitivity of the PD at a wavelength of 1310 nm was measured to be as high as 0.8 A/W, while the 3 dB bandwidth was about 13 GHz.

8.3.4 CMOS ICs

With the miniaturization of the CMOS process technology, not only digital circuits but also analog circuits have become possible to operate at millimeter wave levels. The cut-off frequency (fT) of MOS has exceeded 200 GHz, and is comparable to that of a Si bipolar transistor. An fT of 349 and 242 GHz for n-MOS and p-MOS in

Table 8.4 Target specifications of Tx IC

Parameter	Units	Min	Typ	Max	Comments
Bit rate	Gb/s	25			NRZ Format, Pn: $2^{31}-1$
Differential input voltage swing	mV	600	800	1000	
Differential output voltage swing	V_{pp}	1.5			
Output rise/fall times	ps			18	20%/80%
Cross-point control range	%		75		High
			25		Low
Input return loss	dB	10			F < 20 GHz
Supply voltage	V	0.855	0.9	0.945	
Supply current	mA		125		Including T/A
Operating ambient temperature	°C	−10		85	

Table 8.5 Target specifications of Rx IC

Parameter	Units	Min	Typ	Max	Comments
Bit rate	Gb/s	25			NRZ Format, PRBS: $2^{31}-1$
Small signal transimpendance bandwidth	GHz	12.5			
Input current	μA_{pp}	80		1000	
Transimpedance gain	dBΩ		80		
Differential output voltage swing	mV_{pp}	600	800		
Low cutoff frequency	KHz			100	
Output return loss	dB	10			F < 20GHz
Supply voltage	V	0.855	0.9	0.945	
Power consumption	mW		125		Including Driver IC
Operating ambient temperature	°C	−10		85	

28 nm process technology have been reported, respectively [14]. There are a current-mode logic (CML) and a CMOS inverter in the CMOS circuit design. The CML has the advantages of a differential current operation, a low switching noise, and a low voltage swing for high speed. On the other hand, the CMOS inverter has the advantage of very low static power consumption, a large voltage swing, and a compact size. The power consumption happens during the switching, and the voltage swing is equal to a supply voltage. The authors designed to combine both the CML and CMOS inverter to meet the specifications of Tx and Rx ICs. The target specifications of the Tx and Rx ICs are listed in Tables 8.4 and 8.5. The data rate per channel is from 25 to 28 Gbps. The power consumption per data rate of both ICs is 5 mW/Gbps. The driver IC has a differential input and four differential

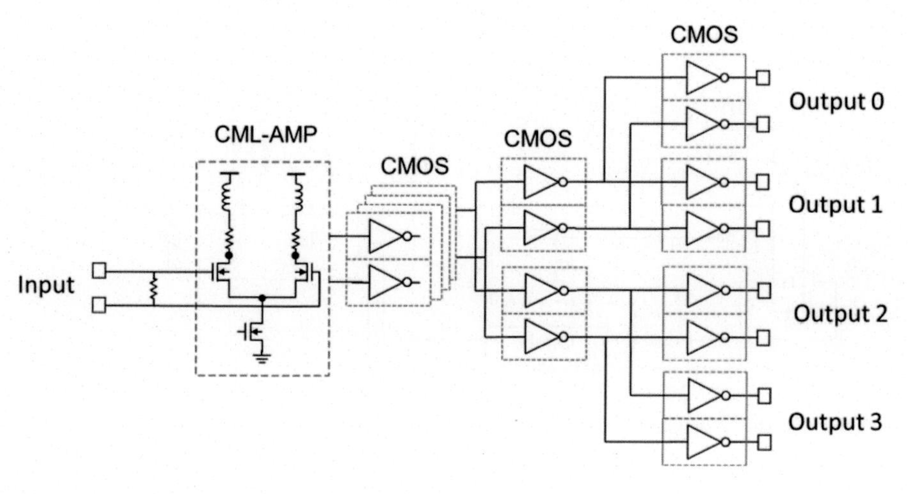

Figure 8.9 Block diagram of the driver IC.

outputs per channel, which drive the four divided segments in the Mach-Zehnder modulator. A differential output-voltage swing of more than 1.5 Vp-p is needed to maintain the optimum optical modulation amplitude from the modulator.

The TIA has a differential output of more than 600 mVp-p. The transimpedance gain is 80 dBΩ. Both ICs have a digital I2C management interface. The management interface is used for the shutdown and configuration of outputs such as voltage swing and cross-point. The authors fabricated the driver IC and the TIA using 28 nm CMOS process technology. The sizes of the driver IC and the TIAs are $1.68 \times 1.07\ \text{mm}^2$ and $1.08 \times 0.93\ \text{mm}^2$, respectively. A block diagram of the driver IC is shown in Fig. 8.9. The driver IC consists of a CML input buffer and CMOS inverters. The CML input buffer is assigned for a 100 Ω-differential termination. CMOS inverters for the outputs are designed to drive capacitances of 0.3 pF in the modulator, and obtain output-voltage swings of more than 1.5 Vp-p. The output impedances are lower than 50 μ. This allowed release from the limit of RC bandwidth in the modulator. The supply voltage is typically 0.9 V. The differential output-voltage swing of the driver was 1.8 Vp-p. A cross point of the optical output from the modulator can be adjusted by the digital I2C management interface. A skew between the differential output signals is fitted in the delay of optical signal in the modulator. The power consumption per data rate is designed to be 3.1 mW/Gbps. A block diagram of the TIA is shown in Fig. 8.10. The TIA consists of a transimpedance amplifier using a CMOS inverter, CML limiting amplifiers, and a CML output buffer. A single-ended signal is converted to a differential signal at the first-stage CML amplifier by a DC feedback loop. The CML amplifiers include compact spiral inductors to enhance the bandwidth of the TIA. The transimpedance gain is 80 dBΩ. The supply voltages of the transimpedance and CML limiting amplifiers are 0.9 V. The differential output-voltage swing could be controlled from 580 to 970 mVp-p, according to application by the digital I2C

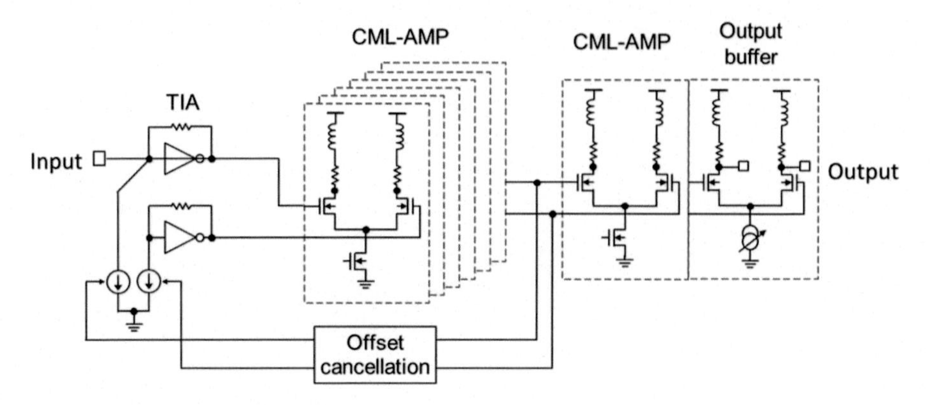

Figure 8.10 Block diagrams of the TIA.

management interface and the supply voltage for the output buffer from 0.7 to 1.1 V. The minimum power consumption per data rate is 1.8 mW/Gbps.

8.3.5 Optical coupling structure

The optical coupling structure of the optical I/O cores connects outer multi-mode optical fibers or waveguides to the I/O interfaces on the Si photonic chip, namely the grating couplers (GCs) for a transmitter (TX), and the photodiodes (PDs) for a receiver (RX). Because the GCs and the PDs are placed close to the driver or transimpedance amplifier (TIA) IC to minimize the form factor of the optical I/O core, it is impossible for the Si photonic chip to connect directly to the optical fiber or waveguide. Considering that the core area of a multi-mode fiber (MMF) is larger than the area of the GC or the PD, control of the spot size of input and output light is also a critical issue. The optical coupling structure, therefore, has to guide the light and control the mode field of the light to take an advantage of large misalignment tolerance. The optical pin is a vertical waveguide structure, which is made from ultra-violet (UV) curable resins and is formed by photolithography [15]. The vertical waveguide structure can move the optical I/O interface from congested Si surfaces to the upper region over the IC. Photolithography has an advantage in adjusting the shape of the waveguide structure by photomask patterns and exposure conditions to control the mode field, as well as formation of high density waveguide structures on the Si surface. From these viewpoints, the optical pin is suitable for use as an optical coupling structure for the optical I/O core. As is the case with planar polymer waveguides, the core shape mainly makes an impact on the transmission properties of the optical pin. The optimum shape can be determined by ray-trace simulation for optical coupling losses from a GC or a PD to a GI50 MMF [16,17]. Fig. 8.11 shows the structural model used in the simulation. The difference in refractive indices of the core and cladding resins used in the following simulation and fabrication is 4.3%, which corresponds to a numerical aperture of 0.466. The height of the simulated

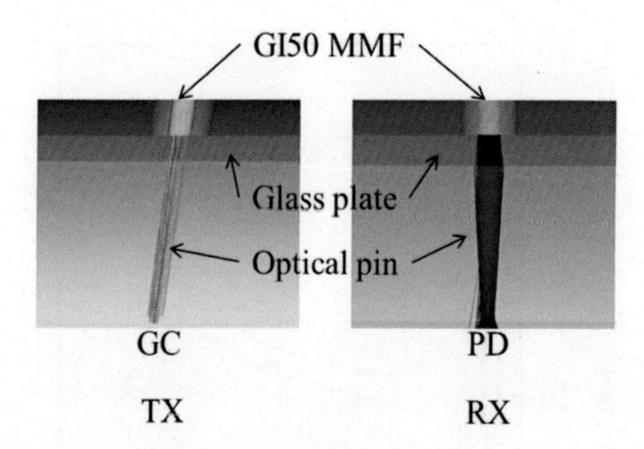

Figure 8.11 Structural model used in misalignment tolerance estimation.

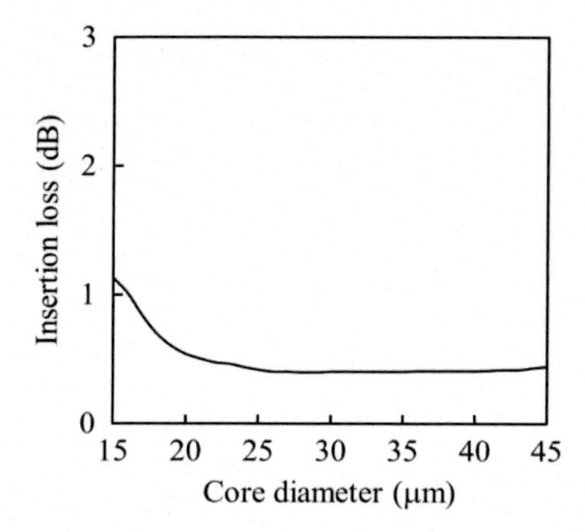

Figure 8.12 Insertion loss as a function of the core diameter for the TX.

optical pin is 300 μm. The optical pin in the TX guides output light from the GC to the MMF. The beam direction from the GC is designed to be inclined at 12.7° from the perpendicular in air. Considering the refractive index of the core material, the core has to be tilted at 8°. The near-field pattern (NFP) and far-field pattern (FFP) of the GC is calculated by finite-difference time-domain (FDTD) simulation. The spot size for the beam from the GC is less than 14 μm. The core diameter dependence for the insertion loss of the TX optical pin is shown in Fig. 8.12. The insertion loss increases with decreasing core diameter when the core diameter is smaller than 25 μm. This is due to an increase in coupling loss at the interface between the GC and the optical pin. Fig. 8.13 shows the allowable misalignment tolerance between the optical pin and an MMF as a function of the core diameter. The allowable misalignment tolerance is defined as a range in the insertion loss smaller than 1 dB. Because the core diameter of a GI50 MMF is larger than that of the simulated optical pin, the allowable misalignment tolerance is estimated to be larger than 20 μm, and decreases with increasing core diameter of the optical pin. From a fabrication point of view, the formation of a cylindrical core by photolithography becomes easier with an increase in the core diameter. As a result, the target core shape for the TX was

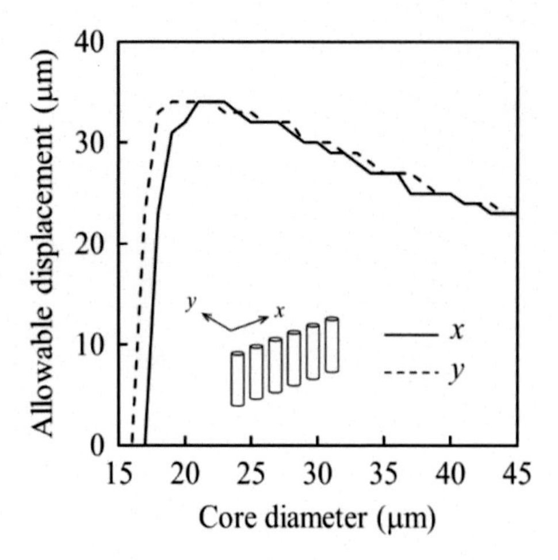

Figure 8.13 Allowable misalignment tolerance between the optical pin and an MMF for the TX.

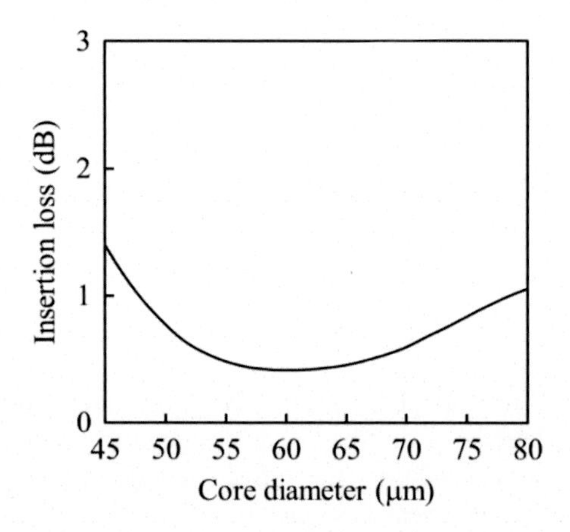

Figure 8.14 Insertion loss as a function of the MMF side core diameter for the Rx.

decided to be a 35 μm diameter 8°-tilted cylinder. The optical pin in the RX has to act as a spot-size converter, because the core diameter of a GI50 MMF is larger than that of the PD, which is 30 μm. For this reason, the core of the RX was designed to be the shape of an inverse truncated cone. In the simulation, the NFP and FFP of the light source into the MMF satisfies the overfilled launch condition. When the diameter of the core at the PD side is 30 μm, the minimum insertion loss is obtained at the core diameter at the MMF side of 60 μm, as shown in Fig. 8.14. Increase in the insertion loss at the range of the core diameter smaller than 55 μm results from an increase in the coupling loss between the optical pin and a MMF. On the other hand, increase of the leak light with increasing the taper angle of the core mainly causes insertion loss in the region of the core diameter larger than 70 μm. As shown

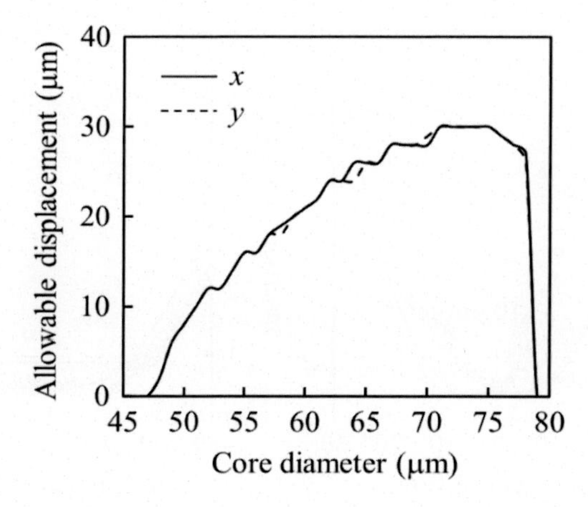

Figure 8.15 Allowable misalignment tolerance between the optical pin and MMF for the Rx.

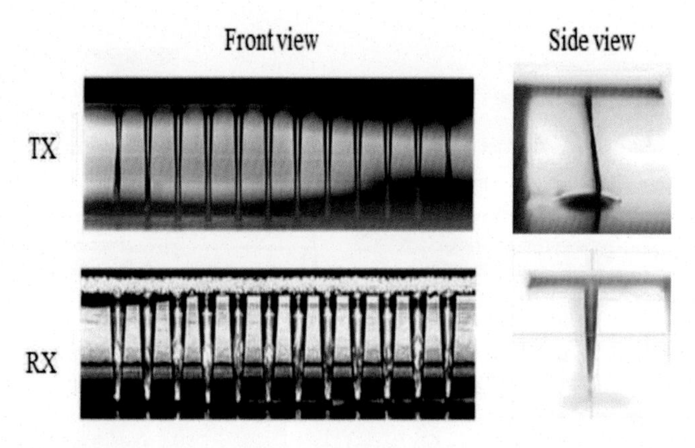

Figure 8.16 Photographs of the typical core arrays: (A) the TX, and (B) the RX.

in Fig. 8.15, the allowable misalignment tolerance increases with increasing core diameter at the MMF side, and abruptly falls at a core diameter of 78 μm. When the core diameter is between 60 and 78 μm, the allowable misalignment tolerance is predicted to be larger than 20 μm. From these results, the target core diameter at the MMF side is decided to be a diameter of 70 μm. Fig. 8.16 shows photographs of the typical core arrays for the TX and RX. The pitch is 125 μm. The fabricated shapes are substantially the same as the design. The measured and calculated insertion losses, depending on the misalignment between an optical pin and a GI50 MMF, are plotted in Fig. 8.17. For the TX, the measured misalignment tolerance is 10 μm, which is smaller than the calculated value, though the measured minimum coupling loss of 0.49 dB is almost the same as the calculated value. In contrast, the measured minimum coupling loss and alignment tolerance is 0.41 dB and approximately 35 μm. These values almost coincide with the calculated values. The measured

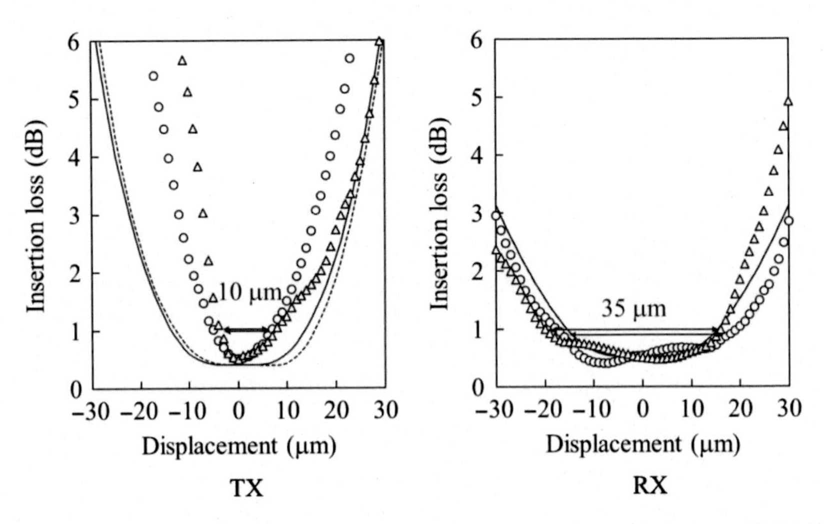

Figure 8.17 Insertion loss depending on misalignment between an optical pin and a GI50 MMF.

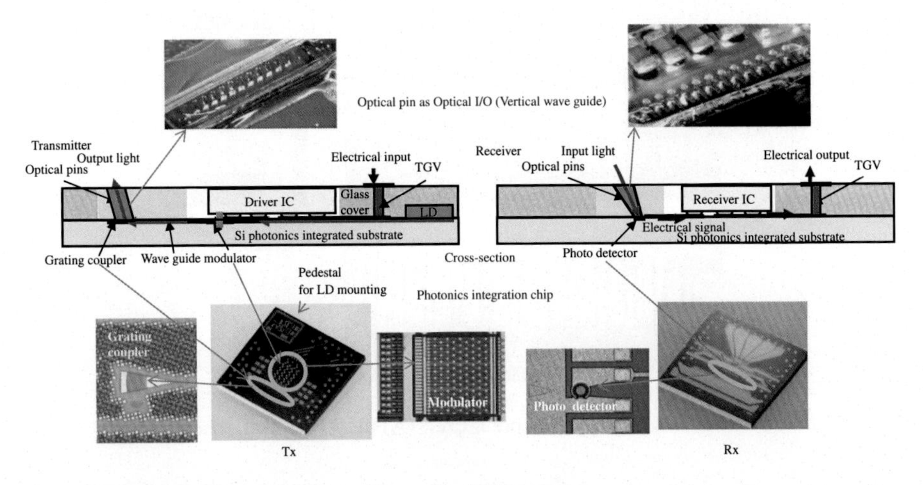

Figure 8.18 Cross-section of Tx and Rx optical I/O cores.

misalignment tolerance values are large enough for handling with commonly used flip-chip mounters, though further improvement of the design and fabrication process would be necessary for the TX. Open circles and triangles are measured values for *x*- and *y*-direction, respectively. Solid and dashed lines are calculated values for *x*- and *y*-direction, respectively. The *x*- and *y*-directions are the same as that shown in Fig. 8.13.

8.3.5.1 Packaging

Packaging of the optical I/O cores is based on a hybrid integration approach. Cross-sections of Tx and Rx optical I/O cores are shown in Fig. 8.18. A Si photonic chip

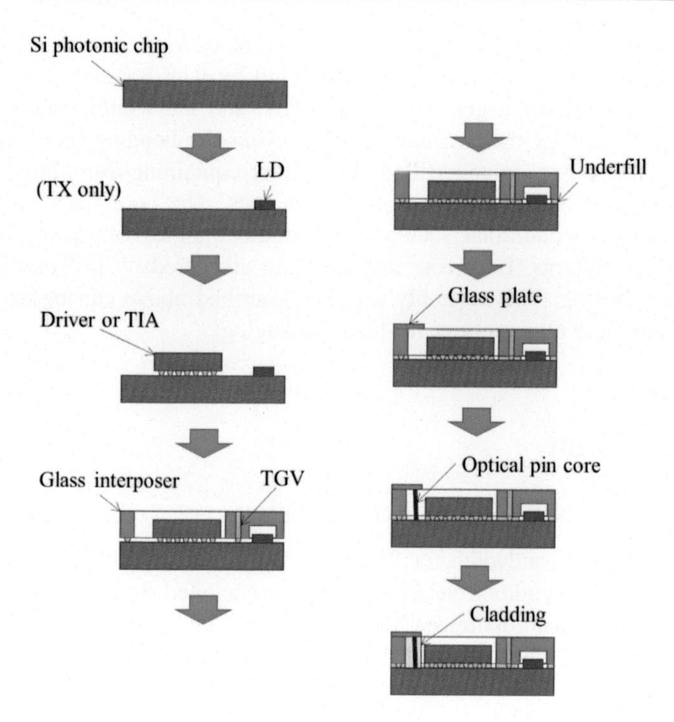

Figure 8.19 Assembly process for the optical I/O cores.

serves as a base for the optical I/O core, and the other components are assembled on the chip. The optical I/O cores comprise a Si photonic chip, driver or TIA, glass interposer, optical pins, and chip capacitors. A Fabry–Perot laser diode (LD) is also mounted in the TX as a light source. The glass interposer has through-glass-vias (TGVs) and Au-coated pads covering either ends of the TGVs, so that the interposer makes connections between the electrical wirings on the Si photonic chip and an outer electrical circuit. The interposer is also designed to surround and mechanically protect the components on the Si photonics chip. The optical I/O structure contains an array of optical pins and a 50 μm-thick glass plate. The surface of the glass plate is the physical interface for the outer optical fibers or waveguides. The optical pins enable the GCs or the PDs to be arranged in the immediate vicinity of the TIA. This layout contributes to reducing signal loss, as well as to downsizing the optical I/O cores. Both optical and electrical I/O interfaces are placed on one side of the optical I/O core. This configuration allows both optical and electrical interconnections to a polymer waveguide-embedded PCB to be made by a single flip-chip bonding process. The overall assembly process for the optical I/O core is illustrated in Fig. 8.19 [18]. An LD is mounted with AuSn solder bumps by flip-chip bonding. To precisely align with Si optical waveguides in the Si photonic chip, horizontal positioning is done by visual alignment, and vertical alignment utilizes Si pedestals [19]. The electrodes of the LD are electrically connected to wiring in the Si photonic chip by wire bonding, and the LD is covered with a transparent gel. The driver, TIA, and glass interposer are bonded with Au stud bumps by thermosonic bonding. After the bonding processes, the IC and the glass interposer are underfilled with a filler-containing resin. Finally, optical pins are fabricated by photolithographic techniques. The core shape can be controlled by exposure conditions, such as dose, incident angle, collimation half angle, and photomask patterns. Both core and cladding are cured by UV irradiation and post-exposure baking. The assembly process described above can be expanded to batch assembly, and ultimately wafer-level packaging.

8.4 Evaluation

8.4.1 Optical coupling (encircled flux)

To secure the MMF bandwidth for 850 nm VCSELs in the 10 Gb/second Ethernet standard, the launch condition determined by the encircled flux (EF) must be satisfied. With a 1.3 μm band optimized MMF, EF within 4.5 μm must be less than 30%, and EF within 19 μm must be more than 86%. Measured EF of the TX optical I/O core is shown in Fig. 8.20. EF was measured under the condition that the GI50 MMF is directly connected with the optical pin. The EFs within 4.5 and 19 μm met the required launch condition. EF tolerance regarding misalignment between the optical pin and MMF was also measured. Dependences of EF within 4.5 and 19 μm for displacements in the (a) *x*- and (b) *y*-directions, respectively, are plotted in Fig. 8.21. EF within 4.5 μm is always less than 30%. Only EF within 19 μm shows a restriction regarding the displacement. The tolerance in the *x*-direction was 44 μm, and that in the *y*-direction was 24 μm. However, these tolerances were wider than the tolerance for the coupling loss. Accordingly, it can be concluded that the misalignment between the MMF and the optical pin was limited by its coupling loss.

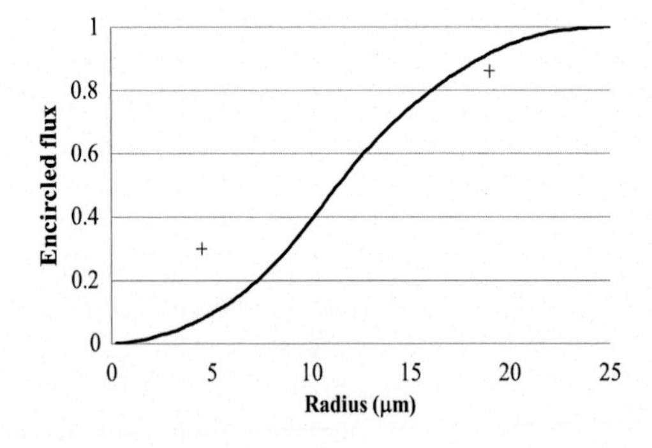

Figure 8.20 Measured EF of TX optical I/O core.

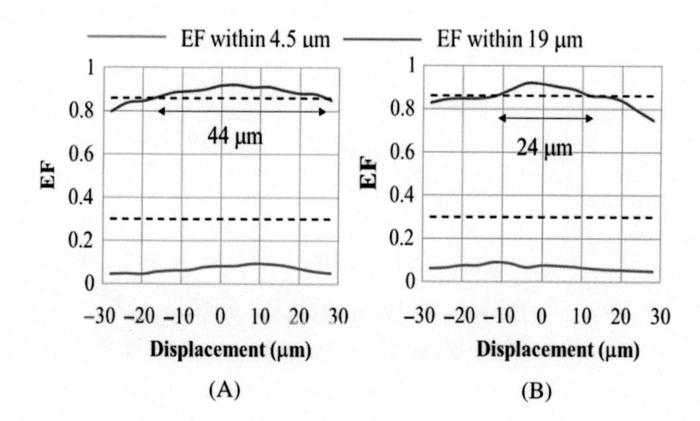

Figure 8.21 EF tolerance of the TX optical I/O core for displacement between MMF and optical pin.

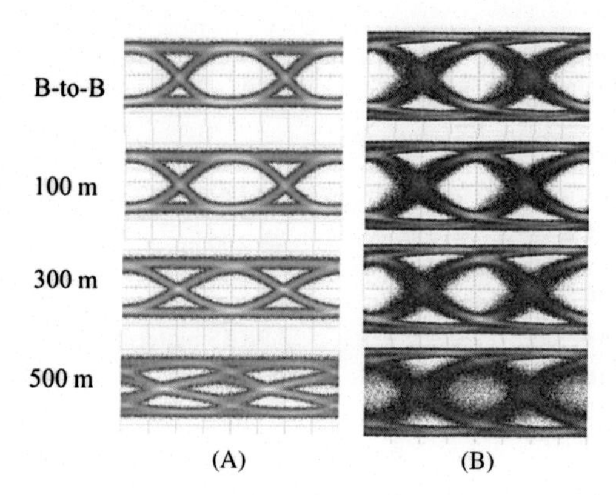

Figure 8.22 Eye pattern: (A) TX optical eye patterns, and (B) RX electrical eye patterns.

8.4.2 Transmission characteristics

To test the performance of the optical I/O cores, the 25 Gbps transmission characteristics with Corning Clearcurve LX multi-mode fiber (MMF) were evaluated. TX and RX optical IO cores of type-B were mounted on printed circuit boards for evaluation. Three MMFs, 100, 300, and 500 m in length, were used for the test. The transmission rate was 25.78125 Gbps. TX optical eye patterns and RX electrical eye patterns after MMF transmission are shown in Fig. 8.22A and B, respectively. A PRBS (pseudo random binary sequence) pattern of $2^{31}-1$ was applied. Clear eye openings were obtained up to 300 m. The TX eye pattern shows large ISI (inter-symbol interference) after 500 m. The extinction ratio of the TX B-to-B (back-to-back) eye pattern was 9.5 dB. The rise time and the fall time at 20–80% were respectively 13.5 ps and 12.3 ps. Dependence of deterministic jitter (DJ) and

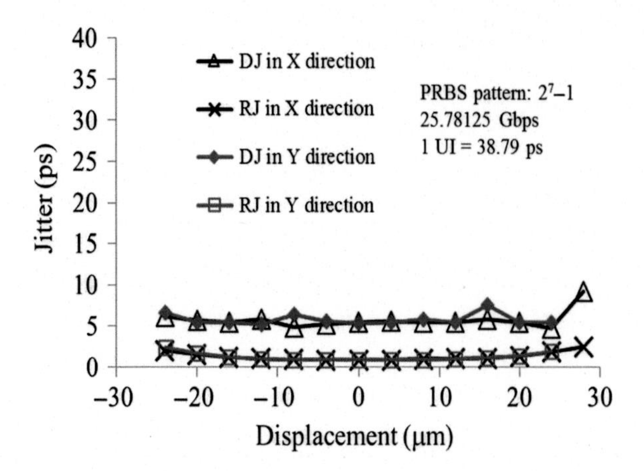

Figure 8.23 Dependence of deterministic jitter (DJ) and random jitter (RJrms) on misalignment between MMF and optical pin.

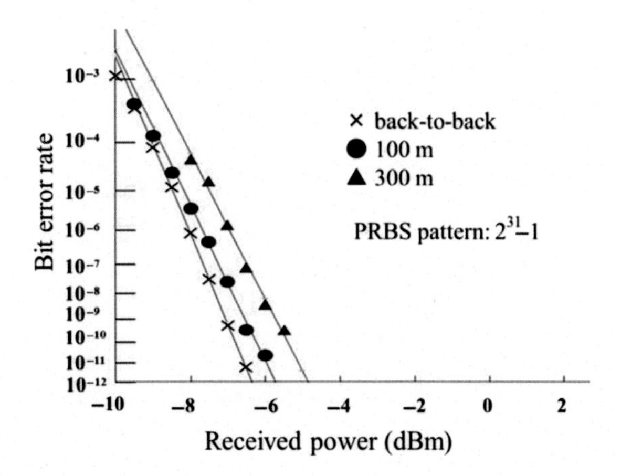

Figure 8.24 Bit error rate characteristics.

random jitter (RJrms) on misalignment between the MMF and the optical pin is plotted in Fig. 8.23. According to the figure, DJ and RJrms are significantly suppressed. This result coincides with the fact that EF had a wide misalignment tolerance. BER characteristics, including B-to-B transmission between the TX and RX, are shown in Fig. 8.24. According to the figure, error-free operation was obtained up to 300 m. The power penalty after 300 m transmission was 1.7 dB. The minimum average received power of B-to-B transmission is −6.3 dBm. These basic performance results indicate that optical I/O cores are applicable to an interconnection covering a distance over 300 m (namely, that needed in a data center).

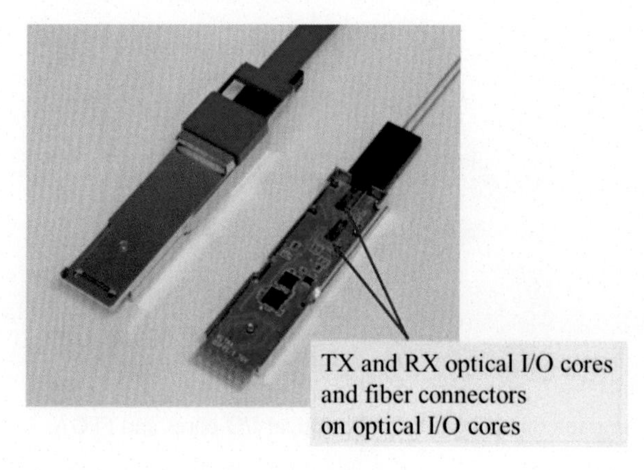

Figure 8.25 Prototype of active optical cable (AOC).

8.5 Application

8.5.1 Transceivers

Development of the optical I/O core will be completed in the near future at the first step. By changing peripheral parts such as the optical/electrical connector or custom design substrates, the optical I/O core can be designed for many applications, not only rack-to-rack, but also optical backplane for board-to-board, built-in systems. A prototype of an active optical cable (AOC) is shown in Fig. 8.25. To achieve a low-cost module, fiber connectors should be mounted on the optical I/O cores by passive alignment or visual alignment. To apply such alignment techniques as a multi-mode waveguide, a transmission medium like a GI50 MMF is indispensable, because it has a large alignment tolerance with respect to the optical components.

8.5.2 System LSI with optical I/Os

The optical I/O core can be designed for many applications, not only rack-to-rack, but also optical backplane for board-to-board, and built-in systems, by changing peripheral parts such as the optical/electrical connector or custom design. To achieve a low-cost module, fiber connectors should be mounted on the optical I/O cores by passive alignment or visual alignment. To apply such alignment techniques as a multi-mode waveguide, a transmission medium like a GI50 MMF is indispensable because it has a large alignment tolerance in regard to the optical components. In the second step, multi-optical I/O cores will be built in an organic LSI interposer, and completed with a middle scale integration of roughly 3 Tbps. An image of an MCM with an ASIC or FPGA chip is shown in Fig. 8.26. The optical I/O core is suitable for mounting on an LSI interposer because of its small footprint and high-density I/O. In this approach, a rich store of knowledge such as power/signal integrity, thermal management, and so on will be gathered. Putting

Figure 8.26 Photograph of MCM with multi-optical I/O cores and FPGA.

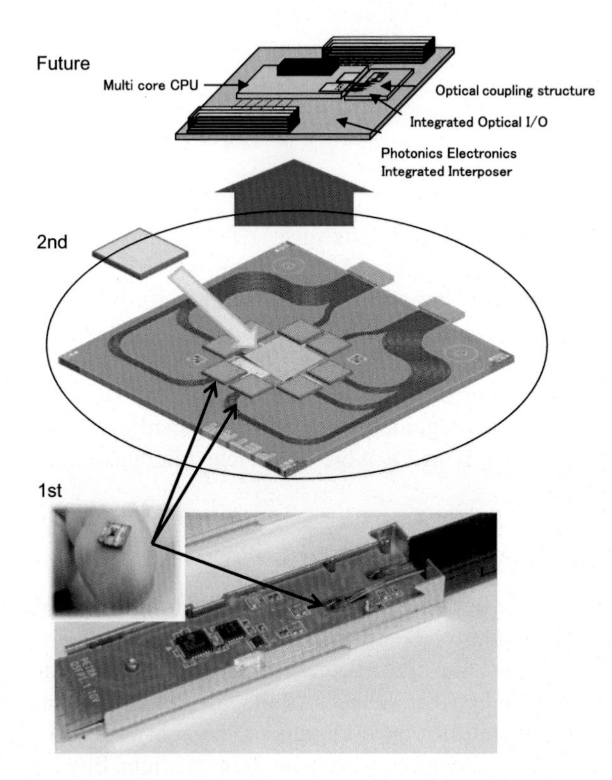

Figure 8.27 Expansion from AOC, MCM to silicon photonics interposer.

this knowledge of large integration to practical use, highly integrated silicon interposer built-in silicon photonics circuits will be designed. In the future, an on-chip server, which consists of CPU and memory mounted on silicon photonics interposer will be realized as shown in Fig. 8.27 [20].

References

[1] Ritter MB, Vlasov Y, Kash JA, Benner A. Optical technologies for data communication in large parallel systems. In: Proc topical workshop on electronics for particle physics; 2010.

[2] De Dobbelaere P, Abdalla S, Gloeckner S, Mack M, Masini G, Mekis A, et al. Si photonics based high-speed optical transceivers. In: ECOC 2012, Paper We.1.E.5; 2012.

[3] Kurata K, Suzuki Y, Kurihara M, Tokushima M, Hagihara Y, Ogura I, et al. Prospect of chip scale silicon photonics transceiver for high density multi-mode wiring system. Opt Commun 2016;362:36–42.

[4] Mogami T, Horikawa T, Kinoshita K, Sasaki H, Morito K, Kurata K. A 300 mm Si photonics platform for multi-applications. In: OECC 2015, paper S3-2, JWe.43; 2015.

[5] Chen X, Bickham SR, Liu H–F, Dosunmu OI, Hurley JE, Li M–J. 25G/s transmission over 820 m of MMF using a multimode launch from an integrated silicon photonics transceiver. Opt Express 2014;22:2070–7.

[6] Yashiki K, Suzuki Y, Hagihara Y, Kurihara M, Tokushima M, Fujikata J, et al. 5 mW/Gbps hybrid-integrated Si-photonics-based optical I/O cores and their 25-Gbps/ch error-free operation with over 300-m MMF. In: Proc. OFC; 2015, Th1G. 1.

[7] Fujikata J, Ushida J, Ming-Bin Y, Yang ZS, Liang D, Guo-Qiang PL, et al. 25 GHz operation of silicon optical modulator with projection MOS structure. In: Proc. OFC; 2010, OMI3.

[8] Shimizu T, Okano M, Hatori N, Ishizaka M, Urino Y, Yamamoto T, et al. Multichannel and high-density hybrid integrated light source with a laser diode array on a silicon optical waveguide platform for interchip optical interconnection. Photon Res 2014;2: A19–24.

[9] Narducci LM, Abraham NB. Laser physics and laser instabilities. Singapore: World Scientific; 1988. p. 285.

[10] Tokushima M, Kamei A, Horikawa T. Dual-tapered 10-μm-spot-size converter with double core for coupling polarization-independent silicon rib waveguides to single-mode optical fibers. Appl Phys Express 2012;5(No. 2):022202.

[11] Tokushima M, Ushida J, Uemura T, Kazuhiko K. Shallow-grating coupler with optimized anti-reflection coating for high-efficiency optical output into multimode fiber. Appl Phys Express 2015;8(9):092501.

[12] Tokushima M, Ushida J, Uemura T, Kazuhiko K. High-efficiency folded shallow-grating coupler with minimal back reflection toward isolator-free optical integration, In: Proc. ECOC; 2015, P.2.20.

[13] Ushida J, Tokushima M, Uemura T, Kazuhiko K. Highly fabrication-tolerant shallow-grating coupler for robust coupling to multimode "optical pin". In: Proc. of international conference on solid state devices and materials; 2015, PS-7-14.

[14] Yang M, et al. RF and mixed-signal performances of a low cost 28 nm low-power CMOS technology for wireless system-on-chip applications. In: 2011 VLSI technology symposium technical digest, pp. 40–1.

[15] Obata Y, Kanda M, Mikami O. Self-written waveguide on a VICSEL-emitting window using photomask transfer method. IEEE Photon Technol Lett 2006;18: 1308–10.

[16] Uemura T, et al. 125-μm-pitch × 12-channel "optical pin" array as I/O structure for novel miniaturized optical transceiver chips. In: Proc. IEEE electronic components and technology conference (ECTC); 2015. pp. 1305–9.

[17] Yashiki K, et al. 25-Gbps/ch error-free operation over 300-m MMF of low-power-consumption silicon-photonics-based chip-scale optical I/O cores. IEICE Trans Electron; in press.
[18] Takemura K, et al. Chip-scale packaging of hybrid-integrated Si photonic transceiver: Optical I/O core. In: Proc. IEEE CPMT symposium Japan (ICSJ); 2015.
[19] Shimizu T, et al. Over-1000-channel hybrid integrated light source with laser diode arrays on a silicon waveguide platform for ultra-high-bandwidth optical interconnections. In: Proc. IEEE CPMT symposium Japan (ICSJ); 2014. pp. 55–8.
[20] Arakawa Y, Nakamura T, Urino Y, Fujita T. Silicon photonics for next generation system integration platform. IEEE Comun Mag 2013;13:72–7.

Scalable three-dimensional optical interconnects for data centers

R. Morris[1], *A.K. Kodi*[1] *and A. Louri*[2]
[1]Ohio University, Athens, OH, United States, [2]George Washington University, Washington, DC, United States

9.1 Introduction

With the growth in handheld devices coupled with the information explosion, there is a trend towards server-side computing rather than client-side computing, leading to the need for massive computing infrastructure called warehouse-scale computers (WSC) or data centers. These data centers are composed of several components: processors, memory, secondary storage, interconnection networks, cooling, power supply units, and many more. Search-based applications (webpages, images) are one class of workload where massive computing infrastructure is required, and data centers would be a good fit for that class of workload. Many companies, such as Amazon Cloud Computing, Google, Microsoft Online Services, etc., provide such Internet-based services to clients by hosting large data centers.

Large portions of hardware resources and software programs must work in cohesion to achieve the desired performance-per-Watt in these data centers. The energy consumed by these data centers is upwards of MWatts, and therefore, there is an urgent need for energy proportional computing to reduce the power consumption of these massive data centers. Several of these companies that are hosting data centers are evaluating application workload, peak and average utilization, and are developing energy-efficient techniques to reduce the total power consumption. In this chapter, we focus on improving the energy-efficiency of the on-chip communication infrastructure called Network-on-Chips (NoCs) that acts like a glue that connects to the cores, caches, and memory modules. The NoC architecture provides the communication infrastructure for multi-cores, which consists of routers for switching, arbitration and buffering, and links for data transfer from source to destination.

Future projections based on the International Technology Roadmap for Semiconductors (ITRS) indicates that complementary metal oxide semiconductor (CMOS) feature sizes will shrink to sub-nanometer levels within a few years. While the NoCs design paradigm offers modular and scalable performance, increasing core counts lead to an increase in serialization latency and power dissipation as packets are processed at many intermediate routers. Although metallic interconnects can provide the required bandwidth due to the abundance of wires in NoCs, ensuring high-speed

Optical Interconnects for Data Centers. DOI: http://dx.doi.org/10.1016/B978-0-08-100512-5.00009-7

inter-core communication within the allocated power budget in the face of technology scaling (and increased leakage currents) will become a major bottleneck for future multi-core architectures which are the backbone of data centers [1,2].

9.2 Photonic and three-dimensional interconnects

Emerging technologies such as silicon photonics and 3D stacking are under serious consideration for meeting the communication challenges posed by multi-cores. Photonic interconnects provide several advantages such as: (1) bit rates independent of distance, (2) higher bandwidth due to multiplexing of wavelengths, (3) larger bandwidth density by multiplexing wavelengths on the same waveguide/fiber, (4) lower power by dissipating only at the endpoints of the communication channel, and many more [3–7]. The recent surge in photonic components and devices such as silicon-on-insulator (SOI) based micro-ring resonators compatible with complementary – metaloxide semiconductor (CMOS) technology that offers good performance in terms of density, power efficiency, and high bandwidth [8] characteristics are generating considerable interest for NoCs [3–5,9–13].

Most photonic interconnects adopt external laser and on-chip modulators, called micro-ring resonators (MRRs). On application of voltage V_{on}, the refractive index of the MRR is shifted to be in resonance with the incoming wavelength of light, which causes a 0 to appear at the end of the waveguide. Similarly, when no voltage is applied, the MRR is not in resonance and a 1 appears at the output. MRRs are used both at the transmitter (modulators) and receiver (filters) sides, and have become a favorable choice due to a smaller footprint (10 μm), lower power dissipation (0.1 mW), high bandwidth (> 10 Gbps), and low insertion loss (1 dB) [14]. Complementary – metal oxide semiconductor (CMOS) compatible silicon waveguides allow for signal propagation of on-chip light. Waveguides with micron-size cross-sections (5.5 μm) and low-loss (1.3 dB/cm) have been demonstrated [14]. Recent work has shown the possibility of multiplexing 64 wavelengths (wavelength-division multiplexing) within a single waveguide with 60 GHz spacing between wavelengths, although the demonstration was restricted to four wavelengths [4,14]. An optical receiver performs the optical-to-electrical conversion of data, and consists of a photodetector, a transimpedance amplifier (TIA), and a voltage amplifier [15,16]. A recent demonstration showed that a Si-CMOS-amplifier has energy dissipation of about 100 fJ/bit with a data rate of 10 Gbps [15]. The thermal stability of MRRs is one of the major challenges causing a mismatch between the incoming wavelength and MRR resonance. Techniques ranging from thermal tuning (more power), athermal tuning (applicable only at fabrication), tuning free-spectral range with backend circuitry (more power), and current injection (smaller tuning range) have been proposed which offer different power consumption levels [17,18].

Similarly, 3D stacking of multiple layers have shown to be advantageous due to: (1) shorter inter-layer channel, (2) reduced number of hops, and (3) increased bandwidth density. A prevalent way to connect 3D interconnects is to use TSVs (through-silicon vias), or micro-bump or flip-chip bonding. The pitch of these vertical vias is very small ($4 \sim 10\ \mu m$), and delays on the order of 20 ps for a 20-layer stack. Jalali's group at UCLA has fabricated SIMOX (Separation by IMplantation of Oxygen) 3D sculpting to stack optical devices in multiple layers [19]. The Lipson group at Cornell has successfully buried an active optical ring modulator in polycrystalline silicon [20]. Moreover, recent work on using silicon nitride has shown the possibility of designing multi-layer 3D integration of photonic layers. Therefore, 3D interconnects could be combined with photonic technology to design scalable, high bandwidth, and low latency communication fabrics for multi-core systems.

To address the requirements of energy-efficient and high-throughput NoCs, we leverage the advantages of two emerging technologies, photonic interconnects and 3D stacking, with architectural innovations to design high-bandwidth, low-latency, multi-layer, reconfigurable networks, called 3D-NoC. A 3D-NoC consists of 16 decomposed photonic interconnect-based crossbars placed on four optical communication layers, thereby eliminating waveguide crossing and reducing the optical power losses. The proposed architecture divides a single large monolithic crossbar into several smaller and more manageable crossbars which reduces the optical hardware complexity, and provides additional disconnected waveguides which provide opportunities for reconfiguration. As the cost of integrating photonics with electronics will be high, statically designed network topologies will find it challenging to meet the dynamically varying communication demands of applications. Therefore, in order to improve network performance, we propose a reconfiguration algorithm whose purpose is to improve performance (throughput, latency) and bypass channel faults by adapting the available network bandwidth to application demand by multiplexing signals on crossbar channels that are either idle or healthy. This is accomplished by monitoring the traffic load and applying a reconfiguration algorithm that works in the background without disrupting the on-going communication. Our simulation results on 64-cores and 256-cores using synthetic traffic, SPEC CPU2006, Splash-2 [21], and PARSEC [22] benchmarks provide an energy saving of up to 6–36% and outperforming other leading photonic interconnects by more than 10–25% for adversial traffic via reconfiguration.

9.3 Optical architecture: three-dimensional-NoC

The proposed 3D-NoC architecture consists of 256 cores in a 64 tile configuration on a 400 mm^2 3D IC. As shown in Fig. 9.1, 256 cores are mapped on an 8×8 network with a concentration factor of four, called a tile [23]. From Fig. 9.1A, the bottom layer, called the electrical die (adjacent to the heat sink), contains the cores, caches, and the memory controllers. Each core has its own private L1 cache and shared-L2 cache which connects four cores together to create a tile. The left inset shown in Fig. 9.1B illustrates a tile. The grouping of cores allows for a reduction in the cost of Nis, as every core does not require lasers attached, and more

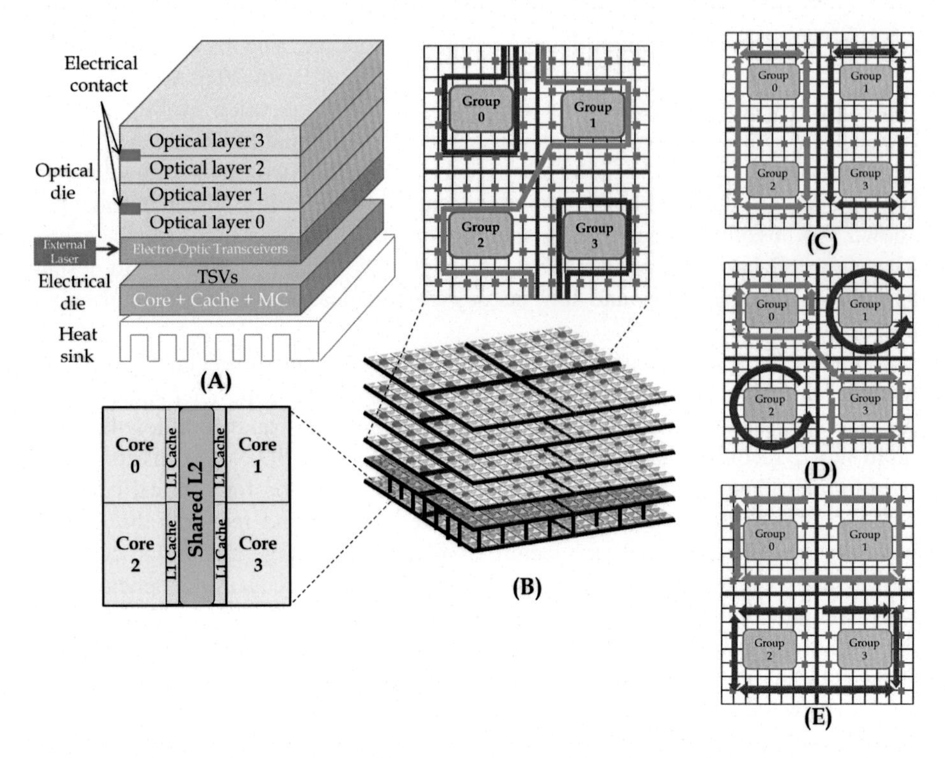

Figure 9.1 Proposed 256-core 3D chip layout. (A) Electrical die consists of the core, caches, the memory controllers, and TSVs to transmit signals between the two dies. The optical die on the lower-most layer contains the electro-optic transceivers and four optical layers. (B) Three-dimensional chip with four decomposed nanophotonic crossbars with the top inset showing the communication among one group (layer 0), and the bottom inset showing the tile with a shared cache and four cores. The decomposition, slicing, and mapping of the three additional optical layers: (C) optical layer 1, (D) optical layer 2, and (E) optical layer 3 [12].
Source: © (2014) IEEE. Reprinted with permission from [12].

importantly, local tile communication is facilitated through a shared-L2 cache. Each tile has a slice of shared-L2 cache along with directory information; memory addresses are interleaved across the shared-L2 cache. For 64-core version, we have 16 memory controllers located within the chip; as we scale to 256 cores, we can increase the number of memory controllers (this has not been modeled, as we assumed synthetic traffic for 256cores). There are two key motivations for designing decomposed NI-based crossbars: first, an optical crossbar is desired to retain a one-hop network; however, a long winding waveguide connecting all the processors increases the signal attenuation (and thereby requires higher laser power to compensate). Second, decomposed crossbars on multiple layers eliminate waveguide crossings; this naturally reduces signal attenuation when compared to 2D networks, where waveguide intersection losses can be a substantial overhead.

9.3.1 Proposed implementation

To utilize the advantage of a vertical implementation of signal routing, we propose the use of separate optical and core/cache systems unified by a single set of connector vias. The upper die, called the optical die, consists of the electro-optic transceivers layer which is driven by the cores via TSVs and four decomposed nanophotonic crossbar layers. To this extent, the electro-optic layer of the optical system contains all the optoelectronic components (modulators, detectors) required for optical routing, as well as the off-chip optical source coupling elements. Layers 0–3 contain optical signal routing elements, composed almost exclusively of MRRs and bus waveguides, with the exception of the electrical contacts required for ring heatings and reconfiguration (these are explained later). The top inset of Fig. 9.1B shows the interconnect for layer 0, whereas Fig. 9.1C–E show layers 1–3. We determine the optimum number of optical layers by analyzing the requirement such that all groups can communicate while preventing waveguide crossing. For example, if Group 0/Group 3 are connected together, then only intra-group communication between Group 1 and Group 2 can take place without waveguide crossing. Therefore, based on this reasoning, we concluded that we will require at least four optical layers. We also provide electrical contact between layers 0/1 and 2/3 to tune ring resonators required for reconfiguration. Vertical coupling of resonators can be very well controlled, as intermediate layer thicknesses can be controlled to tens of nanometers [24].

The core region of the optical layer is composed of ZnO, which is chosen due to its extreme low optical loss in the C-band region, high crystal quality at low deposition temperatures on amorphous substrates [25], high index of refraction ($n \sim 2$ for 1.55 μm), high electro-optical coefficient for efficient modulation [26], and high process selectivity to standard CMOS materials. The fabrication of optical layers 0–3 follows a similar process of PECVD deposition of SiO_2, RF sputtering of ZnO, photolithography and etching to define resonators and waveguides, spin deposition of a planarizing compound (such as SOG (spin on glass), BCB (benzocyclobutene), or Cytops), and an O_2-plasma etchback for planarization. The electro-optic layer will also require the additional steps of e-beam Ge deposition and photolithography for definition of the photodetectors, as well as RF sputtering of indium gallium zinc oxide (IGZO) contacts for both the modulated rings and Ge detectors.

9.3.2 Quantitative comparison: corona, firefly and three-dimensional-NoC

In the proposed 3D layout, we divide tiles into four groups based on their physical location. Each group contains 16 tiles. Unlike the global 64×64 nanophotonic crossbar design in Vantrease et al.[3] and the hierarchical architecture in Pan et al. [7], 3D-NoC consists of 16 decomposed individual nanophotonic crossbars mapped on four optical layers, as shown in Fig. 9.2A–D. Each nanophotonic crossbar is a 16×16 crossbar connecting all tiles from one group to another (12 inter-group and 4 intra-group). It should be noted that the proposed architecture cannot be designed

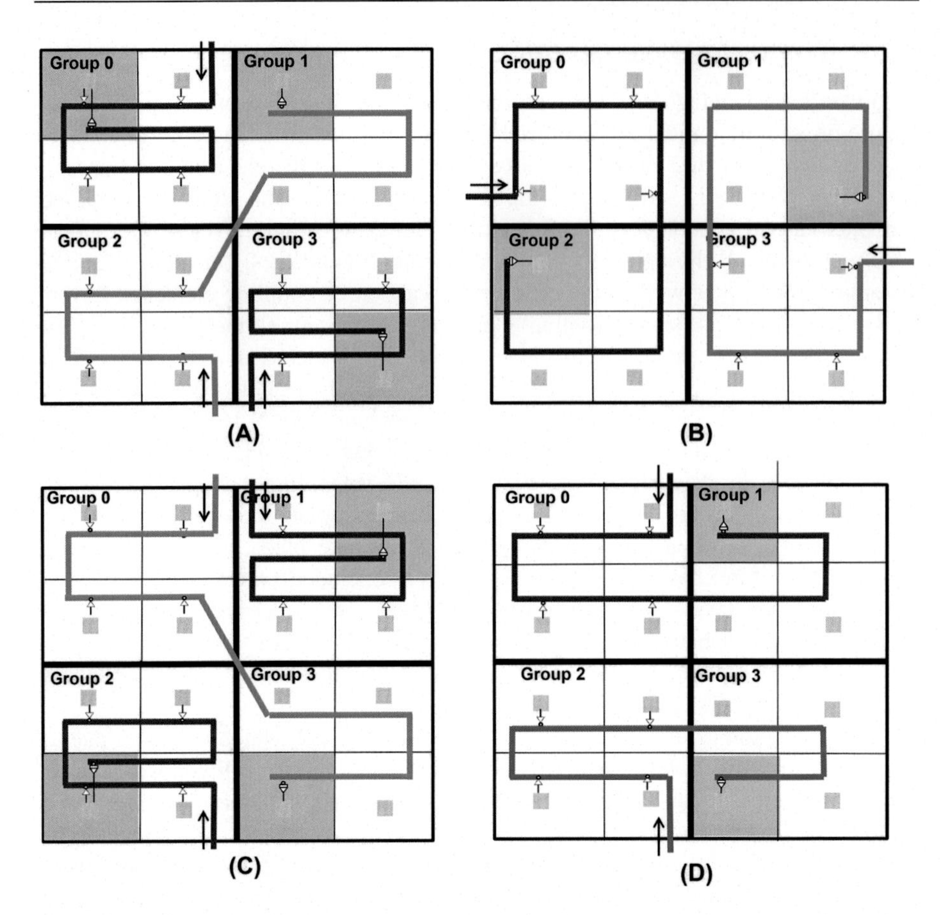

Figure 9.2 The layout of different waveguides for: (A) layer 0 communication, (B) layer 1 communication, (C) layer 2 communication, and (D) layer 3 communication [12].
Source: © (2014) IEEE. Reprinted with permission from [12].

for an arbitrary number of cores; 3D-NoC is restricted to 64 and 256-core versions only. It is composed of many Multiple-WriteSingle-Read (MWSR) nanophotonic channels. A MWSR nanophotonic channel allows multiple nodes the ability to write on the channel, but only one node can read the channel, and therefore requires arbitration. If the arbitration is not fair (early nodes have more priority than later nodes), then latency and starvation could become a problem. On the other hand, a Single-Write-Multiple-Read (SWMR) channel allows only one node the ability to write to the channel, but multiple nodes can read the data, and therefore it requires efficient signal splitters and more power. A reservation-assist sub-network has been proposed in Firefly [7], where MRRs divert light only to those nodes that require the data; thus, SWMR can reduce the power but comes at the price of higher complexity and cost. Therefore, in this work, we adopt MWSR combined with token slot [3] to improve the arbitration efficiency and implement fair-sharing of the communication channels.

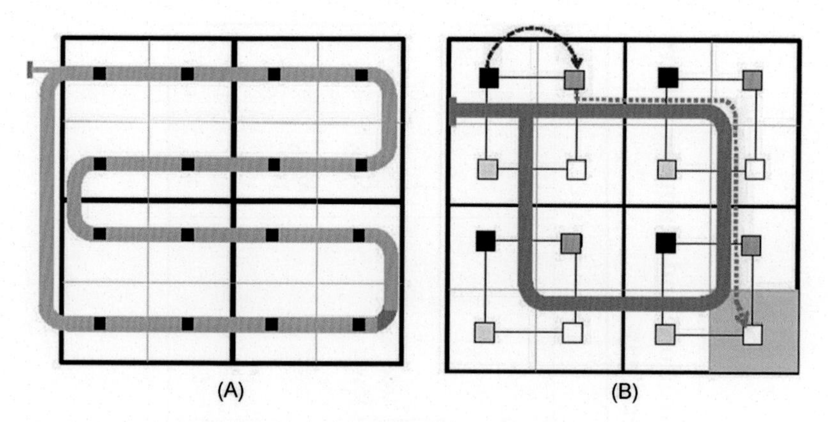

Figure 9.3 The network layout for: (A) Corona, and (B) Firefly [12].
Source: © (2014) IEEE. Reprinted with permission from [12].

To further illustrate the difference between 3D-NoC and other leading nanophotonic networks, Fig. 9.3 illustrates the topologies of Corona and Firefly. In Fig. 9.3A, the optical crossbar topology for a 64-core version of Corona is shown [3]. Each waveguide in Corona traverses around all the tiles, where every tile can write onto a waveguide, but only one tile can read a waveguide. In Fig. 9.3B, the optical topology for a 64-core version of Firefly is shown [7]. Firefly concentrates few local tiles into a group using an electrical mesh network, and then the same numbered tiles in different groups are connected using photonic SWMR interconnects. Table 9.1 shows the optical device requirements for 64- and 256-core versions of Corona, Firefly, and 3D-NoC. As Firefly is an opto-electronic network, the hop count is more than either Corona or 3D-NoC, but it reduces the radix of the network from 8 to 6. Three-dimensional-NoC requires the maximum number of ring resonators due to multiple layers that are designed to prevent waveguide crossings. Firefly requires the most photo-detectors, because each tile can receive data from four other tiles simultaneously. The last rows indicate the performance metric of average hop count and energy from running uniform traffic (more details in Section 9.5). Three-dimensional-NoC and Corona are single hop networks; however, because of the decomposition, 3D-NoC offers lower energy for communication.

9.3.3 Optical communication: intra and inter-group

In the proposed 3D layout, we divide tiles into four groups based on their physical location. Each group contains 16 tiles. Unlike the global 64×64 photonic crossbar design in Vantrease et al.[3], and the hierarchical architecture in Pan et al.[7], 3D-NoC consists of 16 decomposed individual photonic crossbars mapped on four optical layers. Each photonic crossbar is a 16×16 crossbar connecting all tiles from one group to another (Inter-group). It is composed of Multiple-Write-Single-Read (MWSR) photonic channels, which requires less power than the Single-Write-Multiple-Read (SWMR) channels described in Pan et al.[7]. A MWSR photonic channel allows multiple nodes the ability to write on the channel, but only one node can

Table 9.1 Topology comparison for different network sizes, radices, concentration, and diameter where w and k indicate the number of wavelengths and radix of the switch respectively

Component		Corona		Firefly			3D-NoC		
Network size (N)	–	64	256	–	64	256	–	64	256
Network radix (k)	–	4	8	–	4	6	–	4	8
Concentration	–	4	4	–	4	4	1	4	4
Network diameter	1	1	1	(k/2) + 1	3	5		1	1
Wavelengths (w)	–	64	64		64	64	–	64	64
MRRs photo-detectors	$4wk^4$	65K	104K	$4wk^3$	16K	131K	$4wk^4 + 12wk^2$	77K	1097K
	$4wk^2$	4K	16K	$4w(k-1)k^2$	12K	114K	$16wk^2$	16K	65K
Electrical links	–	–	–	k^2	16	64	–	–	–
Bisection	$4wk^2$	4K	16K	$4wk^2$	4K	16K	$16wk^2$	16K	65K
Bandwidth	–	256	256	–	256	256	–	256	256
Bandwidth/channel									
Average hops	–	1	1	–	1.75	3.63	–	1	1
Average energy	–	1.26	1.41		1.28	1.74		1.17	1.25

read the channel. This channel design reduces power, but requires arbitration as multiple nodes can write at the same time. On the other hand, a SWMR channel allows only one node the ability to write to the channel, but multiple nodes can read the data. This channel design reduces latency as no arbitration is required, but requires a source destination handshaking protocol or else the power to broadcast will be higher. We adopt MWSR and Token slot [3] in this architecture to improve the arbitration efficiency for the channel. Each waveguide used within a photonic crossbar has only one receiver which we define as the *home channel*. During communication, the source tile sends packets to their destination tile by modulating the light on the home channel of the destination tile. An off-chip laser generates the required 64 continuous wavelengths, $= \lambda 0, \lambda 1, \lambda 2 \ldots \lambda_{63}$. Fig. 9.1B shows the detailed floor plan for the first optical layer. For optical layer 0, a 32 waveguide bundle is used for communication between Groups 0 and 3, and two 16 waveguide bundles are used for communication within Groups 1 and 2. For inter-group communication between 0 and 3, the first 16 waveguide bundle is routed past Group 0 tiles, so that any tile within Group 0 can transmit data to any destination tile in Group 3. Similarly, the next 16 waveguide bundle is routed past Group 3, so that any tile within Group 3 can communicate with a destination tile located within Group 0. The bidirectional arrows illustrate that light travels in both directions and depends on which group is the source and destination. The remaining two independent waveguide bundles (16 waveguides) are used for intra-group communication for Groups 1 and 2 respectively. Therefore, we require a total of 64 waveguide bundles per layer. A detailed decomposition and slicing of the crossbar on the other three layers is shown in Fig. 9.1C–E.

9.3.4 Router microarchitecture

Fig. 9.4A shows the router microarchitecture in 3D-NoC for tile 0. Any packet generated from the L2 cache is routed to the input demux with the header directed towards RC (routing computation). The two MSBs are used to direct the packet to one of the four sets of input buffers ($IB_0 - IB_3$) corresponding to each optical layer (0–3). For the second set of demuxes, the packet will utilize a unique identifier (that corresponds to the core number) to indicate the source of the packet, to prevent any core from overwhelming the input buffers. Token (Request + Release) ensures that packets are transmitted from the IBs without collision, and the MRRs are used to modulate the signal into the corresponding home channel. At the receiver, the reverse process takes place where the packet from the optical layer is converted into electronics, and according to the unique identifier will find one set of buffers available. Token Control is used to prevent buffer overflow at the home node by checking the number of empty buffer slots. If the number of empty buffer slots falls below a certain threshold, BufT h, then the destination tile will capture the circulating token and will not re-inject the token until the number of free slots increases to the threshold. Furthermore, the receiver of 3D-NoC does not require router computation for an incoming flit of a packet, because flit interleaving does not take place as an optical token is not re-injected until the whole packet is sent. The packets will then compete to obtain the switch (switch allocator (SA)) to reach the L2 cache. It should be noted

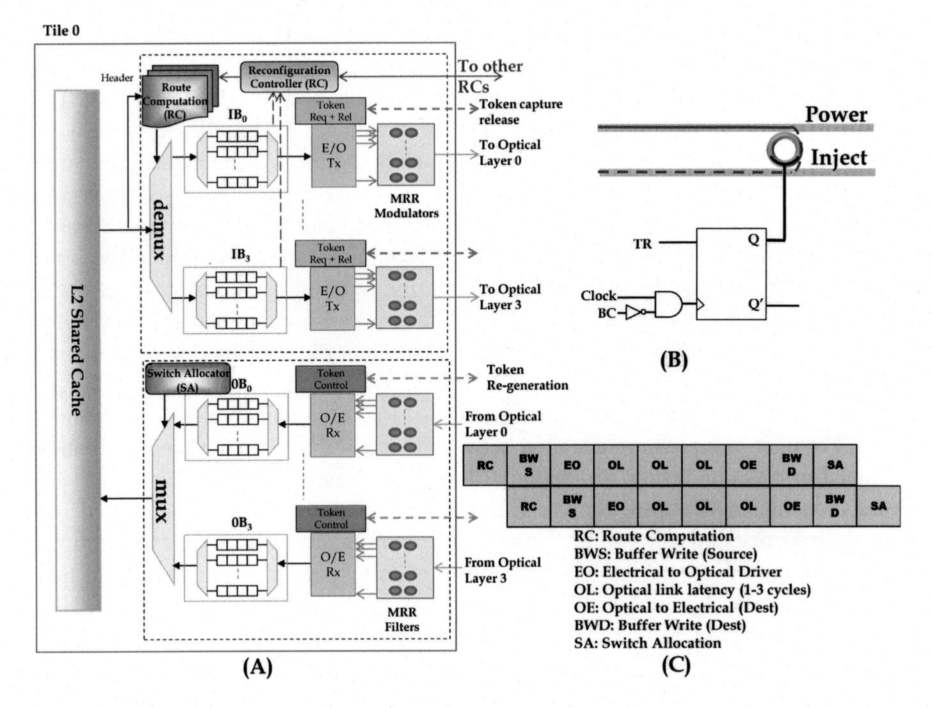

Figure 9.4 (A) Router microarchitecture, (B) token control, and (C) router pipeline [13]. *Source*: © (2012) IEEE. Reprinted with permission from [13].

that the proposed unique identifier is similar to a virtual channel allocator; however, we do not perform any allocation as the decision to enter any buffer is determined on the core number (source or destination). Fig. 9.4B shows the proposed token control block. In the token control block, an optical token is only placed on the token inject waveguide when an optical token is present (high TR signal), and the buffer congestion (BC) signal is low. A low BC signal in this case represents a free buffer slot at the destination tile, and a high BC signal represents that all the buffer slots are full at the destination tile. Fig. 9.4C shows the router pipeline. Routing computation ensures that the packet is directed to the correct output port for both static and reconfigured communication. Busy Write Set (BWS) writes the packet into the buffer slot. The EO conversion takes place with an appropriate buffer chain after the token is received. Optical transmission can take anywhere between 1–3 clock cycles running at 5 Ghz. The OE conversion is repeated at the receiver, BWD writes the packet into the buffer slot, and finally switch allocation (SA) ensures that the packet progresses into the L2 cache.

9.4 Reconfiguration

As future multi-cores will run diverse scientific and commercial applications, networks that can adapt to communication traffic at runtime will maximize the available resources while simultaneously improving the performance. Moreover, faults within the network or the channel can isolate healthy groups of tiles; with the natural redundancy available in the decomposed crossbar, we can take advantage of reconfiguration to overcome channel faults and maintain limited connectivity. To implement reconfiguration, we propose to include additional MRRs that can switch the wavelengths from different layers to create a reconfigurable network. Further, we also propose a reconfiguration algorithm to monitor traffic load and dynamically adjust the bandwidth by reallocating excess bandwidth from under-utilized links to over-utilized links.

9.4.1 Bandwidth reallocation

To illustrate with an example, consider a situation where tiles in Group 0 communicate only with tiles in Group 3. Fig. 9.5 shows the reconfiguration mechanism. The

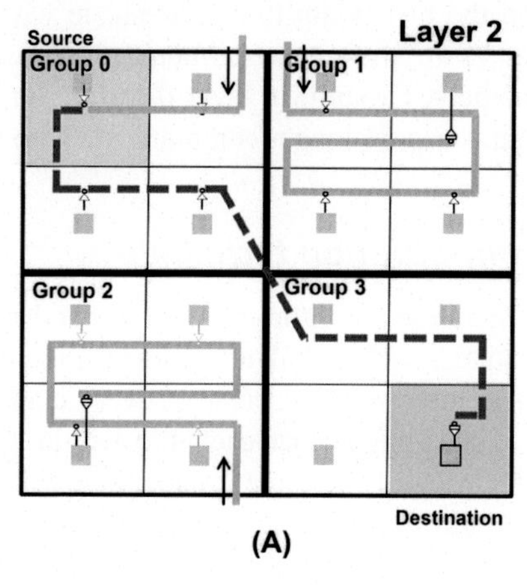

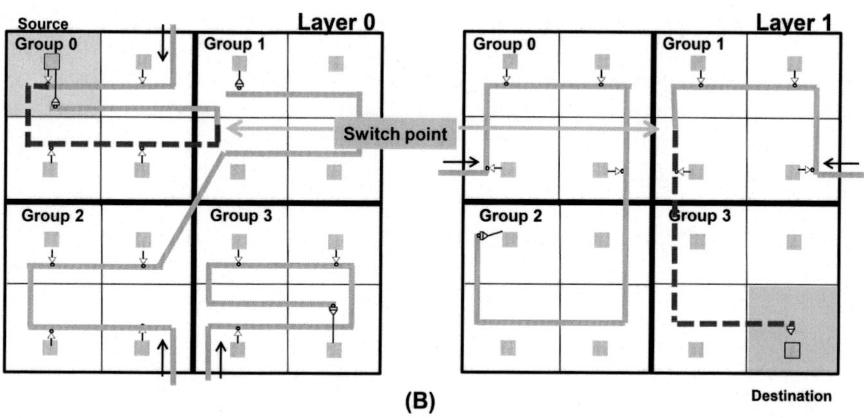

Figure 9.5 (A) Static communication between the source in Group 0 and destination in Group 3. (B) Illustration of reconfiguration between Groups 0 and 3 using partial waveguides from layers 0 and 1 [13].
Source: © (2012) IEEE. Reprinted with permission from [13].

static allocation of channels for communication are in layer 2, as shown in Fig. 9.5A. Suppose no tile within Group 1 (in layer 1) communicates with Group 3, then we can reallocate the bandwidth from Group 1 to Group 0 to communicate with Group 3. To implement reconfiguration, however, we need to satisfy two important requirements: (1) there should be a source waveguide which should be freely available to start the communication on a source layer, and (2) there should be a destination waveguide which also should be freely available to receive the extra packets. As shown in Fig. 9.5B, as the two Groups 0 and 3 talk to each other, we have the first set of waveguides on layer 0 (generally used to communicate within the group) available, therefore this satisfies the first condition. As Group 1 does not communicate with Group 3, we can utilize the destination waveguide available in layer 1, and this satisfies the second condition. The signal originates on layer 0, and switches to layer 1 to reach the destination. Note that this additional channel is available in addition to the layer 2 static configuration, thereby doubling the bandwidth. Therefore, during reconfiguration Group 0 has doubled the bandwidth to communicate with Group 3 via layers 2 (static) and 1 (dynamic). Two different communications are disrupted when the reconfiguration occurs, namely, Group 0 in layer 0 can no longer communicate with itself, and Group 1 in layer 1 can no longer communicate with Group 3.

9.4.2 Dynamic reconfiguration technique

In R-3D-NoC, the reconfiguration algorithm reallocates bandwidth based on historical information. Historical statistics such as link utilization (Linkutil) and buffer utilization (Bufferutil) are collected at the optical receiver of every communication channel by hardware counters [27]. This implies that each tile within a group will have four hardware counters (one for each of the three groups) that will monitor traffic utilization. Both link and buffer utilization are used, as link utilization provides accurate information at low-medium network loads and buffer utilization provides accurate information regarding high network loads [27]. All these statistics are measured over a sampling time window called *Reconfiguration window* or phase, $R_W{}^t$, where t represents the reconfiguration time number, t. This sampling window impacts performance, as reconfiguring finely incurs a latency penalty, and reconfiguring coarsely may not adapt in time for traffic fluctuations. For calculation of $\text{Link}_{\text{util}}$ at configuration window t, we use the following equation:

$$\text{Link}_{\text{util}}^{t} = \frac{\sum_{\text{cycle}=1}^{R_W} \text{Activity(cycle)}}{R_W} \tag{9.1}$$

where Activity(cycle) is 1 if a flit is transmitted on the link, or 0 if no flit is transmitted on the link for a given cycle. For calculation of $\text{Buffer}_{\text{util}}$ at configuration window t, we use the following equation:

$$\text{Buffer}_{\text{util}}^{t} = \frac{\sum_{\text{cycle}=1}^{R_W} \text{Occupy(cycle)}/\text{Total}_{\text{buffers}}}{R_W} \tag{9.2}$$

where Occupy(cycle) is the number of buffers occupied at each cycle, and $Total_{buffers}$ is the total number of buffers available for the given link. When traffic fluctuates dynamically due to short term bursty behavior, the buffers could fill up instantly. This can adversely impact the reconfiguration algorithm as it tries to reallocate the bandwidth faster, leading to fluctuating bandwidth allocation. To prevent temporal and spatial traffic fluctuations affecting performance, we take a weighted average of current network statistics ($Link_{util}$ and $Buffer_{util}$). We calculate the $Buffer_{util}$ as follows:

$$Buffer_w^t = \frac{\sum Buffer_{util}^t \times weight + Buffer_{util}^{t-1}}{weight + 1} \tag{9.3}$$

where weight is a weighting factor, and we set this to three in our simulations [28].

After each $R_W{}^t$, each tile will gather its link statistics ($Link_{util}$ and $Buffer_{util}$) from the previous window $R_W{}^{t-1}$, and send to its local reconfiguration controller (RC) for analysis. We assume that Tile 0 of every group gathers the statistics from the remaining tiles, and this can be a few bytes of information that is periodically transmitted. Next, when each RC_i,($\forall$ $i = 0, 1, 2, 3$), has finished gathering link and buffer statistics from all its hardware controllers, each RC_i will evaluate the available bandwidth for each link depending on the $Link_{util}{}^{t-1}$ and $Buffer_{util}{}^{t-1}$ and will classify its available bandwidth into a select range of thresholds β^{1-4} corresponding to 0%, 25%, 50%, and 90%. We never allocate 100% of the link bandwidth as the source tile may have new packets to transmit to the destination tile before the next R_W. RC_i will send link information (availability) to its neighbor RC_j (j i).If RC_j needs the available bandwidth, RC_j will notify the source and the destination RCs, so that they can switch the MRRs and inform the tiles locally of the availability. Once the source/destination RCs have switched their reconfiguration MRRs, RC_i will notify RC_j that the bandwidth is available for use. On the other hand, if a node within RC_i that throttled its bandwidth requires it back due to an increase in network demand, RC_i will notify that it requires the bandwidth back, and afterwards will deactivate the corresponding MRRs. The above reconfiguration completes a three-way handshake where RC_i first notifies RC_j, then RC_j notifies RC_i that RC_j will use the addition bandwidth, and finally RC_i notifies RC_j that the bandwidth can be used. Table 9.1 shows a psuedo-reconfiguration algorithm implemented in R-3D-NoC. We assume $Link_{util} = 0.0$ to indicate if the link is not being used, $L_{min} = 0.10$ to indicate if the link is under-utilized, $L_{min} = 0.25$ and $B_{con} = 0.25$ to indicate if the link is normal-utilized, and $B_{con} = 0.5$ to indicate that the link is over-utilized [27].

9.4.3 Power reduction technique

To further reduce optical power, we propose a dynamic power reduction technique (PRT) which can regulate the optical laser power when the links are idle. PRT is implemented by augmenting the reconfiguration algorithm by collecting link statistics which have a zero link utilization, and are not required for bandwidth reallocation. From Table 9.1, Step 4 identifies the photonic interconnects which have a zero utilization, indicating idle channels. Step 8 of the reconfiguration algorithm notifies PRT which photonic interconnects are idle, and can be used to deactivate. After PRT collects information about which photonic interconnects are inactive, it will deactivate the unused interconnects, and will also throttle the laser power to the new minimum required power to ensure that the signal can be safely detected. We assume that Vertical Cavity Surface Emitting Laser (VCSEL) arrays are used to generate different wavelengths, and the external laser power can be throttled by reducing the modulation current applied to the VCSEL [27]. VCSELs are chosen due to their low power and direct modulation of the optical signal. The power dissipated by a VCSEL is given as [29]:

$$P_{TX} = V_{DD}(I_v + \gamma I_m) \tag{9.4}$$

where V_{dd} is the supply voltage, I_v is the threshold current, γ is the VCSEL activity factor and I_m is modulation current. By reducing I_m, the total VCSEL power will be reduced to the minimum power needed. Furthermore, if the VCSEL is not required both I_v and I_m can be reduced to zero, which in turn will deactivate the VCSEL. However, completely switching off a VCSEL can also affect the switch-on latency. In our simulation, we account for the switch-on latency when the VCSEL is fully deactivated.

9.5 Performance evaluation

9.5.1 Simulation set up

We first describe the simulation set up of the proposed architecture. Our simulator models in detail the router pipeline, arbitration, switching, and flow control. An aggressive single cycle electrical router is applied in each tile, and the flit transversal time is one cycle from the local core to electrical router [30]. As the delay of Optical/Electrical (O/E) and Electrical/Optical (E/O) conversion can be reduced to less than 100 ps, the total optical transmissions latency is determined by the physical location of the source/destination pair (1–5 cycles), and two additional clock cycles for the conversion delay. In addition, a latency of 1–3 cycles was assumed for a tile to capture an optical token. We assume a input buffer of 16 flits, with each flit consisting of 128 bits. The packet size is 4 flits, which will be sufficient to fit a complete cache line of 64 bytes. We assume a supply voltage V_{dd} of 1.0 V, and a router clock frequency of 5 Ghz [3,7].

We compare 3D-NoC architecture to two other crossbar-like photonic interconnects, Corona [3] and Firefly [7], and two electrical interconnects (mesh and Flattened Butterfly) [31]. We implement all architectures such that four cores (one tile) are connected to a single router. We assume a token slot for both 3D-NoC and Corona to pipeline the arbitration process to increase the efficiency. Multiple requests can be sent from the four local cores to optical channels to increase the arbitration efficiency. We use a Fly Src routing algorithm [7] for Firefly architectures, where intragroup communication via electrical mesh is implemented first, and then inter-group via photonic interconnects. For a fair comparison, we ensure that each communication channel in either the electrical or optical network is 640 Gbps with 64 wavelengths. For closed-loop measurements, we collect traces from real applications using the full execution-driven simulator SIMICS from WindRiver, with the memory package GEMS enabled [32]. We evaluate the performance of 64-core versions of the networks on Splash-2 [21], PARSEC [22], and SPEC CPU2006 workloads, and of the 256-core version on synthetic and workload completion traffic (a mixture of synthetic traces). We assume a 2-cycle latency to access the L1 cache (64 KB, 4-way), a 4-cycle latency to access the L2 cache (4MB, 16-way), a cache line size of 64 bytes and a 160-cycle latency to access the main memory. In addition, there are 16 memory controllers used to access the main memory, and each processor can issue two threads.

9.5.1.1 Splash-2: 64 cores

Fig. 9.6A shows the speed-up for the Splash-2 applications [21]. Three-dimensional-NoC has an average speed-up of about 2.5 for each benchmark over the mesh network. In the water application, 3DNoC has the highest speed-up with a factor of over 3 relative to mesh. This is a result of 3D-NoCs decomposed crossbars allowing for fast arbitration of network resources (less contention), and the reduced hop count relative to the mesh network. In Raytrace and FMM benchmarks, 3D-NoC has the lowest speed-up factor of 2.2, which is contributed to the higher local (few hops) traffic. Nearest-neighbor traffic creates more contention for optical tokens with locally concentrated destinations in 3D-NoC. When 3D-NoC is compared to flattened-butterfly, 3D-NoC has a 25–30% improvement. As flattened-butterfly is a two-hop network, and most traffic under Splash-2 suite is two hops, the intermediate router reduces the throughput of the network. When 3D-NoC is compared to Firefly, 3D-NoC outperforms Firefly by about as much as 38%, which is a result of Firefly routing its traffic through both an electrical and optical network. With R-3D-NoC there is about a 5–12% improvement improvement over 3D-NoC for the select range of Splash-2 traffic traces. For FFT and LU applications, R-3D-NoC has the highest performance improvement over 3D-NoC at about 12%. Both FFT and LU have communication patterns that can take advantage of the reconfiguration algorithm as their communication patterns do not quickly change over time, forcing the network to keep reconfiguring and improving performance. In the other applications (radiosity, raytrace, radix, ocean, rmm, and water), R-3D-NoC has about a 5% increase in performance over 3D-NoC. This is a direct result of Splash-2 traffic traces resembling uniform traffic, which reduces the bandwidth available for reconfiguration.

9.5.1.2 PARSEC and SPEC CPU2006: 64 cores

Fig. 9.6B shows the speed-up for 64 wavelengths for PARSEC and SPEC CPU2006 applications. An average of 2X speed-up is shown by 3D-NoC compared

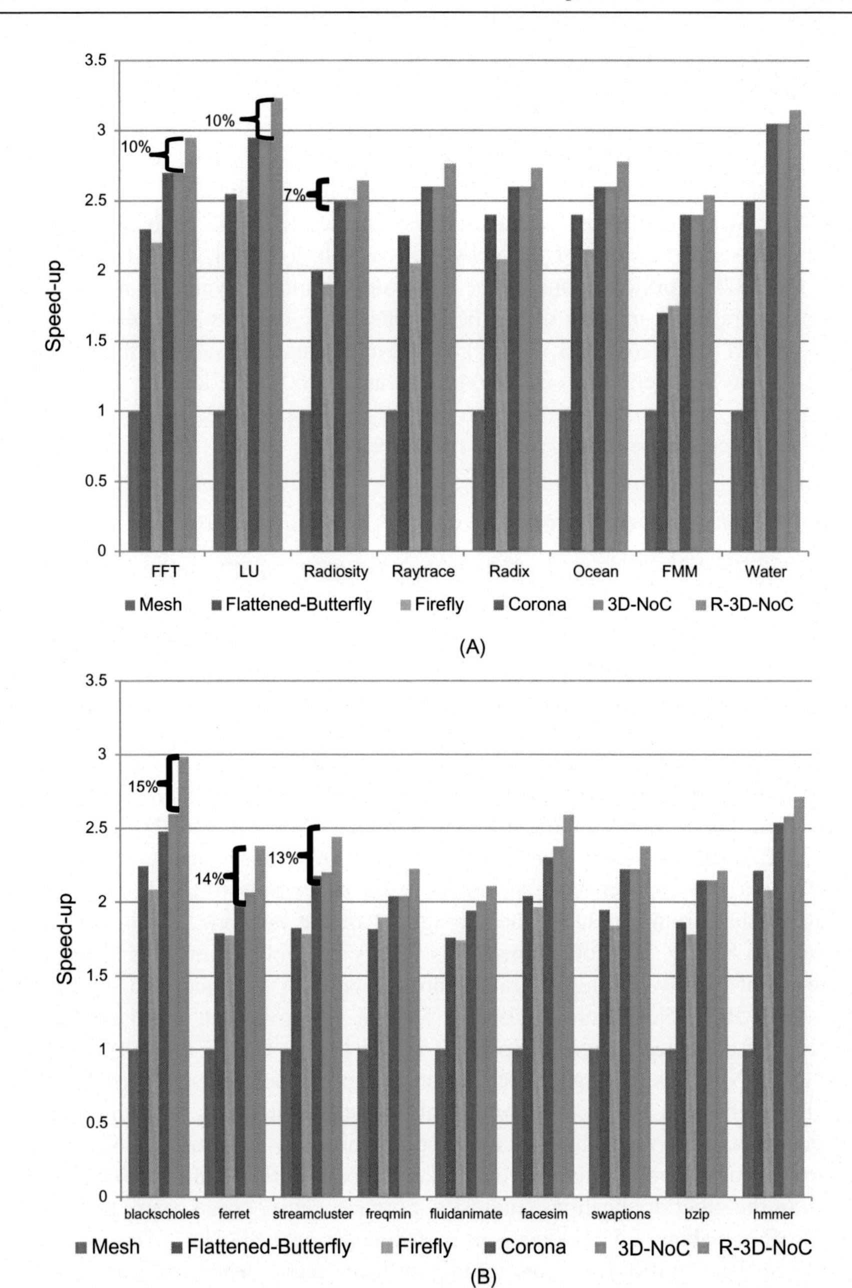

Figure 9.6 Simulation speed-up for 64-core using: (A) SPLASH-2 traffic traces, and (B) PARSEC traffic traces [12].
Source: © (2014) IEEE. Reprinted with permission from [12].

to mesh, and a $10-40\%$ improvement over Flattened-Butterfly and Firefly architectures. When Corona and 3D-NoC are compared to each other, 3D-NoC is able to outperform Corona for most applications except swaptions and bzip. The reason for the improved performance over Corona is primarily due to the communication pattern which makes use of all the four decomposed crossbars simultaneously, thereby sending more data on the network when compared to Corona. For swaption and bzip application traffic, their communication patterns do not take advantage of 3D-NoC decomposed crossbars, and as such there is no significant improvement. Compared to Splash-2 traffic, R-3D-NoC shows a better improvement in performance for PARSEC/SPEC CPU2006 benchmarks. Blackscholes has the largest jump in performance, by almost 20% when compared to Corona and 15% when compared to 3D-NoC. This large increase in performance is attributed to the nature of PARSEC applications which are more communication intensive and therefore, the reconfiguration algorithm maximizes the performance. PARSEC applications emphasize emerging workloads and future shared-memory applications for the study of CMPs, rather than the network.

9.5.1.3 Synthetic traffic: 256 cores

The throughput for all synthetic traffic traces for 256-core implementations are shown in Fig. 9.7, and normalized to mesh network (for Uniform, the mesh has a throughput of 624 GBytes/second). Three-dimensional-NoC has about a $2.5\times$ increase in

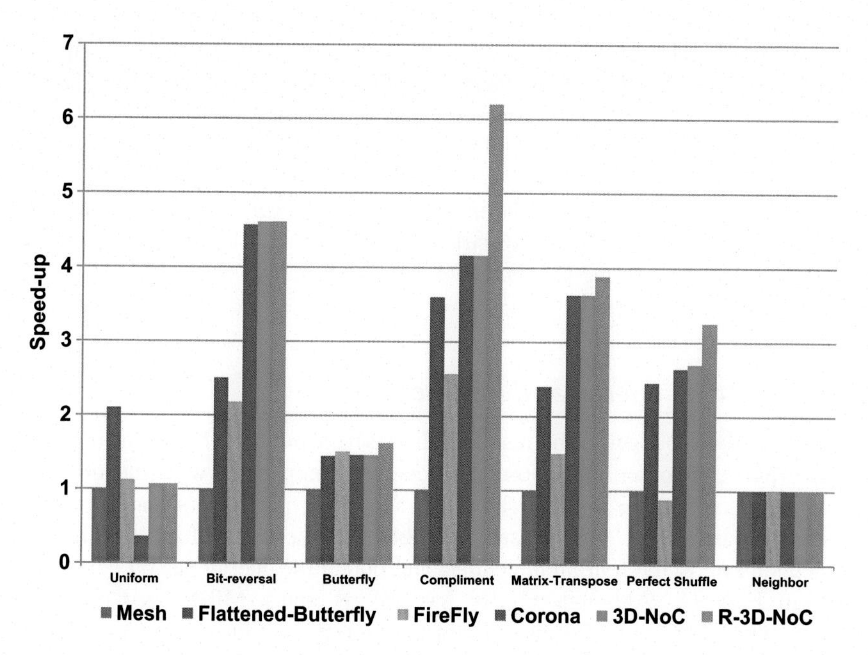

Figure 9.7 Simulation results showing normalized saturation throughput for seven traffic patterns for 256 cores [12].
Source: © (2014) IEEE. Reprinted with permission from [12].

throughput over Corona for uniform traffic, due to the decomposition of the photonic crossbar. The decomposed crossbars allow for a reduction in contention for optical tokens, as now a single token is shared between 16 tiles instead of 64 tiles as in Corona. Firefly slightly outperforms 3D-NoC for uniform traffic, due to the contention found in the decomposed photonic crossbars. Moveover, Firefly uses a SWMR approach for communication which does not require optical arbitration. From the figure, 3D-NoC slightly outperforms Corona for bit-reversal and complement traffic traces. This is due to lower contention for optical tokens in the decomposed crossbars. Three-dimensional-NoC significantly outperforms mesh for the bit-reversal, matrix-transpose, and complement traffic patterns. In these traffic patterns, packets need to traverse across multiple mesh routers, which in turn increases the packet latency and thereby reduces the throughput. The R-3D-NoC is able to outperform 3D-NoC for complement, matrix-transpose, and perfect shuffle traffic traces. These permutation traffic traces exhibit adversial patterns which will benefit R-3D-NoC. In complement traffic, R-3D-NoC has about a 55% increase in performance when compared to 3D-NoC. Complement traffic patterns showcase the best performance, as a single source tile will communicate with a single destination tile, thereby providing opportunities to improve performance via reconfiguration.

9.5.2 Energy comparison

The energy consumption of a photonic interconnect can be divided into two parts, electrical energy and optical energy. Optical energy consists of the off-chip laser energy and on-chip MRRs heating energy. In what follows, we first discuss the electrical energy and then the optical energy consumption. We use ORION 2.0 [33] to obtain the energy dissipation values for an electrical link and router, and modified their parameters for 22 nm technology according to ITRS. We assume all electrical links are optimized for delay and the injection rate to be 0.1. For each optical transmitted bit, we need to provide an electrical back-end circuit for transmitter-end and receiver-end. We assume the O/E and E/O converter energy is 100 fJ/b, as predicted in Krishnamoorthy et al. [34].

9.5.2.1 Optical energy and loss model

The optical power budget is the result of the laser power and the power dissipated for the MRRs. In order to perform an accurate comparison with the other two optical architectures, we use the same optical device parameters and loss values provided in Batten et al. [4], as listed in Table 9.2. We test uniform traffic with a 0.1 injection rate to the four architectures, and obtain the energy per-bit comparison shown in Fig. 9.8. Although Firefly has 1/4 as many MRRs as Corona and 3D-NoC, which results in 1/4 energy consumption per bit on ring heatings, it still consumes more energy per bit than 3D-NoC and CORONA, due to the energy consumption overhead of routers and electrical links. In general, 3D-NoC saves

Table 9.2 Pseudo algorithm used by R-3D-NoC for bandwidth and power regulation

Step 1:	Wait for Reconfiguration window, $R_w{}^t$
Step 2:	RC_i sends a request packet to all local tiles requesting $Link_{Util}$ and $Buffer_{Util}$ for previous $R_W{}^{t-1}$
Step 3:	Each hardware counter sends $Link_{Util}$ and $Buffer_{Util}$ statistics from the previous $R_w{}^{t-1}$ to RC_i
Step 4:	RC_i classifies the link statistic for each hardware counter as: If $Link_{util} = 0.0$ Not-Utilized: Use β_4 & mark this link as inactive (PRT) If $Link_{util} \le L_{min}$ Under-Utilized: Use β_3 If $Link_{util} \ge L_{min}$ and $Buffer_{util} < B_{con}$ Normal-Utilized: Use β_2 If $Buffer_{util} > B_{con}$ Over-Utilized: Use β_1
Step 5:	Each RC_i sends bandwidth available information to RC_j, $(i \neq j)$
Step 6:	If RC_j can use any of the free links then notify RC_i of their use, else RC_j will forward to next RC_j
Step 7a:	RC_i receives response back from RC_j and activates corresponding microrings
Step 7b:	RC_j notifies the tiles of additional bandwidth and RC_i notifies RC_j that the additional bandwidth is now available
Step 8:	Notify PRT of the links not being used (this step is only used by PRT)
Step 9:	Goto Step 1

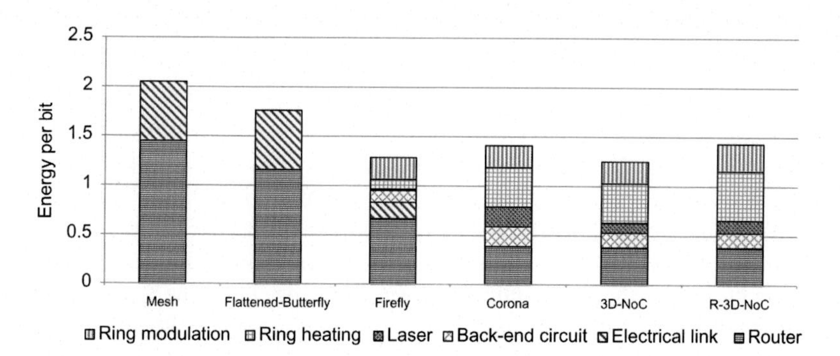

Figure 9.8 Average energy per-bit for electrical and photonic interconnects [13]. *Source*: © (2014) IEEE. Reprinted with permission from [12].

6.5%, 23.1%, and 36.1% energy per bit compared to Corona, Firefly, and Mesh respectively. It should be noted that when the network injection rate increases, 3D-NoC becomes much more energy efficient than other three architectures. There is a slight increase in power dissipation for R-3D-NoC over 3D-NoC, due to the additional MRRs required for reconfiguration.

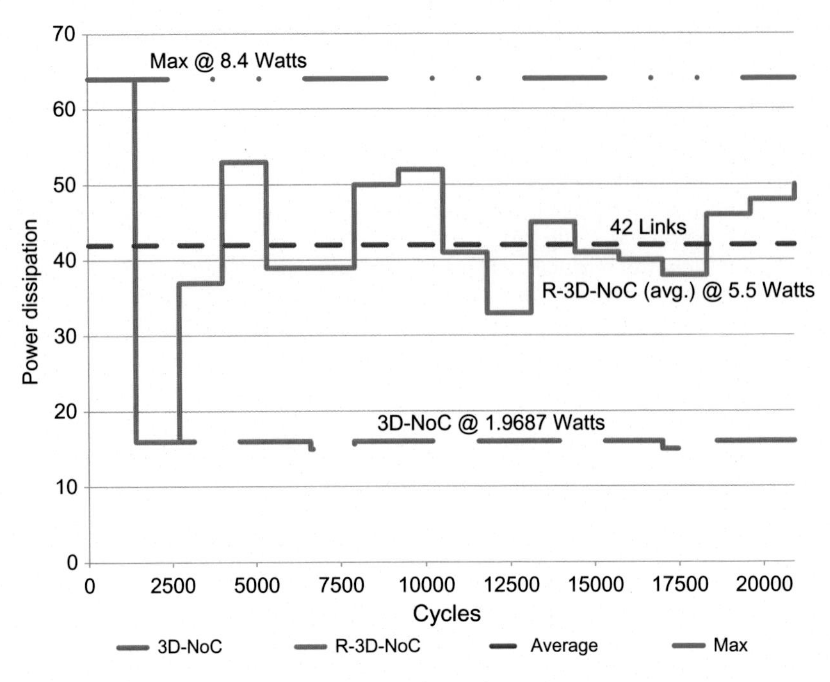

Figure 9.9 Power dissipation when PRT is used for the Blackscholes benchmark from PARSEC suites.

9.5.2.2 Optical power reduction

Fig. 9.9 illustrates the reduction in optical power through deactivation of unused photonic links for 3D-NoC. In Fig. 9.9, the number of required active photonic links for both 64-core versions of 3D-NoC and R-3D-NoC for blacksholes application is shown. During the first R_w (first 1300 cycles), all photonic links in both 3D-NoC and R-3D-NoC are active and the maximum optical power is dissipated (8.4 W). After the first R_w period, the reconfiguration algorithm detects which photonic links are not in use and deactivates those links, along with setting VCSELs modulation current to the new minimum power required. From the figure, the minimum power required for 3D-NoC is now reduced to about 1.96 W, which is a power reduction of about 77%. As for R-3D-NoC, the number of active NI varies during each R_w as the number of reconfigurable photonic links changes dynamically. From the figure, the average number of active NI is 42, which requires an average power of about 5.8 W. This gives R-3D-NoC an optical power reduction of about 31% over the maximum optical power dissipation of 8.4 W. Our proposed reconfiguration technique provides ample opportunity to save power for network links that are not utilized. This is critical in data center and HPC systems that over-provision the network links for bursty traffic; however, the proposed PRT can reduce the average power consumption considerably at periods of low network activity creating a more power-proportional network architecture.

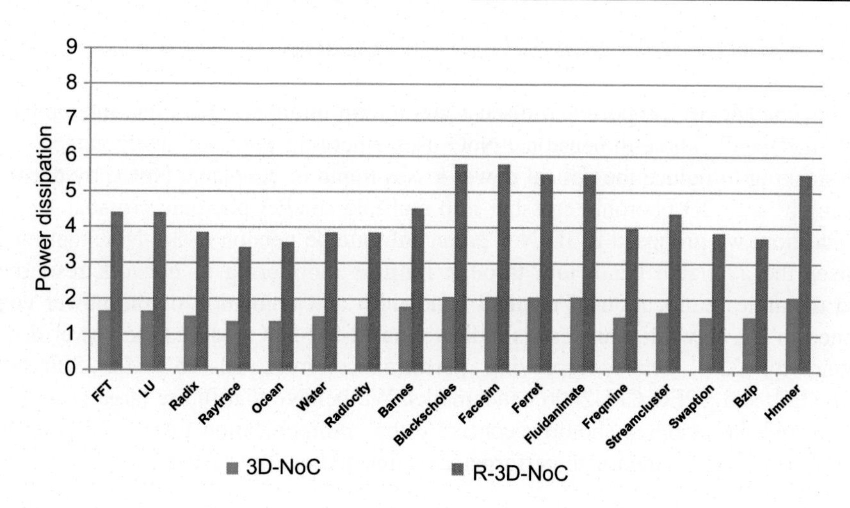

Figure 9.10 Average power dissipation for both 3D-NoC and R-3D-NoC with PRT.

Table 9.3 Electrical and optical power losses for select optical components

Component	Value	Unit
Laser efficiency	5	dB
Coupler (fiber to waveguide)	1	dB
Waveguide	1	dB/cm
Splitter	0.2	dB
Non-linearity	1	dB
Ring insertion and scattering	le-2–le-4	dB
Ring drop	1.0	dB
Waveguide crossings	0.5	dB
Photo detector	0.1	dB
Ring heating (per ring)	26	μW
Ring modulating (per ring)	500	μW
Receiver sensitivity	− 26	dBm

Source: © (2014) IEEE. Reprinted with permission from [12].

Fig. 9.10 shows the average power dissipation for both 3D-NoC and R-3D-NoC when idle photonic links are allowed to be deactivated. From the figure, Raytrace and Ocean benchmarks have the highest power savings of about 80% for 3D-NoC and about 58% for R-3D-NoC. On the other hand, Blackscholes and Facesim benchmarks have the lowest power savings of about 77% for 3D-NoC, and about 31% for R-3D-NoC. As is shown in Fig. 9.10, the reduction in optical power from deactivating photonic links that are not in use allow for a significant reduction in static optical power (Table 9.3).

9.6 Conclusions and future directions

In this chapter, we propose and discuss an on-chip multilayer photonic interconnect called 3DNoC. Three-dimensional-NoC uses emerging photonic interconnects and 3D stacking to reduce the optical power losses found in 2D planar NoCs, by decomposing a large 2D photonic crossbar into multiple smaller photonic crossbar layers. In addition, we proposed R-3D-NoC, a reconfigurable version of 3D-NoC that maximizes the available bandwidth through runtime monitoring of network resources and dynamic reallocation of channel bandwidth and regulation of the power consumed in the network. Our results indicate that R-3D-NoC reduces the power dissipation by 23%, while improving performance from 10–15% for Splash-2, PARSEC, and SPEC CPU2006 benchmarks. We believe that future large-scale networks and data centers should reconfigure the communication fabric and adapt to application load to maximize performance gains and energy savings.

References

[1] Owens JD, et al. Research challenges for on-chip interconnection networks. IEEE Micro 2007;27(5):96–108.

[2] Beausoleil RG, Kuekes PJ, Snider GS, Wang S-Y, Williams RS. Nanoelectronic and nanophotonic interconnect. Proc IEEE 2008;96(2):230–47.

[3] Vantrease D, et al. Corona: system implications of emerging nanophotonic technology. In: Proceedings of the 35th international symposium on computer architecture; 2008. p. 153–64.

[4] Batten C, et al. Building manycore processor-to-dram networks with monolithic silicon photonics. In: Proceedings of the 16th annual symposium on high-performance interconnects; 2008.

[5] Shacham A, Bergman K, Carloni LP. Photonic networks-on-chip for future generations of chip multiprocessors. IEEE Trans Comput 2008;57:1246–60.

[6] Dokania RK, Apsel AB. Analysis of challenges for on-chip optical interconnects. In: GLSVLSI '09: proceedings of the 19th ACM Great Lakes symposium on VLSI; 2009. p. 275–80.

[7] Pan Y, Kumar P, Kim J, Memik G, Zhang Y, Choudhary A. Firefly: illuminating future network-on-chip with nanophotonics. In: The proceedings of the 36th annual international symposium on computer architecture; 2009.

[8] Manipatruni S, Xu Q, Schmidt B, Shakya J, Lipson M. High speed carrier injection 18 Gb/s silicon micro-ring electro-optic modulator. In: IEEE/LEOS 2007; 21–25 Oct 2007.

[9] Kirman N, et al. Leveraging optical technology in future bus-based chip multiprocessors. In: Proceedings of the 39th international symposium on microarchitecture; 2006.

[10] O'Connor I. Optical solutions for system-level interconnect. In: Proceedings of the 2004 international workshop on system level interconnect prediction. ACM; 2004. p. 79–88.

[11] Gunn C. CMOS photonics for high speed interconnects. IEEE Photon Technol Lett 2006;26:58–66.

[12] Morris R, Kodi AK, Louri A, Whaley R. 3D stacked nanophotonic network-on-chip architecture with minimal reconfiguration. IEEE Trans Comput 2014;vol. 63:243–55.

[13] Morris R, Kodi AK, Louri A. Dynamic reconfiguration of 3D photonic networkson-chip for maximizing performance and improving fault tolerance. In: Proceedings of the 2012 45th annual IEEE/ACM international symposium on microarchitecture, ser. MICRO '12, Washington (DC, USA); 2012. p. 282–93.

[14] Sherwood-Droz N, Preston K, Levy JS, Lipson M. Device guidelines for wdm interconnects using silicon microring resonators. In: Workshop on the interaction between nanophotonic devices and systems (WINDS), co located with Micro 43; 2010. p. 15–8.

[15] Zheng X, Liu F, Lexau J, Patil D, Li G, Luo Y, et al. Ultra-low power arrayed CMOS silicon photonic transceivers for an 80 Gbps WDM optical link. In: Optical fiber communication conference; 2011.

[16] Koester SJ, Schow CL, Schares L, Dehlinger G. Ge-on-SOI-detector/Si-CMOS amplifier receivers for high-performance optical-communication applications. J Lightwave Technol 2007;25(1):46–57.

[17] Binkert NL, Davis A, Jouppi NP, McLaren M, Muralimanohar N, Schreiber R, et al. The role of optics in future high radix switch design. In: ISCA; 2011. p. 437–48.

[18] Georgas M, Leu J, Moss B, Sun C, Stojanovic V. Addressing link-level design tradeoffs for integrated photonic interconnects. In: CICC; 2011. p. 1–8.

[19] Koonath P, Jalali B. Multilayer 3-d photonics in silicon. Opt Express 2007;15:12686–91.

[20] Preston K, Manipatruni S, Gondarenko A, Poitras CB, Lipson M. Deposited silicon high-speed integrated electro-optic modulator. Opt Express 2009;17:5118–24.

[21] Woo SC, Ohara M, Torrie E, Singh JP, Gupta A. The SPLASH-2 programs: characterization and methodological considerations. In: Proceedings of the 22nd annual international symposium on Computer architecture (ISCA '95), New York, NY, USA: ACM; 1995. p. 24–36. http://dx.doi.org/10.1145/223982.223990.

[22] Bienia C, et al. The parsec benchmark suite: characterization and architectural implications. In: Proceedings of the 17th international conference on parallel architectures and compilation techniques; 2008.

[23] Dally WJ, Towles B. Principles and practices of interconnection networks. San Fransisco, CA: Morgan Kaufmann; 2004.

[24] Little BE, Chu ST. Microring resonators for very large scale integrated photonics. In: IEEE LEOS annual meeting conference proceedings; 1999. p. 487–8.

[25] Chen K, Chiang KS, Chan HP, Chu PL. Growth of c-axis orientation ZnO films on polymer substrates by radio-frequency magnetron sputtering. Opt Mater 2008;30:1244–50.

[26] Zhang XY, Dhawan A, Wellenius P, Suresh A, Muth JF. Planar ZnO ultraviolet modulator. Appl Phys Lett 2007;91.

[27] Chen X, Peh L-S, Wei G-Y, Huang Y-K, Pruncal P. Exploring the design space of power-aware opto-electronic networked systems. In: 11th international symposium on high-performance computer architecture (HPCA-11); 2005. p. 120–31.

[28] Soteriou V, Eisley N, Peh L-S. Software-directed power-aware interconnection networks. ACM Trans Archi Code Optim 2007;4:1 Article 5 (March 2007). http://dx.doi.org/10.1145/1216544.1216548.

[29] Apsel A, Andreou A. Analysis of short distance optoelectronic link architectures. In: Proceedings of the international symposium on circuits and systems; 2003.

[30] Kumar A, Kundu P, Singh AP, Peh L-S, Jha NK. A 4.6 tbits/s 3.6 ghz single-cycle noc router with a novel switch allocator in 65 nm CMOS. In: ICCD 2007; 2007.

[31] Kim J, Dally WJ, Abts D. Flattened butterfly: cost-efficient topology for high-radix networks. In: Proceedings of 34th annual international symposium on computer architecture(ISCA); 2007. p. 126–37.
[32] Martin M, Sorin D, Beckmann B, Marty M, Xu M, Alameldeen A, et al. Multifacet's genreal execution-driven multiprocessor simulator (GEMS) toolset. In: ACM SIGARCH computer architecture news, no. 4; 2005. p. 92–9.
[33] Kahng AB, et al. Orion 2.0: a fast and accurate noc power and area model for early-stage design space exploration. In: Proceedings of DATE; 2009. p. 423–8.
[34] Krishnamoorthy AV, Ho R, Zheng X, Schwetman H, Lexau J, Koka P, et al. Computer systems based on silicon photonic interconnects. Proc IEEE 2009;97(7):1337–61.

Electronic drivers/TIAs for optical interconnects

10

J. Bauwelinck and X. Yin
Ghent University - iMinds - imec, Ghent, Belgium

Editors

The editors will be the team coordinating the EU's PhoxTrot Project on optical data interconnects:

- Dr. Tolga Tekin, Fraunhofer IZM, Germany
- Dr. Nikos Pleros, Aristotle University of Thessaloniki, Greece
- Dr. Richard Pitwon, Xyratex, United Kingdom
- Dr. Andreas Hakansson, Fraunhofer IZM, Germany

Rationale

High-speed electronic circuits are crucial to the success of optical interconnects, not only to generate, process, and store huge amounts of data, but also to interface between purely digital devices (such as network processors) and other analog electro-optic (E/O) and opto-electronic (O/E) devices, such as lasers, optical modulators, and photodiodes. These high-speed electronic front-end circuits, as the driver in the transmitter and the transimpedance amplifier (TIA) in the receiver, are the main topics of this chapter.

The functionality of the driver and TIA are illustrated in Fig. 10.1. The driver chip receives the data to be transmitted from a digital CMOS chip, which typically includes a "clock and data recovery" (CDR) circuit to synchronize the different data streams, and a serializer to combine the multiple data streams into one high-speed data signal for transmission. Due to the high bit rates, the interface between the digital chip and the driver chip is typically a low-voltage differential digital signal, e.g., using current-mode-logic (CML). The driver chip then converts the incoming data voltage waveform into appropriate current or voltage pulses of a particular amplitude, which are eventually superimposed on a certain bias current or bias voltage to optimize the operating point of the optical laser or modulator. The driver circuit is designed such that the resulting optical waveform is an accurate representation of the digital data. Realizing such high-quality optical modulated waveforms is one of the main challenges of driver design, despite the E/O transfer function of the laser/modulator and the parasitics of the electrical interface between driver and laser/modulator. Moreover, as the characteristics of optical devices

Optical Interconnects for Data Centers. DOI: http://dx.doi.org/10.1016/B978-0-08-100512-5.00010-3

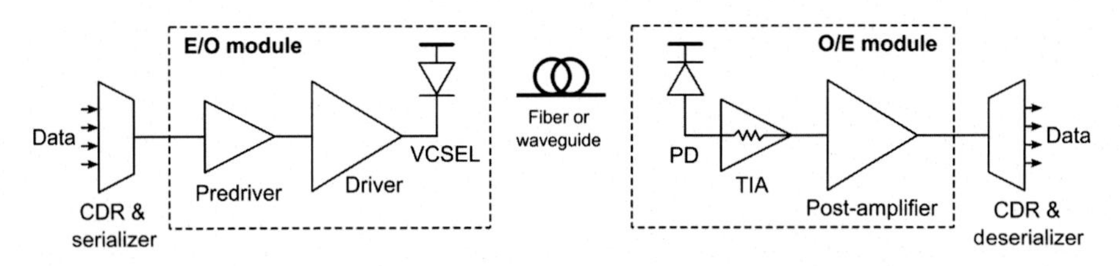

Figure 10.1 Typical E/O and O/E front-end circuits for optical interconnects.

depend generally on the temperature, the driver functionality must be adapted accordingly for best performance. When the E/O bandwidth is marginal for the targeted bit rate, equalization needs to be introduced in the driver or receiver circuitry as well. Excessive parasitics, on the other hand, contributed by interconnections, can degrade the modulating waveform due to reflections, overshoot, ringing, etc.

At the other side of the optical fiber, the receiver must be capable of reconstructing the digital data from the optical waveform. The optical signal is typically attenuated and eventually distorted by dispersion in the fiber. As the received optical signal is typically weak due to losses in the optical interfaces (and in the optical modulator when external modulation is applied), a low-noise TIA is needed to convert the weak photodiode current into a voltage waveform which is strong enough for further processing. The TIA output voltage is then further amplified in a post-amplifier (PA), and the data is then extracted by CDR circuitry. The particular TIA design challenges will be discussed in Section 10.3, considering sensitivity, dynamic range, offset compensation, and automatic gain control (AGC). Notwithstanding that optical interconnects are short, the required receiver sensitivity is a considerable challenge, because the transmitted optical power levels are constrained due to power consumption constraints, technology limitations, or due to coupling losses. Moreover, the fastest optical transmitters today are limited in extinction ratio, i.e., the ratio between high and low modulation levels, making the reception even more difficult in combination with crosstalk.

In data centers, multiple transceivers are applied in parallel to realize higher bit rate point-to-point connectivity inside racks or between racks. This could be implemented using fiber ribbons, wavelength division multiplexing (WDM), or using multi-core fibers. In future, multiple modes may be considered as well, but today, it seems that data center infrastructure will migrate towards single-mode fibers to be future proof, while multi-mode technology is showing very interesting progress as well. Irrespective of the optical multiplexing technology, applying multiple data streams in parallel requires multi-channel driver and receiver electronics. This brings additional challenges to the electronics design, on top of the need for ever faster and lower power transceiver circuitry. Multi-channel devices need to fit in specific packages, requiring a tight integration of electronics and photonics. This tight integration not only brings challenges for the assembly, but restricts power consumption and introduces additional issues such as thermal crosstalk from electronics to photonics, mutual parasitics, electrical crosstalk, power/ground integrity ..., making the design of data center transceivers particularly challenging.

As transceiver circuits for optical interconnects need to be low-cost, low-power and small in footprint to fit in small packages, the circuit complexity should find a good balance between modulation efficiency and implementation efficiency of the analog/digital circuits. For this reason, the basic modulation schemes NRZ (2-level), duobinary (3-level), and PAM-4 (4-level), illustrated in Fig. 10.2, are the most practical options. Whether one or the other is better, depends on the channel attenuation and frequency response, including fiber and the E/O devices. More complex schemes typically require a high-speed digital-to-analog converter (DAC), and a high-speed analog-to-digital converter (ADC) in combination with high-speed digital signal processing (DSP). For long reach applications, this complexity is justified by the higher spectral efficiency and higher throughput. However, for data center applications, complexity is a competitive threshold.

Today, 25 Gb/s VCSEL links are commercially widely available, using NRZ modulation and multi-mode mode fibers up to a few hundred meters. As long as sufficient bandwidth is available, NRZ, eventually assisted by equalizer circuits, will be preferable because of its minimum complexity and low latency. Now, moving towards 50 Gb/s line rates, or even 100 Gb/s, puts traditional NRZ signaling under pressure. As a result, PAM4 is widely being adopted for achieving higher serial data speeds (50 Gbps and higher) with a lower relative bandwidth. Meanwhile, a powerful alternative, duobinary, is often overlooked or misunderstood. However, scientific discussions and experiments in the literature recognize that duobinary actually provides more performance in most cases [1]. The main criticism against duobinary is that duobinary circuits need to operate at the full baud rate; however, this is no problem

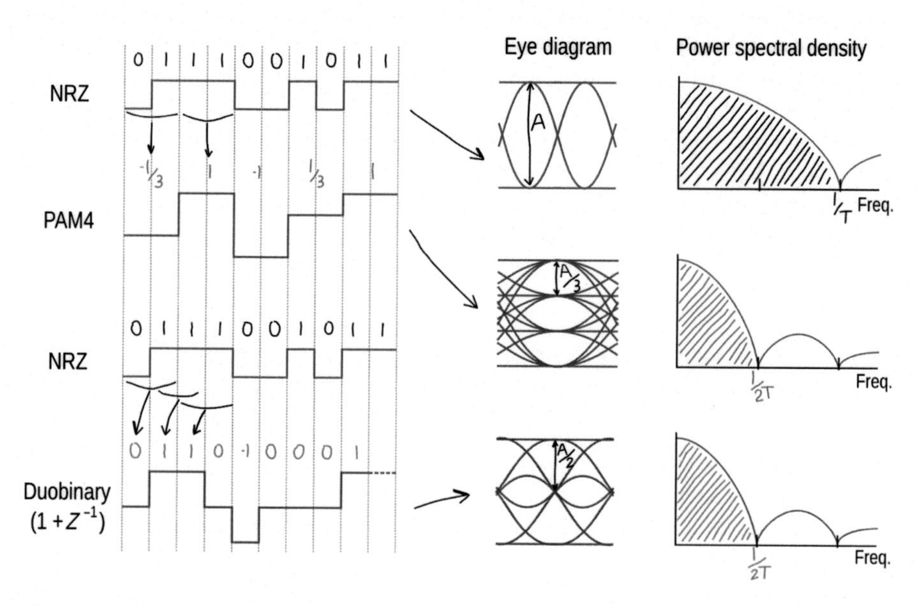

Figure 10.2 Simplified comparison of the basic modulation schemes: NRZ, PAM4, and duobinary.

as demonstrated in Van Kerrebrouck et al. [2] at 100 Gb/s using custom transceiver chips in a mature 130 nm SiGe BiCMOS process.

In this section, we will briefly introduce the concept of these basic modulation schemes. Extensive comparisons of these schemes are presented in Van Kerrebrouck et al. [2], considering electrical interconnections inside a rack, and in Jensen and Monroy [1] considering short-reach optical links. In general, the goal of most advanced modulation schemes is to limit the needed bandwidth in order to achieve higher serial data rates. In this process, SNR is typically traded for bandwidth. When using PAM4, this is done by combining two NRZ bits into a single PAM4 symbol, so that the same throughput is achieved at half the rate, resulting in about half the required bandwidth. As PAM4 combines two bits in the amplitude of one symbol, four different levels can be obtained, leading to an SNR which is theoretically three times smaller compared to NRZ, assuming the same maximum swing.

Duobinary limits the bandwidth in a completely different way. By adding the previous bit to the current NRZ bit, a stream of duobinary symbols is created (function: $1+z^{-1}$). This leads to a three level signal with a symbol rate identical to the original NRZ bit rate. However, the bandwidth (shown by means of the power spectral density (PSD) in Fig. 10.2) is compressed by the addition of bits that make it impossible to have high-speed signal transitions. For example, +1 can never go directly to −1 in one bit period, but takes at least two bit periods. SNR is thus traded for bandwidth compression by going to three level symbols. Note that the SNR penalty of duobinary is lower compared to PAM4. Moreover, duobinary can be created very easily, by sending the NRZ stream through a low pass filter which adds intersymbol interference (ISI) corresponding to $1+z^{-1}$. This can be done by using the channel response of the transmission medium together with some equalization at either the transmitter or the receiver. In this way, part of the channel loss is exploited to create the duobinary stream from an NRZ signal, which means this loss does not need to be compensated. This allows a simple NRZ transmitter providing backwards compatibility to NRZ systems to be used, but requiring less equalization. However, despite its advantages, duobinary is not (yet) included by the main standardization bodies, so in the near future we will mainly see NRZ and PAM4 transceiver products for optical interconnects in data centers.

10.1 Co-design and co-simulation of electronics and photonics

Due to the very high bit rates in today's optical interconnects, the design and simulation processes of optical transmitter and receiver front-ends need to include accurate models for the electronic circuits, the on-chip and off-chip interconnects, the electrical parasitics of the photonic devices, and the electro-optic response of the photonic devices. Such a co-simulation approach allows for a real co-design of electronics, interconnects, and photonics in order to identify the tradeoffs between various parameters and to predict the system-level performance. For example, considering a conventional positive-intrinsic-negative (PIN) photodiode (PD), a larger active area will capture more photons, so a larger PD will have a higher responsivity, which is beneficial on one hand, but a larger PD junction will also show a higher capacitance, reducing the receiver bandwidth. So a co-design approach can yield the required insight to quantify the relevant tradeoffs, and to find a good balance between conflicting requirements in order to derive an overall optimum considering various specifications.

In practice, the "design space" or degrees of freedom are often restricted somewhat by the chosen suppliers or technology platform, and in the worst case, it may not be possible to change anything in the photonic devices, e.g., when the design of laser, modulator, and PD are fixed. In such case, the co-optimization process focuses on the electronic circuit topologies and the layout of the interconnections.

It is clear that for the co-simulation process, one needs to bring together the models of the different parts in one simulation bench. For this purpose, advanced circuit design tools such as Cadence Virtuoso or ADS are well suited, because the most complex circuits to be considered for the transceiver design are in the electronic circuitry and, in the end, the electronic circuitry largely determines the performance (power consumption, bandwidth, sensitivity, signal quality...) for the given photonic devices. Moreover, such professional electronic IC design tools can mix transistor level circuitry with different kinds of models to combine subsystem and system simulations, and various simulators are available in these tools to efficiently simulate different aspects (transient, S-parameter, harmonic balance, noise, stability...).

To derive a model for the interconnects and the photonic devices, one typically starts from S-parameter measurements. S-parameters are measured in the frequency domain under small-signal conditions. The interconnects are electrically "passive", so these can be considered as linear elements, but one has to be aware of crosstalk. It is possible to include S-parameter models in a circuit simulation, but often it is more efficient (shorter simulation time) to first derive an equivalent circuit description. When the interconnect is short compared to the wavelength of the highest frequency involved, then a lumped equivalent circuit is used. However, for increasing frequencies, the lumped approximation is not accurate anymore, so then multiple sections or distributed elements (transmission lines) need to be used. For more complex devices, besides circuit equivalents, it is often very convenient to use a hardware description language such as VerilogA, e.g., to include a fiber model, the rate equations of a laser, or to model an optical modulator.

To illustrate such an S-parameter fit, Fig. 10.3 shows the measured S-parameters of the fastest C-band single-mode VCSEL to date, at a bias of 6 mA, showing 22 GHz of bandwidth [3]. From the measured S-parameters, a small-signal equivalent model was developed that includes the electrical response (determined by parasitics) and the electro-optic response. Such a simulation model allows co-simulation of the driver circuitry with the VCSEL response to tradeoff eye quality, equalizer complexity, and power consumption.

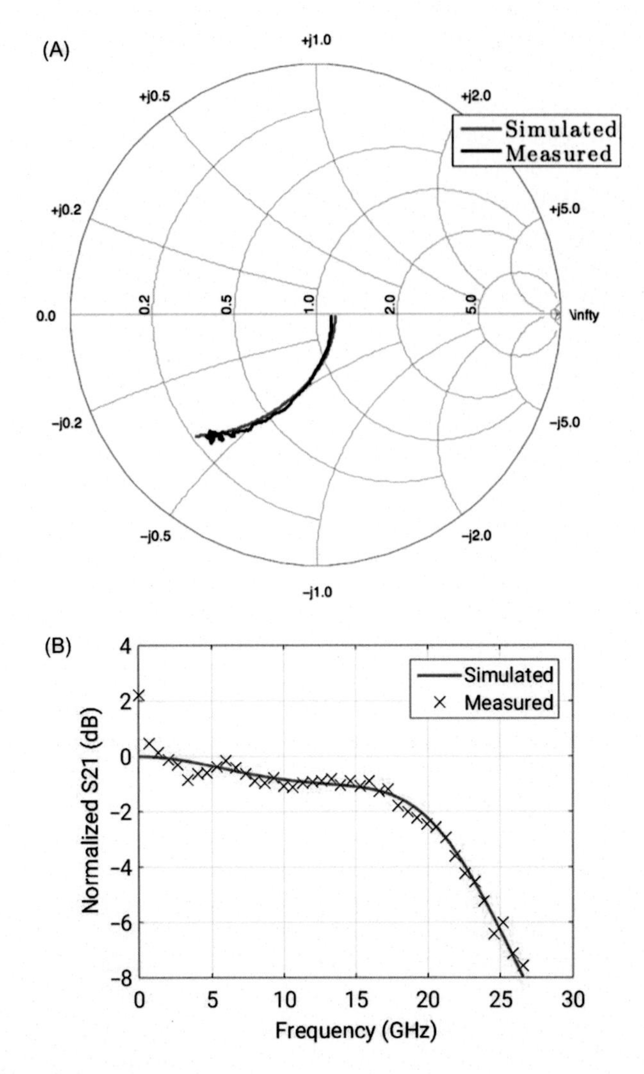

Figure 10.3 VCSEL model versus VCSEL measurements: (A) S11 parameter, (B) Normalized S21 parameter.

10.2 Electronic drivers

10.2.1 VCSEL drivers

In the PhoxTroT project we focused on single-mode transmission at 1550 nm, however, the VCSEL technology (see chapter: 13) can be optimized for 1310 nm operation as well. The main rationale for applying long-wavelength VCSELs is the compatibility with silicon photonic integrated circuits, and to realize optical data center links beyond 500 m which is not possible with 850 nm multi-mode

VCSELs. Without doubt, multi-mode devices are showing very interesting progress as well, although single-mode is considered to be the most future proof. It is, however, somewhat more challenging to develop driver electronics for 1310 or 1550 nm VCSELs, due to some key differences. First of all, long-wavelength VCSELs have a lower bandwidth. Today, the fastest long-wavelength VCSEL achieves 22 GHz of bandwidth [3], which makes it suitable for 28 and 40 Gb/s NRZ applications. For 56 Gb/s interconnects, a multi-level scheme such as duobinary or PAM4 will likely be more efficient than NRZ, but by applying strong equalization as demonstrated by IBM and Vertilas it is still possible to realize 56 Gb/s NRZ with a 18 GHz VCSEL [4]. However, equalization costs power, and operational margins are needed for products to anticipate process variations, temperature variations, end-of-life degradation, etc. A second key difference is that long-wavelength VCSEL arrays typically have a common anode, instead of the typical common cathode configuration for short-wavelength VCSELs. As a consequence, the long-wavelength VCSEL driver typically needs two supply voltages because the VCSEL anode needs to be connected to a higher supply voltage providing sufficient headroom for the driver output circuit, while a common-cathode array can connect its cathode to ground. The last, but also important, difference is that long-wavelength VCSELs typically require a higher drive current compared to short-wavelength VCSELs, but still considerably lower than DFB lasers of similar bandwidth.

Fig. 10.4A and B illustrate the basic architectures of a VCSEL driver for NRZ, duobinary, and PAM4. In both schemes, a bias current I_{bias} is used to operate the VCSEL above its threshold to obtain a high electro-optic conversion bandwidth. The high-speed modulation current is then added to the DC bias current to alternate the VCSEL optical output, as shown in Fig. 10.5 for NRZ and PAM4 modulation. In the case of NRZ the drive current can take on two different values (I_{bias} and $I_{bias} + I_{mod}$) corresponding to a logical “0” or “1” bit. Duobinary can be obtained from the same structure, by tuning the bandwidth to introduce ISI according to $1 + z^{-1}$. For PAM4, two NRZ drivers, with binary weighted drive currents ($I_{msb} = 2I_{lsb}$) are combined as shown in Fig. 10.4B, so that the total drive current can take on four different values (I_{bias}, $I_{bias} + I_{lsb}$, $I_{bias} + I_{msb}$ and $I_{bias} + I_{lsb} + I_{msb}$), corresponding to logical symbols “00”, “01”, “10”, and “11”.

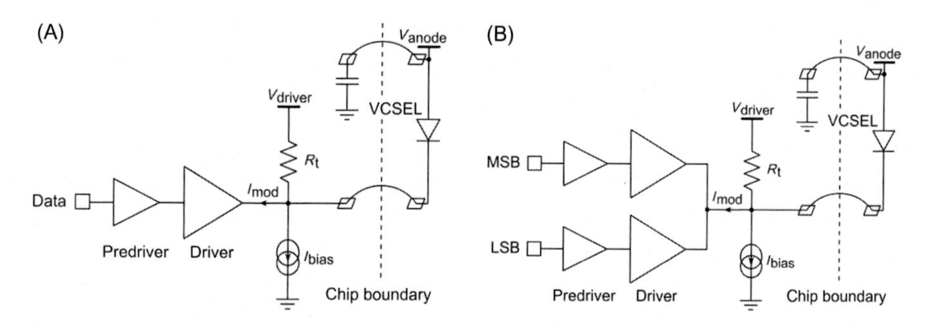

Figure 10.4 (A) Simplified VCSEL driver for NRZ or duobinary. (B) Simplified VCSEL driver for PAM4.

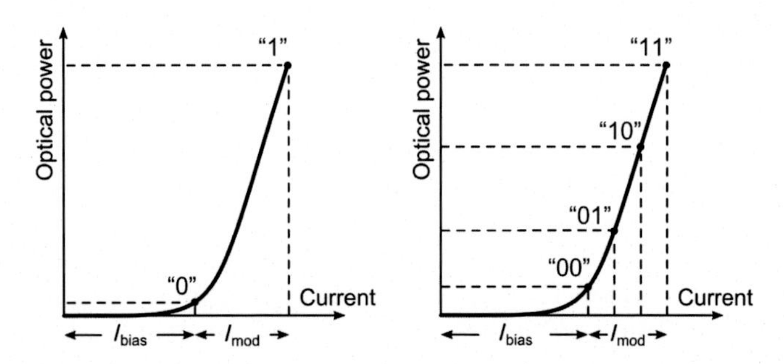

Figure 10.5 Power-current (PI) VCSEL characteristic and its application for NRZ and PAM4 modulation.

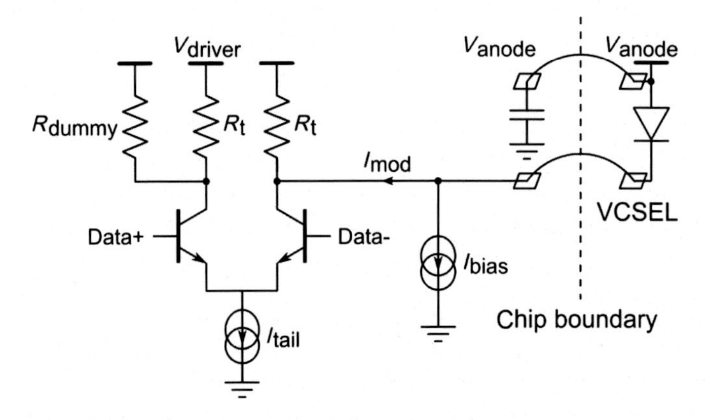

Figure 10.6 Basic VCSEL driver circuit for a common anode VCSEL.

A basic current switching VCSEL driver stage without equalization is shown in Fig. 10.6. The DC current source I_{bias} is used to operate the VCSEL above its threshold; however, as a consequence, the optical "0" level will not be completely "dark", restricting the extinction ratio between optical "1" and "0" pulses. The high-speed modulation current is generated from the DC tail current source by switching with a fast differential pair. While the VCSEL is a single-ended device, the driver circuitry on-chip is typically differential because differential circuits such as amplifiers, buffers, flip-flops ... can achieve higher speeds than their single-ended counterparts, and the incoming data, delivered through transmission lines on a PCB, is also often differential to optimize signal integrity. Moreover, differential circuits are generally robust against common-mode disturbances and power supply noise. In addition, less "switching noise" is generated on the power and ground nets, because the total supply current remains more or less constant during switching, improving crosstalk performance in multi-channel devices.

Depending on the input data signal polarity, the tail current is steered to the left or the right branch. When the tail current flows through the right transistor, it will be divided among the VCSEL and the back termination resistor R_t. So, a part of the tail current is dissipated in the back termination resistor; however, at a high bit rate,

R_t is needed to minimize the degradation of signal quality due to, e.g., reflections or ringing. So the selection of the output resistance R_t is an important design choice and it depends on the VCSEL parasitics, the bit rate, and the interconnection (wire bonding or flip-chip bonding). A high R_t leads to lower power consumption, but also influences the bandwidth and high-speed characteristics of the interface. At the left side of the differential pair, a dummy resistor is typically added, to make the circuit more symmetrical, as the (single-ended) VCSEL is added at the right side. The basic topology is the same for bipolar or CMOS implementations. In a practical realization, the performance of the driver circuit can be improved by various circuit techniques such as inductive peaking, cascode transistors, etc.

Today, however, increasing the data rate often relies on equalization techniques as the need for data outgrows the available bandwidth. In a driver circuit, equalization is often implemented using feed-forward equalization (FFE). In the case of NRZ and duobinary, FFE is applied to the incoming data, and the resulting pre-emphasized waveform is applied to the VCSEL, so that, after all O/E conversions and the fiber channel, the receiver captures a signal with sufficient quality. For PAM4, one could in principle apply the same architecture thereby requiring a PAM4 input data signal and a linear FFE stage. As the data format in processors, memories, networking chips ... is binary, one should at some point in the communications path translate the binary NRZ data into PAM4. For this purpose, the architecture in Fig. 10.4B is much more useful. In Fig. 10.4B, two binary NRZ streams are combined into one PAM4 stream as in Soenen et al. [5]. As each incoming stream is binary, the FFE design is much easier compared to a fully linear FFE. To make sure that the three eyes in the resulting PAM4 signal are properly synchronized, a clock signal is used to synchronize the MSB and LSB bits in both paths.

To introduce FFE functionality in the driver, one can combine multiple basic driver stages, with different programmable tail current sources. By driving these stages with a delayed data signal, one actually obtains a finite impulse response filter (FIR), as illustrated in Fig. 10.7.

Recently, in the EU FP7 Phoxtrot project, a two-channel and an improved four-channel VCSEL driver array have been developed, optimized for Vertilas 1550 nm BTJ VCSELs operating at 28–40 Gb/s NRZ. The block diagram of the two-channel VCSEL driver, designed in 130 nm SiGe BiCMOS technology, is illustrated in Fig. 10.8. The incoming differential data signal is terminated in a 100 Ω differential impedance and buffered towards the predriver. The predriver in conjunction with the main driver delivers the modulation and bias current to the VCSEL. Since the response of the VCSEL strongly depends on the average operating current, the driving capability of the output stage covers a broad range without sacrificing signal integrity. For that reason, the gain of the predriver is dynamically adjusted according to the output current. A two-tap fractionally spaced feed-forward equalizer (FFE) is implemented in the driver chip to compensate the bandwidth limitation of the VCSEL in order to achieve 40 Gb/s NRZ operation.

This driver concept was then further improved to lower the power consumption by 50%, now estimated at 4.5 pJ/bit at 40 Gb/s, while extending the drive current range and supporting four-channel VCSEL arrays with a common anode. This functionality is unique, as most VCSEL drivers are designed for common cathode

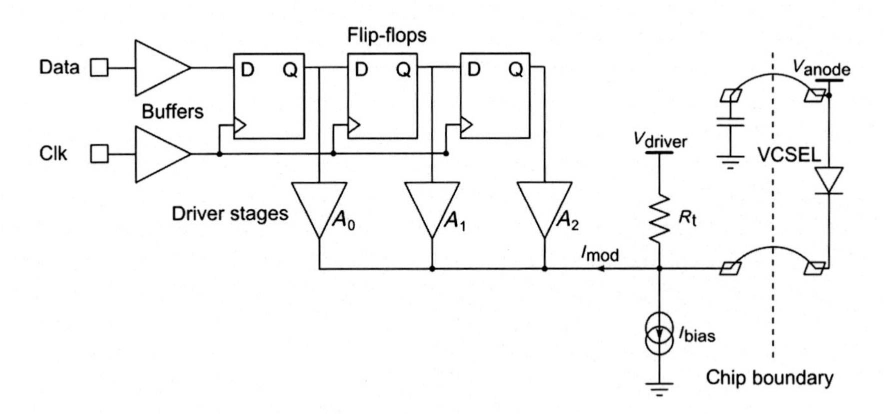

Figure 10.7 Simplified VCSEL driver with three-tap symbol-spaced FFE.

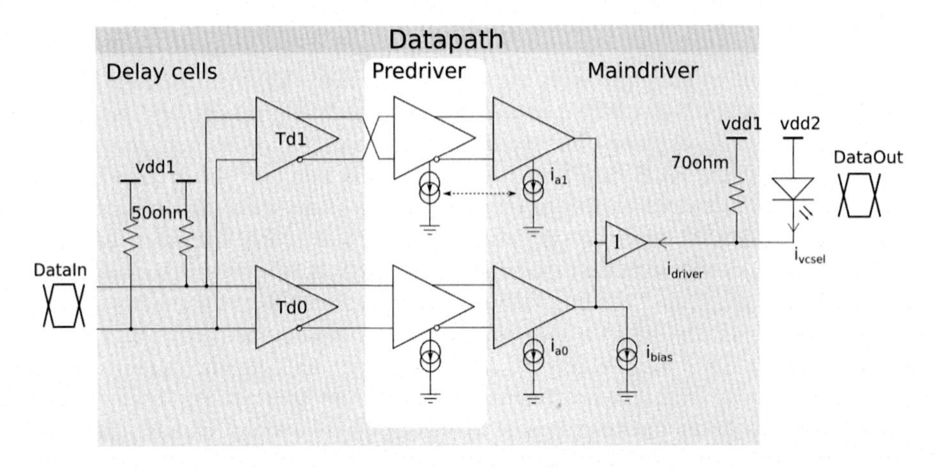

Figure 10.8 First Phoxtrot 2 × 40 Gb/s NRZ VCSEL driver.

VCSELs. In the EU FP7 Mirage project, we developed PAM4 VCSEL drivers for 1550 nm VCSEL arrays, fabricated by TUM. The results of the first Mirage PAM4 VCSEL driver chip are presented in Soenen et al. [5], whereas the optimized PAM4 VCSEL driver is showing good optical performance at 56 Gb/s (not yet published).

10.3 Transimpedance amplifiers

The receiver circuit design, as mentioned earlier, is quite different from the circuit design in optical transmitters. In particular, noise performance is rarely a limitation in transmitters, but is a very common and important aspect for the receiver front-end. Typically, the first stage in an optical receiver is a transimpedance amplifier (TIA) for converting a small input photo-current, provided by a photodiode (PD), into an output voltage signal V_{OUT}, for further processing.

The main design parameters of a TIA are the transimpedance gain, given by V_{OUT}/I_{PD}, the bandwidth, and the dynamic range determined by the sensitivity and the overload levels. The sensitivity of an optical receiver is the weakest optical power required to obtain a specified bit-error rate (BER) performance, while the strong signal handling capability is governed by the TIAs overload optical level.

Since a TIA can be considered in general as a current-to-voltage converter, the simplest form of TIA is just a load resistor R_L, as shown in Fig. 10.9A, which converts the photo-current I_{PD} to the output voltage V_{OUT} simply using Ohm's law. The simple resistive TIA topology shows some fundamental tradeoffs, as both the transimpedance gain and input impedance equal the resistor R_L. In this case the lower limit of the transimpedance gain is determined by the current noise of the resistor R_L: $\overline{i^2_{n,R_L}} = \frac{4kT}{R_L}$. In order to achieve a high sensitivity, the current noise must be low, so for this simple circuit the load resistance R_L must be high. However, a high resistance decreases the receiver bandwidth, since the resistor is directly loaded by the photodiode capacitance and the capacitance of the output network. The high load resistance R_L causes another problem concerning the input overload current: as the output signal swing V_{OUT} is proportional to $R_L.I_{PD}$, a high current will "overload" the receiver, resulting in a significant disturbance in photodetector's reverse bias condition. Ideally, creating a virtual ground at the TIAs input is preferable for reducing this overload effect.

It is clear that the basic TIA circuit with a simple resistive load has several shortcomings. Much better performance can be achieved by introducing amplification and feedback in the TIA circuitry. So far the most common TIA configuration is the shunt – shunt feedback topology, where a negative feedback network shunts both the input and output of the main amplifier, sensing the output voltage and feeding back a proportional current to the input. This type of feedback provides a convenient low-impedance input node for the photodiode current I_{PD}, and also ensures a small output resistance for better output voltage drive capability. Fig. 10.9B shows a basic shunt – feedback TIA configuration, where a feedback resistor R_F is connected across a voltage amplifier with gain $A(s) = A_0/(1 + s\tau_A)$, input resistance R_A, and input capacitor C_A. The transfer function of the basic shunt-feedback TIA can be expressed using:

$$H(s) = \frac{-R_T}{1 + s/(Q\omega_0) + (s/\omega_0)^2}$$

where:

$$R_T = \frac{R_F}{1 + (R_A + R_F)/(A_0 \cdot R_A)}$$

$$Q = \sqrt{\frac{(A_0 + 1) \cdot \tau_{in} \cdot \tau_A}{\tau_{in} + \tau_A}}$$

$$\omega_0 = \sqrt{\frac{A_0 + 1}{\tau_{in} \cdot \tau_A}}$$

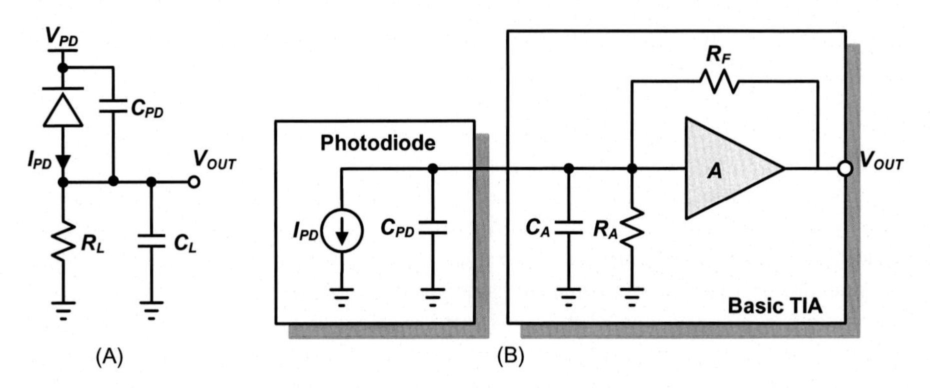

Figure 10.9 Basic TIA: (A) a simple resistive load, (B) a shunt-feedback TIA.

in which $\tau_{\text{in}} = R_F(C_{PD} + C_A)$ is the input node time-constant, Q is the quality factor, and ω_0 is the angular natural frequency. If $A_0{}^*R_A/(R_A + R_F) >> 1$, we find that the trans-resistance gain R_T reduces to a well-known form $\approx R_F$. As R_F ideally determines the trans-resistance gain R_T and the TIA current noise, increasing R_F has a two-pronged effect on the noise at the output [6]: by reducing the bandwidth and quality factor of the transfer function, more noise is filtered out, and it decreases the input-referred noise spectral density. The quality factor Q provides another degree of freedom in TIA design, which is the gain near the angular natural frequency. For high values of Q, the maximum frequency response becomes larger than R_T, resulting in ringing and overshoot in the time-domain response. Therefore, the largest Q before onset of frequency peaking is particularly interesting: for $Q = 1/\sqrt{2} \approx 0.707$ we obtain the so-called Butterworth or maximally flat response, which has a good design balance between bandwidth and group-delay characteristics.

Fig. 10.10A shows a typical self-biased shunt-feedback TIA implementation with a common-emitter input stage [7]; a similar topology can be applied to a common-source input stage when using CMOS technologies. As shown in Fig. 10.10A, Cascode Q_0 protects Q_1 from excessive collector-emitter voltage and reduces its Miller capacitance contribution to the input capacitance. In addition, it provides a convenient low-impedance input for current injection, which can be used to provide an extra bias current to Q_1. Q_1 has a large emitter area to reduce its base resistance and the associated thermal noise. This leads to an increased base-emitter junction capacitance. In turn, this requires a higher bias current in order to reduce the transition time through the base and to improve the high-frequency response. Note that the output has been taken from the collector of the cascade transistor Q_0, which grants better headroom for the subsequent stage compared to the conventional output at the Q_2 emitter. Noise introduced by feedback resistor R_F and input transistor Q_1 are the two dominant sources of TIA noise. Taking into account tradeoffs between sensitivity, power consumption, and bandwidth, detailed analysis and a design optimization scheme have been explained and proposed in Moeneclaey et al. [6].

One alternative is to replace the common-emitter/source amplifier with a parallel nMOS/pMOS gain stage, as shown in Fig. 10.10B. Compared to the topology with

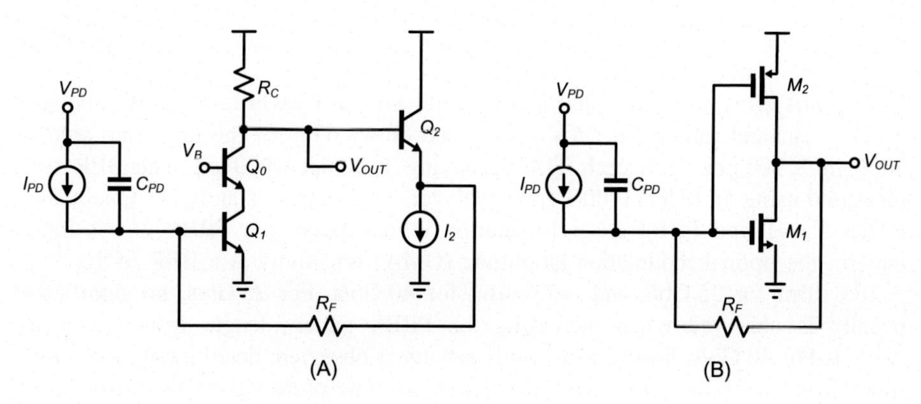

Figure 10.10 (A) Common-emitter/source shunt-feedback TIA, (B) complementary shunt-feedback TIA.

an active pMOS load [8], the complementary stage has a larger overall trans-conductance with the same bias current. The main problem of the complementary shunt-feedback TIA lies in the worse performance of pMOS transistors compared to nMOS transistors, resulting a larger pMOS transistor size (i.e., larger capacitance and lower speed) to achieve the same parallel trans-conductance. However, in more recent technologies below 45 nm, pMOS transistors become comparable to nMOS transistors, and thus using a complementary topology is gaining popularity as it can operate at a lower supply voltage [9]. For design purposes, the complementary stage can be treated as a single G_m cell and loaded directly by the feedback resistor R_F. One important drawback of this scheme is that the output impedance of this G_m cell is not buffered. Thus, direct feedthrough from the feedback network must be analyzed and designed properly, or a voltage buffer may be needed to isolate the complementary stage and the feedback network.

Finally, instead of using a shunt − shunt feedback, we may consider adding a current buffer in front of the load trans-resistance R_L in Fig. 10.9A. A current buffer can be implemented by a common-base or a common-gate stage. More advanced implementations use local feedback to improve the current buffer performance [10] especially for the common-gate topology (due to a smaller G_m). This allows us to isolate the photodiode capacitance and the load resistance, keeping a stable TIA frequency response. The main drawback is the noise performance of the common-base/gate amplifier which is in practice worse than the common-emitter/source stages, although it has been shown suitable to achieve very high bandwidth operation [11].

Recently, in the EU FP7 Phoxtrot project [13], a 2 × 40 Gb/s TIA array has been designed for use in high-speed short-range communication systems in data centers [6]. The data path consists of a TIA input stage, a main amplifier, and an output stage. The TIA input stage is a shunt − shunt feedback TIA optimized for high bandwidth and low noise performance. A control loop is formed using the balancing error integrator, which removes the dc-offset between both output signals by adjusting the dc-voltage at the inverting input of the main amplifier. This creates a low-frequency high-pass pole in the data path at 550 kHz. The TIA controller enables digital adjustment of the gain and bandwidth of the data path stages and is programmed via an external SPI interface.

The TIA array was fabricated in a 0.13 um SiGe BiCMOS technology. The single TIA channel runs off a 2.5 V supply and draws 63 mA. The total chip area is 3000 μm × 900 um, with each TIA occupying 1100 um × 900 um. The BER was measured using an SHF 11100B error analyzer. The input was an NRZ signal using PRBS 2^7-1 and PRBS $2^{31}-1$ patterns. When applying a PRBS 2^7-1 input pattern, the optical modulation amplitude (OMA) sensitivity at a BER of 10^{-12} is −10.6 dBm for 25 Gb/s and −7.6 dBm for 40 Gb/s. For 25 Gb/s, no significant penalty is observed when increasing the PRBS pattern length from 2^7-1 to $2^{31}-1$. For 40 Gb/s, however, a 1.2 dB penalty is observed, deteriorating the sensitivity from −7.6 to −6.4 dBm. Compared to state-of-the-art TIAs operating at 40 Gb/s, the presented TIA is characterized by a low power consumption while maintaining a competitive sensitivity.

As an alternative to NRZ, in the EU FP7 MIRAGE project, PAM4 is investigated for single-mode point-to-point links in data centers. PAM4 makes it possible to reach a higher bit rate, however, at the cost of an increased linearity requirement. In Moeneclaey et al. [12] we presented a linear optical receiver, operating with PAM4 signals at line rates ranging from 50 to 64 Gb/s. In the TIA input stage, the feedback resistor is implemented as an nMOS transistor biased in the linear region in order to control the transimpedance gain. The gain stages after the TIA input block utilize a degenerated differential pair topology for the required linearity.

The receiver was evaluated at 25, 28 and 32 Gbaud. The output electrical eye diagrams, acquired with the equivalent-time oscilloscope, are shown in Fig. 10.11A–C, while the distribution of PAM4 symbols is depicted in the respective histogram of Fig. 10.11D–G, after signal acquisition in the real-time oscilloscope. It can be observed that the eye diagram is still quite open up to 32 Gbaud, whereas the clear separation of the symbol distributions in the corresponding histograms implies sufficient reception performance. The power consumption of the TIA channel after optimizing the receiver settings was 165 mW, yielding an energy consumption of 2.6 pJ/bit at 64 Gb/s. BER measurements were performed to the digitized signal at the real-time oscilloscope after offline clock recovery, re-sampling and automatic thresholding for symbol detection. As can be observed at 25 Gbaud and an average input power range between −2 and 0.4 dBm, the received signal exhibits zero errors, which corresponds to an upper 95% confidence limit of 2.9×10^{-7}. Operation below the FEC limit was achieved in all cases, proving the suitability of the linear receiver for future interconnect standards with serial line rates above 50 Gb/s. In a very recent experiment (not yet published), by further optimizing the TIA chip and set up, we have been able to improve the sensitivity from −3 to −7 dBm at 64 Gb/s for FEC limit of 1×10^{-3}. The new PIN − TIA experiment also achieved an excellent dynamic range of 8.8 dB by switching between a high and low gain mode. To our knowledge, this is the best sensitivity and dynamic range performance reported at 56–64 Gb/s PAM4 for PIN − TIA modules (Fig. 10.11).

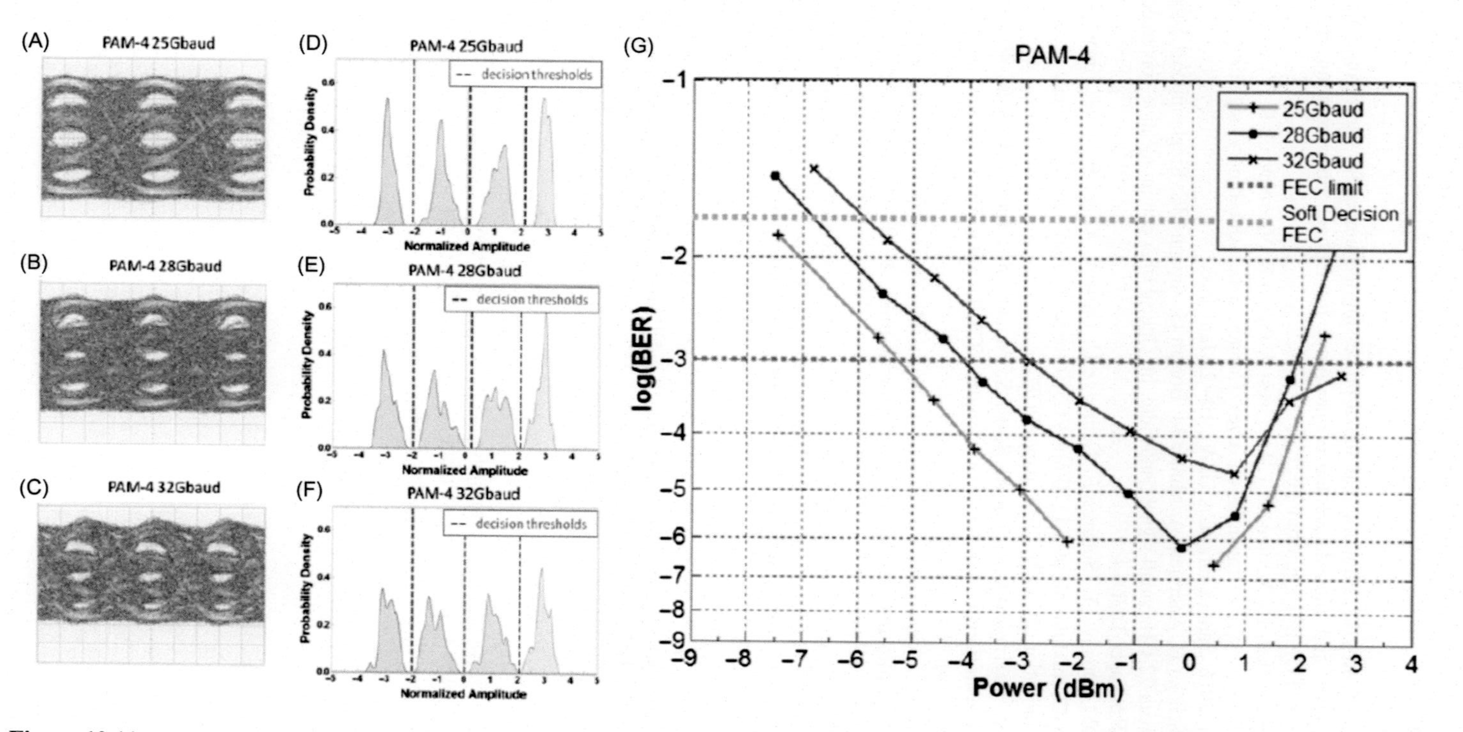

Figure 10.11 Receiver output eye-diagrams at: (A) 25, (B) 28, and (C) 32 Gbaud. Symbol histograms at: (D) 25, (E) 28, and (F) 32 Gbaud, and BER measurement results (G).

References

[1] Jensen JB, Monroy IT. Photonic and signal processing techniques for short range high capacity intra- and inter-datacenter connectivity. In: Optical fiber communication conference (OFC); 2016, tutorial Tu2J. 1.

[2] Van Kerrebrouck J, De Keulenaer T, De Geest J, Pierco R, Vaernewyck R, Vyncke A, et al. 100 Gb/s serial transmission over Copper using duo-binary signaling. In: Designcon 2016, Santa Clara (CA USA); 19–21 Jan 2016.

[3] Spiga S, Schoke D, Andrejew A, Müller M, Boehm G, Amann MC. Single-mode 1.5-μm VCSELs with 22-GHz small-signal bandwidth. In: Optical fiber communication conference (OFC); 2016, paper Tu3D. 4.

[4] Kuchta DM, Doany FE, Schares L, Neumeyr C, Daly A, Kögel B, et al. Error-free 56 Gb/s NRZ modulation of a 1530 nm VCSEL link. In: European conference on optical communication (ECOC); 2015, paper PDP.1.3.

[5] Soenen W, Vaernewyck R, Yin X, Spiga S, Amann MC, Kaur KS, et al. 40 Gb/s PAM-4 transmitter IC for long-wavelength VCSEL links. IEEE Photon Technol Lett 2015;27(4):344–7.

[6] Moeneclaey B, Verbrugghe J, Blache F, Goix M, Lanteri D, Duval B, et al. A 40 Gb/s transimpedance amplifier for optical links. IEEE Photon Technol Lett 2015;27(13):1375–8.

[7] Verbrugghe J, Vaernewyck R, Moeneclaey B, Yin X, Maxwell G, Cronin R, et al. Multi-channel 25 Gb/s low-power driver and transimpedance amplifier integrated circuits for 100 Gb/s optical links [invited]. J Lightwave Technol 2014;32(16):2877–85.

[8] Razavi B. Design of integrated circuits for optical communications. 2nd ed. Hoboken, NJ: Wiley; 2012, Chapter 4.

[9] Liu FY, Patil D, Lexau J, Amberg P, Dayringer M, Gainsley J, et al. 10-Gbps, 5.3-mW optical transmitter and receiver circuits in 40-nm CMOS. IEEE J Solid-State Circuits 2012;47(9):2049–67.

[10] Sackinger E, Guggenbuhl W. A high-swing, high-impedance MOS cascode circuit. IEEE J Solid-State Circuits 1990;25(1):289–98.

[11] Han J, et al. A 20-Gb/s transformer-based current-mode optical receiver in 0.13 um CMOS. IEEE Trans Circuits Syst II: Express Briefs 2010;57(5):348–52.

[12] Moeneclaey B, Kanakis G, Verbrugghe J, Iliadis N, Soenen W, Kalavrouziotis D, et al. A 64 Gb/s PAM-4 linear optical receiver. In: OFC 2015, vol. M3C.5, Los Angeles (CA, USA), March 20–24, 2015. pp. 1–3.

Part III

Circuit Boards

Electrical and photonic off-chip interconnection and system integration

M. Zia, C. Wan, Y. Zhang and M. Bakir
Georgia Institute of Technology, Atlanta, GA, United States

11.1 Introduction

11.1.1 Moore's law and off-chip I/O trends

With the increased number of parallel processing architectures and the complexity of integrated circuits, there is an overburdening demand being put on the I/O interface so as to allow performance benefits at the system level; the ability to sustain high-bandwidth low-loss communication between chips is limiting system performance, despite the increased computing power of individual chips. Fig. 11.1 shows the total off-chip bandwidth trends, along with the number of package pins [1]. As seen from the figure, the total off-chip I/O bandwidth is consistently increasing, with an ever-increasing mismatch between the required off-chip bandwidth and the increase in the number of off-chip I/O pins. Although an increase in channel data rate can allow for an increase in total aggregate bandwidth, the increased need for equalization and noise cancellation requires additional power and area, deeming it a less attractive option [2].

11.1.2 Packaging in retrospect: current challenges and needs

Packaging plays a critical role in determining overall system performance; current packaging schemes and interconnect methodologies are quickly becoming incapable of sustaining the high-bandwidth demand. For example, motherboard-based interconnects (Fig. 11.2A) have been widely used for relatively long range ($\sim$30−100 cm) [3] interconnects between modules. However, with the required I/O bandwidth between chips and modules increasing drastically, motherboard-based electrical interconnects have not been able to keep up with the bandwidth demand. This is primarily because of the inherent limit of the minimum channel pitch achievable on the motherboard; a high density differential stripline pair typically has a pitch in the order of $\sim$600 μm [3]. Assuming a 10 Gb/second channel data rate would imply that the bandwidth density achievable utilizing motherboard-based electrical interconnects is in the order of $\sim$16 Gb/second per mm. Operating channels at a higher frequency can result in a higher aggregate bandwidth between modules; however, higher channel frequencies are limited by channel dispersion limits, the need

Optical Interconnects for Data Centers. DOI: http://dx.doi.org/10.1016/B978-0-08-100512-5.00011-5

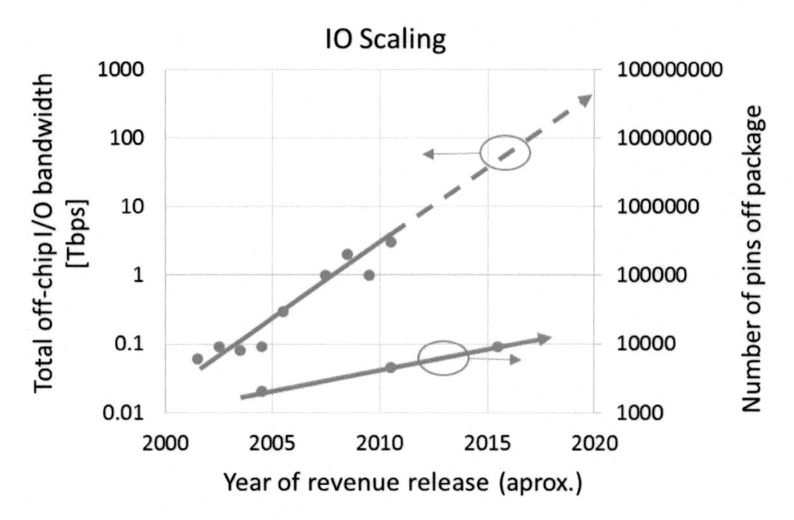

Figure 11.1 The scaling of processor I/O [1].

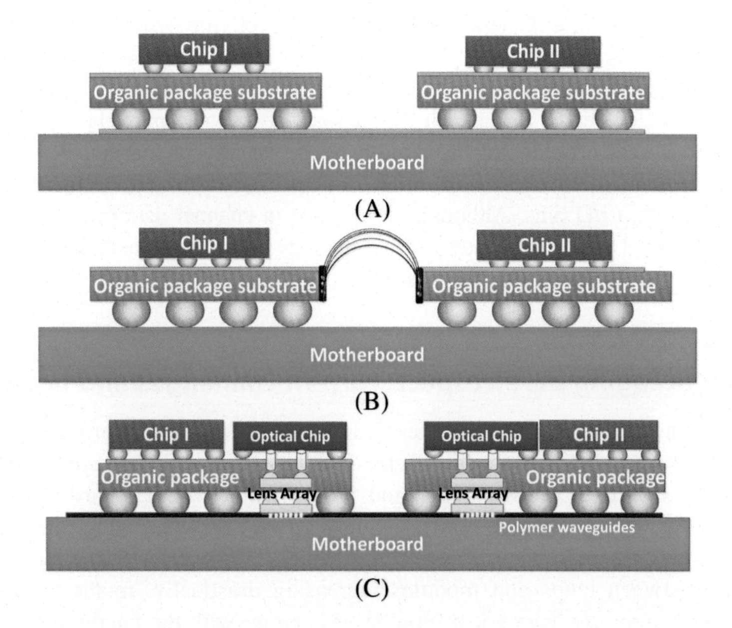

Figure 11.2 Conventional interconnect methodologies: (A) electrical interconnects on a motherboard, (B) flex electrical connectors, and (C) optical interconnects on a motherboard.

for a higher number of equalization taps, and a consequent increase in power dissipation [2]. Fig. 11.2B shows flex connectors as an alternative to connecting package modules via a motherboard. High-speed connectors are used to directly connect the package substrate, thereby bypassing the conventional socket and motherboard. This can allow a three times increase in raw bandwidth, and the ability to transmit higher data rates over longer distances as compared to FR4 boards [4].

Optical interconnects have also been explored for low-loss long-range chip-to-chip interconnects. The ability to send multiple wavelengths in the same channel using wavelength division multiplexing (WDM) allows higher bandwidth communication between modules. However, incorporating the electrical-to-optical conversion overhead and the laser efficiency increases the total energy per bit expended, and hence limits their utilization for short distances for which electrical interconnects expend less overall energy. Furthermore, the pitch of the waveguides on the board is typically fabrication limited, and hence cannot be scaled to very fine dimensions. Fig. 11.2C shows the schematic of terabus architecture with polymer waveguides at the board level fabricated at a 62.5 μm pitch [5,6]. A silicon-based "optical chip" converts the electrical signals to optical signals, which are then relayed to the motherboard via a lens array and optical couplers. The terabus has been shown to provide a bidirectional aggregate data rate of 360 Gb/second bandwidth over 24 transmitter and 24 receiver channels with a polymer waveguide on the optical PCB, with each channel operating at 15 Gb/second and a link EPB of 9.7 pJ/bit [6]. This translates to a bandwidth density of ~240 Gb/second per mm.

11.2 Emerging electrical and photonic interconnects

11.2.1 Silicon interposer technology: "the good and the bad"

With performance scaling bottlenecks becoming more and more packaging centric, there is ever growing research being carried out in packaging for next generation computing platforms. In particular, solutions that can provide high-bandwidth and low-energy communication between chips are being extensively explored. In recent years, silicon interposers have been explored for various benefits including higher bandwidth density, heterogeneous integration, and reduction in form factor [7]. The interposer, as shown in Fig. 11.3A, inserts as an intermediate tier between the VLSI chips and the package substrate, and provides the capability of high density wiring between the VLSI chips. This dense wiring on silicon interposers allows higher aggregate bandwidth between chips. For example, assuming a stripline differential pair pitch of 22 μm [8] for interconnectson a silicon interposer, and a channel data rate of 10 Gb/second, the bandwidth density achievable is ~450 Gb/second per mm. The energy efficiency of such a link of length 4 cm is 5.3 pJ/bit [8]. While EPB is lower than the optical PCB case discussed earlier, the EPB for interposer-based interconnects is a strong function of length and would quickly surpass that of the optical PCB case for longer interconnects. Based on the 2.5D integration platform, there have also been different topologies presented leveraging the key benefits of interposer technology while addressing cost, energy, or packaging challenges. For example, Intel's embedded multi-die interconnect bridge (EMIB) packaging [9] reduces the overall silicon area on the package (compared to using a silicon interposer), eases the assembly process, and reduces the overall cost when compared to conventional interposer technology.

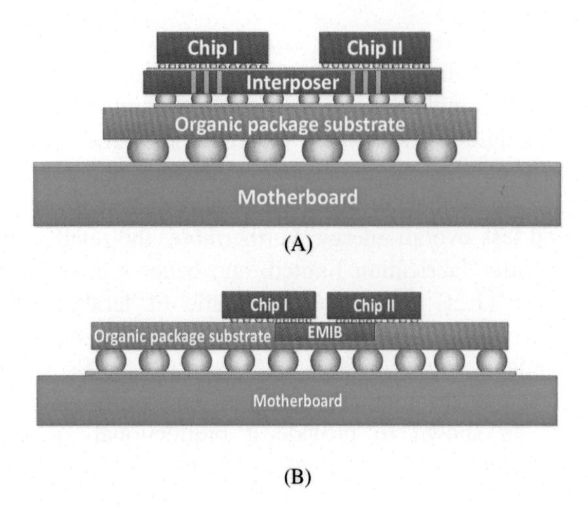

Figure 11.3 Silicon-based: (A) electrical interposer, and (B) Intel's embedded multi-die interconnect bridge (EMIB)—fine pitch wires achievable on silicon enable high density communication.

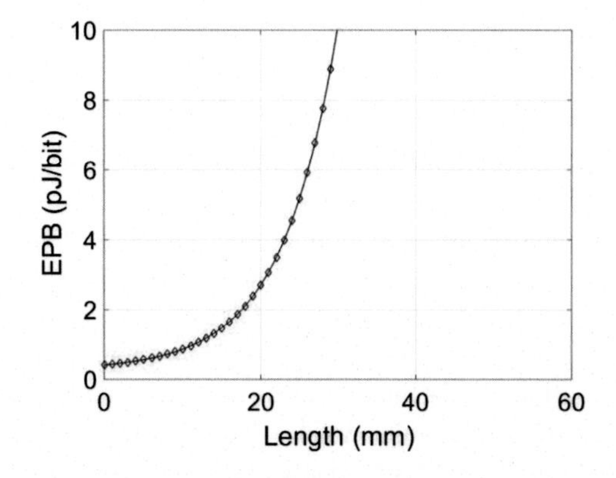

Figure 11.4 Energy per bit expended for a differential copper transmission line on silicon.

However, as the system size and the number of VLSI chips in a system continue to grow, there is a growing demand for high bandwidth, energy efficient communication over longer distances (>5 cm). Although this can be achieved by reducing wire pitch or increasing data rate per channel, having long, fine pitch wires on silicon carrying data rates >10 Gbps would make the interconnects extremely lossy. Fig. 11.4 shows the EPB of a differential copper transmission line on silicon with a 2.5 μm wire width and a 12.5 μm channel pitch as a function of length. The EPB for the electrical interconnects is calculated assuming a differential voltage mode signaling scheme [10]; the geometric dimensions of the transmission line are used to extract the R,L,C, and G parameters; the propagation constant is then determined based on these parameters. Using the receiver noise condition for a maximum BER of 10^{-12}, the minimum driving current is calculated which is then used to find the total loss in the transmission line [10]. As seen from the graph, as the length of the interconnect increases, the EPB expended rises exponentially. This limits the reach of high bandwidth density interconnects due to energy budget constraints.

11.2.2 Silicon photonics and optical coupling

With the range of electrical interconnects being limited by the energy budget for chip-to-chip interconnection coupled with a continuous increase in the required off-chip bandwidth, silicon photonics-based optical interconnects have emerged as a lucrative alternative to conventional electrical interconnects. Compared to electrical interconnects which suffer from increased latency, and power consumption, as well as decreased bandwidth as they scale, photonic interconnects exhibit a completely different signal propagation mechanism allowing them to achieve ultra-high-speed [11]. Therefore, photonics has become a promising solution for high-bandwidth, high-density, and low power-consumption communications with applications in datacom, access networks, I/O for bandwidth-intensive electronics [12], and chip-to-chip interconnects [13].

Silicon-based photonic systems have already gained a foothold in the photonic industry. Most silicon photonics are demonstrated on the CMOS-compatible silicon-on-insulator (SOI) platform, due to its ability to integrate both electronics and photonics on the same chip. The high index contrast between the silicon core and the cladding material (air, silicon dioxide, etc.) allows for tight confinement of light in the core. Moreover, it is easy to fabricate submicron structures such as single-mode waveguides, resonators, photonic crystals, etc. [14].

The integration of dense electrical and photonic wires is required to meet the ever growing needs of terabits per second data rates driven by modern computing systems. When photonics are used at the chip level, optical connectors are necessary in order to couple optical signals in and out of waveguides. Furthermore, optical couplers might also be utilized for interlayer coupling within the system. Thus, the coupling loss can become a significant contributor to the total energy expended and requires efficient coupling schemes.

The most favorable approach for optical coupling is using diffractive gratings. Grating coupling has demonstrated many benefits, such as using CMOS-compatible fabrication, eliminating facet cleaving, polishing or additional processes, allowing different mode coupling, as well as enabling wafer-scale optical testing and low-cost packaging. Grating couplers can be applied in either fiber-to-chip coupling or chip-to-chip coupling. In the case of chip-to-chip coupling, a pair of diffractive gratings can be used as output/input couplers to achieve interlayer optical communications.

The diffractive grating consists of a periodic variation in the surface profile, called the surface-relief grating(Fig. 11.5A), or in the refractive index of the material, called the volume grating (Fig. 11.5B). Surface-relief grating can be surface-corrugated (Fig. 11.6A), substrate-corrugated (Fig. 11.6B), or double-corrugated (Fig. 11.6C). Compared with volume gratings which are mostly polymer-based,

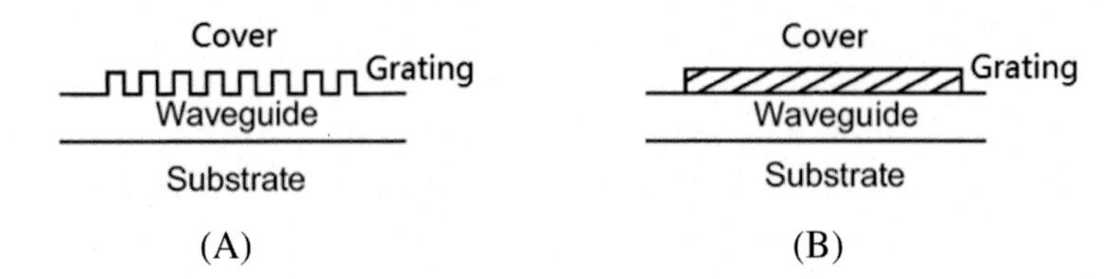

Figure 11.5 Two types of diffractive gratings: (A) surface-relief grating, and (B) volume grating.

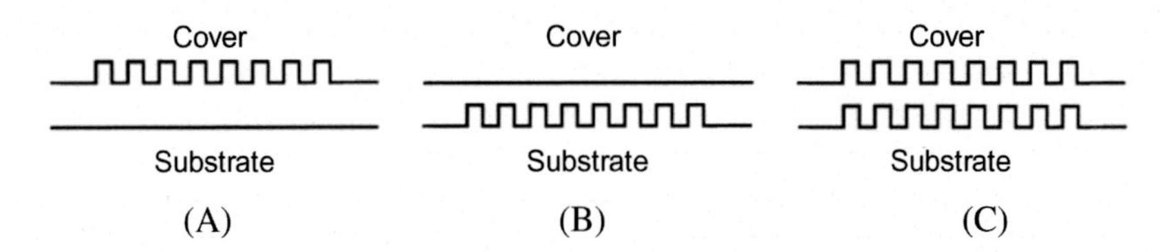

Figure 11.6 Three types of surface-relief gratings: (A) surface-corrugated, (B) substrate-corrugated, and (C) double-corrugated.

surface-relief gratings have reduced thicknesses, thus becoming a popular solution in compact interconnect technologies. Even more, surface-relief gratings can be fabricated using Si-based materials, and thus can be easily integrated in the CMOS platform.

High efficiency grating couplers can reduce losses at the connections, and thus extend the scale of the photonic system. Whether it is fiber-to-chip coupling or chip-to-chip coupling, improving the diffraction efficiency of a single grating plays a vital role. Several approaches, with the goal of enhancing radiation directionality, have been proposed to improve grating diffraction efficiencies.

The radiation directionality is defined as the ratio of radiated power into the cover P_c to the total radiated power $P_c + P_s$, where P_s indicates the radiated power into the substrate. The grating can be made intrinsically directional with $P_c >> P_s$. The first strategy is to define an asymmetric grating profile. This is due to the fact that gratings with symmetric profiles (e.g., sinusoidal and rectangular) radiate the incident power (or guided power) almost equally into the cover and the substrate, resulting in a diffraction efficiency close to 50%, provided the refractive index of the cover is comparable to that of the substrate. On the contrary, gratings with asymmetric profiles (e.g., blazed (trapezoidal, sawtooth or triangular) and parallelogramic) emit more than 90% of the radiated power to the cover [15–18]. It is reported that the blazed grating can direct 97% of the total radiated power to the desired angle [19]. Nevertheless, the radiation factor α (defined as the radiated power per unit grating length) of the blazed grating is very small, and thus a longer grating is required to radiate the same amount of incident (or guided) power. Researchers have found that a parallelogramic grating provides both a larger radiation factor and a higher radiation directionality than a grating with other tooth profiles [20–22]. Therefore, it offers the benefits of simultaneously obtaining high

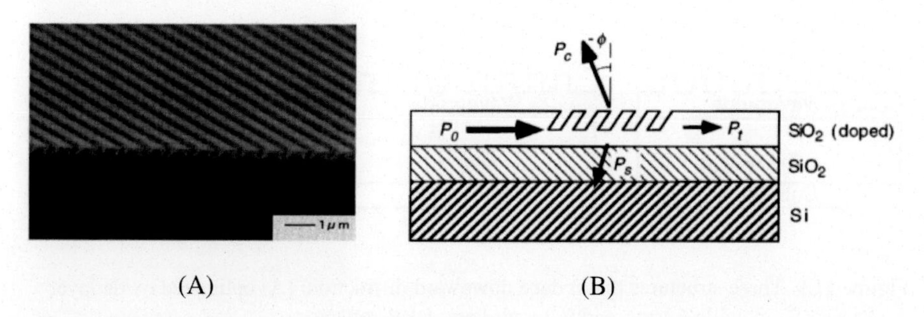

Figure 11.7 (A) SEM image of a parallelogramic grating in SiO_2 with grating period 0.8 μm; (B) a schematic diagram of a parallelogramic grating on a SiO_2/Si planar waveguide [21].

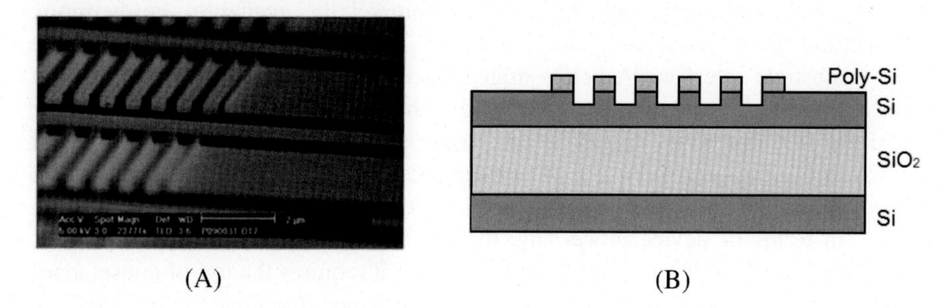

Figure 11.8 (A) SEM image of an SOI grating with poly-Si overlay [24]; (B) a schematic diagram of an SOI grating with poly-Si overlay.

coupling efficiency and high device compactness. The SEM image of a parallelogramic grating is shown in Fig. 11.7A, and its corresponding structure is schematically shown in Fig. 11.7B, where P_0, P_c, P_s, and P_t indicate the guided power, the radiated power into the cover, the radiated power into the substrate, and the transmitted power, respectively.

The second strategy is to define a high-refractive index overlay structure prior to the etching of the grating. The overlay structure enhances the directionality to the cover by facilitating the constructive interference between the diffracted light propagating in the cover (or grating grooves), and the light propagating in the overlay region. As a result, the upward-propagating waves are intensified while downward-propagating waves are suppressed. Usually, a poly-Si overlay is applied on the Si-on-insulator (SOI) grating structure [23]. This directionality is thus a function of the waveguide thickness, the etch depth in the waveguide, and the silicon overlay thickness. Experimentally, by using the overlay structure (Fig. 11.8), −1.6 dB (69.18%) fiber-to-chip coupling efficiency is obtained for TE polarized light [23].

Another approach to enhance directionality into the cover is to reduce the diffracted power into the substrate P_s. A simple method to accomplish this is to optimize the buried oxide (BOX) layer thickness (Fig. 11.9A) so that light diffracted to

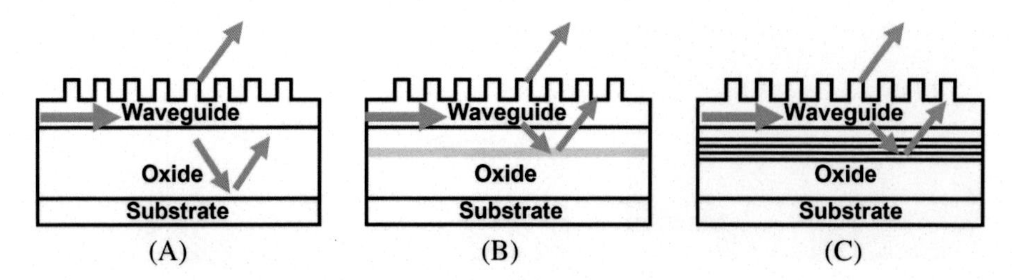

Figure 11.9 Three structures that reduce downward diffraction: (A) optimized oxide layer thickness, (B) metallic mirror, and (C) multilayer DBR reflector.

the substrate is reflected at the bottom of the BOX layer, and constructively interferes with light diffracted to the cover. However, this method requires a custom SOI waveguide structure which is not commercially available. Furthermore, this method only moderately improves the directionality, depending on the oxide – substrate interface. A better strategy is to apply a bottom mirror to reflect close to 100% of the downward diffracted light. This can be achieved by either depositing metallic layers or distributed Bragg reflectors (DBR) at the bottom of the waveguide, as shown in Fig. 11.9B and C, respectively. This approach still requires careful optimization of the distance between the bottom mirror and the grating. In terms of device processing, the fabrication process is complicated and not directly transferable to a CMOS process, since it requires the use of nonstandard bonding processes. As examples, high-efficiency grating structures for the 1550 nm wavelength were realized using a gold bottom mirror [25–27], and using a two-pair Si/SiO_2 DBR stack [28].

Other design considerations include reducing undesirable interference from back-reflection or other diffraction orders [29–31], improving mode matching between waveguide and fiber [32–36], and introducing subwavelength structures to suppress diffraction effects (e.g., scattering loss and large mode mismatch) common in conventional grating [37].

11.3 Large-scale interconnected system using a "silicon bridging" concept

Integration technologies at 2.5D and 3D , including silicon photonics-based interposers, discussed earlier, enable high-bandwidth low-energy communication between chips. However, there is some tradeoff necessary in lieu of the benefits that the technology brings. While EMIB technology enables high-bandwidth communication between chips connected via the bridge, it increases the package complexity and substrate processing. Furthermore, EMIB technology is specific for chips that are spatially in close proximity in the package, and high density communication is only between the adjacent chips in the package. Moreover, since the platform does not support optical communication, it cannot take advantage of WDM to achieve higher aggregate bandwidths.

Both electrical and optical silicon interposers allow high-bandwidth low-energy communication between chips that are even spatially further away. While electrical interconnects on silicon interposers are range limited due to the exponential increase in EPB, large optical interposers pose assembly and packaging challenges on top of higher costs.

Likewise, 3D integration poses a unique set of challenges introduced by stacking chips; thermal management of stacked dice, especially with high power dice such as processors present in the stack, becomes a critical issue. Also, the wafer-to-wafer bonding and sequential testing of stacked dice poses additional challenges that need to be overcome before large-scale adoption of the technology occurs [38–40].

The shortcomings of these technologies are exacerbated by the fact that the number of VLSI chips needing to be integrated in a system are continuously increasing. Thus, incorporating all of these chips on an interposer would require a very large silicon interposer, which would pose mechanical handling and cost challenges. Similarly, having a 3D stack with an ever-increasing number of stacked chips would further amplify the current challenges. Thus, there is still a need for a large-scale silicon system that allows high density electrical and optical communication between chips, and extends beyond the physical limits of an interposer. The next subsection describes one approach to realize this large-scale silicon system, along with its enabling technologies.

11.3.1 System overview

The discussion thus far motivates the utilization of interposers with electrical and optical interconnects, while eliminating the package layer, as it can provide the highest bandwidth density and efficient energy utilization compared to other methodologies discussed. It also motivates an integration solution that allows extension of bandwidth and energy benefits of an interposer beyond its physical and practical limits. In this context, Yang et al. [41,42], have proposed using silicon bridges to bridge adjacent interposer "tiles". Fig. 11.10 shows the overview of a silicon-bridged multi-interposer system platform using a two tile example. The interposer tiles are directly mounted onto the FR4, thereby eliminating package substrate; four positive self-alignment structures (PSAS) are fabricated on the FR4 corresponding to the inverse pyramid pits on the interposer tiles to provide a low-cost high accuracy alignment. Dense mechanically flexible interconnects (MFIs) are utilized to provide reliable connections to the motherboard, as well as the adjacent interposer via a silicon bridge while overcoming CTE mismatch and surface variations [43]. The silicon bridge allows one to extend the high-bandwidth fine pitch connections possible on an interposer beyond the physical limits of the interposer. Thus, the silicon-bridged multi-interposer system emulates a contiguous piece of large silicon allowing high density communication between interposers.

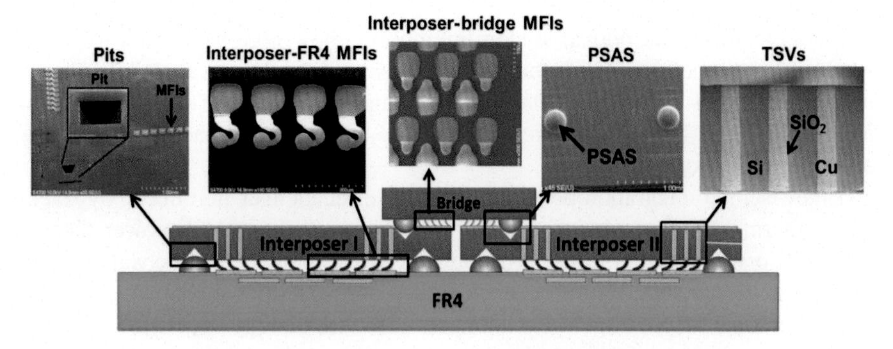

Figure 11.10 Two interposer tiles mounted on FR4 bridged using silicon bridge.

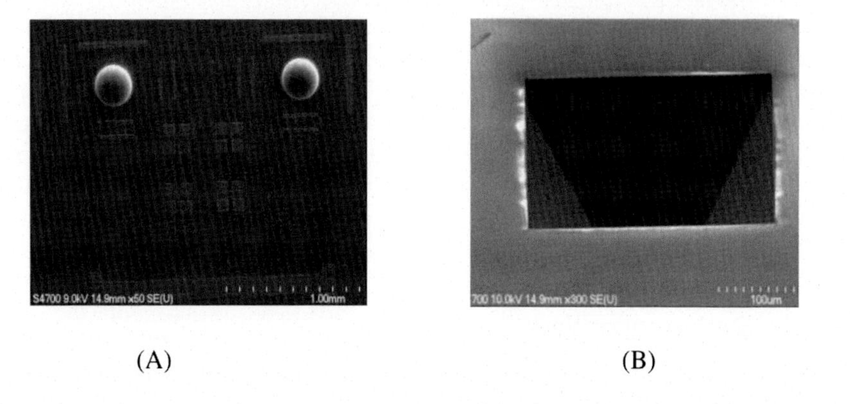

(A) (B)

Figure 11.11 (A) PSAS on bridge; (B) inverse pyramid pit.

11.3.2 Self-alignment devices and performance

Another key enabling technology for the silicon-bridged multi-interposer system is the self-alignment utilizing PSAS and inverse pyramid pits, as shown in Fig. 11.11. The inverse pyramid pits are fabricated using KOH etch of silicon using a nitride mask. The dimensions of the pits on both sides of the interposer are kept at $300\ \mu m \times 300\ \mu m$. The gap between the substrates is controlled by tuning the height of the PSAS [44]. PSAS are fabricated by precise reflow of photoresist. The PSAS − pit duo, shown in Fig. 11.11, enables precise control over the gap between substrates, which is a critical factor determining the optical coupling loss. The measured silicon-to-silicon and silicon-to-FR-4 alignment accuracy using vernier scale patterns is summarized in Table 11.1. Fig. 11.12 also shows the profilometer scan of the fabricated PSAS. As seen from the scans, the precise reflow of the photoresist results in a perfect truncated sphere and, along with the pit, allows sub-micron alignment accuracy and precise control over the gap between substrates.

Table 11.1 Alignment accuracy (μm) using PSAS

Regions	Silicon–Silicon		Silicon–FR4	
	X	Y	X	Y
Bottom left	< +1	< +1	+4.4	+2.0
Bottom right	+1	+1	+3.2	−3.2
Top left	< +1	+1	−1.6	−3.2
Top right	< +1	< +1	−2.8	+2.4

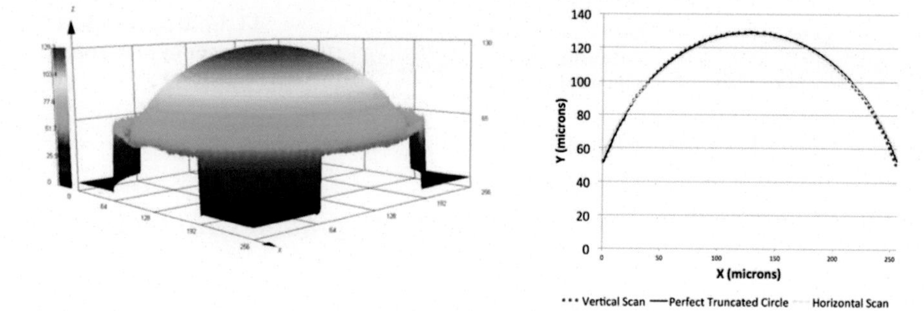

Figure 11.12 Profilometer scan data for fabricated PSAS [42].

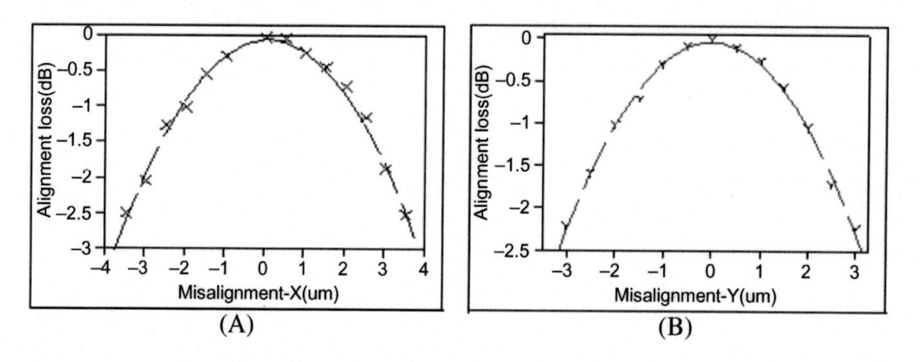

Figure 11.13 Excess coupling loss due to x and y misalignment [43].

Accurate alignment is also imperative to ensure coupling loss is at a minimum. Specifically for the silicon-bridged multi-interposer system, the alignment accuracy plays an even more critical role, as two optical couplings are required to optically connect the two adjacent interposers. As seen from Fig. 11.13, the coupling loss increases by 2 dB for a misalignment of 3 μm.

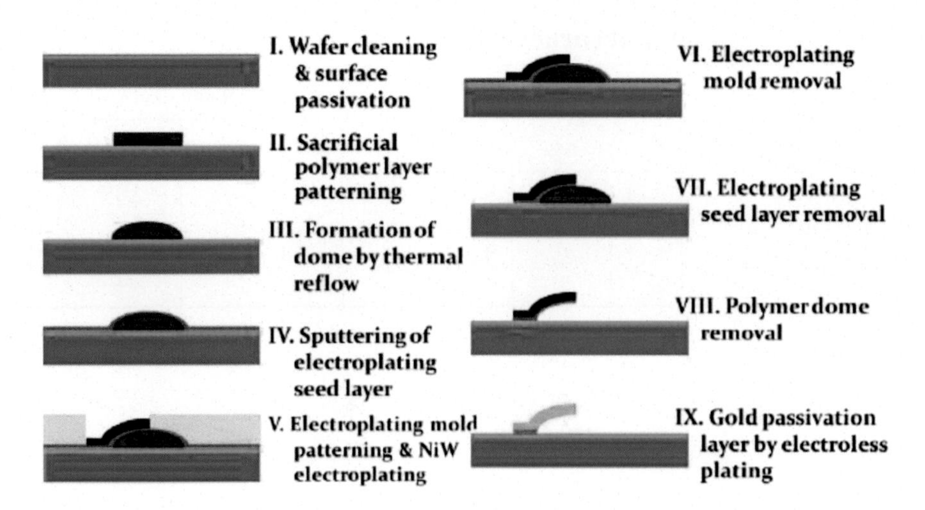

Figure 11.14 MFI fabrication flow [45].

11.3.3 Electrical flexible interconnect integration with photonic interconnects

MFIs are an integral enabling technology for the realization of the system; Fig. 11.14 shows the fabrication flow for MFIs. The MFIs are fabricated using NiW; the higher yield strength of NiW, as compared to copper [46], allows a greater range of elastic motion for the fabricated MFIs. The electroless gold plating passivates the MFIs and prevents any oxidation, ensuring a good contact. Fig. 11.15 shows the optical and SEM images of the fabricated gold passivated NiW MFIs [45]. The vertical stand-off height for these MFIs is 65 μm.

Indentation tests for mechanical characterization were performed using the Hysitron Triboindentor. Each cycle of indentation consisted of a downward motion, in which the indentor head deforms the MFI to a specified depth, and an upward motion in which the indentor head returns to starting height. A force–displacement graph is then plotted for the indentation cycle. Fig. 11.16 shows the compliance measurements performed on a single gold passivated NiW MFI, along with compliance measurements of a Cu MFI. As is evident from the measurements, the NiW MFIs show elastic behavior even after 100 indentations to a vertical depth of 65 μm. On the other hand, Cu MFIs quickly go into plastic deformation and lose vertical height. This can be seen from Fig. 11.16, where the sudden decrease in compliance at the 10th indentation can be attributed to the fact that the indentor head comes in contact with the stage on which the MFI sample is placed. These results show that the NiW MFIs can recover to the original height if they, e.g., go through warpage cycles in a system due to CTE mismatch. Also, the elastic behavior ensures a good electrical contact with time.

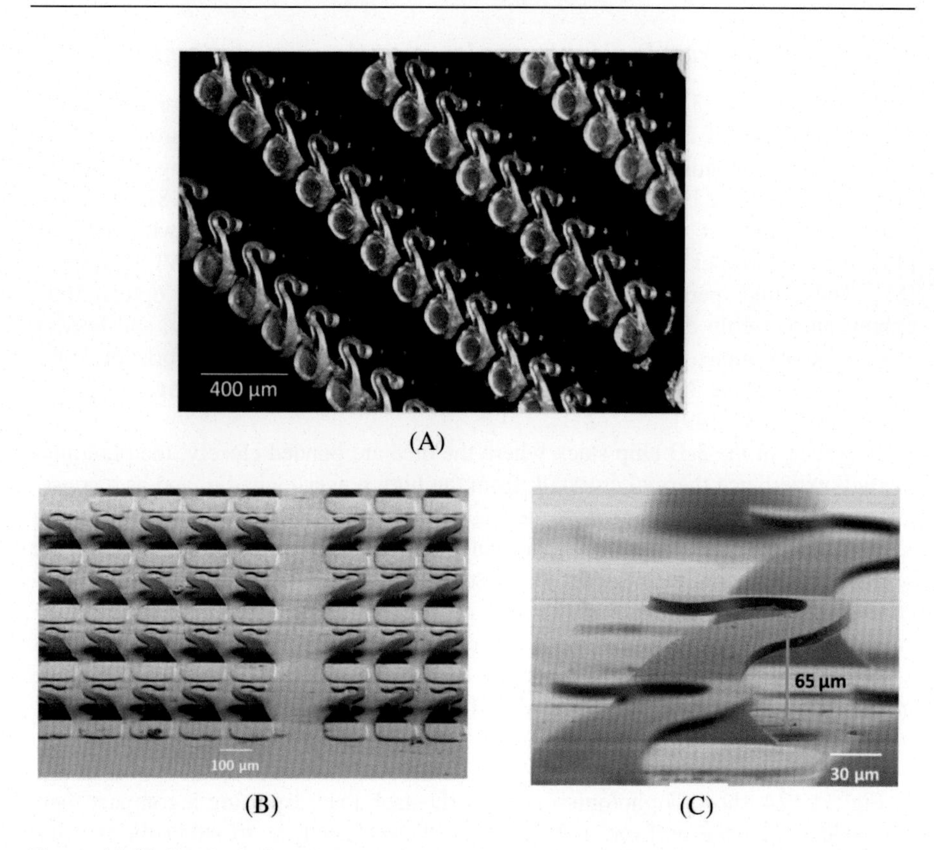

Figure 11.15 (A) Optical, and (B and C) SEM images of the MFIs. The vertical stand-off height is 65 μm [45].

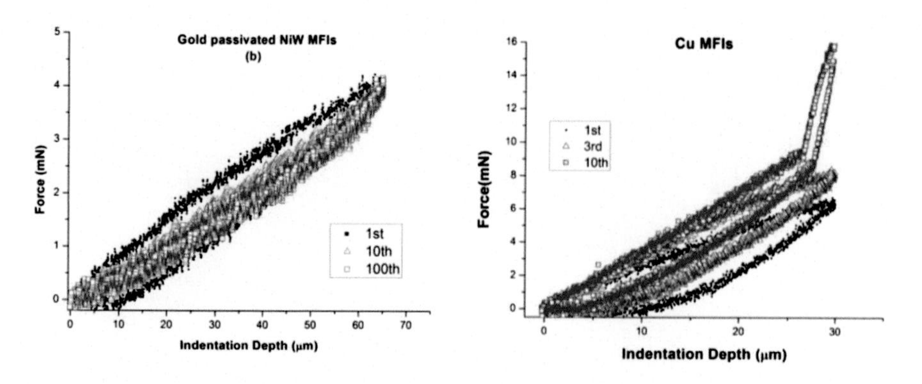

Figure 11.16 Compliance measurements for gold passivated NiW and Cu MFIs [45].

11.3.4 3D stacked logic, photonics, and thermal challenges

Three-dimensional (3D) integration is an emerging technology to address the bandwidth challenges of conventional nanoelectronics [44]. The use of through silicon vias (TSVs) or monolithic inter-die vias enable heterogeneous stacking of multiple dice with very large bandwidth [47] and low energy communication [48]. There has been recent interest to stack the silicon-compatible photonics die with memory [49], logic controller [50], and logic dice [51] to build large-scale VLSI systems [52]. These high performance computing blocks connected through optical links exhibit chip-to-chip communication advantages in bandwidth, density, and latency [53]. As an example, the circuit-switched optical interconnection networks on logic, can not only reduce 35% of the memory access latency, but also improve the power efficiency by 28% [49].

However, in the 3-D chip stack where the dice are bonded closely, the photonics die will experience thermal crosstalk from the high power logic die, and as a consequence, the thermal fluctuation of the logic die will be "mirrored" into the photonics die. But in the domain of silicon nanophotonics, a number of components are sensitive to temperature variation. For instance, the temperature sensitivity of a silicon based 10 μm-diameter electro-optic ring modulator is 0.11 nm/°C [54]. Assuming 64 channels of WDM sharing the working band, i.e., bands of 1530−1625 nm, each wavelength channel is 1.48 nm wide [55]. If we use the above modulators in the two ends (the send and receive), there will be a complete wavelength mismatch when the temperature drifts only by 14°C (wavelength drift is 1.54 nm > 1.48 nm).

Fig. 11.17A shows a photonic-on-logic 3D stack [51]. By using a compact thermal model [56], we perform a steady state thermal analysis to study the thermal coupling phenomenon mentioned above. The chip size is assumed to be 1 cm × 1 cm. The power map of the logic die is based on the Intel Core i7 microprocessor [57], and is shown in Fig. 11.17B. The total power is 74.49 W. The photonics die is assumed to consume 5 W with a uniform power distribution [51, 58]. The ambient temperature is assumed to be 38°C, and the thermal resistance of air

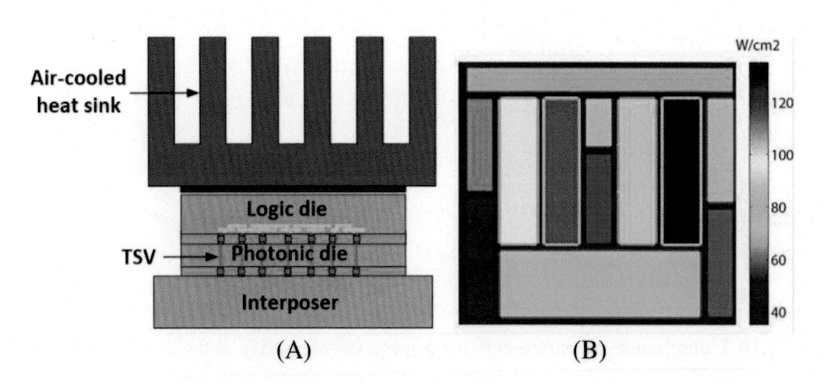

Figure 11.17 (A) A photonic-on-logic 3-D stack; (B) Power map of the logic die.

Table 11.2 The specification of simulated stack

	Conductivity (W/K · m)	Thickness (μm)
TIM	3	20
Photonics die	149	5 [58]
Underfill layer	0.9	40
Processor die	149	50
Micro-bump	60	40
Interposer	149	200

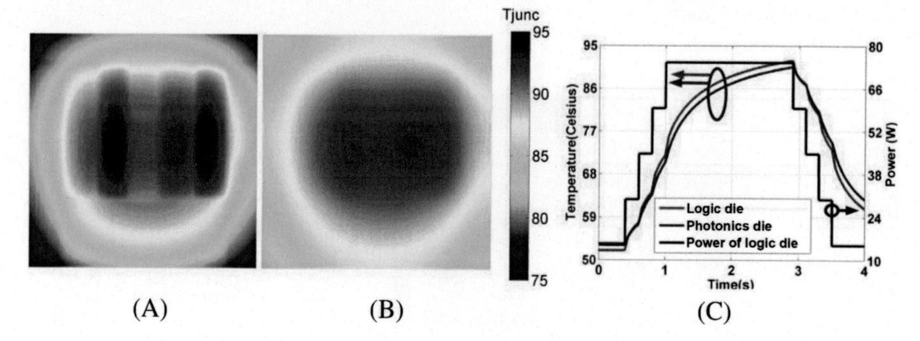

Figure 11.18 Thermal profiles: (A) Logic die: 94.60°C; (B) photonic die: 93.35°C; (C) Transient analysis results. The maximum temperature of each die is plotted.

cooled heat sink is assumed to be 0.5 K · cm^2/W. The thermal property and thickness of each layer are shown in Table 11.2.

The thermal profiles of both dice are shown in Fig. 11.18A and B, respectively. Firstly, we observe that although the photonic die dissipates a relatively low power (5 W), its maximum temperature is 93.35°C. Secondly, the photonic die thermal map is a mirror image of that of the processor, which reveals the strong thermal coupling between the two dice. The intra-die temperature variation is 12.7°C. From this simulation, we conclude the spatial temperature variation of processor die will be reflected into the photonics die.

To investigate the time domain thermal coupling between the logic and the photonic die, we perform transient thermal analysis [59] on the 3D stack, as shown in Fig. 11.17A. The logic is assumed to experience an on − off cycle, as plotted in Fig. 11.18C. The initial power of the logic is 15 W. Next, the logic power gradually increases to a peak power state (75 W), and remains for 1.9 seconds. The logic eventually returns to its initial state. From Fig. 11.18C, we observe the temperature of the photonic die always follows the logic die temperature, and the total temperature change is 37.0°C. This time domain thermal coupling phenomenon poses an unstable thermal environment for the temperature-sensitive photonic die, and will result in a communication error or power overhead of the thermal stabilizer circuit [60].

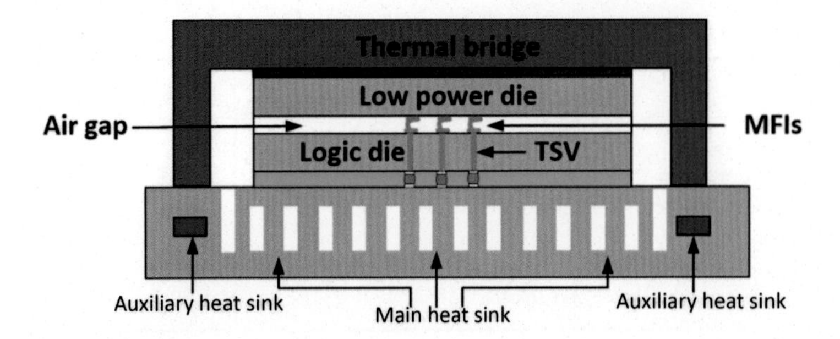

Figure 11.19 Proposed architecture with air gap, clustered MFIs, and separate cooling for logic and low power die.

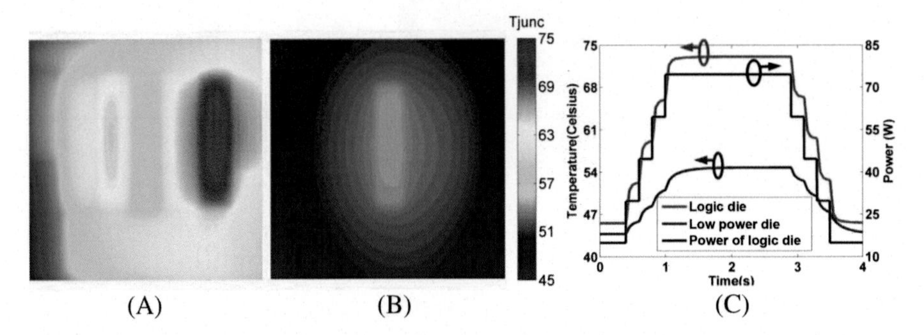

Figure 11.20 Thermal profiles: (A) Logic die: 73.12°C; (B) low power die: 54.79°C; (C) Transient analysis results of the proposed architecture. The maximum temperature of each die is plotted.

To reduce the thermal coupling, a novel architecture with a set of cooling and isolation technologies has been proposed, and is shown in Fig. 11.19 [56]; the figure shown is generic and not customized to a particular application or ICs. There are three key features. Firstly, interposer embedded microfluidic cooling is used. The main heat sink is designed for the high power logic die, and the auxiliary one is designed for cooling the thermal bridge, which is an extended heat spreader used for cooling the low-power die. Secondly, the I/Os between the two dice are clustered in the center following the specification of wide I/O technology. Therefore, the thermal coupling is limited to a specific area, instead of occurring throughout the whole chip. Lastly, MFIs are used to provide mechanical support and eliminate the use of underfill. Moreover, the MFI dimension can be carefully selected to enhance thermal isolation.

Figure 11.20A and B shows the simulated thermal profiles of the proposed architecture. The low-power die temperature is 19°C lower than the logic die. Moreover, the spatial temperature variation is only 7°C. Fig. 11.20C shows the transient analysis results. Although the logic die experiences a temperature change of 27.7°C, the low-power die temperature varies by 11.0°C. Compared to the conventional case shown in Fig. 11.18C, the temperature variation reduces by 26.0°C.

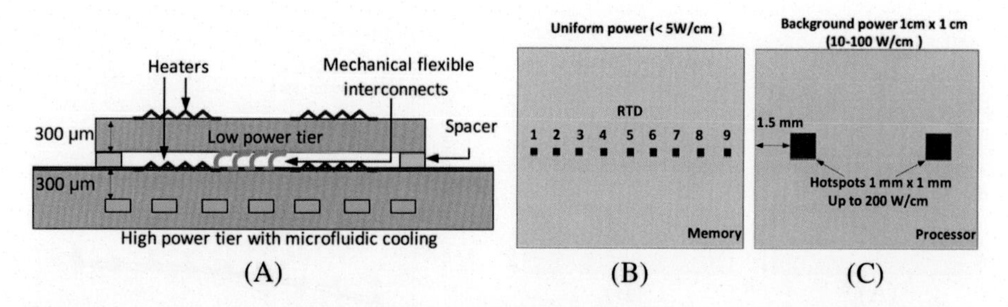

Figure 11.21 (A) Schematic of the designed testbed for evaluation of the proposed thermal isolation technologies. (B) The top (low-power), and (C) bottom tier (high-power tier).

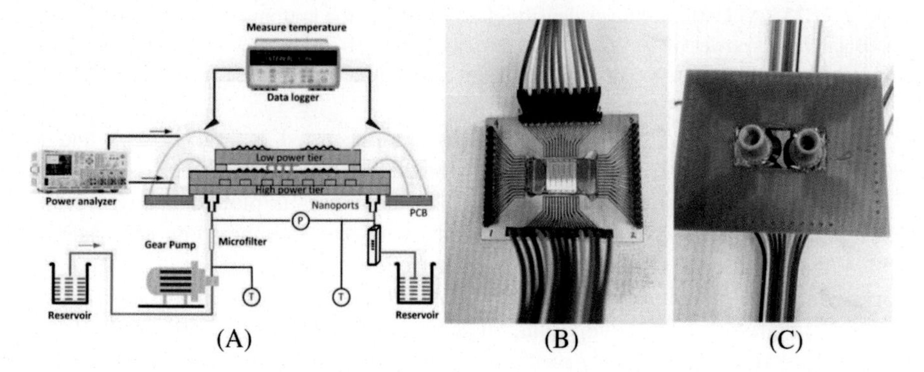

Figure 11.22 (A) Microfluidic test set up to evaluate the thermal isolation technologies. (B) Top and (C) bottom view of the stack assembled to a PCB board using wire bonding.

Guided by the above modeling and analysis, a thermal testbed is designed, as shown in Fig. 11.21A, to explore thermal coupling and solutions [61]. Fig. 11.21B shows the power map and temperature sensor designs for the low-power tier. The low-power tier dissipates a uniform power ≤ 5 W. A spiral heater is formed over a 1 cm × 1 cm area. Nine resistance temperature detectors (RTDs) are inserted along the middle of the chip in order to measure temperature along the length of the chip. Fig. 11.21C shows a schematic illustration of the high-power tier. The chip area is 1 cm × 1 cm. There are two hotspots on the chip, each measuring 1 mm × 1 mm. The two chips are interconnected with an array of NiW MFIs. The array contains 12 × 100 MFIs yielding a total of 1200 MFIs. This number is chosen based on the wide I/Os specifications [62]. The MFI design has a pitch of 75 μm × 100 μm. The entire MFI array is 9940 μm × 870 μm.

The microfluidic test set up is shown in Fig. 11.22A. The top tier is bonded to the bottom tier through MFIs that are located in the center region. An Agilent DC power analyzer is used to source current into the on-chip heater/RTDs on both tiers. The data logger is used to measure the resistance of the RTDs on the top and bottom tiers and extract the junction temperatures [63]. Fig. 11.22B and C show the assembled top and bottom die in a PCB using wire bonding.

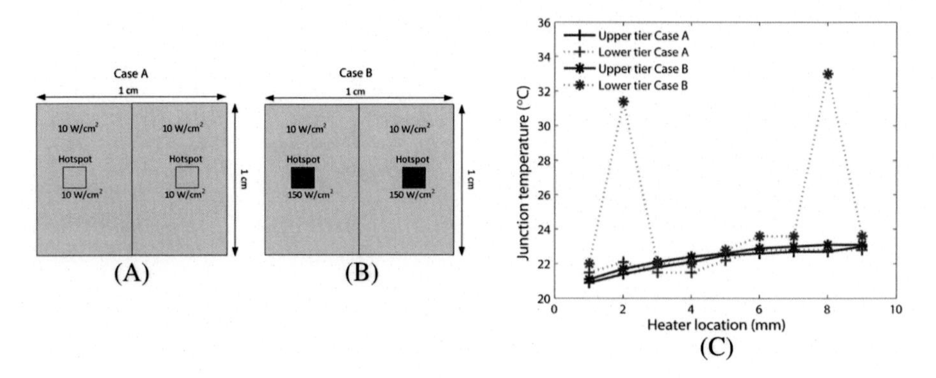

Figure 11.23 (A) Uniform power density of 10 W/cm^2 in the bottom tier (Case A), (B) Background power of 10 W/cm^2 plus two hotspots each dissipating 150 W/cm^2 (Case B), and (C) Junction temperature fluctuation of top and bottom tiers in Case A and Case B.

The power maps of the simulated cases are illustrated in Fig. 11.23A and B, respectively. In Fig. 11.23A, the bottom tier dissipates 10 W/cm^2 across the chip. The junction temperature for each location on both tiers is plotted in Fig. 11.23C (Case B). Next, (Case B), the power density of the two hotspots increases to 150 W/cm^2, while the background power remains unchanged (Fig. 11.23B). The corresponding temperature of each chip is plotted in Fig. 11.23C (Case B). In Case B, the temperature is relatively flat, indicating uniform temperature without hotspots. When the power density of the hotspot increases, one obvious observation is that there are two peak temperatures that occur in the bottom die. This is expected because of the large power density of the hotspots. The two peak temperatures are 31.4°C and 33.0°C, respectively. However, also in Case B, there are no obvious hotspots in the upper tier. The temperature of the upper tier gradually increases from 21.1°C to 23.1°C. This demonstrates that the proposed thermal isolation concept effectively minimizes the hotspot coupling between the stacked tiers.

11.4 Conclusion

With the perpetual increase in bandwidth requirements of chip-to-chip communication, it is imperative that an all-encompassing solution be devised—one that allows high-bandwidth low-energy communication for both short and long-reach interconnects. Conventional interconnects through the motherboard suffer from coarse channel pitch and the need for higher equalization taps for longer range. While flex-based interconnects bypass the package and motherboard, the channel pitch still limits the bandwidth density achievable. Optical interconnects at motherboard level are a promising alternative; however, the relatively coarse waveguide pitch and added package complexity makes them less attractive. Silicon interposer-based electrical interconnects allow high-bandwidth communication owing to the fine pitch achievable. However, long fine pitch wires on silicon quickly become very lossy and expend very high EPB. This has motivated efforts in the optical interposer regime which allow ultra-high bandwidth communication owing to the fine waveguide pitch and WDM. However, as the link EPB is dominated by the transmitter and receiver loss and the laser wall plug efficiency, the silicon nanophotonic links expend far more EPB at shorter lengths than the electrical interconnects. Although having a physically large interposer allows integration of multiple chips on to the same substrate, enabling high density communication, the cost and mechanical handling of such a large interposer are not feasible. The silicon-bridged multi-interposer system provides a viable alternate that allows the extension of the high-bandwidth density of an interposer beyond the practical limits. It further enables both electrical and optical communication for short and long-reach interconnects. Self-alignment technology allows submicron alignment accuracy that ensures low optical coupling loss, which is imperative for nanophotonic link integration. Three-dimensional integration [64] also aims to increase the bandwidth density and reduce the interconnect loss by stacking chips, thereby reducing the interconnect length. However, the thermal challenge is a potential show stopper for the 3D photonics-on-logic stack, despite the superior performance using vertical interconnections. Thermal coupling from a time-varying logic die leads to spatial and time domain temperature variation in the logic die, where many temperature sensitive components are integrated. The simulation results with the novel air isolation technology discussed show that it reduces the spatial thermal variation from 12.7°C to 7°C, and time domain variation from 37°C to 11°C. The experimental testbed fabricated to demonstrate the technology shows a spatial temperature reduction of 10.3°C.

With the demand for off-chip bandwidth projected to continue to grow, large-scale silicon systems, like the silicon-bridged multi-interposer system, are likely to come into more and more use. It is also foreseeable that coexistence of electrical and nanophotonic interconnects would be critical in achieving low energy interconnects for short and long-reach chip-to-chip interconnects.

References

[1] Zheng X, Krishnamoorthy AV. Si photonics technology for future optical interconnection. In: Communications and photonics conference and exhibition. Asia: ACP; 2011. p. 1–11, 13–6.

[2] Kam DG, et al. Is 25 Gb/s on-board signaling viable?. IEEE Trans Adv Packag 2009;32 (2):328–44.

[3] Kuwahara T, et al. A study of high-density differential transmission line of the package board based on crosstalk reduction. In: 2015 international conference on Electronics packaging and iMAPS all Asia conference (ICEP-IACC); 2015. p. 878–81.

[4] Braunisch H, et al. High-speed flex-circuit chip-to-chip interconnects. IEEE Trans Adv Packag 2008;31(1):82–90.

[5] Schares L, et al. Terabus: terabit/second-class card-level optical interconnect technologies. IEEE J Sel Top Quantum Electron 2006;12(5):1032–44.

[6] Doany FE, et al. Terabit/s-class optical PCB links incorporating 360-Gb/s bidirectional 850 nm parallel optical transceivers. J Lightwave Technol 2012;30(4):560–71.

[7] Kim N, et al. Interposer design optimization for high frequency signal transmission in passive and active interposer using through silicon via (TSV). In: 2011 IEEE 61st Electronic components and technology conference (ECTC); May 31, 2011–June 3, 2011. p. 1160, 1167.
[8] Dickson TO, et al. An 8 × 10-Gb/s source-synchronous I/O system based on high-density silicon carrier interconnects. IEEE J Solid-State Circuits 2012;47(4):884–96.
[9] Braunisch H, Aleksov A, Lotz S, Swan J. High-speed performance of silicon bridge die-to-die interconnects. In: 2011 IEEE 20th conference on Electrical performance of electronic packaging and systems (EPEPS); 2011. p. 95–8.
[10] Cho H, et al. Power comparison between high-speed electrical and optical interconnects for inter-chip communication. In: Proceedings of the IEEE 2004 international Interconnect technology conference; June 7–9, 2004. p. 116, 118.
[11] Arakawa Y. Silicon photonics for next generation system integration platform. IEEE Commun Mag 2013;51(3):72.
[12] Bogaets W. Design challenges in silicon photonics. IEEE J Sel Top Quantum Electron 2014;20(4):8202008.
[13] Young IA, et al. Optical technology for energy efficient I/O in high performance computing. IEEE Commun Mag 2010;48(10):184.
[14] Xu XC, et al. Complementary metal-oxide-semiconductor compatible high efficiency subwavelength grating couplers for silicon integrated photonics. Appl Phys Lett 2012;101(3):031109.
[15] Peng ST, Tamir T. Directional blazing of waves guided by asymmetrical dielectric gratings. Opt Commun 1974;11(4):405–9.
[16] Streifer W, Burnham RD, Scifres DR. Analysis of grating-coupled radiation in GaAs: GaAlAs lasers and waveguides-II: blazing effects. IEEE J Quantum Electron 1976;12 (8):494–9.
[17] Marcuse D. Exact theory of TE-wave scattering from blazed dielectric gratings. Bell Syst Tech J 1976;55(9):1295–317.
[18] Chang KC, Tamir T. Simplified approach to surface-wave scattering by blazed dielectric gratings. Appl Opt 1980;19(2):282–8.
[19] Aoyagi T, Aoyagi Y, Namba S. High efficiency blazed grating couplers. Appl Phys Lett 1976;29(5):303–4.
[20] Li M, Sheard SJ. Experimental study of waveguide grating couplers with parallelogramic tooth profiles. Opt Eng 1996;35(11):3101.
[21] Li M, Sheard SJ. Waveguide couplers using parallelogramic-shaped blazed gratings. Opt Commun 1994;109(3–4):239.
[22] Liao TD, Sheard S, Li M, Zhu JG, Prewett P. High-efficiency focusing waveguide grating coupler with parallelogramic groove profiles. J Lightwave Technol 1997;15:1142–8.
[23] Roelkens G, Van Thourhout D, Baets R. High efficiency silicon-on-insulator grating coupler based on a poly-silicon overlay. Opt Express 2006;14(24):11622–30.
[24] Roelkens G, Vermeulen D, Selvaraja S, Halir R, Bogaerts W, Van Thourhout D. Grating-based optical fiber interfaces for silicon-on-insulator photonic integrated circuits. IEEE J Sel Top Quantum Electron 2011;17(3):571.
[25] Van Laere F, Roelkens G, Ayre M, Schrauwen J, Taillaert D, Van Thourhout D, et al. Compact and highly efficient grating couplers between optical fiber and nanophotonic waveguides. J Lightwave Technol 2007;25(1):151–6.
[26] Kang JH, Atsumi Y, Hayashi Y, Suzuki J, Kuno Y, Amemiya T, et al. Amorphous-silicon inter-layer grating couplers with metal mirrors toward 3-D interconnection. IEEE J Sel Top Quantum Electron 2014;20(4):1.

[27] Yao Y, et al. Low loss optical interlayer coupling using reflector-enhanced grating couplers. In: 2013 IEEE optical interconnects conference; 2013, p. 31–2.
[28] Selvaraja SK, Vermeulen D, Schaekers M, Sleeckx E, Bogaerts W, Roelkens G, et al. Highly efficient grating coupler between optical fiber and silicon photonic circuit. In: Lasers and electro-optics/international quantum electronics conference, Munich, Germany; 2009.
[29] Zhang Z, Huang B, Zhang Z, Cheng C, Chen H. Monolithic integrated silicon photonic interconnect with perfectly vertical coupling optical interface. IEEE Photon J 2013;5. pp. 6601711-6601711.
[30] Werquin S, Vermeulen D, Bienstman P. Implementation of surface gratings for reduced coupling noise in silicon-on-insulator circuits. IEEE Photon Technol Lett 2014;26 (16):1589.
[31] Li YL, Li LY, Tian B, Roelkens G, Baets RG. Reflectionless tilted grating couplers with improved coupling efficiency based on a silicon overlay. IEEE Photon Technol Lett 2013;25:1195–8.
[32] Shi R, Guan H, Novack A, Streshinsky M, Lim AEJ, Guo-Qiang L, et al. High-efficiency grating couplers near 1310 nm fabricated by 248-nm DUV lithography. IEEE Photon Technol Lett 2014;26:1569–72.
[33] Song JH, Budd RA, Lee BG, Schow CL, Libsch FR. Focusing grating couplers in unmodified 180-nm silicon-on-insulator CMOS. IEEE Photon Technol Lett 2014;26:825–8.
[34] Na N, Frish H, Hsieh IW, Harel O, George R, Barkai A, et al. Efficient broadband silicon-on-insulator grating coupler with low backreflection. Opt Lett 2011;36:2101.
[35] Antelius M, Gylfason KB, Sohlstrom H. An apodized SOI waveguide-to-fiber surface grating coupler for single lithography silicon photonics. Opt Express 2011;19:3592–8.
[36] Liu L, Zhang J, Zhang C, Wang S, Jin C, Chen Y, et al. Silicon waveguide grating coupler for perfectly vertical fiber based on a tilted membrane structure. Opt Lett 2016;41 (4):820–3.
[37] Chen X, Li C, Fung CKY, Lo SMG, Tsang HK. Apodized waveguide grating couplers for efficient coupling to optical fibers. IEEE Photon Technol Lett 2010;22:1156–8.
[38] Lau JH. Evolution, challenge, and outlook of TSV, 3D IC integration and 3d silicon integration. In: 2011 international symposium on Advanced packaging materials (APM), 2011. p. 462–88.
[39] Mourier T, et al. 3D integration challenges today from technological toolbox to industrial prototypes. In: 2013 IEEE international Interconnect technology conference (IITC), 2013. p. 1–3.
[40] Ko CT, et al. Wafer-level 3D integration with Cu TSV and micro-bump/adhesive hybrid bonding technologies. In: 2011 IEEE international 3D systems integration conference (3DIC), 2012. p. 1–4.
[41] Yang HS, et al. In: 2014 IEEE Optical interconnects conference, May 4–7, 2014. p. 71–2.
[42] Yang HS, et al. Self-alignment structures for heterogeneous 3D integration. In: 2013 IEEE 63rd Electronic components and technology conference (ECTC); 2013. p. 232–9.
[43] Yao J, et al. Grating-coupler based low-loss opticl interlayer coupling. In 2011 8th IEEE international conference on Group IV photonics (GFP); Septemeber 14–16, 2011. p. 383, 385.
[44] Meindl JD. Interconnect opportunities for gigascale integration. IEEE Micro 2003;23 (3):28–35.
[45] Zhang C, et al. Highly elastic gold passivated mechanically flexible interconnects. IEEE Trans Compon Packag Manuf Technol 2013;3(10):1632–9.

[46] Slavcheva E, et al. Electrodeposition and properties of NiW films for MEMS application. Electrochim Acta 2005;50(28):5573−80.
[47] Kang U, Chung H-J, Heo S, Ahn S-H, Lee H, Cha S-H, et al. 8 Gb 3D DDR3 DRAM using through-silicon-via technology. In: IEEE international solid-state circuits conference; 2009, p. 130−1.
[48] Zhang T., Wang K, Feng Y, Chen Y, Li Q., Shao B, et al. A 3D SoC design for H.264 application with on-chip DRAM stacking. In: 3D systems integration conference (3DIC); 2010, p. 1−6.
[49] Brunina D, Liu D, Bergman K. An energy-efficient optically connected memory module for hybrid packet- and circuit-switched optical networks. IEEE J Sel Top Quantum Electron 2013;19(2):3700407.
[50] Udipi A, Muralimanohar N, Balasubramonian R, Davis A. Jouppi N. Combining memory and a controller with photonics through 3D-stacking to enable scalable and energy-efficient systems. In: Proc. of international symposium on computer architecture (ISCA); 2011.
[51] Demir Y, Hardavellas N. Parka: thermally insulated nanophotonic interconnects. In: Proc. 9th international symposium on networks-on-chip (NOCS '15). New York (NY): ACM, 2015.
[52] Ho R, Amberg P, Chang E, Koka P, Lexau J, Li G, et al. Silicon photonic interconnects for large-scale computer systems. IEEE Micro 2013;33(1):68−78.
[53] Krishnamoorthy AV, et al. Computer systems based on silicon photonic interconnects. Proc IEEE 2009;97(7):1337−61.
[54] Manipatruni S, Dokania R, Schmidt B, Sherwood-Droz N, Poitras C, Apsel A, et al. Wide temperature range operation of micrometer-scale silicon electro-optic modulators. Opt Lett 2008;33:2185−7.
[55] Li Z, Mohamed M, Chen X, Dudley E, Meng K, Shang L, et al. Reliability modeling and management of nanophotonic on-chip networks. IEEE Trans Very Large Scale Integr Sys 2012;20(1):98−111.
[56] Zhang Yang, Zhang Yue, Bakir MS. Thermal design and constraints for heterogeneous integrated chip stacks and isolation technology using air gap and thermal bridge. IEEE Trans Compon Packag Manuf Technol 2014;4(12):1914−24.
[57] Intel, [online]. Available: http://newsroom.intel.com.
[58] Sun C, Wade MT, Lee Y, Orcutt JS, Alloatti L, Georgas MS, et al. Single-chip microprocessor that communicates directly using light. Nature 2015;528(7583):534.
[59] Zhang Y, Sarvey TE, Bakir MS. Thermal challenges for heterogeneous 3D ICs and opportunities for air gap thermal isolation. In: Proc. IEEE 3D systems integration conference (3DIC), Ireland; December 2014.
[60] Padmaraju K, Logan DF, Shiraishi T, Ackert JJ, Knights AP, Bergman K. Wavelength locking and thermally stabilizing microring resonators using dithering signals. J Lightwave Technol 2014;32(3):505−12.
[61] Zhang Yue, Zhang Yang, Sarvey T, Zhang C, Zia M, Bakir MS. Thermal isolation using air gap and mechanically flexible interconnects for heterogeneous 3D Ics. IEEE Trans Compon Packag Manuf Technol 2016;6(1):31−9.
[62] WIDE I/O SINGLE DATA RATE (WIDE I/O SDR), JEDEC Standard JESD229, Dec. 2011. [Online]. Available: http://www.jedec.org/standards-documents/docs/jesd229.
[63] Zhang Y, Dembla A, Joshi Y, Bakir M. 3D stacked on-demand microfluidic cooling for high performance 3D ICs. In: Proc. IEEE electronic components and technology conference; 2012, p. 1644−50.
[64] Tekin T. Review of Packaging of Optoelectronic, Photonic, and MEMS Components. IEEE Journal of Selected Topics in Quantum Electronics May-June 2011;17 (3):704−19. Available from: http://dx.doi.org/10.1109/JSTQE.2011.2113171.

Electro-optical circuit boards with single- or multi-mode optical interconnects

12

L. Brusberg[1]*, M. Immonen*[2] *and T. Lamprecht*[3]
[1]Fraunhofer Institute for Reliability and Microintegration, Berlin, Germany, [2]TTM Technologies Inc., Turku Area, Finland, [3]Vario-optics AG, Heiden, Switzerland

12.1 Motivation and classification of optical interconnects at the board level

Optical interconnects for data transmission at board level offer a significant reduction in power consumption, increased energy efficiency, system density, and bandwidth scalability compared to purely copper driven systems. So far, such embedded optical architectures do not exist in data center and network systems. However, there is a clear need to replace the electrical signal lines with optical interconnects for increased high-speed data transmission, due to the higher bandwidth by length product of optical interconnect structures in the backplane and line cards [1]. The system enclosure consists of different peripheral line cards that are plugged into an electro-optical backplane where signals are routed across. The integration of optics to the line cards can be divided into three possible configurations, as shown schematically in Fig. 12.1. Line card A has an optical engine located close to the integrated circuit in order to convert high-speed electrical signals into optical signals for data transmission through the system by flexible optical signal links. On line card B, the flexible optical links are replaced by rigid integrated waveguide layers. Line card C shows a configuration in which the electrical IC and optical engine are merged to a photonic IC. Such a configuration can be feasible using silicon photonic ICs with single-mode waveguides working at 1310/1550 nm at IC level.

The requirements for an electro-optical circuit board (EOCB) are the following:

- Low material absorption at all key wavelengths (850/980/1310/1550 nm)
- Tight waveguide bends
- Waveguide crossovers
- Low signal dispersion
- Termination with standard interfaces (e.g. MT)
- Thermal stability
- Volume production

EOCBs can be classified by the material in which the optical light is propagating, like optical glass fibers, polymers, and multicomponent glasses. Glass fibers will be embedded in polymer and terminated with MT connectors for fiber-based

Optical Interconnects for Data Centers. DOI: http://dx.doi.org/10.1016/B978-0-08-100512-5.00012-7

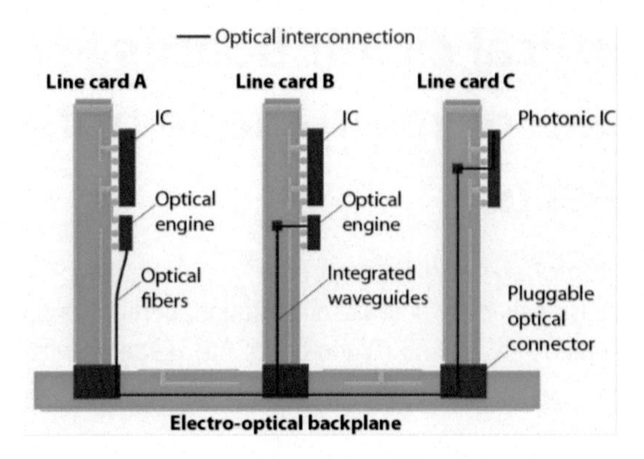

Figure 12.1 Example of a board-level optical interconnection (*red line*) which consists of an electro-optical backplane with integrated waveguides, pluggable optical board-to-board connectors, and three possible line card configurations with flexible optical fiber or integrated waveguide links.

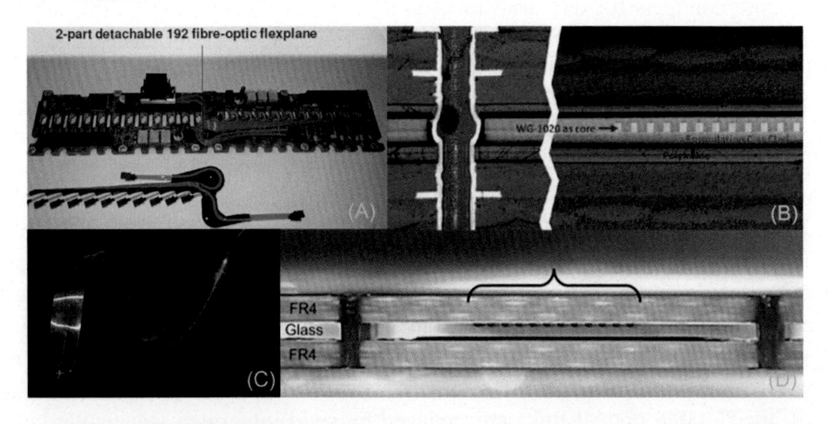

Figure 12.2 Classification of electro-optical circuit boards in: (A) optical fiber shuffles or fiber-optic flex plane [1], (B) planar integrated polymer waveguide in PCB stack-up [10], (C) polymer flex plane [11], and (D) planar integrated glass waveguide in PCB stack-up [12].

EOCBs, as shown in Fig. 12.2A. It seems to be the first step in implementing optical interconnects in commercial applications without the need to change the printed circuit board (PCB) stack-up concept by embedding additional optical layers. The benefits of optical fiber shuffles are: commercial availability [2,3], variety of termination solutions to industry-standard connectors like MT and others, very low propagation loss with smaller than 0.0023 dB/km for all key wavelengths, tight bends in the case of optimized waveguide profiles with high NA, and low dispersion in the case of single-mode and graded-index multi-mode fibers. The drawbacks are increasing layer thickness for crossings, and limitation of point-to-point links.

Polymer waveguide layers can be fabricated as rigid (Fig. 12.2B), flex (Fig. 12.2C), in combination flex-rigid-flex for embedding in EOCBs, or as flex on top with terminated connectors comparable to optical fiber shuffles. The benefit of polymer is the potential for low-cost high-volume waveguide fabrication with complex layouts including splitters, combiners, crossings, and tight bends [4]. Low propagation loss is limited to a wavelength range of around 850 nm, because of high bond vibration at telecom wavelength except with perfluorinated polymers. Recent planar polymer single-mode [5] and graded-index [6,7] waveguide development work addresses low signal dispersion.

Glass waveguides can be planar integrated in commercially available display glass which will be embedded as an optical layer in the PCB, as shown in Fig. 12.2D. Complex waveguides including splitters, combiners, crossings, and tight bends are achievable similar to polymer waveguides. For glass, the two sophisticated waveguide technologies are laser direct writing [8] and ion-exchange [9]. Ion-exchanged glass waveguide characterizes single- and multi-mode graded-index profiles for low signal dispersion and low propagation loss at all key wavelengths.

The fabrication of EOCBs is dependent on applied waveguide technology. The selection of suitable waveguide technology is dependent on application wavelength, layout complexity, connector interface, reliability requirements, and costs.

12.2 Manufacturing of integrated planar polymer waveguides

This section deals with the integration of an optical polymer waveguide layer into a stack of electrical layers to eventually form the EOCB. It discusses the general approach and the resulting requirements which apply for EOCB manufacturing. Finally, an industrially manufactured, customer-driven technology demonstrator for optical on-board communication is reported.

12.2.1 Basic concept of EOCBs

The cross-section illustrated in Fig. 12.3 shows the conceptual stack-up of an EOCB. The main component of an EOCB is an optical polymer layer, consisting of

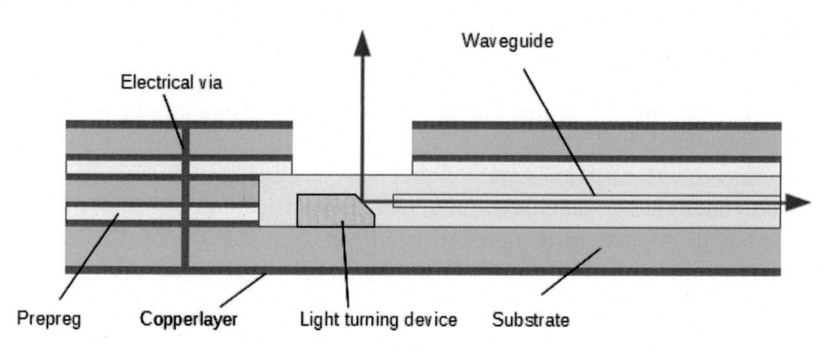

Figure 12.3 Principle stack-up of an electro-optical circuit board.

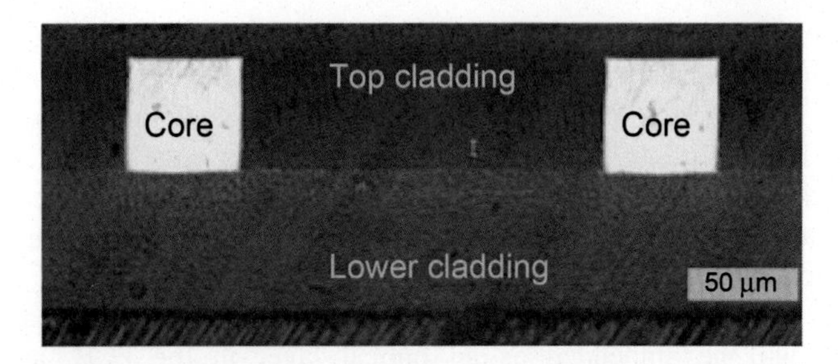

Figure 12.4 Cross-section of polymer waveguides with 50 μm core structures.

the waveguide core structures embedded in the polymer cladding layer. Core and cladding material have different refractive indexes in order to enable guiding of the light. A cross-section is shown in Fig. 12.4. Usually, the refractive index contrast of the polymer material is adjusted to the numerical aperture of the optical fiber. The light turning device enables vertical optical coupling to adjacent photonic components, i.e., photon emitting and receiving elements. In general, the optical layer is embedded within several layers of electrical circuits to form the EOCB. The stack-up sequence thereby follows conventional PCB fabrication. An electrical stack consists of alternating layers of dielectric material and a patterned electrically conductive layer, usually copper. The vertical connection between these individual copper layers is ensured by electrical vias.

Various reports on planar polymer waveguide technologies [13–15] show low insertion losses ≤ 0.05 dB/cm at wavelengths in the 850-nm range. Thereby, they demonstrate the general applicability of this technology for short reach multi-mode optical communication such as on-board and intra-rack optical links.

Different optical coupling schemes exist to bring the optical signals into and out of the waveguides. A possible integration approach of such a light turning device is illustrated in Fig. 12.3. It is embedded in the optical layer and allows light to couple vertically from VCSELs or photodiodes to the waveguides. At the board edge butt-coupling or end-firing approaches, e.g., to connectors [16,17], are usually implemented.

12.2.2 Manufacturing of an electro-optical PCB

The basic approach in EOCB manufacturing is to use conventional PCB fabrication processes, sequences, and standards in order to expedite technological acceptance. However, due to the required level of precision for optical waveguides, such as sidewall roughness in the submicron range and geometrical tolerances in the micron range, one has to implement waveguide-specific processes and tools for the optical layer itself.

The layout data for the optical and electrical layers are co-designed using a conventional EDA (electronic design automation) software tool. Each layer, the optical and the conventional electrical inner layers, is then manufactured separately, as described in Section 12.2.4. These layers are then assembled into a stack which eventually forms the EOCB, see Section 12.2.4.

A general waveguide fabrication approach based on photoexposure is described here. Processes and treatments related to specific optical materials are omitted. The detailed properties of various optical polymer materials, which are patternable by photoexposure and are applicable for electro-optical PCBs, have been discussed in depth in other publications [13–15,18,19], or are available from the material suppliers directly.

The PCB industry provides a wide variety of dielectric materials which can be incorporated in EOCBs. FR4 (flame retardant 4) is the workhorse among the family of woven-glass reinforced laminates. It has good availability, can be obtained with application-specific properties such as glass transition temperature (T_g) or glass fabric, and it is suitable for price-sensitive applications. Woven-glass dielectrics for high-frequency applications are made of specific matrix materials to provide low dielectric loss and low dielectric constant. They are more expensive and often require specific processing. Dielectric layers without woven glass reinforcement are available based on polyimide, PEEK, LCP, and other polymers. EOCBs are intended to send high-speed signals over the optical path, thus reducing the amount of expensive and difficult-to-process high-frequency dielectrics necessary in the stack-up. In principle, all the above dielectrics can be, and are actually used for EOCBs.

12.2.3 Manufacturing of planar polymer waveguides

Different methods exist for patterning planar polymer waveguides, which include: photoexposure, laser ablation, dry etch, and photobleaching [20]. Within this chapter, waveguide core formation based on photoexposure, in particular photomask lithography and laser direct imaging (LDI), is described [16]. Both methods make use of UV-photons in order to selectively cure the polymer and produce waveguide core structures with a high-quality sidewall topography. This is required to minimize optical propagation losses induced by light scattering. Photolithography and LDI are complementary technologies, whereby each method provides application-specific advantages.

The starting point is a substrate, preferably a dielectric with a patterned copper layer. It features various reference markers and also electrical traces when appropriate. The lower cladding layer, a liquid polymer, is deposited onto the substrate. Depending on the application, the desired layer thickness, and the topography of the substrate, different deposition tools can be deployed. Doctor blading is used for a wide variety of applications which require layer thicknesses in the range of several tens of micrometers. Doctor blading is very material efficient, and allows fabrication of layers with good homogeneity of layer thickness. For applications requiring thinner layers, one can apply conventional spin-coating. The lower cladding is finally cured by photolithography, in case a structure is required, or UV flood exposure for a continuous cladding layer.

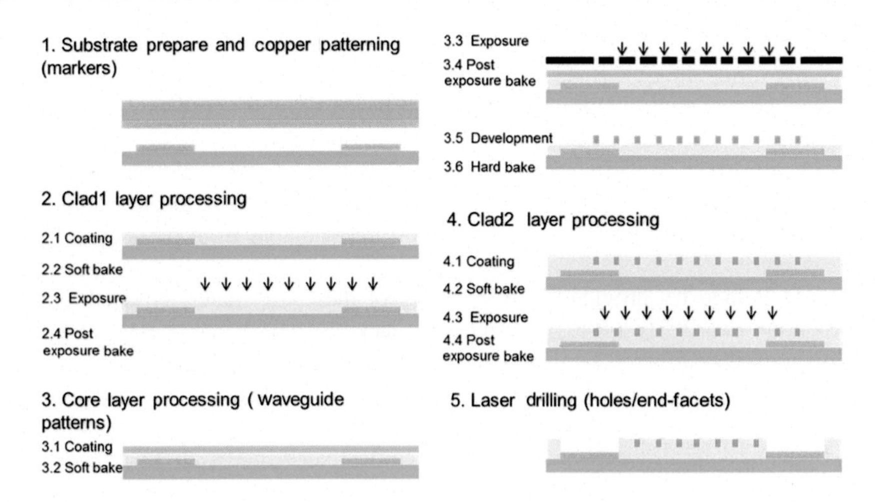

Figure 12.5 Process flow to fabricate one optical layer on a PCB.

In the next step the liquid polymer which eventually forms the core layer is deposited. The material is then patterned by photoexposure and subsequent removal of uncured polymer, the so called development step. The waveguide layer fabrication flow is illustrated in Fig. 12.5.

LDI allows direct patterning of the waveguide core structure via photo-induced curing of the polymer. This method is very flexible and cost-effective for manufacturers with a large product mix. The implemented vision system maps the distortion, which can easily reach a hundred micrometers, for each substrate individually. The controller then exactly maps the optical layout onto the distorted substrate. This correction allows the electrical layer and the optical layer to be precisely aligned to each other, which is particularly important in the coupling region. The LDI system currently installed at vario-optics is capable of exposing an area of $530 \times 610\ \text{mm}^2$, which is compliant with a mid-size standard PCB panel.

Mask exposure of optical waveguides requires high-quality chromium masks. The high cost and effort for large format masks precludes their utilization for large EOCB products, such as backplanes. However, smaller EOCB products, such as sensor applications, can be economically manufactured by applying step and repeat mask exposure, e.g., the mask is exposed multiple times on the same panel.

Once the waveguide cores are developed, they are covered with the upper cladding layer, using the same process sequence previously used for the lower cladding. By patterning the upper cladding, one gets access to the surface of the lower cladding and the dummy waveguide sidewalls as precise mechanical datum in the vertical and lateral dimensions, respectively [21,22].

Based on this manufacturing sequence, waveguide core structure dimensions ranging between 5–500 μm, thus covering a range of two magnitudes in core dimension, have been manufactured. Their applications range from single-mode and multi-mode optical communication, respectively, to sensor applications with lasers and LEDs [17,20,23].

12.2.4 Integration of planar polymer waveguides

After fabrication of the different electrical and optical layers, they are stacked together. The PCB industry conventionally follows a sequential stack-up starting from the inside, e.g., the optical layer. The individual layers are bonded together by prepregs (pre-impregnated composite fibers) during a lamination process. Prepregs are partially cured (B-stage) dielectric layers. They are usually matched, in terms of mechanical and dielectric properties, to the dielectric laminate of the previously manufactured electrical layers (Fig. 12.6).

During a common PCB lamination process, pressures of up to 15 bar and temperatures up to 190°C are applied for several hours, therefore curing the prepregs and eventually bonding the stack together (see Fig. 12.7).

Throughout EOCB fabrication, one has to pay attention to the distortions which occur during several process steps, in particular during the lamination process. The layout of certain electrical layers is adjusted to ensure that all layers, electrical and optical ones, are aligned to each other within the respective tolerance.

Electrical vias are used to connect the various electrical layers in a vertical direction. To form an electrical via, a hole is drilled and subsequently the sidewall of the hole is electrochemically plated with copper after adequate chemical treatment.

To avoid any design restrictions, e.g., in the case of an optical layer in the center of the EOCB, the delicate polymer waveguide structures have to withstand the harsh conditions occurring during lamination, drilling, and plating. The mechanical, and in particular the optical properties must be maintained. A key aspect is therefore the application-specific selection of the appropriate optical and dielectric materials. In Lamprecht et al. [15] such an EOCB, based on an optical silicone material from Dow Corning, was successfully manufactured and tested. It contains six electrical layers and one optical layer, with waveguides having a 50 μm core size.

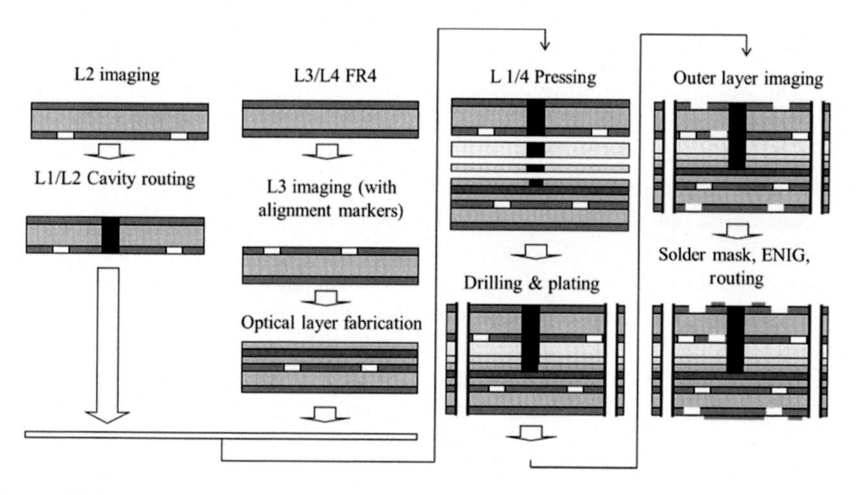

Figure 12.6 Example of the process flow to fabricate one optical layer (TTM Technologies).

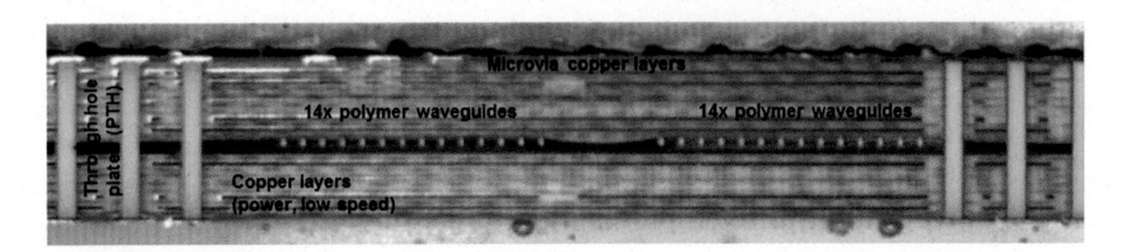

Figure 12.7 Cross-section of an EOCB with electrical copper layers, one optical polymer layer, and electrical vias penetrating the optical layer.

Another critical aspect is the registration between optical and electrical layers across the EOCB stack-up, particularly when the optical layer is located in the inner layers. In this case, the optical emitter/detector devices on the optical-electrical (O/E) ball grid array (BGA) packages are aligned to the optical waveguides located inside the PCB stack. Although micro-optics are utilized in the relatively long optical path, the EOCB must be designed to minimize dimensional and registration inaccuracies inherent in PCB materials and board fabrication processes. Registration errors of the contact copper layer (O/E package land pattern) to the optical layer corresponds to $X - Y$ misalignment in the optical path from the optical emitter/detector and optical waveguide input/output. Modern fabrication processes can achieve ± 0.075 to ± 0.05 mm (special) registration of layers in high layer count products with an overall thickness of up to 3 mm. In Immonen et al. [4], TTM Technologies reported the successful development of a modified process flow for achieving higher-than-standard alignment accuracy between the contact copper layer (L1) and the optical inner layer. Microscopic investigation revealed <20 μm top copper layer to waveguide accuracy across an approximate distance of 1 mm (960 μm copper core) (Fig. 12.8).

12.2.5 Technology demonstrator: on-board optical communication

An industrially manufactured EOCB with embedded polymer waveguides and electrical channels has been produced as a technology demonstrator by an EOCB manufacturer (vario-optics ag). It is made out of common high-frequency dielectrics (RO4000 from Rogers Corporation), and features 96 optical interconnects made out of silicone-based optical polymers [15]. The eight electrical layers plus one optical layer EOCB features reflow soldered coaxial connectors (Huber + Suhner) and electrical vias, as well as optical waveguide crossings and MT-based optical connectivity (Huber + Suhner). This demonstrator shows the technological readiness level required for commercial applications of on-board optical communication based on embedded polymer waveguides.

Another example of an EOCB is a demonstrator by TTM Technologies and Seagate (Fig. 12.9) developed under the PhoxTrot project (code: PhoxDem09.01) [4,24]. The demonstrator's top and bottom copper layers include complex routed high-speed layouts extracted from the existing design for the Seagate

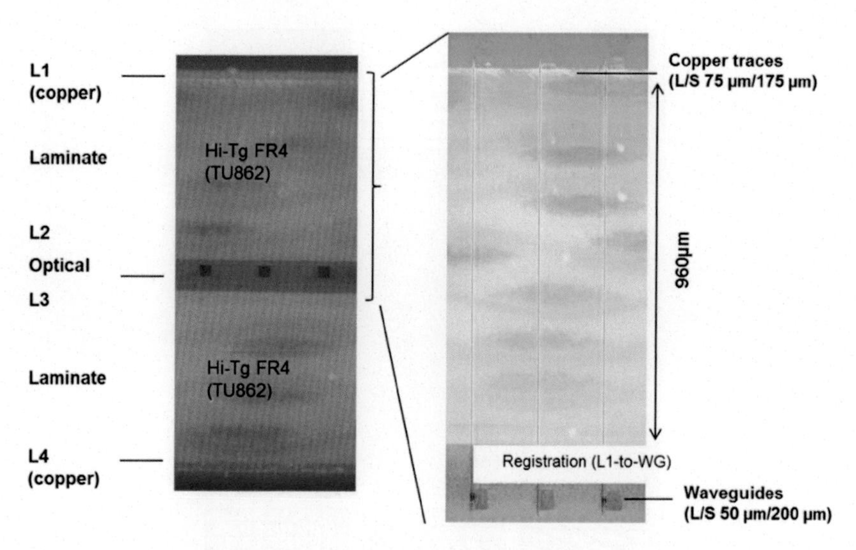

Figure 12.8 Cross-section of the 4L + 1Opt. board stack. Registration of the signal copper layer to waveguide core layer is <20 μm (TTM Technologies).

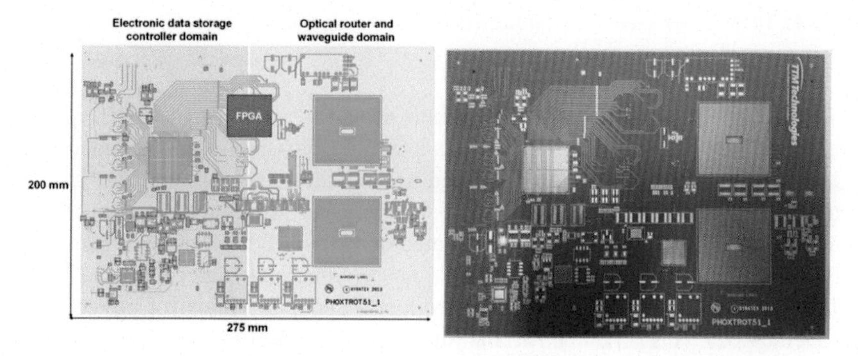

Figure 12.9 Technology demonstrator of an electro-optical PCB card layout containing 209 polymer optical waveguide test channels.

ThunderValley2 controller cards used in the optically enabled data storage platform demonstrator. The purpose of these complex routed copper layers is to enable the crucial demonstration of the fabrication challenges inherent in electro-optical PCBs with multiple complex routed copper layers and one or more optical layers. The two BGA sites with electrical land patterns matching O/E router (Compass Networks) layout with approximately 4000 I/O are designed for optical chip-to-chip traffic. The optical window under BGA is designed to accommodate the microlens prism component to couple light in/out from the VCSEL/PDs assembled on the BGA package.

The PhoxDem09.01 optical layer accommodates numerous waveguide test-groups exhibiting a full range of geometric configurations including: varied cross-

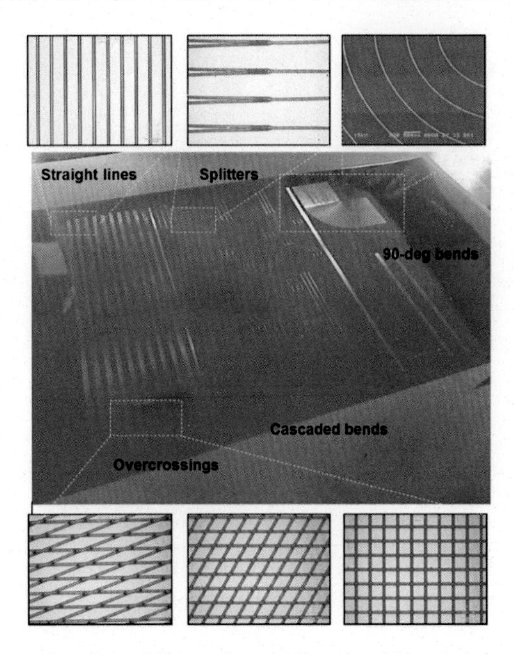

Figure 12.10 Micrographs of the fabricated optical waveguide structures in a 16″ × 20″ production panel.

sectional sizes (30–70 μm), 90 degree bends of varying radii (2–40 mm), cascaded bends with varying radii (10–20 mm), waveguide crossings with varied crossing angles (90–20 degree), splitters, tapered waveguides, and waveguides interconnected to midboard interface slots (Fig. 12.10). The form factor of the polymer waveguide PCB is 275 × 200 mm^2 fabricated on a 16″ × 20″ production panel.

The third EOCB demonstrator example is a commercial router product by TTM Technologies and Compass Networks developed under the Phoxtrot project (code: PhoxDem09.02) (Fig. 12.11). In the demonstrator, chip-to-chip intra card and ASIC-to-fiber traffic is realized using embedded optical waveguide links. The card shows, for the first time, a high layer count PCB product, wherein a multi-mode polymer waveguide signal layer is embedded into a complex PCB stack with 16 functional copper layers.

The results have shown the successful fabrication of optical waveguides with optical losses of ~0.1 dB/cm at 1000 nm, the operational wavelength of the electro-optical router. The waveguide-to-chip coupling losses were measured at about −3 dB, which is in line with the simulation results, and inter-waveguide crosstalk was measured at <25 dB.

12.2.6 Polymer waveguides on flexible substrates

As data rates increase annually, optical fiber communication is required for shorter distances. The use of fibers is nowadays common not only between, but also within

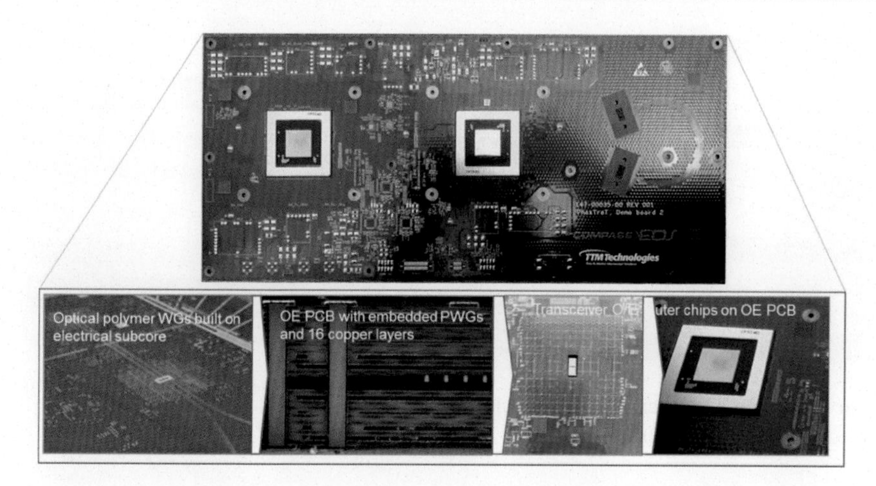

Figure 12.11 A high layer count OE router product with optical chip-to-chip and chip-to-fiber links embedded on board (16 electrical + 1 optical).

racks. New concepts include star, matrix, or even all-to-all communication, with the consequence of a very high number of high-speed communication channels required.

It is obvious that the assembly effort is directly related to the number of fibers. Another problem that occurs is that the space occupied by such fiber-based systems can reach a significant volume, with other negative consequences such as disturbance of the air cooling system. A first step to reduce the space occupied by fibers is to connect them to fiber ribbons of typically 12, 24, or 48 fibers for point-to-point connections. With fiber cable harnesses the architecture can be supported to a certain extent, but they still require significant space. A further step in the integration is the use of fiber shuffles. Single fibers are fixed on flexible substrates, allowing very complex routings.

12.3 Integrated glass waveguide based EOCBs

12.3.1 Basic concept

The glass-based EOCB consists of one or multiple glass waveguide panels which are embedded as inner layers in the core of the PCB. The glass panel is fully processed before embedding. A simple stack-up of one glass layer and two electrical layers is sketched in Fig. 12.12. The electrical package can be made of multiple copper and FR4 prepreg layers fabricated with standard PCB technologies (lamination, drilling, plating, etching), thus allowing advanced high-density electrical PCB designs. These electrical packages will be laminated from both sides to an optical package, which consists of a single glass or a stack of glass waveguide panels, as shown in the schematic in Fig. 12.12.

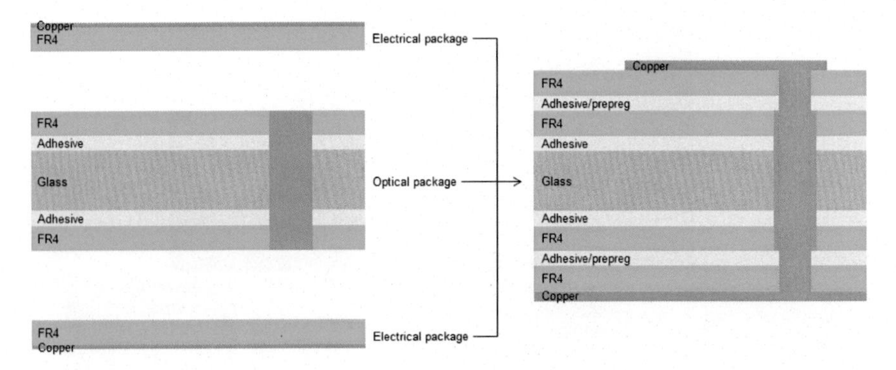

Figure 12.12 EOCB stack-up concept with a pre-processed glass layer which will be embedded in the core of the PCB.

12.3.2 Glass waveguide panel fabrication

The display industry produces alkali-borosilicate and alkali-aluminosilicate display glass sheets to which standard subsequent finishing processes like thin film deposition, lithographic processing, stacking, and cutting have been up-scaled. Spray coating and dip coating are common technologies, and sputter machines can handle large substrate sizes with sufficient homogeneity. In recent years a lot of laser processing has been developed and introduced into mass production. For instance, mask-less lithography using LDI technology avoids high mask costs while supplying highest process accuracies and flexibilities. For glass cutting, a CO_2 laser process in combination with cold jetting to thermo-mechanically scribe and mechanically break the glass achieves high-quality optical end-faces. The glass waveguide panel process line is shown in Fig. 12.13.

The glass waveguide panel process consists of a two-step thermal ion-exchange technology suitable for large formats and low-cost batch processing. Thin glass panels developed for surface strengthening by potassium ion-exchange containing monovalent alkali ions (Na^+) are commercially available for touch display applications. Refractive index modulation of display glass can be done by silver ion-exchange between salt melt and glass, which increases the refractive index sufficiently. The waveguide layout is defined by an aluminum thin film mask on the glass surface, as shown in Fig. 12.14A. The mask opening width directly influences the lateral waveguide dimensions. Then the masked glass is dipped into a hot salt melt bath (300–400°C) which serves as the ion source for the silver diffusion process (Fig. 12.14B). The ion-exchange between the sodium of the glass and the silver of the salt melt starts immediately because of the concentration difference in the area of the mask openings between the glass and salt melt. Silver ions diffuse into the glass and form an isotropic graded refractive index profile in the area of the mask opening. The refractive index maximum is located on the glass surface at the mask opening, and decreases to the bulk refractive index by moving deeper into the glass and under the mask. The maximum refractive index rise and diffusion

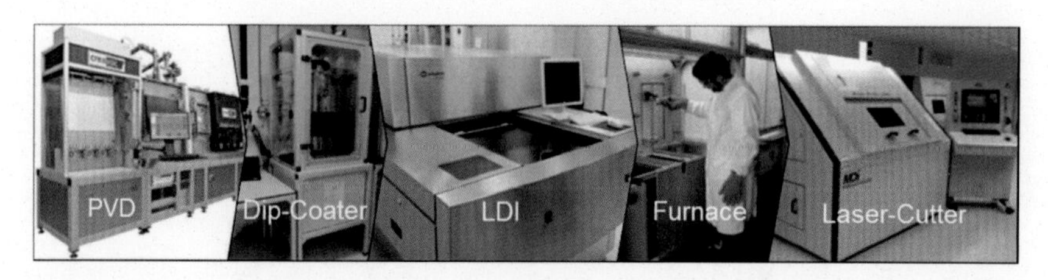

Figure 12.13 Fraunhofer's unique glass waveguide panel process line.

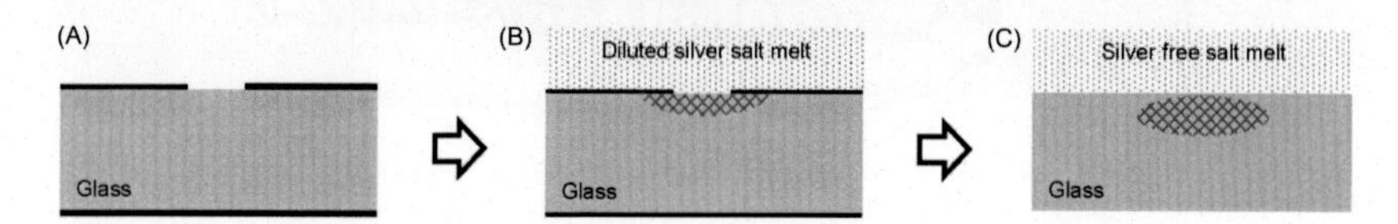

Figure 12.14 Ion-exchange waveguide process flow. (A) AI thin film mask on glass. (B) Thermal silver ion-exchange. (C) Mask removal & thermal ion-exchange.

depth can be adjusted by the silver salt melt concentration, process duration, and temperature control. The diffusion process stops when the masked glass is removed from the salt melt.

After removing the mask layer by chemical wet etching, a second thermal ion-exchange in silver-free salt melt is performed, as shown in Fig. 12.14C. A reverse silver diffusion out of the glass decreases the refractive index at the glass surface, and the position of the refractive index maximum is shifted from the glass surface deeper into the glass. In that process step the final waveguide profile is defined for high coupling efficiency to optical glass fibers. Single-mode and multi-mode waveguides can be manufactured by applying the introduced ion-exchange waveguide process in commercially available display glass by adaptation of the process parameters like mask opening width, temperature, duration, and salt melt concentration.

Single-mode waveguides have been fabricated in thin glass located 5 μm below the glass surface. The graded refractive index profile has a maximum index difference of $\Delta n = 0.003$. The measured mode field diameter is 19 μm in the horizontal and 12 μm in the vertical direction, because of the elliptical waveguide shape. The vertical mode field diameter matches almost perfectly to standard single-mode fibers, and the coupling loss in the best case without misalignment is only 0.3 dB from the fiber into the waveguide. The propagation loss of the single-mode waveguide is below 0.1 dB/cm at a wavelength of 1550 nm [24]. Multi-mode waveguides have also been fabricated in thin glass. The maximum refractive index difference of $\Delta n = 0.03$ is located at a depth of 30 μm below the glass surface. The propagation loss was measured at 0.13 and 0.05 dB/cm for 850 and 1310 nm, respectively. Such multi-mode glass waveguides which show very low bend losses and tight waveguide bends with a radius of ≥ 11 mm can be fabricated. A waveguide insertion loss of 90 degree arc waveguide group was characterized, and the results are plotted in Fig. 12.15 [25].

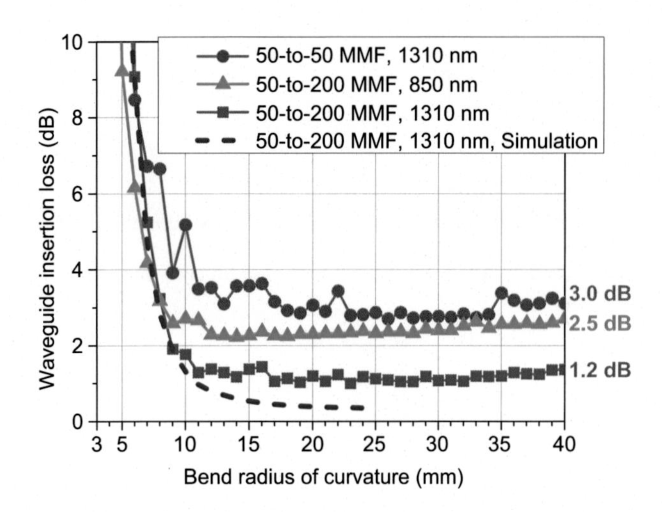

Figure 12.15 Insertion loss of 90 degree arc bend waveguide characterized with different measurement set-ups. The launching fiber is a 50 μm core GI-MMF, the collecting fiber is either a 50 μm core GI-MMF, or a 200 μm core SI-MMF [25].

12.3.3 Integration of glass waveguide panels in PCB stack-ups

Novel lamination techniques were developed to allow glass waveguide panels to be reliably integrated into a conventional electronic multilayer PCB. The optical package fabrication has been approved using two different concepts: lamination of different glass layers in a conventional multilayer press using adhesive sheet layers [26], and lamination of different glass layers with a cold lamination technique far below a temperature of 100°C using adhesive sheet layers [27]. A crucial point is the alignment of the glass layer to other layers during preparation of the first lamination process. One way is to align the glass layers and adhesive sheets by drilling precise holes for pin registration. Alternative alignment by marks was applied using visual alignment. The lamination cycle and parameters, especially ramp-up setting of press pressure, were adjusted to avoid glass cracking. The cut-outs in the glass are protected with inlets. In a parallel run, the electrical package is fabricated from a stack of FR4 and copper layers. The areas of optical windows in the FR4 layers for access to the glass for optical coupling were totally opened after the EOCB processing was completed. During processing, the cut-out areas have to be covered to protect the optical glass waveguide interface quality. Previously, lamination cavities have been milled into the electrical package to about half of the thickness in the areas of the optical windows. This supports the opening process from the outer side after lamination of the entire stack-up. The glass is protected during all wet-chemical processing. After final lamination of the outer layers, electrical vias were drilled and the board proceeded as a conventional rigid PCB. A crucial requirement is void free and reliable copper deposition inside the holes. After applying a solder mask layer and finishing the solder pads with, e.g., ENIG (electroless nickel/immersion gold), the optical windows were opened with depth controlled milling. An EOCB with four glass layers is shown in Fig. 12.16.

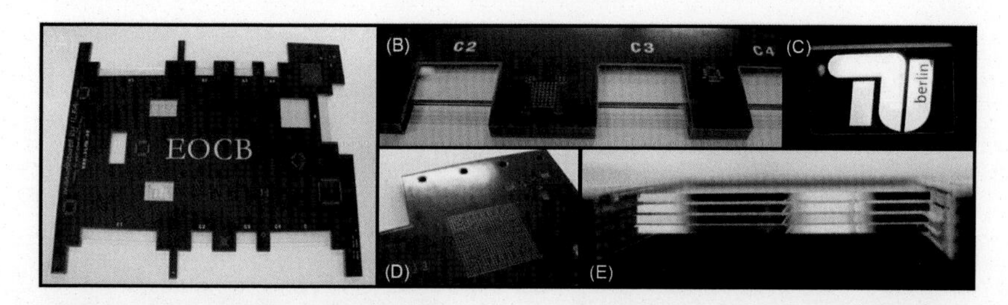

Figure 12.16 Fabricated EOCB (A), with access on the edges to glass waveguides (B), illuminated partner logo patterned on glass (C), electrical BGA pad array (D), and embedded four double-sided glass waveguide panels (E).

It was found that both pure laser drilling, as well as pure mechanical drilling in the overall hybrid composites (glass and FR4), have major drawbacks that need to be evaluated as a "show stopper" for glass-based EOCB fabrication technology. A laser drilling approach in combination with material plugging of the through glass vias and mechanical drilling was successfully developed [26]. Glass layers are drilled before lamination. Then the best drilling technologies for through vias can be applied for each material. For glass, different drilling technologies are suitable like powder blasting, laser drilling, etc., [28] in contrast to FR4. A solid state laser (Nd:YVO4) with optical output power of 20 W in combination with an optical galvo scanning system was successfully selected for cylindrical through-glass via drilling. The laser beam is focused on the glass panel back-side and starts drilling from there by moving the laser beam focus position continuously up to the glass surface. The system was chosen because drilling and free-form cutting can be realized easily using the same machine. The smallest via diameter of 0.5 mm in glass can be achieved with such a system. After drilling the glass layer with an adequate technology, the holes have to be filled with a dielectric plugging material. The glass layer will be aligned by registration marks to the other FR4 or epoxy layers and then laminated together. Using a high-flow prepreg which flows into the pre-drilled through glass vias can substitute the mentioned additional filling step if prepreg resin fills them up completely. Afterwards, high-speed mechanical drilling can be applied through the pre-drilled glass without contacting the glass itself. The resulting coaxial through vias are compatible with standard via metallization processing because of polymer passivation which ensures high copper adhesion inside the vias. The same concept is applied for the optical mid-board waveguide interface in the glass layer. A glass cut-out will be implemented in the glass layer before the lamination process. The glass cut-out has at least one edge of optical end-face quality for low-loss optical signal coupling into the planar integrated waveguides. A post-processing of the optical waveguide end-face can be avoided when the optical waveguide end-face will be covered during standard PCB processing against pollution and damage. An EOCB with 1301 electrical vias drilled through a hybrid stack-up was fabricated as shown in Fig. 12.17. An electrical daisy chain with 302 vias was successfully electrically tested without any failures.

Different reliability tests for glass-based EOCBs have been successfully performed during development and process optimization. A FR4-stack-up with

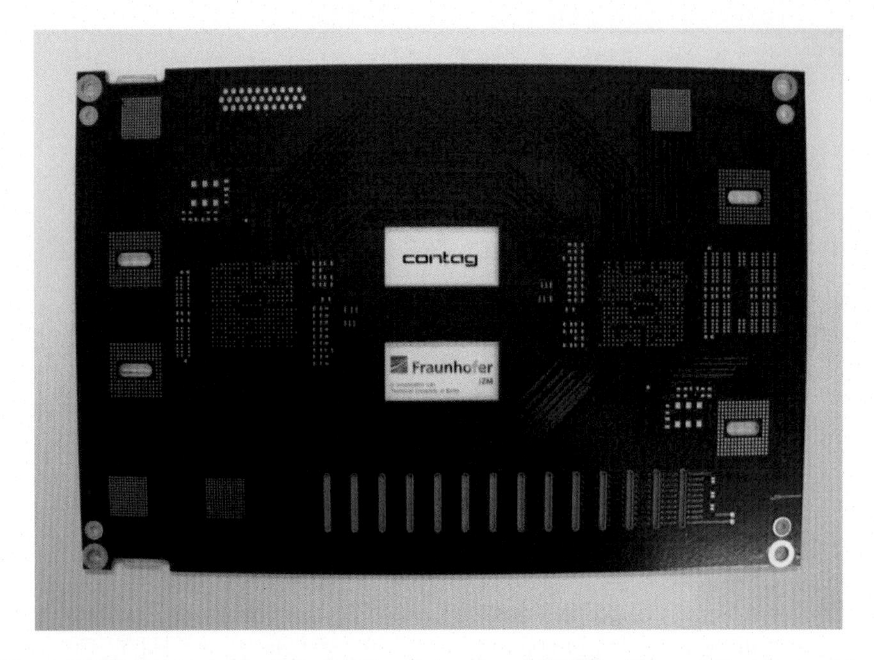

Figure 12.17 Top view of a fully functional glass-based EOCB with one optical layer and two electrical layers, an area of 270 × 184 mm^2, thickness of 1.9 mm, and 1301 electrical through vias.

multiple adhesive foil layers was thermally shock tested for 10 seconds with a peak temperature of 288°C. A thermal transition test with 100 cycles over a temperature range of +125°C and −40°C was also performed without impact. Electrical components were soldered on the multi-mode glass-based EOCB and optical performance did not change [27]. For single-mode glass waveguides the impact of high temperature bonding has been evaluated in detail. A glass waveguide sample laminated between FR4 prepreg layers was thermally cycled three times between temperatures of 100 and 300°C by a defined heat profile (100°C→3.3 K/second→300°C for 15 second→ − 1.6 K/second→100°C). No delamination or change in insertion loss could be observed after the thermal stress test [29]. Because of the mismatch of coefficient of thermal expansion (CTE) between FR4, glass, and FR4 prepreg, the stack-up has to be symmetrical, avoiding board wrapping as result of thermal impact during PCB manufacturing processes or post-processes like soldering with higher temperatures.

12.3.4 Fiber-to-board and chip-to-board coupling interface

For optical fiber-to-board coupling an optical edge connector and receptacle was assembled in a cut-out of the PCB directly on the glass edge of the glass waveguide panel. The connector design has been adapted to meet the opto-mechanical waveguide requirements of the glass foils, with embedded planar waveguides as

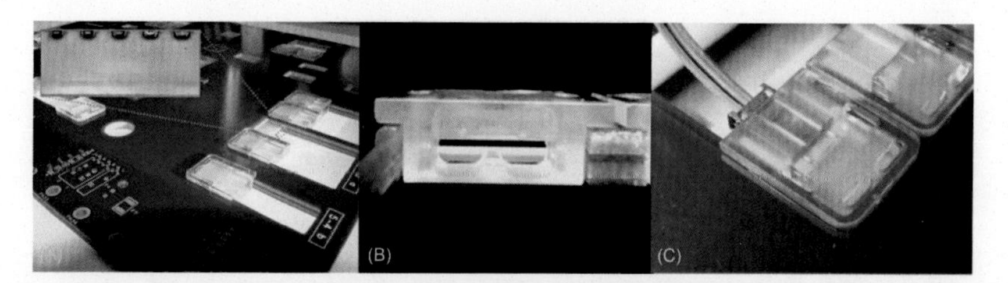

Figure 12.18 MT receptacle mounts assembled on glass-based EOCB (A), MT receptacle mount (B), and plugged MT fiber ribbon patch cord (C) [28].

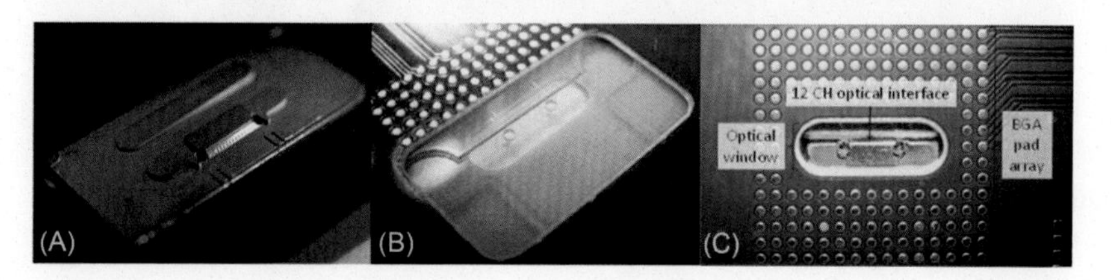

Figure 12.19 Coupling element with 12 spherical mirrors for beam deflection (A), detailed back-side view of the EOCB demonstrator with glass cut-outs and waveguide interfaces for coupling element assembly (B), and (C) top view on the BGA pad array and the optical window with the 12-channel optical out-of-plane glass waveguide coupling interface.

deployed within a multilayer electronic PCB stack-up. The receptacle with plugged MT fiber patch cord is shown in Fig. 12.18 [28].

For chip-to-board coupling a micro-optical coupling element will be placed in a cut-out of the glass panel from the opposite side, as the electro-optical chip will be mounted on the EOCB. The concept was demonstrated with a 12-channel spherical mirror coupling element fabricated by PMMA hot-embossing, laser-separation, and Au-sputtering (see Fig. 12.19A) [26]. For mounting the mirrors of the optical coupling element a cut-out in the glass waveguide panels is cut by laser. Then the pre-processed glass waveguide panel is embedded in the PCB stack-up and windows in PCBs layers are implemented for coupling element assembly from the back-side and optical out-of-plane coupling to the front-side (see Fig. 12.19B). The coupling element was mounted by UV curing adhesive after finishing EOCB fabrication. A top view of the BGA array for optical engine assembling on the top-side of the EOCB is shown in Fig. 12.19C. The distance between the top-side of the EOCB demonstrator and the mirror is about 1.5 mm.

12.3.5 Optical backplane demonstration

A fully integrated connection platform comprising a $281 \times 233\ \text{mm}^2$ multilayer electro-optical backplane with integrated planar glass waveguides, a pluggable connector system, and slots for five pluggable test cards are demonstrated [28] in Fig. 12.20. Both on-card and externally generated 850 and 1310 nm optical test

Figure 12.20 Optical backplane demonstrator platform with embedded multi-mode glass waveguide layer [28].

data were conveyed through the connector and waveguide system and characterized for in-system and system-to-system optical connectivity at data rates up to 32 Gb/second with a test signal conveyed at 850 nm over an electro-optical backplane with open eye diagrams and BER $<10^{-13}$ [27,28].

Due to its optical transparency and low-loss waveguide performance in the telecom wavelength range, it seems straightforward to propose the possibility of EOCBs made from thin glass layers as packaging and interconnection platforms for silicon photonics board-level packaging and single-mode optical backplanes. Ongoing work for single-mode fabrication in glass, fiber-to-board, and chip-to-board coupling presents very promising results [30].

12.4 Mass production and reliability

Optical waveguide loss and overall EOCB yield is very sensitive to process-induced errors and parameter deviations. In optical structures, errors such as coating defects, film non-uniformities, imaging defects, development residues, contaminants, or humidity cause degradation of physical and functional characteristics, or even total failure of the device [31,32]. Controlling process deviation is very challenging, particularly in the product development stage when production quantities are relatively small, design changes are frequent, and the number of different development test vehicles is relatively high (low-volume high-mix highly complex products). To meet highly repeatable EOCB production, PCB manufacturers have to develop a process that utilizes standard fabrication procedures in key steps. In Pitwon et al. [33], for the first time ever, a high-volume PCB fabricator reported the development of a standardized fabrication process to embed polymeric

Figure 12.21 Production equipment at TTM for EOCB fabrication. Example of a fabricated EOCB product.

waveguide layers as part of rigid PCBs. EOCB fabrication is conducted using standard PCB facilities and equipment (Fig. 12.21). Each optical layer (cladding/core/cladding) is realized by subsequent coating, imaging, and curing cycles to form polymeric higher index trace patterns (cores) fully surrounded by a lower index optical material (cladding). After waveguide fabrication, the optical layer is laminated as part of the PCB stack-up using a standard press-lamination process. Subsequent process steps after stacking optical and electrical layers in the board stack-up include various drilling, metallization, curing, and cleaning cycles, which were investigated separately for optical material compliance and optimum processing conditions.

References

[1] Pitwon R, et al. Demonstration of fully-enabled data centre subsystem with embedded optical interconnect. Proc SPIE 2014;8991.

[2] Data sheet. LIGHTRAY OFX optical fiber circuit assemblies, <www.te.com>.

[3] Data sheet. FlexPlane™ optical circuitry, <www.molex.com>.

[4] Immonen M, Wu J,Yan HJ, Zhu LX, Chen P, Rapala-Virtanen T, Development of electro-optical PCBs with embedded waveguides for data center and high performance computing applications. In: Proc. SPIE. 8991, Optical interconnects xIV, 899113; 2014.

[5] Zgraggen E, Soganci IM, Horst F, Porta AL, Dangel R, Offrein BJ, et al. Laser direct writing of single-mode polysiloxane optical waveguides and devices.. J Lightwave Technol 2014;32(17):3036–42.

[6] Ishigure T. Multimode/single-mode polymer optical waveguide circuit for high-bandwidth-density on-board interconnects. In: SPIE OPTO. International Society for Optics and Photonics; 2015.

[7] Soma K, Ishigure T. Fabrication of a graded-index circular-core polymer parallel optical waveguide using a microdispenser for a high-density optical printed circuit board. IEEE J Sel Top Quantum Electron 2013;19(2) 3600310-3600310.

[8] Lapointe J, Gagné M, Li MJ, Kashyap R. Making smart phones smarter with photonics. Opt Express 2014;22(13):15473–83.

[9] Tervonen A, West BR, Honkanen S. Ion-exchanged glass waveguide technology: a review. Opt Eng 2011;50(7) 071107-071107

[10] John RSE, Amb CM, Swatowski BW, Weidner WK, Halter M, Lamprecht T, et al. Thermally stable, low loss optical silicones: a key enabler for electro-optical printed circuit boards. J Lightwave Technol 2015;33(4):814–19.

[11] <http://opticalinterlinks.com/products.html>.

[12] Schröder H, Brusberg L, Arndt-Staufenbiel N, Richlowski K, Ranzinger C, Lang KD. Advanced thin glass based photonic PCB integration. In: Electronic components and technology conference (ECTC), 2012 IEEE 62nd. IEEE. p. 194–202.

[13] Dangel R, Bapst U, Berger C, Beyeler R, Dellmann L, Horst F, et al., Development of a low-cost low-loss polymer waveguide technology for parallel optical interconnect applications. In: Biophotonics/optical interconnects and VLSI photonics/WBM microcavities, 2004 digest of the LEOS summer topical meetings. presented at the biophotonics/optical interconnects and VLSI photonics/WBM microcavities, 2004 digest of the LEOS summer topical meetings, IEEE; 2004. <http://dx.doi.org/10.1109/LEOSST.2004.1338689>.

[14] Immonen M, Wu J, Kivilahti J. Fabrication of polymer optical waveguides with integrated micromirrors for out-of-plane surface normal optical interconnects. In: 4th IEEE international conference on polymers and adhesives in microelectronics and photonics, 2004. POLYTRONIC 2004. Presented at the 4th IEEE international conference on polymers and adhesives in microelectronics and photonics, 2004. POLYTRONIC 2004; 2004. p. 206–10. <http://dx.doi.org/10.1109/POLYTR.2004.1402762>.

[15] Lamprecht T, Halter M, Meier D, Beyer S, Betschon F, John R, et al.. Highly reliable silicone based optical waveguides embedded in PCBs. In: Optical fiber communications conference and exhibition (OFC), 2014. Presented at the optical fiber communications conference and exhibition (OFC); 2014. p. 1–3. <http://dx.doi.org/10.1364/OFC.2014.Th4J.5>.

[16] Bierhoff T, Schrage J, Halter M, Betschon F, Duis J, Rietveld W. All optical pluggable board-backplane interconnection system based on an MPXTM-FlexTail connector solution. In: 2010 IEEE photonics society winter topicals meeting series (WTM). Presented at the 2010 IEEE photonics society winter topicals meeting series (WTM); 2010. p. 91–2. http://dx.doi.org/10.1109/PHOTWTM.2010.5421981.

[17] Schmidtke K, Flens F, Worrall A, Pitwon R, Betschon F, Lamprecht T, et al. 960 Gb/s optical backplane ecosystem using embedded polymer waveguides and demonstration in a 12G SAS storage array. J Lightwave Technol 2013;31:3970–5. Available from: http://dx.doi.org/10.1109/JLT.2013.2278401.

[18] Anzures E, Dangel R, Beyeler R, Cannon A, Horst F, Kiarie C, et al. Flexible optical interconnects based on silicon-containing polymers. In: Proceedings of SPIE. Presented at the photonics packaging, integration, and interconnects IX, San Jose, CA; 2009, p. 72210I–72210I–12. http://dx.doi.org/10.1117/12.808396

[19] Swatowski BW, Amb CM, Breed SK, Deshazer DJ, Weidner WK, Dangel RF, et al. Flexible, stable, and easily processable optical silicones for low loss polymer waveguides. In: Tabor CE, Kajzar F, Kaino T, Koike Y, editors; 2013, p. 862205. http://dx.doi.org/10.1117/12.2007419.

[20] Ma H, Jen AK-Y, Dalton LR. Polymer-based optical waveguides: materials, processing, and devices. Adv Mater 2002;14:1339–65. Available from: http://dx.doi.org/10.1002/1521-4095(20021002)14:19<1339::AID-ADMA1339>3.0.CO;2-O.

[21] Krähenbühl R, Lamprecht T, Zgraggen E, Betschon F, Peterhans A. High-precision, self-aligned, optical fiber connectivity solution for single-mode waveguides embedded in optical PCBs. J Lightwave Technol 2015;33:865–71.
[22] Park S, Lee J-M, Ko SC. Fabrication method for passive alignment in polymer PLCs with U-grooves. IEEE Photon Technol Lett 2005;17:1444–6. Available from: http://dx.doi.org/10.1109/LPT.2005.848287.
[23] Betschon F, Lamprecht T, Halter M, Beyer S, Peterson H. Design principles and realization of electro-optical circuit boards; 2013, p. 86300U–86300U–12. doi:10.1117/12.2006379.
[24] Immonen M, Wu J, Yan HJ, Zhu LX, Chen P, Rapala-Virtanen T. Electro-optical backplane demonstrator with multimode polymer waveguides for board-to-board interconnects. In: Proc. of IEEE 5th electronics system-integration technology conference (ESTC), Helsinki, Finland, 16–18 Sept 2014.
[25] Brusberg L, Schröder H, Herbst C, Frey C, Fiebig C, Zakharian A, et al. High performance ion-exchanged integrated waveguides in thin glass for board-level multimode optical interconnects. In: Proc. of European conference on optical communication (ECOC), Valencia, Spain, 27 Sept–1 Oct 2015.
[26] Brusberg L, Schröder H, Ranzinger C, Queisser M, Herbst C, Marx S, et al. Thin glass based electro-optical circuit board (EOCB) with through glass vias, gradient-index multimode optical waveguides and collimated beam mid-board coupling interfaces. In: Proc. of 65th electronic components and technology conference (ECTC), San Diego, CA, 26–29 May 2014.
[27] Brusberg L, Schröder H, Pitwon R, Whalley S, Miller A, Herbst C, et al. Electro-optical backplane demonstrator with gradient-index multimode glass waveguides for board-to-board interconnection In: 64th electronic components and technology conference (ECTC), 27–30 May 2014. p. 1033–41.
[28] Pitwon RCA, Brusberg L, Schröder H, Whalley S, Wang K, Miller A, et al. Pluggable electro-optical circuit board interconnect based on embedded graded-index planar glass waveguides. J Lightwave Technol 2014;Vol. 33(No. 4):741–54.
[29] Brusberg L, Manessis D, Neitz M, Schild B, Schröder H, Tekin T, et al. Development of an electro-optical circuit board technology with embedded single-mode glass waveguide layer. In: Proc. of 5th electronics system-integration technology conference (ESTC), Helsinki, 16–18 Sept 2014.
[30] Brusberg L, Manessis D, Herbst C, Neitz M, Schild B, Schröder H, et al. Single-mode board-level interconnects for silicon photonics. In: Optical fiber communication (OFC) conference, Los Angeles, CA, 22–26 March 2015.
[31] Immonen M, Tian D, Junnila S, Rapala-Virtanen T. Emerging from Lab to LAP: development of large area processing technologies for low-cost photonic interconnects on PCBs. In: Proc. of IEEE 1st electronics system-integration technology (ESTC), Dresden, 5–7 Sept 2006. p. 1347–55.
[32] Immonen M, Yan HJ, Chen P, Xu JX, Rapala T. Development of electro-optical PCBs with polymer waveguides for high-speed intra-system interconnects. Circuit World 2012;38(3):104–12.
[33] Pitwon R, Immonen M, Wang K, Itoh H, Shioda T, Wu J, et al. International standards for optical circuit board fabrication, assembly and measurement. J Opt Commun Elsevier 2016;362:22–32.

International and industrial standardization of optical circuit board technologies

13

R. Pitwon[1,8], M. Immonen[2], J. Wu[3], L. Brusberg[4], H. Itoh[5], T. Shioda[6], S. Dorresteiner[7], H.J. Yan[3], H. Schröder[4], D. Manessis[4], P. Schneider[7] and K. Wang[1]
[1]Seagate, Langstone Road, Havant, Hampshire, United Kingdom, [2]TTM Technologies, Salo, Finland, [3]TTM Technologies, Shanghai, P.R. China, [4]Fraunhofer Institute for Reliability and Microintegration, Berlin, Germany, [5]Industrial Technology Center of Tochigi Prefecture, Tochigi, Japan, [6]Mitsui Chemicals, Inc., Chiba, Japan, [7]TE Connectivity, s'Hertogenbosch, The Netherlands, [8]University of St Andrews, St Andrews, KY, Scotland

13.1 Introduction

Although the prospects for commercial proliferation of optical circuit board (OPCB) technology at the turn of the century were crippled by the rapid slow-down in the telecoms sector following the stock market crash in 2001, embedded waveguide-based OPCB technology continued to advance steadily, albeit at a slower pace than anticipated at the turn of the century. By 2016 substantial improvements in the three crucial areas of optical waveguide material, waveguide fabrication, and connectorization had finally allowed polymer waveguide cable products to emerge with comparable loss and dispersion in the operational wavelength range around 850 nm to optical fiber.

Now optical materials are available which offer the required resilience to thermal cycling and humidity to allow them to be integrated into printed circuit board (PCB) fabrication and assembly processes. A variety of waveguide fabrication techniques have evolved, which lend themselves to high-volume production (photolithography, batch-processing) or low-volume or prototype development (laser direct imaging). Graded index waveguide profiles can now be fabricated in both glass [1] and polymer [2] to offer reduced modal dispersion, crosstalk, and radiative losses though the side walls. Waveguide connectorization, considered the final technical barrier to OPCB commercialization, had reached the stage by 2015 where low termination losses have been demonstrated both in glass [3] and in polymer [4].

The deployment of optical interconnect into low-cost, high-volume data communication enclosures is now made possible by the emergence of a new technology eco-system, including board-mountable optical transceivers and very high density parallel optical interfaces [5].

Major international collaborative research and development initiatives driving this migration of optical interconnect into mainstream data communication environments

Optical Interconnects for Data Centers. DOI: http://dx.doi.org/10.1016/B978-0-08-100512-5.00013-9

include the European PhoxTrot project [6], the European Nephele project [7], and the US HDPuG Optoelectronics project [8].

In order for OPCB technology to become commercially viable, industrial processes for high-volume OPCB fabrication, assembly, and quality assurance must be established and advanced to the point that multiple global PCB vendors can provide marketable products.

Three main varieties of OPCB technology exist today, which differ strongly from each other in terms of their optical channel technology and market readiness: (1) fiber optic circuits, (2) polymer waveguides [9], and (3) planar glass waveguides.

In this chapter we will discuss emerging industrial, high-volume PCB manufacturing process standards for the fabrication of OPCBs based on fiber optic circuit laminates, embedded planar polymer waveguide layers, and embedded planar glass waveguide layers.

The most challenging aspect of OPCB assembly is the provision of reliable technologies and processes for high-yield assembly of connecting interfaces to planar optical waveguides. Substantial progress has been made in terminating MT compliant ferrules onto single or multirow waveguides for both multimode and single-mode waveguides [10,11]. In this chapter, we will review the emerging international standards which track these achievements, including the first MT compliant optical connector standard for polymer waveguides.

Furthermore, measurement results of planar optical waveguides have historically been notoriously inconsistent from one tester to another, raising strong questions about the validity of reported waveguide measurements in general. This is due to a lack of harmonized measurement conditions, which would yield a predicted margin of measurement repeatability and, as such, would serve as the basis for an OPCB quality assurance process.

In this chapter, we set out recommended measurement conditions for OPCBs and describe how both the application of a measurement identification system and recommended reference measurement conditions can bring about a dramatic improvement in results consistency, which can reduce variation in measurement results to less than 10%. These will serve as the basis for the first international measurement standards for OPCB technologies (Fig. 13.1).

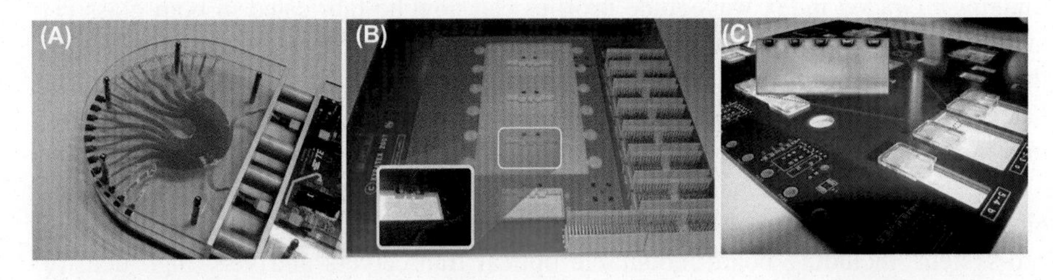

Figure 13.1 OPCB varieties (A) fiber optic laminate, (B) polymer waveguides and (C) planar glass waveguides.

13.1.1 Diversity in OPCB Types

There are three main varieties of circuit board embedded optical interconnect, which have emerged at different stages over the past 20 years and differ strongly from each other in terms of their material composition, waveguide profile, channel and performance characteristics, fabrication process, and maturity. These varieties include:

1. Fiber optic circuit laminates [12],
2. Embedded planar polymer waveguides [13,14],
3. Planar glass waveguides [15,16].

13.1.1.1 Fiber optic circuit laminates

Laminated fiber optic circuits in which optical fibers are pressed and glued into place on a substrate benefit from the reliability of conventional optical fiber technology. However, these circuits cannot accommodate waveguide crossings in the same layer, i.e., fibers must cross over each other and cannot cross through each other. Also, with each additional fiber layer, backing substrates must typically be added to hold the fibers in place, thus significantly increasing the thickness of the circuit. This would limit the long-term usefulness of laminated fiber optic circuits in PCB stack-ups. At best they can be glued or bolted onto the surface of a conventional circuit board.

13.1.1.2 Polymer waveguides

The absorbtion characteristics of most modern optical polymer materials would make polymer waveguides unsuitable to convey longer operational wavelengths (1060 nm, 1310 nm, or 1550 nm) over longer distances due to higher intrinsic losses, though this can be mitigated in some perfluorinated polymer formulations [17]. However, they would be suitable for very short reach, versatile, low-cost links such as inter-chip connections on a board, or board-to-board connections within an enclosure. Indeed, in spite of the higher absorption at longer wavelengths, single--mode polymer waveguides are emerging as an efficient means of optically connecting photonic integrated circuit (PIC) devices to fiber optic channels through evanescent adiabatic coupling mechanisms, providing an efficient alternative to current grating coupler approaches [18–21].

They would also be suitable for applications in which certain properties of the polymer, such as thermo optic, electro optic or strain optic coefficients could be used to support advanced integrated functions such as Mach-Zehnder switches or long-range plasmonic interconnects [22,23].

13.1.1.3 Planar glass waveguides

Planar glass waveguide technology could combine some of the performance benefits of optical fibers, such as lower material absorption at longer operational wavelengths and lower modal dispersion with the ability to fabricate dense complex optical circuit layouts on single layers and integrate these into PCB stack-ups. The Fraunhofer Institute of Reliability and Microintegration (Fraunhofer IZM) in Germany are producers of planar glass waveguide-based OPCBs [3].

13.1.1.4 Free space optics

With the proliferation of expanded optical beam technologies, free space optical interconnect represents a viable emerging solution for in-system interconnect applications, as the internal misalignment tolerances inherent to such systems can be more easily accommodated by expanded beam technologies. Target applications include server backplane interconnectivity [24,25].

13.1.1.5 Target applications

The target application defines the tradeoff space constraining the selection of waveguide type and thus plays an important role in deciding the most suitable choice of OPCB. For example, in data center applications cost would be the dominant factor, while in high performance computers or supercomputers emphasis is placed on performance optimization.

13.2 Industrial manufacturing processes for OPCBs

13.2.1 Fiber optic circuit laminates

13.2.1.1 Introduction

The increase in optical transmission requirements in data center environments is putting a great strain on the architectures of the fiber-optic infrastructure, with fiber management both outside and inside the enclosure becoming more challenging than ever before.

The need for complex fiber switching and cross connect, as well as the increasing density of the individual fiber management systems, add to this burden. The increasing number of patch cords, and fan-out cables required to connect higher bandwidth density infrastructures is becoming problematic not only in terms of cable management, but also from a cooling perspective.

One solution to this problem is a low-cost fiberinterconnect system that can handle high density complex optical cross connects in a reduced space. This space reduction is essential as higher processing and communication data rates go hand in hand with higher processor power consumption, and cooling these processors becomes critical to maintain the system reliability and performance.

Fiber optic circuit laminates or flex foils offer a solution to this problem. The fiber optic circuits comprise a number of individual fibers, which have been precisely laid into a predetermined form or pattern, and fixed into place using a special coating designed to bond the fibers together as they are laid down. In some cases, a part of the construction may be routed onto a sheet of thermally stable material (e.g., Kapton), allowing these parts to withstand a temperature range of −40 to +85 °C. These thin foils are very efficient in terms of space saving and structured fiber management. The fiber layout has very little restriction, and therefore offers a solution to almost every fiber management problem. The most common application will be fiber on the board, board-to-board as well as backbone cross connections for data communications and telecommunications applications, but harsh military and aerospace applications will also be strong contenders.

13.2.1.2 Fiber laying technology

For both the fiber laying and the fiber coating, TE Connectivity located in the Netherlands has developed two different machines capable of handling the manufacturing process of precision routing of the fibers, as well as the coating process (Fig. 13.2).

13.2.1.3 Fiber optic circuit laminates manufacturing process

The manufacturing of fiber optic circuit laminate products consists of two major process steps: fiber routing on disposable/permanent adhesive foils, and application of conformal coating by spray coating (Fig. 13.3).

For fiber routing two types of adhesive tapes are used, a disposable tape and a permanent tape (Fig. 13.4).

During the fiber routing process the fiber is placed on adhesive tapes by a guiding needle at a given force and speed. The fiber routing pattern is designed in CAD software and converted by CAM into a CNC program that controls the path of the needle. Fiber is routed on a disposable tape and on a permanent tape. The main function of the disposable tape is to maintain position of the fibers for straight

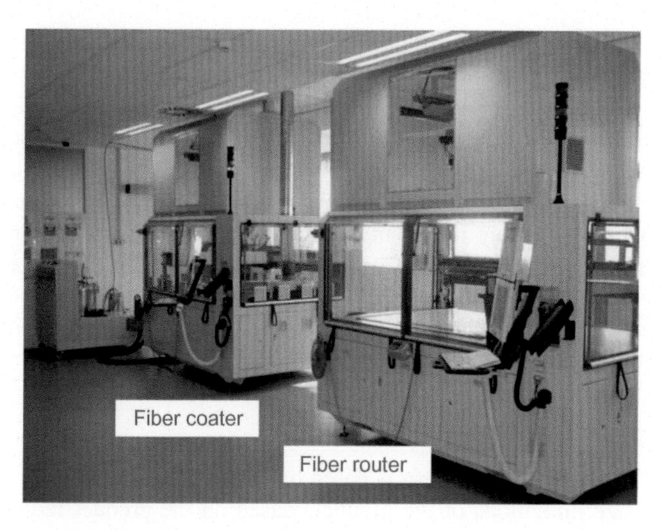

Figure 13.2 Fiber coating and routing machines.

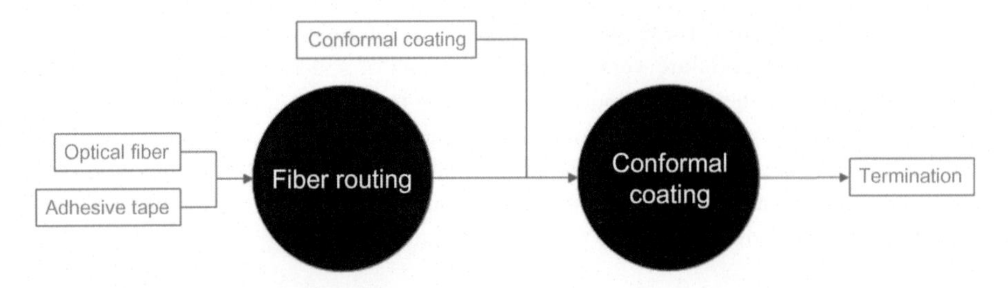

Figure 13.3 Fiber routing process.

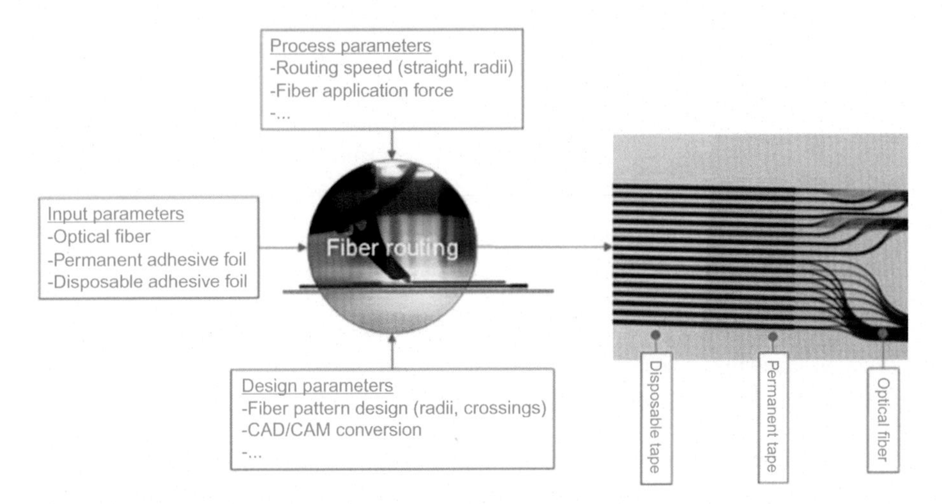

Figure 13.4 Fiber routing process components.

tails and positioning of the permanent adhesive foil. Typically, the adhesive layer of the disposable tape is thin and sufficiently strong the maintain fiber position in straight lines. After completion of the OFX product the fibers and permanent adhesive foil must be peeled of with little effort. The permanent tape is used to route fibers in crossings and radii. The permanent tape is either PET or Kapton material with a thick (100–140 μm) and strong adhesive layer (Figs. 13.5, 13.6, and 13.7).

For areas where fibers are routed in circular paths and where fibers are crossed, a strong adhesion layer is required. Currently a permanent foil material is selected, a PET foil with an adhesive layer. The adhesive layer is a firm acrylic pressure-sensitive solvent resistant adhesive system. It features high ultimate bond strength with excellent high temperature performance and excellent solvent resistance. Bond strength for this specific tape increases substantially with natural aging.

Product requirements

The design of a fiber routing pattern is based on the wiring diagram supplied by the customer and size limitations of the product. Based on the product requirements, a pattern is designed using the design boundary conditions which are currently

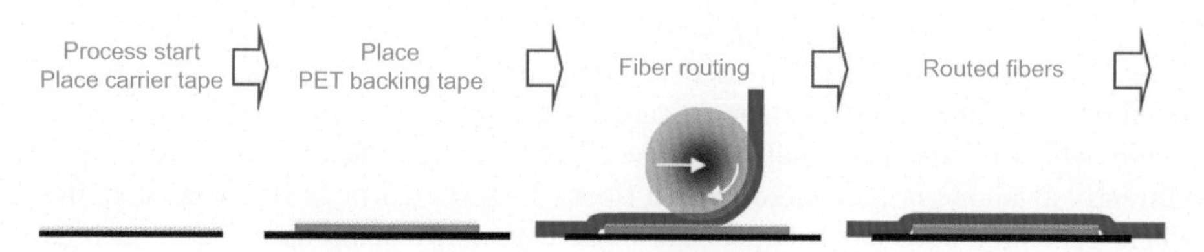

Figure 13.5 Fiber routing process steps.

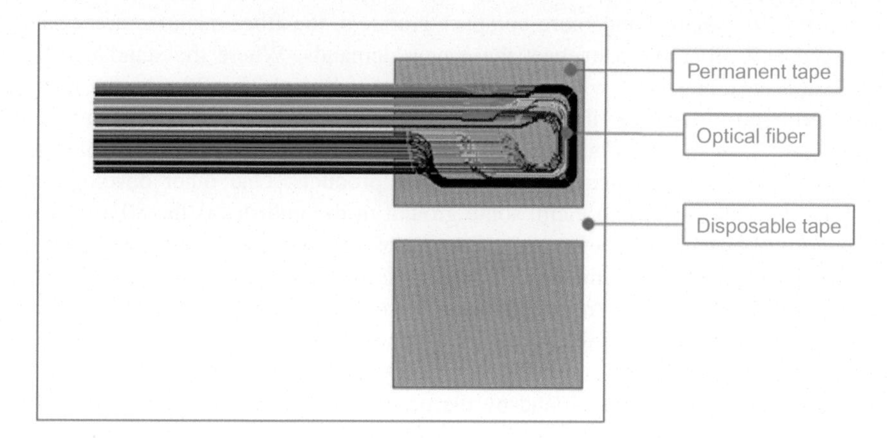

Figure 13.6 Positioning of materials on to the vacuum bed of the fiber routing machine.

Figure 13.7 Vacuum bed of the fiber routing machine. The white surface is the carrier.

available. Most important design boundaries are the minimum bend radius either given by fiber type or by the used adhesive layer and the number of crossings required. The fiber router system is capable in routing a minimal bend radius of 4 mm, which is typically smaller than the accepted bend radius for most fiber types currently available in the market. When fiber optics started in its first inception, the fibers generally were much larger than those commonly being used today, such as the multimode 100/140 fiber. The industry standard has since then moved on, and much has been standardized to a 125 μm glass fiber with a coating of 250 μm. This fiber is available in various types, such as single-mode having a 9 μm core (9/125), and multimode fiber having a 50 μm or 62.5 μm core (50/125, 62.5/125). Due to the demand for smaller and more compact products, the fiber manufacturers have developed new fiber types to meet these new demands. Where the standard fiber types could only be routed with a minimum bend radius of 25 – 30 mm, the newly developed bend insensitive fibers can cope with much smaller radii, down to as little as 5 mm. This smaller fiber bending radius helps the designers of fiber management systems greatly reduce the size of their products. One other development which is only just starting to claim some ground in the industry is the 80 μm fiber. This 80 μm fiber has a protective outer coating of 125 μm, and is being designed in both single-mode 9/80 and multimode 50/80. As one can imagine, by halving the outer diameter of the fiber from 250 μm down to 125 μm, the footprint of optical products can be reduced significantly. This certainly applies when the 80 μm fiber is combined with the bend insensitive technology. Bend radius is in general limited by the fiber specification itself, and by the performance of the adhesive type. The challenge faced for fiber routing is to maintain the position of fibers directly after routing and after coating. Position of the fibers is especially critical during the fiber routing itself, the system cannot route where a fiber is placed. Fig. 13.8 shows an example of a multilayer fiber flexplane circuit.

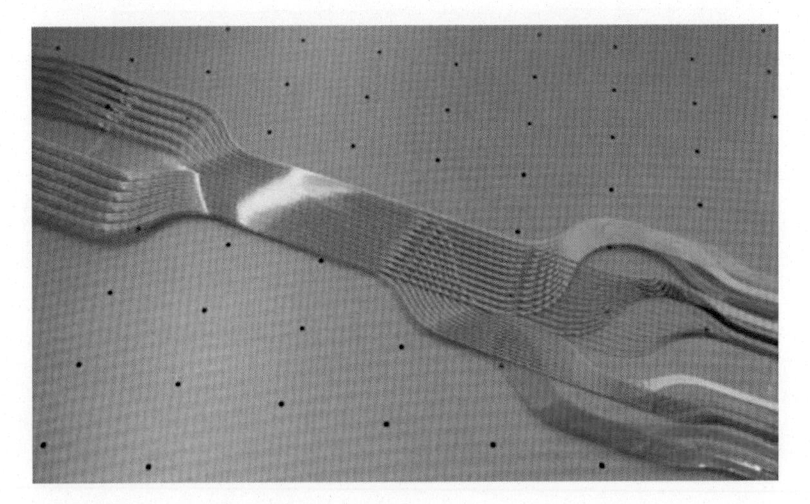

Figure 13.8 Flat plane routing with multiple layers.

13.2.1.4 Fiber coating

Coating process

After the routing process, the entire product will be transferred to a second machine that has been designed to coat the fibers and the adhesive foil with a conformal coating. This conformal coating serves several purposes. This coating is sprayed onto the surface of the adhesive carrier material, and onto the optical fibers. This spray can be controlled very accurately in order to maintain an exact layer of coating over the entire surface. This may vary depending on the application of the product.

1. The entire surface of the permanent carrier foil is covered with the conformal coating. This serves to protect the fibers, and improves the adhesion of the fibers to the base carrier material. Furthermore, the remaining surface area of the permanent foil is covered with the conformal coating to cover the exposed adhesive.
2. The fiber tails on the temporary foil are covered with the same base material conformal coating in order to form the loose fibers into a ribbon. This ribbon is required for further processing of the optical flex foil in the further termination process (Fig. 13.9).

Once the surface has been coated, the product will be left to air cure. After this curing process the temporary foil will be removed from the product, leaving a product that can be terminated with connectors, or possibly spliced directly into the end application (Fig. 13.10).

Coating material

The conformal coating is a silicon-based material, which is commonly used in the electronics industry to protect electrical circuit boards from the elements. In this application, the silicon material provides a very strong adhesion to the optical fibers, giving a very durable product that can withstand harsh environmental conditions (Fig. 13.11).

13.2.1.5 Future outlook for fiber optic flexplane technologies

Most recent developments for fiber optic circuits are progressing in several directions, complex multilayer flex foils, very small size flex foils, lamination of the optical circuit in PCB, and ruggedized solutions where the flex is applied in harsh environment applications. Examples are shown in Fig. 13.12.

The fiber optic circuit laminate products can be used in a wide range of current and future applications. These products are a very versatile and cost-effective product process that can be used in a vast number of applications.

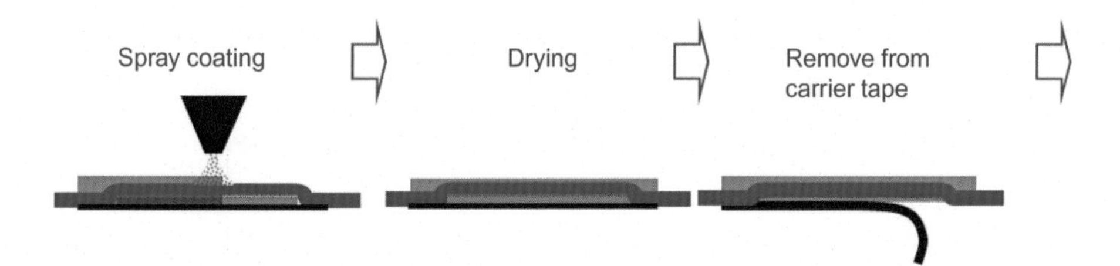

Figure 13.9 Fiber coating process steps.

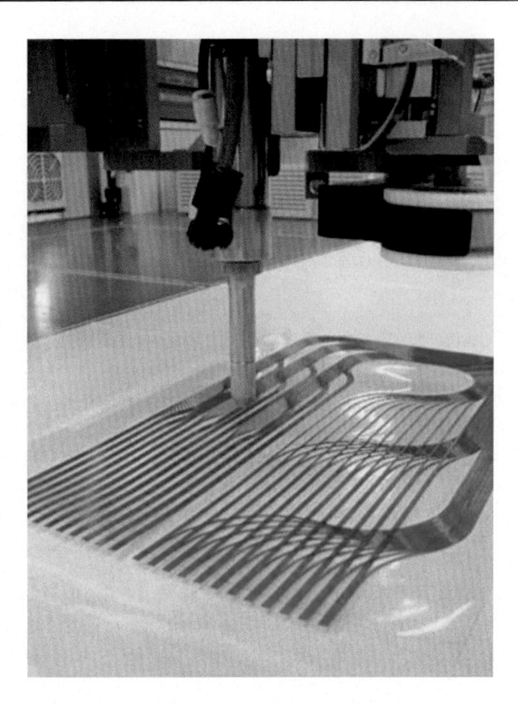

Figure 13.10 Fiber coating unit in its spray ready position.

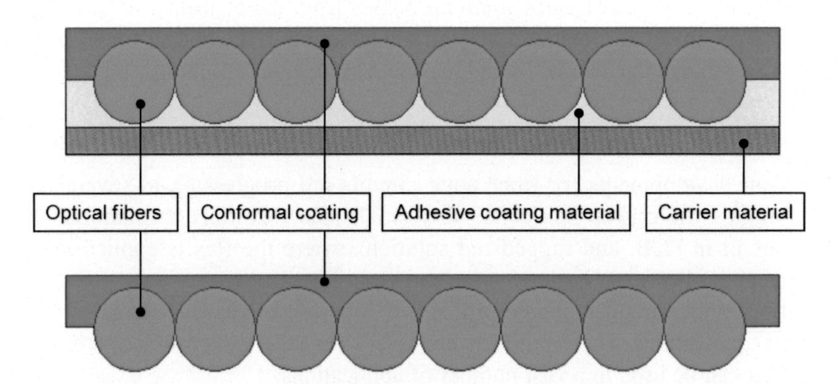

Figure 13.11 Top: Graphic showing the fibers embedded in the adhesive layer, and covered with the conformal coating. Bottom: Graphic showing the fibers with the adhesive backing material removed, leaving a ribbon style cable.

13.2.2 Polymeric planar embedded optical waveguides

OPCB with embedded polymer waveguides is a multisignal circuit construction that has both optical and electrical interconnect functions. In typical designs, copper traces provide the majority of the interconnect structures, especially lower speed signals and power networks. Polymer waveguides are designed for selected high

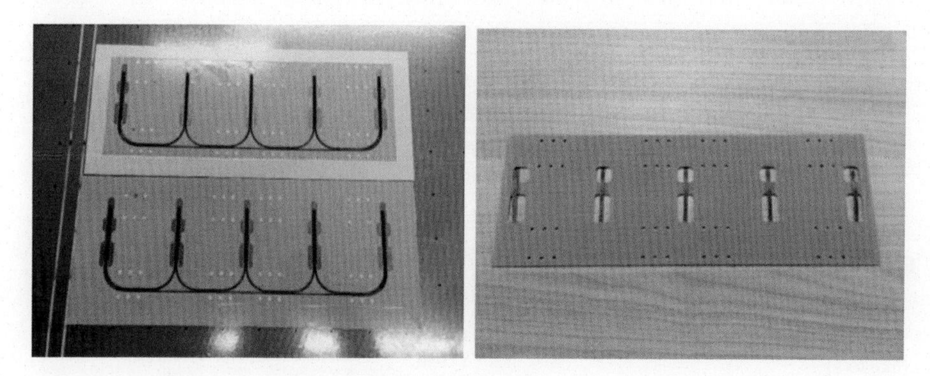

Figure 13.12 Examples of fiber optic circuit laminates.

I/O chip-to-chip and chip-to-module links where parallel optics enable higher bandwidth density and improved board real-estate utilization. Due to resulting hybrid stack-up, optical waveguide layer and optical materials have to withstand traditional PCB manufacturing processes, including press-lamination, drilling, desmear, metallization, and surface finishing processes.

TTM Technologies has evaluated various materials to manufacture optical waveguides in a high-volume production, as well as their compliance to standard PCB processes and requirements [26]. Silicones [27], polysiloxanes [28], and silsequioxanes [29] representing hybrid silicon-polymer classes have been deemed attractive material candidates for polymer waveguides due to their easy processability, very low optical losses (with the state-of-the-art being 0.031 dB/cm at 850 nm [30]), and high thermal, mechanical, and chemical stability during conventional PCB processes and board assembly. More recently, optical losses of 0.29 and 0.45 dB/cm at 1310 m and 1550 nm respectively, have been reported for a modified polysiloxane [31], further expanding end use applications for this material class to cover telecom wavelengths besides datacom applications. Silicones are inorganic polymers with a Si-O backbone. The structure of the polymer can be tuned between an inorganic glass, where each Si atom is bonded to only oxygen atoms, and flexible rubber where two oxygen atoms are substituted by organic functional groups [30]. Polynorborene is another state-of-the-art polymer class with very low loss and potential for polymer waveguides in PCBs [4].

Controlling process deviation in the product development stage is critical and particularly challenging when production quantities are relatively small, design changes are frequent, and the number of different development test vehicles is relatively high (low-volume, high-mix, high-complex products). To meet highly repeatable EOPCB production TTM has developed a process that utilizes standard fabrication procedures in key steps.

The main steps of TTM's EOPCB fabrication and polymer waveguide fabrication processes are illustrated in Figs. 13.13 and 13.14, with key process equipment shown in Fig. 13.15.

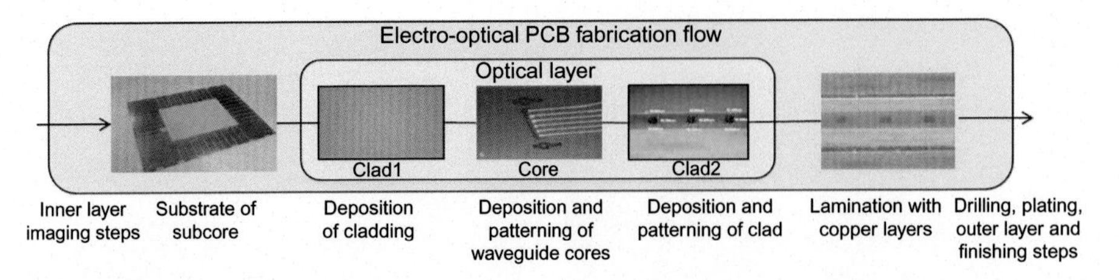

Figure 13.13 Electro-optical PCB fabrication flow with embedded polymer waveguides.

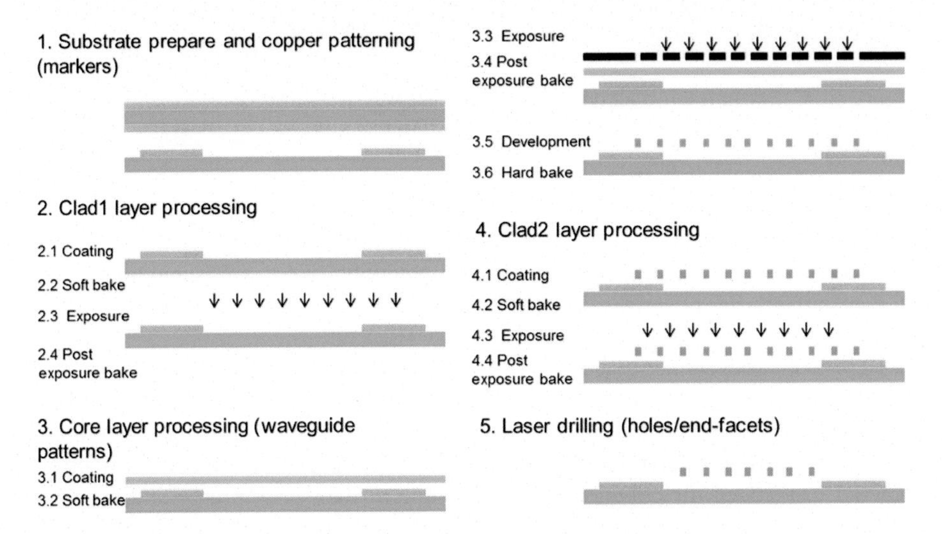

Figure 13.14 Process flow to fabricate one optical layer on PCB.

Figure 13.15 Process equipment to fabricate optical polymer waveguides and OE PCBs at TTM.

In the waveguide manufacturing process three optical layers—bottom cladding, core, and top cladding are fabricated on substrates or subcores by a sequential build-up process. Each optical layer is passed through subsequent coating, imaging, and curing cycles. As a result, polymeric higher index trace patterns (optical cores) are fully surrounded by lower index optical material (optical cladding) to form the waveguiding structures and the polymeric optical waveguide layer, which is part of the PCB stack. Liquid optical material is deposited on the substrate using doctor blading or slot die coating methods (Fig. 13.14, Steps 2.1, 3.1, 4.1). These coating methods enable high process throughput, scalable layer thickness (the core height) from 10 to 100 μm, panel scale fabrication up to 510 mm × 610 mm, and high thickness uniformity of ± 2 μm. Aforementioned characteristics are critical to meet controlled and repeatable OPCB fabrication in a production environment. After coating, the panels are soft-baked in a production convection oven to drive out solvent from the coated layer.

Optical polymer traces are formed into the material layer by UV initiated cross-linking followed by post-exposure bake (PEB) and a wet-chemical etching step (Figure 13.14, Steps 3.3 – 3.5). The optical polymer materials scoped for PCB and packaging level waveguide applications have been designed to be photo-imageable with conventional i-line ($\lambda = 365$ nm) photo-tools. UV imaging machines are widely available in PCB inner layer, outerlayer, and solder mask processing steps, including semi- or fully-automatic contact exposure systems (e.g., Adtec Adex-series), laser direct imaging (i.e., Orbotec Paragon-series), and direct imaging tools (e.g., ORC or Adtec) are selected depending on available lamp characteristics relative to the photo-initiator in question. TTM's optical waveguides are patterned either by contact imaging with glass artworks or by maskless direct imaging depending on features and other optical design details.

After waveguide fabrication, the optical layer is laminated as part of the PCB stack-up using a standard press-lamination process. Subsequent process steps after stacking optical and electrical layers into the PCB stack-up include various drilling, metallization, curing, and cleaning cycles, which were investigated separately for optical material compliance and optimum processing conditions. An example of a finished EOPCB with 16 copper and one embedded optical layer is illustrated in Fig. 13.16.

13.2.3 Glass waveguide OPCBs

The emerging requirement for optical interconnect at the longer wavelengths (O, C, L band) in future data center environments is being driven by global investment into affordable, mass-producible, PIC technologies, primarily silicon photonics

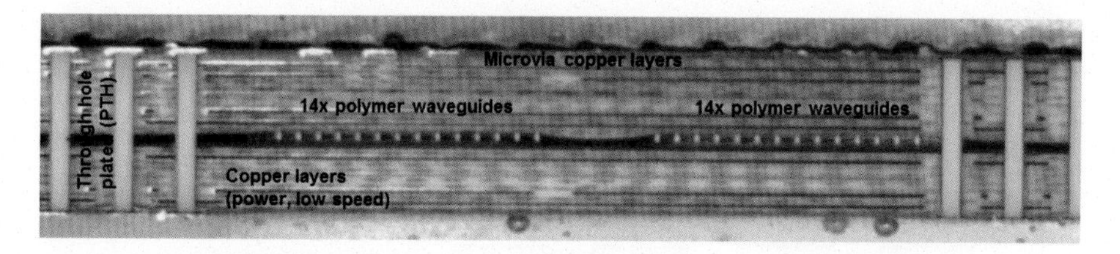

Figure 13.16 Example of a high layer count printed circuit board (HLC-PCB) with 16 copper and one embedded optical polymer layer (16L + 1Opt). Besides 14 + 14 TX + RX optical signal traces, all electrical function layers (S,P,G), and microvia and PTHs for layer-to-layer connections are provided in the stack.

transceivers and integrated switches. While embedded polymer waveguides are an eminently suitable solution for OPCBs supporting lower-cost, multimode transceiver interconnect, based on shorter wavelength 850 nm VCSEL technologies, polymer becomes a less suitable medium at longer wavelengths due to higher intrinsic material absorbtion. Though formulations have been developed in which absorbtion losses at longer wavelengths can be reduced, for instance through perfluorinated polymer blends [17], reported losses at the O, C, and L band wavelengths are still substantially higher than in conventional optical fibers.

A preferred medium for longer wavelength transmission would be glass. The increasing requirement for thermal and mechanical stability in PCBs, make lamination of thin glass foils within conventional PCB substrate stack-ups an attractive option for embedding optical interconnect functionality at the PCB level.

This would require the embedded glass foils to be equipped with optical waveguides, deflection elements, and cavities. Thin film metallization on glass and electrical vias would also enable direct assembly of electronic and optoelectronic devices on glass.

This opens up the possibility of OPCBs made with either glass interconnect layers, polymer interconnect layers, or even both.

High-volume manufacture of glass-based OPCBs can be achieved by leveraging state-of-the-art display materials and display processing technologies such as thin film metallization on glass, ion-exchange, and laser-cutting. One viable approach involves pre-fabrication of glass waveguide panels and lamination of those panels into a PCB stack-up using conventional PCB fabrication processes. Issues such as CTE mismatch can be accommodated through temperature control of the lamination process and materials.

A proven method of patterning planar waveguides into glass foils, is through the use of a two-step thermal ion-exchange process as reported by Brusberg et al. [32]. Definition of the required waveguide layout onto the glass panel can be achieved by patterning an aluminum mask layer deposited on the glass foil. The process flow is shown in Fig. 13.17. The glass substrate should contain an adequate concentration of ions that are exchangeable with counterpart ions in the salt melt to induce a change in the refractive index. Preferable choices are fusion-drawn alkali-aluminosilicate or borosilicate thin-sheet glass with a composition designed for ion-exchange. This glass is commonly used to realize the transparent

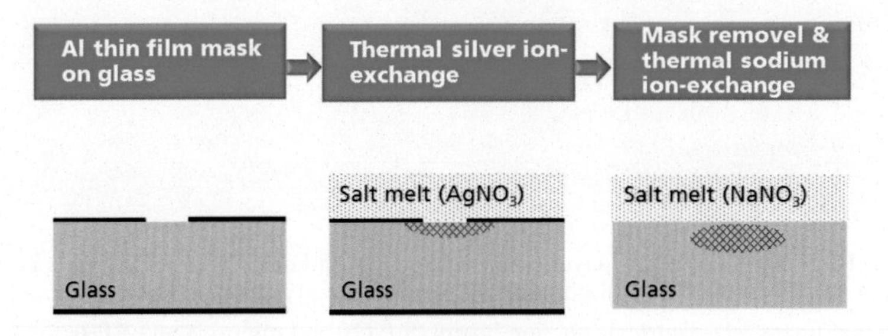

Figure 13.17 Process flow for glass waveguide panel fabrication.

cover plate for handheld mobile telephones and similar devices: when used in that application, the glass is usually chemically strengthened by a potassium-for-sodium ion exchange process [32].

The display industry continues to drive requirements for increased glass panel sizes. In 2015, glass panel areas of 2.94 m × 3.37 m were announced for a Gen 10.5 LCD production line [33]. Such LCD production includes subsequent finishing processes like thin film deposition, lithographic processing, stacking, and glass cutting. For those large panels having a thickness between 0.5 and 0.7 mm, all processes are typically automated. In order to structure the large panel substrates, resist coating and metallization by sputtering is essential and can be realized in an automated process. Spray coating and dip coating are common technologies and sputter machines can handle large substrate sizes with sufficient homogeneity. Fraunhofer IZM has developed processes for patterning an aluminum diffusion mask on glass panels and subsequent waveguide fabrication using ion exchange. First, the diffusion barrier is deposited on both glass surface sides. Fraunhofer IZM deposits an aluminum layer of 400 nm thickness, which is DC-sputtered by Creavac Creamet 600 physical vapor deposition (PVD) equipment. Then lithography and wet-chemical processing is required to pattern the aluminum mask on the front-side of the panel. A dip coating step covers both sides of the glass panel with photoresist. The exposure of the front-side optical trace geometry and alignment marks is carried out on an Orbotec Paragon Ultra 200 laser direct imaging (LDI) system transferring the diffusion mask layout to the photoresist layer. The LDI exposed glass panel is developed and layout successfully transferred to the photo resist layer, as shown Fig. 13.18 for a glass panel with area of 440 × 570 mm^2. Then the aluminum layer on the front-side is structured with acid treatment and the photoresist is completely removed.

Planar optical waveguides will then be formed through an ion exchange process, which offers excellent waveguide performance, process compatibility, and packaging possibilities. In the exchange process, ions in the glass matrix are exchanged with ions from a hot salt melt, activated by thermal diffusion. As a result, the ion-

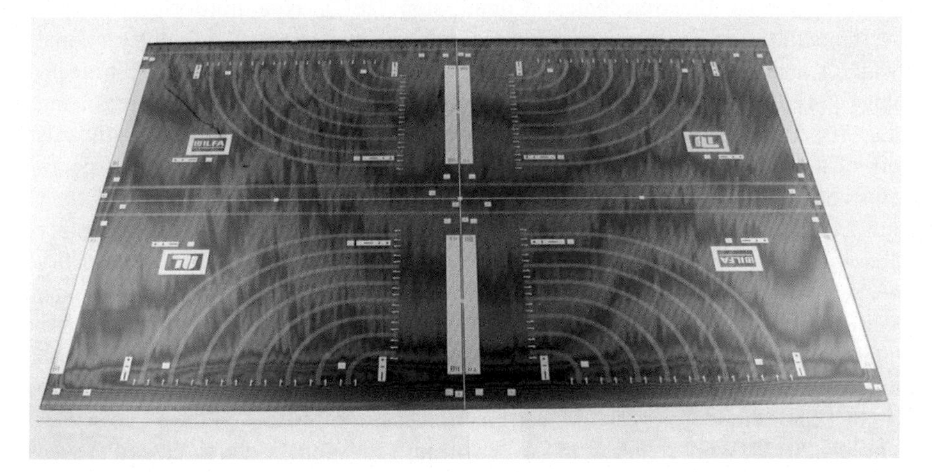

Figure 13.18 Photoresist mask layer on glass (440 × 570 mm^2).

exchange locally increases the refractive index and forms optical waveguides in the glass. The key benefits of using such thermal processes are: (1) waveguide fabrication is largely independent of substrate size, and (2) high throughput by batch processing. This capability is already demonstrated commercially in state-of-the-art processes for surface strengthening of cover glass panels on large volume industrial equipment. After waveguide integration the panel will be cut by laser processing for separation and achieving optical end-face quality on the glass edge. Fraunhofer IZM prefers a CO_2 laser in combination with cold jetting to thermo-mechanically cleave glasses with thicknesses of typically 150 μm to 5 mm with edges that can be directly used as optical interfaces at minimal roughness. Additional laser processing can be used to selectively drill holes or cut-outs into the glass panel prior to applying mechanical drilling through the entire PCB after the glass panel has been embedded [1]. A pre-processed glass panel with alignment marks, optical waveguides, thin film metallization, holes, and cut-outs can be embedded in the stack-up by standard PCB equipment.

13.3 International standardization of OPCBs

While standardization is not necessarily a prerequisite to commercial adoption of a technology, it can be a very effective means of accelerating a technology to market, provided that the standardization does not in any way constrain innovation required to further improve the technology. Areas in which standardization can be useful are measurement standards, to ensure test and measurement of a technology is carried out in a mutually agreed way, and performance benchmarks for the technology. Section 13.4 provides a detailed discussion of a measurement standard for OPCBs.

13.3.1 International Electrotechnical Commission (IEC)

The International Electrotechnical Commission (IEC), is a not-for-profit, non-governmental organization, founded in 1906, which develops International Standards and operates conformity assessment systems in the fields of electrotechnology [34]. Some 174 TCs (Technical Committees) and SCs (Subcommittees), and about 700 Project Teams (PT)/Maintenance Teams (MT) carry out the standards work of the IEC [35,36]. The 509 Working Groups (WGs) prepare standards documents under TC or SC. Technical Committee 86 (TC86) develops standards for fiber optic technologies, and Technical Committee 91 (TC91) for electronics assembly technology. In 2006, an IEC working group, IEC/TC86/JWG9 (Optical functionality for electronic assemblies), was established to prepare international standards and specifications for OPCBs, intended for use with opto-electronic assemblies [37]. As technology for OPCB is an interdisciplinary field between optical interconnection and electronic packaging, this group was established as a joint working group (JWG) of both TC86 (Fiber optics) and TC91 (Electronics assembly technology).

In July 2015, JWG9 comprised 31 experts from 12 National Committees with seven international standards published to date. In July 2015, the convenor and secretary of JWG9 are Hideo Itoh (Japan) and Richard Pitwon (Great Britain), respectively.

Currently, international standards for OPCB interconnect technologies have been restricted to optical fiber applications, such as optical fiber flexes, as this is a mature technology with a supporting commercial eco-system [38,39]. However, given the substantial progress made over the past decade in embedded waveguide technologies, in particular polymer waveguides, new revisions of these standards will now accommodate polymer and glass waveguides.

13.3.2 Polymer MT connector standard

One important example of how this group is driving commercialization of OPCB technology is the development of the standard for an MT compliant optical connector, which can be assembled onto planar polymer waveguide structures. Fig. 13.19a shows the construction and assembly of the polymer waveguide MT (PMT) connector. The PMT connector is comprised of a ferrule, cover, and boot section, the external dimensions of which, once assembled match those of existing optical fiber based MT ferrules. The optical waveguide attachment region should have a maximum width of 3 mm and minimum length of 16 mm to fit within the ferrule and boot section. The waveguide strip should contain a centered group of 12 waveguides with a center-to-center pitch of 250 μm in direct accordance with the existing standards for single row MT fiber interfaces.

The PMT ferrule is aligned to the optical waveguide interface on either a flexible or rigid optical waveguide strip such that the relative alignment of the waveguide core centers to the centers of the MT pin slots on the ferrule are in compliance with existing MT interface standards. The connector is typically fixed with quick-drying glue after which the connector facet must be polished to ensure the waveguide end facets are in the same plane as the ferrule connecting surface.

Fig. 13.19b shows an example of a commercial PMT connector for multimode polymer waveguides manufactured by Hakusan Mfg. Co. Ltd. [40] to allow direct connection to standard optical fiber MT connectors. This PMT connector was developed according to the requirements of the JPCA (Japan Electronics Packaging and Circuits Association) standard, JPCA-PE03-01-07S.

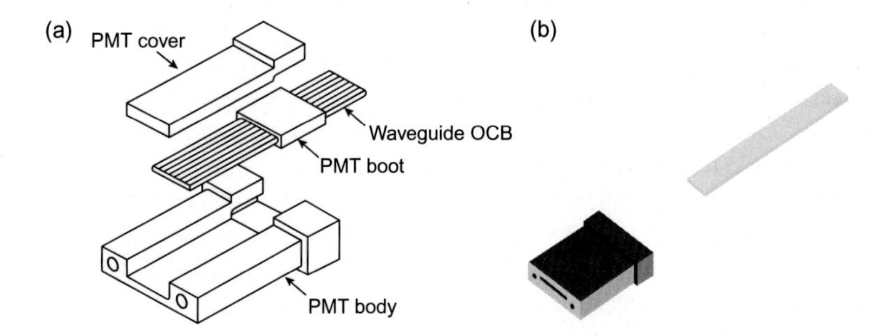

Figure 13.19 Polymer MT (PMT) ferrule standard for the termination of optical waveguides (a) PMT connector components, (b) Sumitomo Bakelite PMT connector assembly prior to waveguide ribbon attachment.

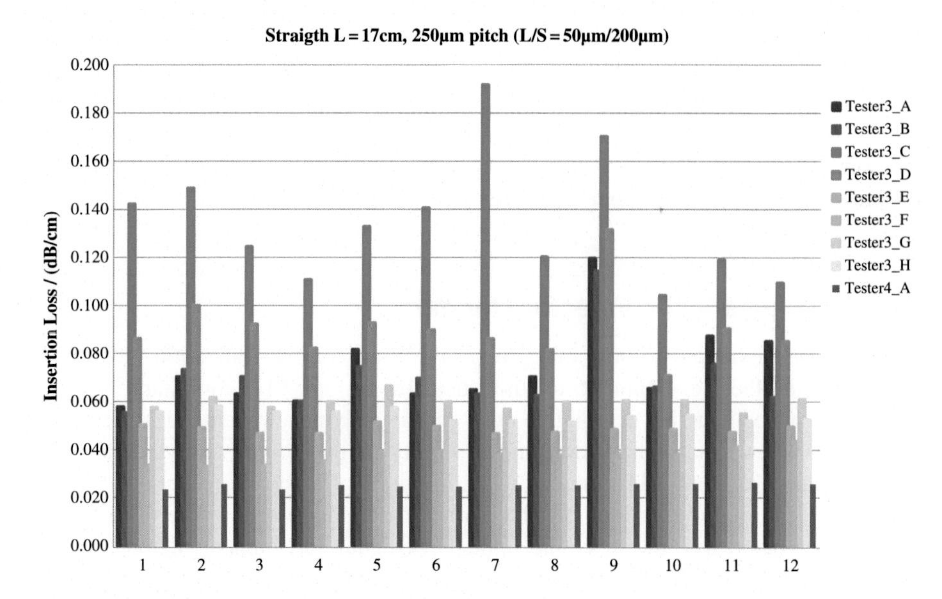

Figure 13.20 Insertion loss (dB/cm) results measured of 12-ch set of straight waveguides by multiple testers (IDs 0 − 4) using different methods.

13.4 OPCB measurement

A serious and common problem with the measurement of optical waveguide systems has been both a lack of proper definition of the measurement conditions for a given test regime and unsuitable measurement conditions, giving rise to strong inconsistencies in the results of measurements by different parties on the same test sample. This is clearly evident from the collection of insertion loss measurements on a given waveguide sample by different testing organizations [8] shown in Fig. 13.20, in which measurement results are shown to vary by over 400%.

13.4.1 Measurement definition system for OPCBs

Independent repeatability of waveguide measurements is very difficult to achieve if there is lack of clarity between testing organizations on how measurement conditions are specified. One important prerequisite to the commercial adoption of OPCBs is therefore a reliable test and measurement definition system, which is agnostic to the type of waveguide system under test and therefore can be applied to different OPCB technologies, as well as being adaptable to future variants. Such a definition system must properly capture sufficient information about the measurement conditions to ensure that the results of measurement on an identical OPCB sample by independent parties will be consistent within an acceptable margin of error.

Given the large number of measurement parameter permutations possible, the amount of information required to sufficiently describe the measurement conditions is prohibitive. It would be impractical for testers to provide a full textual description for each type of measurement, especially in situations where OPCBs are subjected to a variety of different measurement regimes, for instance, as part of a comprehensive quality assurance regime in a commercial OPCB foundry.

13.4.1.1 Measurement definition system requirements

A measurement definition system should satisfy the following core requirements:

Accuracy

The measurement definition system should capture sufficient information to ensure that the variability in independent measurement results falls within an acceptable margin, preferably less than 10%.

Accountability

The measurement definition system would force testers to be accountable to provide sufficient information about the measurement conditions. The system would therefore comprise a formalized framework to capture the required amount of information about the measurement conditions.

Efficiency

The measurement definition system should allow the entirety of the measurement condition information to be abbreviated into a measurement identification code (MIC) such that it can be contained within a short amount of text.

Convenience

The MIC should be easy to construct and deconstruct using the references look-up tables in an international standards document.

Agnostic

The measurement definition system should be agnostic to the type of OPCB under test in order to accommodate different varieties of optical interconnect, e.g., fiber, glass waveguide, or polymer waveguide. To this end, the type of optical channel under test is itself not captured in the information to be specified, it will be treated as a "black box" bounded by the input facet and output facet of the optical channel under test.

Customizable

Where the parameters of a measurement condition are not explicitly provided in the corresponding look-up tables, the MIC should be extendable to accommodate user-defined parameters.

The measurement definition system should give preference to measurement configurations which are:

- Accessible, favoring the use of available and affordable equipment,
- Viable, favoring measurements which can be easily carried out by most organizations without the requirement for specialized or restricted equipment or expertise.

13.4.2 Measurement definition system criteria

The measurement definition system should provide information on the following five critical aspects of the measurement environment:

- Source characteristics
- Launch conditions
- Input coupling conditions
- Output coupling conditions
- Capturing conditions.

13.4.2.1 Source characteristics

Typical sources for common measurements on OPCB channels include LEDs, laser diodes, and white light sources, while less common sources include amplified spontaneous emission devices. In order to accommodate a comprehensive range of available source types and characteristics, the measurement identification system will define most sources in terms of permutations of key properties including wavelength and spectral width. Source optical power or modal profile need not be specified, as only the optical power and modal profile at the launch facet need be specified as part of the launch conditions.

Modulated sources

According to the standard the source, amplitude and phase is considered to be unmodulated. Optical modulation is a large and complex area, with many possible permutations of modulation type, duty cycle, and data characteristics. Modulation schemes would include standard On-Off Keying (OOK) and multilevel modulation schemes such as PAM, in-phase and quaternary (IQ) modulation schemes such as Quadrature Phase Shift Keying (QPSK), multilevel Quadrature Amplitude Modulation (nQAM), multipulse modulation schemes, and discrete multitone (DMT). Data characteristics would include PRBS (pseudo random binary sequence) data with various correlation lengths, as well as test data associated with real data transmission protocols. Modulation will not be included in the measurement definition system described in this standard. In the event of a modulated source, the modulation characteristics will have to be explicitly stated.

Wavelength division multiplexed sources

According to this standard the source is considered to be centered on a single wavelength with varying spectral widths or white, which is consistent with the use of common commercial sources including laser diodes, LEDs or Amplified Spontaneous Emission devices. It may be desirable to characterize the performance of the channel under test with wavelength division multiplexed (WDM) light, in which multiple wavelengths are superposed onto the launch conduit in accordance with various WDM schemes including coarse WDM, whereby of the order of 10 separate wavelength encoded light streams can be superposed, and dense wavelength division multiplexing (DWDM), in which of the order of hundreds of wavelength encoded light streams can be superposed onto the same channel.

WDM sources are not included in this standard as the possible permutations would be prohibitively complex. In the event of a wavelength division multiplexed source, the wavelength division multiplexing characteristics must be explicitly stated. Preferably, if convenient, each wavelength encoded channel can be uniquely specified using the measurement identification system outlined in this standard.

13.4.2.2 Launch conditions

Launch conditions have the greatest effect on variability of measurement results on OPCB channels, it is therefore crucial that these be sufficiently defined.

Launch conditions must include the following information, which determines how light propagates through the optical channel under test and therefore determines the independent reproducibility of the measurement:

1. Launch facet size and shape, which is typically defined by the core of the launch conduit. For a standard fiber, it would be sufficient to specify the fiber type.
2. Total optical power amplitude at the launch facet.
3. Spatial (near field) and angular (far field) optical power distribution of light at the launch facet. The launch conditions for multimode conduits should preferably comply with both encircled flux (EF) requirements as defined in IEC 62614, and encircled angular flux (EAF) requirements as defined in IEC 61300-3-53. Such launch conditions can be reliably achieved by deploying appropriate mode filtering equipment around or in-line with the launch conduit. The launch conditions for single mode launch conduits should comply with IEC 61300-3-53.

13.4.2.3 Input coupling conditions

Input coupling conditions provide information on how the launch conduit is connected to the input facet of the optical channel under test, e.g., through butt-coupling or imaging through a lens system, and whether or not the input facet is treated with refractive index matching or damping materials to mitigate scattering losses.

Compliant and fixed refractive index matching or damping materials

It is common practice to apply a refractive index matching or damping material to the input and/or output facet of the channel under test in order to mitigate Fresnel reflection and scattering effects caused by the roughness of the input and/or output facet surface. The refractive index material can be in the form of a liquid or gel, which will provide a compliant buffer and is best suited to measurement whereby the launch facet is butt-coupled in direct contact or within a few microns of the input facet such that the liquid or gel completely fills the gap between the launch facet and the input facet of the channel under test. The use of liquid or gel would not be suitable in the case of a free space projection of light onto the input facet of the channel under test (such as imaging of the output of a fiber facet onto the input facet of the channel under test using a lens assembly), as the surface tension of the compliant material would cause it to form a boundary of unpredictable geometry around the input facet of the channel under test. The alternative to using a compliant refractive index matching or damping material is to use a fixed refractive index matching or damping material with a defined flat surface, such as a thin film. This is useful when the input facet of the channel under test has high roughness, but a free space launch is used.

Polarization dependent input coupling conditions

One important possible measurement parameter for single-mode launch conditions is the fast or slow polarization axis of the launch facet relative to the channel under test. For example, this would be required to characterize the polarization dependent loss or the birefringence of the channel under test. For this purpose a top axis of the input channel under test must be defined by the first tester or test sample creator and clearly marked on the sample containing the one or more channels under test or otherwise described in accompanying literature.

In measurements requiring a defined polarization, the single-mode fiber could be a polarization maintaining fiber as defined in IEC TR 62349:2014. Polarization maintaining fibers are fabricated with structures to impart an asymmetric stress profile across the fiber, enhancing the birefringence of the fiber between the axes normal to and parallel to the imparted stress. The birefringence is such that it prevents the coupling of light propagating in one polarization axis to the orthogonal polarization axis.

The optical power exiting a polarization maintaining fiber will be divided between the fast axis and the orthogonal slow axis. The ratio of optical power contained in the fast axis to the optical power contained in the slow axis depends on a number of conditions, including how the power was launched into the fiber.

In the event that a polarization maintaining fiber or other speciality fiber is used in which the optical power contained in the two orthogonal polarization axes of the launch facet is a required measurement parameter, the measurement will be defined using two instances of the measurement identification system outlined in this standard. In this way, a single measurement will be treated as two separate measurements, each defining one of the polarizations on its own.

13.4.2.4 Output coupling conditions

Output coupling conditions provide information on how the light is coupled out of the output facet of the optical channel under test to the capturing conduit, e.g., through butt-coupling or imaging through a lens system, and whether or not the output facet is treated with refractive index matching or damping materials to mitigate scattering losses.

As with the input coupling conditions, provision should also be made to define capturing using a polarization maintaining conduit, such as a single-mode polarization maintaining fiber. Again, the angle will be defined as part of the measurement identification system outlined in this standard.

13.4.2.5 Capturing conditions

The capturing conditions include information on both the capturing conduit used to extract the optical signal from the optical channel under test and basic information on the measuring element, such as a photodetector or CCD camera. However, it must be noted that, all other conditions being equal, the response of the measurement equipment itself will vary from device to device, so it will be a requirement of this standard that the measurement equipment be explicitly specified as well.

13.4.3 Standardized definition of waveguide position

The position of the launch facet relative to the input facet of the channel under test, and the position of the capturing facet relative to the output facet of the channel under test are critical parameters. In waveguide measurements, the standard procedure is to adjust the launch and capturing axes around the input and output facets of a channel under test respectively to achieve the maximum transmittance or minimum insertion loss. The respective positions of launch and capturing axes relative to the input and output facets of the channel under test required to achieve maximum transmittance, depend strongly on the geometry of the channel under test, and rarely coincide with the exact center of the input and output facets. Indeed, it would be prohibitively difficult for testers to identify the exact geometric center of the input and output facets and align the centers of their launch and capturing facets to them.

The most viable and repeatable measurement position on a channel under test is therefore through identification of the point of maximum transmittance achievable by adjusting the launch and capturing elements within the regions of the input and output facets of the channel under test. This assumes the testers are competent to properly adjust their launch and capturing devices to identify this, but this is the most reproducible approach.

Thus, the actual geometric center of the input and output facets of the channel under test are irrelevant.

13.4.4 Recommended measurement launch conditions

Different types of light sources produce very different launch conditions. For example, a light emitting diode (LED) will tend to "overfill" a multimode fiber, coupling light to a higher number of mode groups, while a laser will tend to "underfill" a multimode fiber, coupling light to fewer mode groups. Overfilling a fiber tends to produce insertion loss measurements that are too high, while underfilling a fiber tends to produce insertion loss measurements that are too low. Quality assurance or certification testing on OPCBs involving underfilled launches could therefore obscure high-loss events such as misaligned connectors or excessive waveguide wall roughness, leading in turn to false "pass" results.

13.4.4.1 Standardized optical modal profiles

As a result of the need for more reproducible multimode attenuation measurements, international standards have been developed which define a narrow range of acceptable optical power or modal distribution launch profiles from an optical channel interface.

Historically, it was sufficient to describe methods of arranging a channel such that the distribution of optical power between the modes of the channel reached a quasi-steady state known as "equilibrium modal distribution," but this did not provide a definition of optical modal distribution and as such was not sufficient to enable repeatable measurements.

New definitions for optical modal distribution in optical channels have since been devised and divided into two categories:

1. Near field optical power distribution, which equates to the spatial distribution of optical power at the fiber interface facet and is defined by a parameter known as EF.
2. Far field optical power distribution, which equates to the angular distribution of optical power after exiting the fiber facet and is defined by a parameter known as EAF.

Equilibrium modal distribution

After propagation through a sufficient length of unchanging optical fiber, a quasi-stable distribution of optical power across the optical modes of the fiber can be achieved. This distribution was known historically as the equilibrium modal distribution (EMD). It should be noted that equilibrium modal distribution does not mean a state whereby the optical power in each mode stays constant. In practice, this does not occur in multimode optical waveguides, as power continuously couples from one mode to another causing power in each mode to fluctuate. Rather, EMD refers to a state whereby the fluctuating optical power in each mode has reached a constant maximum and minimum value respectively and so, as the power in any mode fluctuates, the peak and trough power will not change unless there is change in the channel parameters. In particular, it relates to the state whereby excessive optical power in higher order modes will have been coupled to the radiation band through the channel side walls, and peak optical power in those modes will have reduced to a steady value.

EF: nearfield optical modal distribution

The near field optical modal distribution is framed by a metric known as EF, which defines the circular area integral of optical power over the channel (e.g., fiber) aperture with increasing radius from the channel center. EF defines a near field optical power distribution in terms of an EF template, as shown in Fig. 13.21. The vertical axis EF defines the total amount of optical power contained within a circle of given radius from the channel center, as given by the horizontal axis. The EF envelope or template sets defined limits on the amount of optical power included within these circles.

An underfilled channel would have more power nearer the channel center within a given radius, so it becomes the upper bounding curve of the EF template shown.

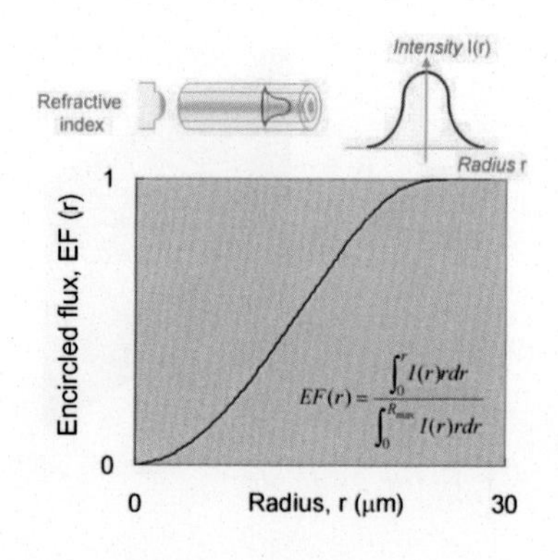

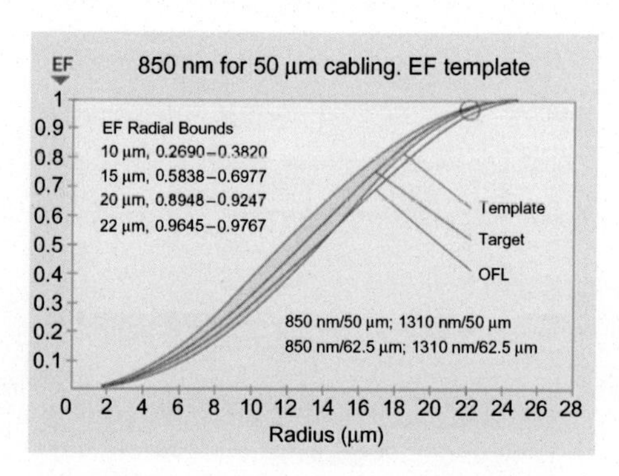

Figure 13.21 Encircled flux template as defined by IEC 61280-4-1 Ed. 2.0-Fiber-optic communication subsystem test procedures—Part 4-1: Installed cable plant— Multimode attenuation measurement.

An overfilled launch (OFL) would have less power nearer the center and more power in the outer regions of the core, so it becomes the lower OFL curve.

In order for the launch aperture of an optical channel to produce launch conditions, which meet this EF standard, the measured value of EF for each radius value from the channel center needs to fall within the upper and lower radial bounds defined by the EF template.

EF has now become an accepted part of many multimode fiber testing standards replacing previous less reproducible methods.

Both IEC and TIA standards bodies have documents that describe the requirements for EF, specifically:

- *IEC 61280-4-1* Ed. 2.0: Fibre optic communication subsystem test procedures—Part 4-1: Installed cable plant—Multimode attenuation measurement.
- This standard defines four different compliance templates for 850 and 1300 nm for both 50 and 62.5 μm cabling.
- *TIA-526-14-B* Optical Power Loss Measurements of Installed Multi-Mode Fiber Cable Plant.

EF represents a substantial improvement on previous launch condition standards based on mode power distribution (MPD), which was the metric used to describe launch conditions within the International Organization for Standardization and International Electrotechnical Commission's standard ISO/IEC 14763-3 Information Technology—Implementation and Operation of Customer Premises Cabling—Testing of Optical Fibre Cabling. The MPD metric was superseded by EF because it was found that channel or link loss consistency using sources compliant with MPD were insufficient to meet the tight loss budgets for Gigabit Ethernet.

EAF: farfield optical modal distribution

The far field optical distribution of an optical channel aperture is defined by the angular distribution of optical power measured far enough from the aperture to meet the far field condition whereby the aperture can be treated as a point source.

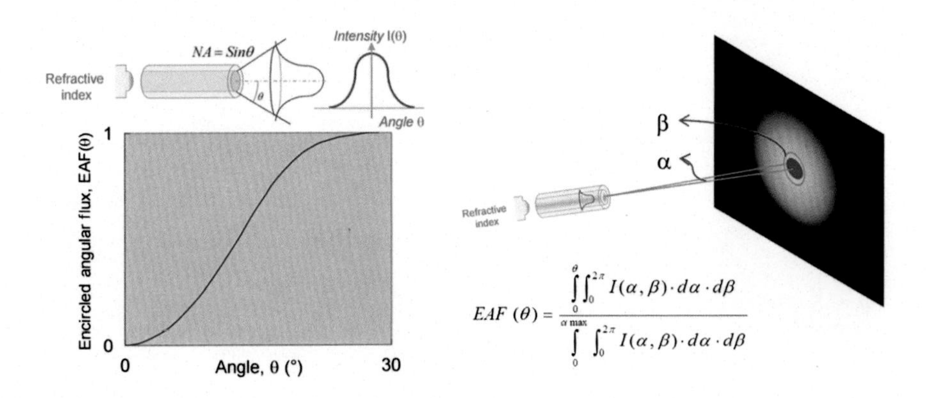

Figure 13.22 Encircled angular flux template as defined by IEC 61300-3-53.

The far field distribution is framed by a metric known as EAF, which is the fraction of the optical power radiating from an optical channel core within a certain solid angle, centered on the optical axis of the channel, measured as a function of the numerical aperture full angle. EAF thus defines a far field optical power distribution in terms of an EAF template, as shown in Fig. 13.22.

EAF is still a very new concept, with one published international standard in existence as of 2015: "IEC 61300-3-53:2015—Fibre optic interconnecting devices and passive components—Basic test and measurement procedures—Part 3-53: Examinations and measurements—EAF measurement method based on two-dimensional far field data from step index multimode waveguide (including fibre)."

This standard is currently limited to step-index waveguides; however, efforts are underway in the IEC to extend the standard or develop related standards, which include EAF definitions for 850 and 1300 nm for both 50 and 62.5 μm cabling, thus bringing them into line with the existing compliance standards for EF.

13.4.4.2 Mode filters and conditioners

Mode filters are used to selectively remove higher order modes in order to bring about an equilibrium modal distribution (EMD). Mode filters are generally most effective when the fiber is modally overfilled, which can be achieved by coupling the fiber input to a broadband source such as an LED.

One preferred, low-cost method of achieving an equilibrium mode distribution is by wrapping the fiber around a mandrel. By defining both the mandrel diameter and number of fiber turns around the mandrel, the higher order modes can be stripped in a controlled manner.

The TIA/EIA-568-B standard includes a specification for mandrel geometries and number of turns around the mandrel suitable for 50/125 and 62.5/125 μm fibers with either 3 mm jacketed cable or 900 μm buffered fiber.

Alternative methods of mode filtering include passing the fiber through to a series of S bends, either through specially formed fiber, fiber flexplane with the proper geometry, or a series of pin structures in a fiber holding plate.

More advanced mode conditioners are required, which can cause the modal optical distribution at the output to match defined EF or EAF profiles, such as those specified in the standard IEC 61280-4-1 Ed. 2.0. Certain suppliers can already provide more advanced modal filters or modal conditioners including the provision of EF preserving cables [41].

Bend insensitive fibers

Bend insensitive fibers are a variant of commercial fibers with a cross-sectional refractive index profile, enhanced to better confine higher order modes. Such fibers are becoming increasingly popular for use inside buildings where tight bends may be needed. As a consequence of their enhanced refractive index profile, they are also less susceptible to modal filtering and are therefore not recommended for use as a launch conduit.

13.4.4.3 Recommended launch conditions

Table 13.1 defines key recommended launch profiles including underfilled profiles, various mode filtered multimode profiles, and overfilled profiles, as well as recommendations on how to reproduce some of these modal profiles.

13.4.4.4 Recommended single-mode fiber launch measurement setup

The recommended measurement set-up for single-mode fiber launch conditions is shown in Fig. 13.23. A single-mode optical source should be connected with a single-mode optical fiber, first through a single-mode optical isolator to shield the source from unwanted back-reflections occurring at different interfaces further on down the test link, especially the interface between the launch facet and the input facet of the channel under test. The output from the optical isolator should then be connected through a variable single-mode optical attenuator. This will allow the tester to adjust the optical power at the launch facet to match the required optical power as defined in the MIC. This can alternatively be achieved by using a power tuneable source.

13.4.4.5 Recommended multimode fiber launch measurement setup

The recommended measurement setup for multimode fiber launch conditions is shown in Fig. 13.24. A single-mode or multimode optical source should be connected with a single-mode or multimode optical fiber, first through a single-mode or multimode optical isolator to shield the source from unwanted back-reflections occurring at different interfaces further on in the test link, especially the interface between the launch facet and the input facet of the channel under test. The output from the optical isolator should then be connected with single-mode or multimode fiber to the input of a variable single-mode or multimode optical attenuator. This will allow the tester to adjust the optical power at the launch facet to match the

Table 13.1 **Recommended modal launch profiles**

Designation	Modal distribution at launch facet	Recommended measurement set-up to achieve modal distribution
Single-mode launch		
L1	***UF*** Underfilled launch complies with single-mode launch requirements in IEC 61300-1, Annex B.2.2	Preferably, optical isolator between source and OS1 launch conduit 2-m long OS1 single-mode fiber (SMF) provides a single-mode launch profile
Multimode launch		
L2[a]	***EF/EMD*** Complies with EF requirements in IEC 61280-4-1	The source is passed into a 5 m graded index multimode fiber (GI-MMF) which is wrapped 20 times around a 38 mm diameter mandrel. The output of the mandrel is then passed through a mode controller/filter producing a mode filtered optical intensity profile, which complies with EF requirement of IEC 61280-4-1. This is then used as the input to a 5 m GI-MMF, which is wrapped 20 times around a 38 mm diameter mandrel to produce a mode-stripped optical intensity profile at the GI-MMF launch facet
L3[a]	***EF*** Complies with EF requirements in IEC 61280-4-1	5 m graded index multimode fiber (GI-MMF) is passed through a mode controller/filter producing a mode filtered optical intensity profile at the GI-MMF launch facet, which complies with EF requirement of IEC 61280-4-1
L4[a]	***EMD*** Equilibrium modal distribution	5 m 50 μm graded index OM3 multimode fiber (GI-MMF) is wrapped 20 times around a 38 mm diameter mandrel to produce a mode-stripped optical intensity profile at the GI-MMF launch facet

(*Continued*)

Table 13.1 (Continued)

Designation	Modal distribution at launch facet	Recommended measurement set-up to achieve modal distribution
L5[a]	*OF* Overfilled distribution—uniform near-field optical intensity distribution	5 m 105 μm step index multimode fiber (SI-MMF) is wrapped 20 times around a 38 mm diameter mandrel to create a mode-scrambled, overfilled optical intensity profile at the SI-MMF launch facet
L6[a]	*VOF/EAF* Very overfilled distribution Complies with EAF requirements in IEC 61300-3-53	5 m 200 μm core step-index fiber (SI-MMF) is passed through a mode controller producing a mode filtered optical intensity profile at the launch facet, which complies with the EAF requirement of IEC 61300-3-53

[a]Bend insensitive fiber is not recommended for MM or SM test leads.

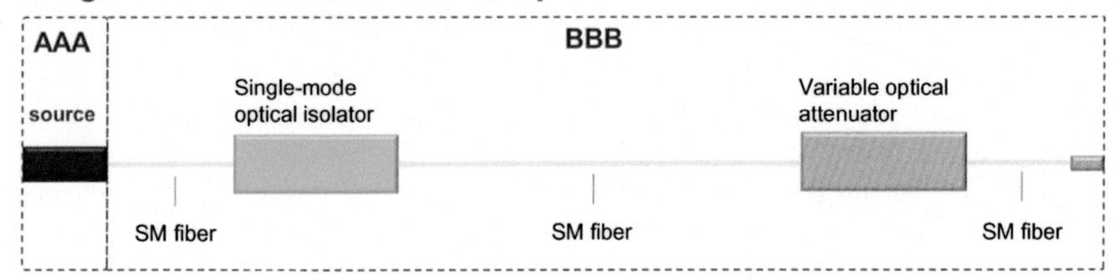

Figure 13.23 Recommended test setup for single-mode fiber launch conditions.

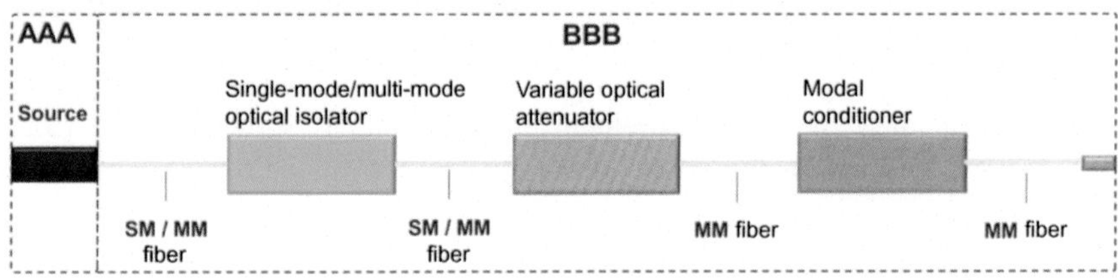

Figure 13.24 Recommended test setup for multimode fiber launch conditions.

required optical power as defined in the MIC. This can alternatively be achieved by using a power tuneable source. The output of the variable optical attenuator will then be connected with non-bend insensitive multimode fiber to the input of a modal conditioning or filtering system, the output of which will be connected with multimode fiber to the launch facet. The purpose of the modal conditioning or filtering system will be to ensure that the modal profile of the launch facet is defined according to L2, L3, L4, L5, or L6 in Table 13.1.

L2 is the preferred launch condition, in which the modal profile is created from an Encircled Flux (EF) compliant profile as defined in IEC 61280-4-1, which is then injected into a fibre mandrel.

13.4.5 Measurement identification code

"IEC 62496-2—General guidance for definition of measurement conditions for optical characteristics of optical circuit boards" is an international conformity standard designed to: (1) force testers to sufficiently specify the measurement conditions for a given optical channel under test, and (2) provide a set of recommended reference measurement conditions, in order to ensure consistency of measurement results within an acceptable margin. Given the variety in properties and requirements for different OPCB types, some test environments and conditions would be more appropriate than others for a given board. The proposed standard provides a measurement definition system to specify the conditions for measurement of OPCBs, whereby a MIC is constructed through the use of reference look-up tables to identify different aspects of the measurement environment. The MIC system thus captures sufficient information about the measurement conditions to ensure consistency of independently measured results within an acceptable margin, thus forcing testers to be accountable for providing sufficient information about the measurement conditions.

All attributes of the optical channel being measured are considered the "channel under test," which is effectively treated as a closed system or "black box." The measurement standard serves to properly capture and define the measurement conditions around the channel under test, which is assumed to have an input facet and an output facet. The measurement standard does not extend to specify the attributes that the channel under test should have, such as end facet surface quality, as this is considered part of the closed system. The idea is that multiple testers may take the same sample, whatever its attributes, and measure in the same way to ensure a like-for-like comparison.

The MIC is comprised of five three-digit numerical coordinates, which represent the five critical areas of the measurement environment, as shown in Fig. 13.25, whereby each coordinate value is obtained from the corresponding reference look-up table in the standards document.

13.4.6 AAA: source characteristics

Coordinate AAA contains the information about the source used to stimulate the optical channel, whereby sources are defined in terms of permutations of their key properties including wavelength, spectral width, and output power.

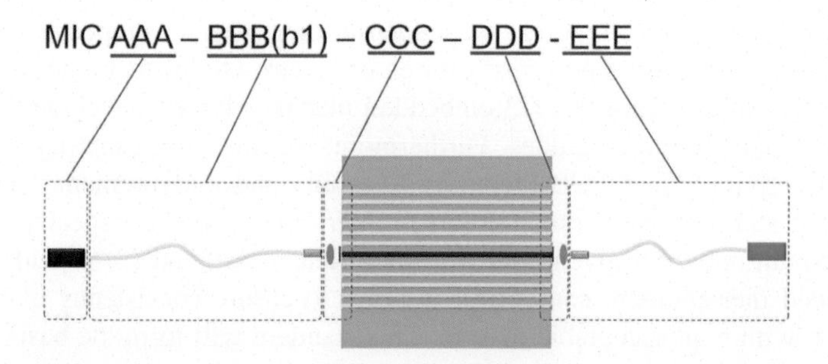

Figure 13.25 Measurement identification code construction.

13.4.7 BBB(b1): launch conditions

Coordinate BBB contains information about the conduit used between the light from the source and the input point of the optical channel under test, and includes fiber types and mode filtering mechanisms. The parameter b1 is mandatory and defines the total optical power measured at the launch facet.

13.4.8 CCC: input coupling conditions

Coordinate CCC contains information about how light is coupled from the launch conduit to the input point of the optical channel under test and includes index boundary treatment.

13.4.9 DDD: output coupling conditions

Coordinate DDD contains information about how light is coupled out of the output facet of the optical channel under test to the capturing conduit or element.

13.4.10 EEE: capturing conditions

Coordinate EEE contains information about the capturing conduit (if any) between the output facet and the optical measurement device.

The MIC method was first applied on polymer waveguide-based OPCBs developed on the PhoxTrot project to evaluate its effectiveness [42].

13.5 Conclusion

The commercial adoption of electro-optical printed circuit board (EOCB) technology will be accelerated by the development of industrial and conformity standards for high-volume fabrication, connector assembly, and waveguide measurement.

In this chapter we have reported on industrial processes developed for the high-volume fabrication of three principal classes of electro-OPCB technology, namely: (1) fiber optic circuit laminates, (2) embedded planar polymer waveguides, and (3) embedded planar glass waveguides. Furthermore, we introduced international standardization activities for OPCB including assembly and measurement. The OPCB measurement standard will both: (1) force testers to sufficiently specify the measurement conditions for a given optical channel under test, and (2) provide a set of recommended measurement conditions, in order to ensure consistency of measurement results within an acceptable margin. This standard will form the basis for validated waveguide measurements in future.

References

[1] Brusberg L, Schroder H, Ranzinger C, Queisser M, Herbst C, Marx S, et al. Thin glass based electro-optical circuit board (EOCB) with through glass vias, gradient-index multimode optical waveguides and collimated beam mid-board coupling interfaces. In: 2015 IEEE 65th electronic components and technology conference (ECTC); 2015, p. 789–98.

[2] Soma K, Ishigure T. Fabrication of a graded-index circular-core polymer parallel optical waveguide using a microdispenser for a high-density optical printed circuit board. IEEE J Sel Top Quantum Electron 2013;19(2). p. 3600310–3600310.

[3] Pitwon R, Brusberg L, Schroeder H, Whalley S, Wang K, Miller A, et al. Pluggable electro-optical circuit board interconnect based on embedded graded-index planar glass waveguides. J Lightwave. Technol 2015;33(4):741–54.

[4] Ishigure T. Graded-index core polymer optical waveguide for high-bandwidth-density optical printed circuit boards: fabrication and characterization. Opt. Interconnects Xiv 2014;8991:899102.

[5] Doany FE, Schow CL, Member S, Lee BG, Budd R a, Baks CW, et al. Terabit/s-class optical PCB links incorporating optical transceivers. Lightwave 2012;30(4):560–71.

[6] Tekin T. PhoxTrot project. [Online]. Available: http://www.phoxtrot.eu/ [accessed 16.11.15].

[7] National Technical University of Athens, "Nephele Project." [Online]. Available: http://www.nepheleproject.eu [accessed 08.12.15].

[8] HDPuG Optoelectronics. [Online]. Available: http://hdpug.org/content/optoelectronics [accessed 16.11.15].

[9] Ishigure T, Shitanda K, Kudo T, Takayama S, Mori T, Moriya K, et al. Low-loss design and fabrication of multimode polymer optical waveguide circuit with crossings for high-density optical PCB. In: 2013 IEEE 63rd Electron. Components Technol. Conf.; 2013, p. 297–304.

[10] Krahenbuhl R, Lamprecht T, Zgraggen E, Betschon F, Peterhans A. High-precision, self-aligned, optical fiber connectivity solution for single-mode waveguides embedded in optical PCBs. J Lightwave Technol 2015;33(4):865–71.

[11] Pitwon R, Worrall A, Stevens P, Miller A, Wang K, Schmidtke K. Demonstration of fully enabled data center subsystem with embedded optical interconnect. Opt Interconnects Xiv 2014;8991:899110.

[12] Berger C, Kossel M, Menolfi C, Morf T, Toifl T, Schmatz M. High-density optical interconnects within large-scale systems. Proc SPIE 2003;4942(2003):222–35.

[13] Pitwon R C a, Wang K, Graham-Jones J, Papakonstantinou I, Baghsiahi H, Offrein BJ, et al. FirstLight: pluggable optical interconnect technologies for polymeric electro-optical printed circuit boards in data centers. J Lightwave Technol 2012;30 (21):3316–29.
[14] Schares L, Kash J a, Doany FE, Schow CL, Schuster C, Kuchta DM, et al. Terabus: terabit/second-class card-level optical interconnect technologies. IEEE J Sel Top Quantum Electron 2006;12(5):1032–43.
[15] Schroder H. Glass panel processing for electrical and optical packaging. In: ... (ECTC), 2011 IEEE ...; 2011, p. 625–33.
[16] Pitwon R, Schröder H, Brusberg L, Graham-Jones J, Wang K. Embedded planar glass waveguide optical interconnect for data centre applications. In Optoelectronic interconnects Xiii, Proc. SPIE, vol. 8630; 2013, p. 86300Z.
[17] Takenobu S, and Okazoe T. Heat resistant and low-loss fluorinated polymer optical waveguides at 1310/1550 nm for optical interconnects. In: 37th Eur. conf. expo. opt. commun. 1, p. We.10.P1.31; 2011.
[18] Roelkens G, Dave UD, Gassenq A. Silicon-based photonic integration beyond the telecommunication wavelength range. IEEE J Sel Top Quantum Electron 2014;20(4).
[19] Barwicz T, Taira Y. Low-cost interfacing of fibers to nanophotonic waveguides: design for fabrication and assembly tolerances. IEEE Photon J 2014;6(4):1–18.
[20] Hsu H-H, Nakagawa S. Dry-film polymer waveguide for silicon photonics chip packaging. Opt Express 2014;22(19):23379–84.
[21] Dangel R, Hofrichter J, Horst F, Jubin D, La Porta A, Meier N, et al. Polymer waveguides for electro-optical integration in data centers and high-performance computers. Opt Express 2015;23(4):4736.
[22] Huang C-C. Ultra-long-range symmetric plasmonic waveguide for high-density and compact photonic devices. Opt Express 2013;440(7083):2333–42.
[23] Zografopoulos DC, Beccherelli R. Liquid-crystal tunable long-range surface plasmon polariton directional coupler. Mol Cryst Liq Cryst 2013;573(1):70–6.
[24] Szymanski T. "Architecture of a terabit free-space intelligent optical backplane". J Parallel Distrib Comput 1998;55(1):1–31.
[25] Fey D, Erhard W, Gruber M, Jahns J, Bartelt H, Grimm G, et al. Optical interconnects for neural and reconfigurable VLSI\narchitectures. Proc IEEE 2000; 88(6):838–48.
[26] Immonen M, Wu J, Yan HJ, Zhu LX, Chen P, Rapala-Virtanen T. Development of electro-optical PCBs with embedded waveguides for data center and high performance computing applications. In: SPIE photonics west 2014-OPTO: optoelectronic devices and materials, vol. 8991; 2014, p. 899113.
[27] Dow Corning. Polymer waveguide silicones. [Online]. Available: http://www.dowcorning.com/content/electronics/electronicsmarkets/polymer-waveguide-silicones.aspx [accessed 07.12.15].
[28] Microresist Technology Gmbh. Ormocore and Ormoclad. [Online]. Available: www.microresist.com [accessed 07.12.15].
[29] Microchem. LightLink waveguide materials. [Online]. Available: http://www.microchem.com/Prod-Light-Link.htm [accessed 07.12.15].
[30] Swatowski B, Amb C, Breed SK, Deshazer DJ, Ken WW. Flexible, stable, and easily processable optical silicones for low loss polymer waveguides. In: SPIE ...; 2013, p. 11.
[31] Sato T. Novel organic-inorganic hybrid materials for optical interconnects. Proc SPIE 2011;7944. p. 79440M–79440M–8.

[32] Brusberg L, Schröder H, Herbst C, Frey C, Fiebig C, Zakharian A, et al. High performance ion-exchanged integrated waveguides in thin glass for board-level multimode optical interconnects. In: Proc. European conference on optical communication (ECOC); 2015.
[33] BOE's Gen 10.5 Display Equipment Is A Pie In The Sky For Korean Equipment Companies ETNews, [Online]. Available: http://english.etnews.com/20150710200003 [accessed 25.11.15].
[34] International Electrotechnical Commission—Who we are. [Online]. Available: http://www.iec.ch/about/profile/ [accessed 07.07.15].
[35] International Electrotechnical Commission—Management Structure. [Online]. Available: http://www.iec.ch/dyn/www/f?p = 103:63 [accessed 07.07.15].
[36] IEC TC/SCs IEC Technical Committees & Subcommittees—About TC/SCs. [Online]. Available: http://www.iec.ch/dyn/www/f?p = 103:62:0::::fsp_lang_id:25 [accessed 07.07.15].
[37] International Electrotechnical Commission TC 86 Fibre optics JWG 9 Convenor & Members. [Online]. Available: http://www.iec.ch/dyn/www/f?p = 103:14:0::::FSP_ORG_ID,FSP_LANG_ID:2775,25 [accessed 07.07.15].
[38] Mikawa T. Low-cost, high-density optical parallel link modules and optical backplane for the last 1 meter regime applications. Proc SPIE 2008;6897:XI−XX.
[39] Masuda H, Saito K, Suzuki S, Kinoshita M, Ibaragi O, Okada Y, et al. High-density optical backplane using small diameter optical fiber. J JAPAN Inst Electron Packag 2006;9(4):289−95.
[40] PMT connector, Hakusan Mfg. [Online]. Available: http://www.hakusan-mfg.co.jp/productinfo/ict/optcom/pmt-connector.html [accessed 07.07.15].
[41] Arden photonics EF preserving cables.
[42] Pitwon R, Wang K, Immonen M, Wu J, Zhu LX, Yan HJ, et al. International standardisation of optical circuit board measurement and fabrication procedures. In: Optical interconnects Xv, vol. 9368; 2015, p. 93680W.

Requirements for process automation of optical interconnect technologies

14

I. Piacentini and T. Vahrenkamp
ficonTEC Service GmbH, Achim, Germany

14.1 An introduction and a list of issues

This chapter will concentrate on the need for automation of the packaging of PICs, and will not address the system-level integration of silicon photonics at board level/backplane/enclosures/inter-rack connectivity [1], partly because PIC packaging is where the authors are more knowledgeable, but also because it is felt that automation of the PIC packaging processes—and the inherent cost savings—is and will be an important first step, and will be largely responsible for widespread acceptance of PICs.

From wafer level to a fully packaged photonics device the path is fairly long and requires the assembly, alignment, testing, and integration of several elements/process steps, but essentially packaging should take care of the following:

- Complete the photonics circuit with the additional elements not integrated at wafer-level, typically with a flip-chip bonding process;
- Insert the chip into a "box" (the package) that provides mechanical and environmental protection to the device (bath-tub metallic packages, ceramic or organic/plastic packages;
- Provide electrical connectivity (wire-bonding inside the package, pins outside the package);
- Provide optical connectivity (single/multiple fiber optics pig-tailing);
- Ensure thermal contact/thermal interfacing for proper power dissipation.

The path is equally long also from the manual assembly of complex PIC devices in the labs that relies on very skilled and patient personnel, with a single device assembly time that can span over several hours with uncertain yields, to a full industrial production targeting minutes or possibly tens of seconds per device, with extremely high yields and very repetitive and accurate processes.

14.1.1 Let us pick an example

The example in Fig. 14.1 is taken from Doe [2], and shows a single photonic chip produced at wafer level with well proven CMOS foundry technology. An external CW infrared laser source is mounted on the diced chip, with the beam facing down and coupled via a grating structure to the optical circuitry of the chip. A bundle of

Optical Interconnects for Data Centers. DOI: http://dx.doi.org/10.1016/B978-0-08-100512-5.00014-0

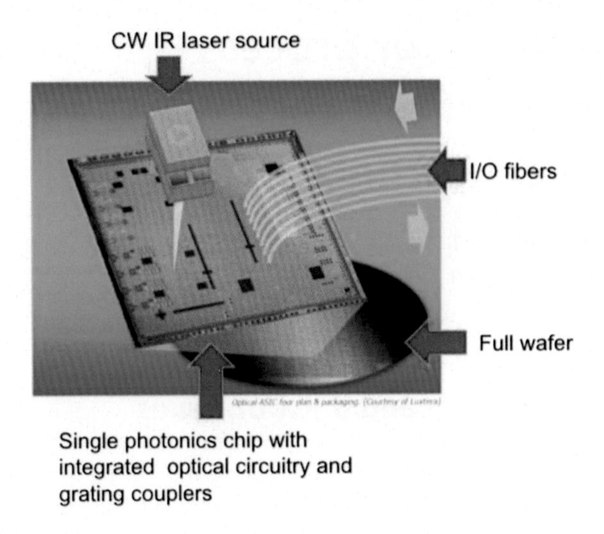

Figure 14.1 An example of a photonics chip assembly.
Source: Courtesy of Luxtera.

fibers carrying the input and output signals is also shown. The fibers will typically be assembled in a ribbon and pre-spaced on a glass block. The laser source and the fiber block have dimensions of a few millimeters, and require positioning with sub-micron accuracy. Epoxy glue bonding is used to fix these components to the silicon substrate. These kinds of operations represent the core of automated PIC packaging/assembly.

It should be noted that the example above does not show an actual outer package and more conventional process steps, for instance wire-bonding.

14.1.2 Complexity levels in packaging processes and the evolution of assembly machines

The level of complexity of PICs can cover a very broad range, from fairly simple TOSA/ROSA to a multi element free-space optics hybrid assembly of a full transceiver, as illustrated in the following figures. In a similar way, dedicated assembly machines have evolved from R&D laboratory equipment or compact demonstrators to fully-fledged automated industrial equipment.

The assembly in Fig. 14.2 shows a single fibers aligned to a photo-detector and held in position by a fairly large "glob" of laser-induced soldering. This fixing method requires a metallized fiber, and it is no longer in use today. This assembly can hardly qualify as a PIC as it is really an assembly of parts, including SMT passive components, on what looks like a ceramic substrate.

A more recent PIC assembly is shown in Fig. 14.3: a chip is held in a chuck and is being contacted by an electrical probe-card on the left, while a fiber-array block is being positioned from the right.

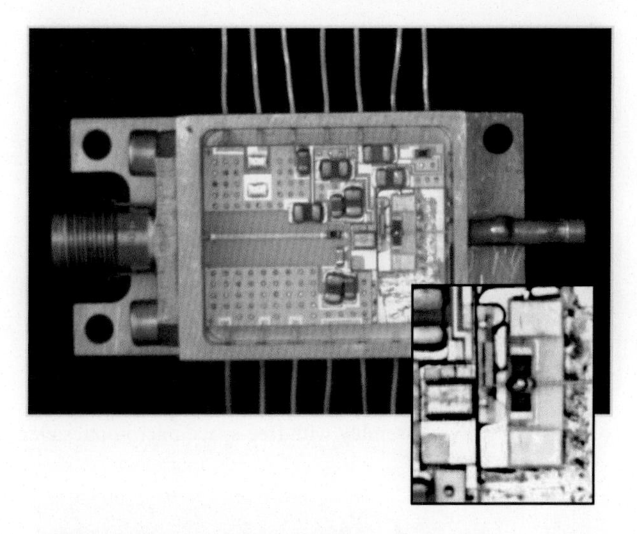

Figure 14.2 An optical receiver from the mid-1990s.

Figure 14.3 A photonics chip being coupled to a fiber-array block.

A device with a much larger complexity, a transceiver that also requires the bonding of free space micro-optical elements, is illustrated in Fig. 14.4. The actual device is held on the chuck between the two electrical probe cards and micro-optical elements are aligned on its right-hand side.

To address the evolution of machines, a few examples are illustrated below, starting with a picture from 2001. While it is true that the motion mechanics has also progressed since then, the wide availability and cost reduction of "accessory" technologies like machine vision, nanometer-resolution of position sensors, industrial computer processing power, and tools for faster software development, is largely responsible for the transition to fully automated industrial machines.

Figure 14.4 A complex transceiver assembly with free-space micro-optics elements.

Figure 14.5 An active alignment demonstration unit developed by miCos GmbH and National Instruments Italy in 2001.

What we see in Fig. 14.5 above is a table-top five-axis mechanical set up targeted at demonstrating fast active alignment. It is complete with high-magnification cameras and is fully controlled by the PXI-based industrial controller on the left, housing also the motion control, image frame-grabbers, and data acquisition boards.

A more complex set up with a total of 14 mechanical axes is shown in Fig. 14.6, with the 3D mechanical layout and a detail of the fiber-arrays block held on electrically heated grippers (for thermal epoxy bonding). The primary goal was active dual-sided fiber optic pig-tailing of a new type of modulator with edge-coupling optical input output. With a typical coupling area of 1.5×5 μm, the required accuracy to minimize losses was in the order of 100 nm, with a minimum controlled

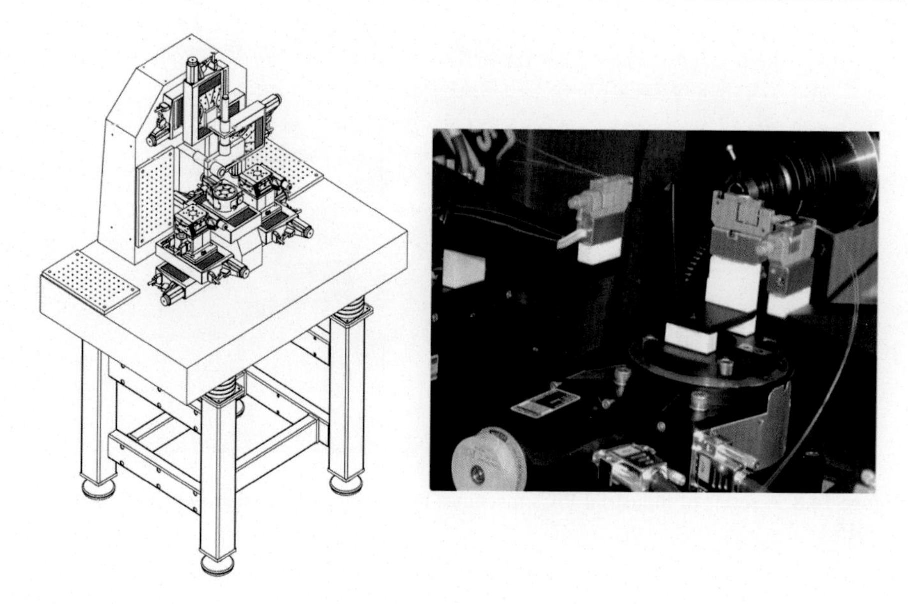

Figure 14.6 A 14-axis dual-sided FO pig-tailing machine at Pirelli Labs in 2001.

Figure 14.7 A pre-production PICs assembly machine designed and built in 2013.
Source: Courtesy of PI-miCos.

step of the alignment mechanics of 25 nm. Synchronizing power meter data acquisition and motion control was challenging at that time, and it is further explained in Section 14.2.

In Fig. 14.7 we jump ahead some 12 years. The machine depicted was designed with a strong accent on flexibility, as it was targeting the pre-production of a limited batch of PICs and would then serve as a generic assembly platform at CNIT-Inphotec in Pisa, Italy [3].

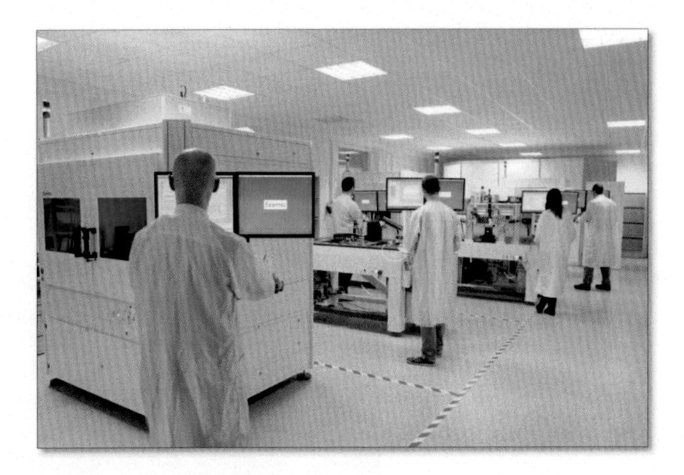

Figure 14.8 A series of industrial PICs machines being commissioned at ficonTEC GmbH in 2015.

Within this long period of time two things are worth remarking on: vertical optical coupling based on grating structures had become widely used in PICs, and an alternative to the stack of axes in the assembly machines had emerged in the form of parallel kinematic hexapod-like 6 DOF mechanical devices. On this machine, ancillary operations like the pick and place of the components to be assembled, as well as epoxy-glue dispensing and UV curing, are performed by an of-the-shelf 6-axis anthropomorphic robot.

As the demand for volumes increased more functionalities were added into the machines, with fully automated long sequences of operations, including multiple-components pick and place, epoxy dispensing or laser welding/soldering, in-line testing, etc. The look of the machines also changed, with full outer enclosures, larger buffers/loaders for parts, and a top-mounted tri-color semaphore signaling machine status. A series of such machines being commissioned prior to delivery can be seen in Fig. 14.8.

The key "ingredients" of these machines are further presented and discussed within thischapter.

14.1.3 The main steps of a complex packaging process

In the world of conventional semiconductor devices, the terms front-end/back-end usually refer to wafer-level processing and the following processes to package the semiconductor die. Some of the semiconductor fabrication steps are also common to photonics, and are simply adopted for the packaging of PICs, but not all the steps need to be included in a PICs assembly machine. Electrical wire-bonding is a good example: this process has reached such a level of optimization that it would not make sense to include it in machines dedicated to PICs. Furthermore,

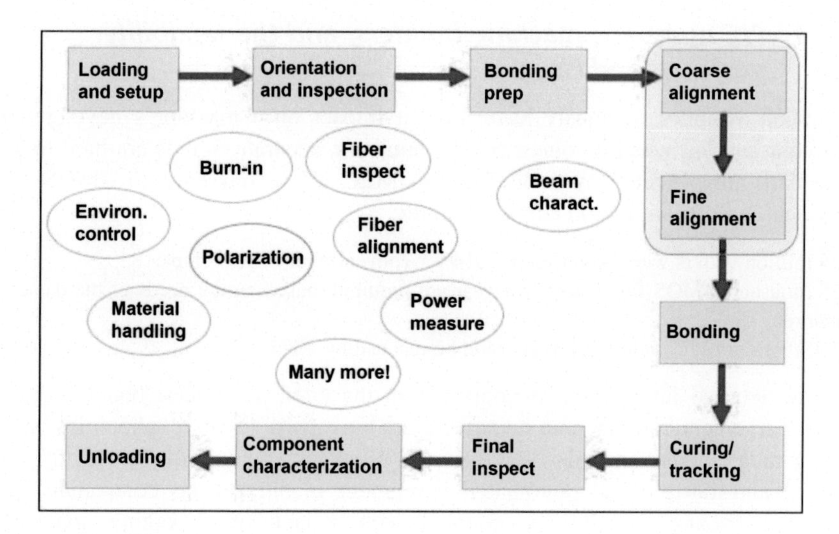

Figure 14.9 The steps of a typical assembly process from a 2002 presentation.

specific equipment is offered from reputable suppliers with very good price/performance ratios, only achievable when technologies have stabilized for decades. The sub-components of PICs are in most cases delivered as pre-diced wafers on what is commonly termed as blue-tape, in gel-packs, waffle-packs, or ad-hoc loading trays.

The actual sequence of process steps is invariably specific to a given PIC device, requiring some form of sequencing flexibility/programmability to be designed into the assembly machine. Fig. 14.9 recycles an old slide of a speech given in 2002 by one of the authors during a workshop at Flextronics, in Cork, Ireland, but it is still very much applicable today. The square boxes list the actual process steps from the loading of parts to the unloading of the completed assembly, while the bubbles add additional/auxiliary topics of the assembly process. Coarse and fine alignment are in today's jargon often replaced with passive/active alignment, as discussed in Section 14.2.

Some form of flip-chip bonding is often part of the process in order to add additional components to the optical circuitry chip, ranging from complex CMOS driver circuits for multiple MZ modulators, making use of micro-copper pillars [4] or other "bumps" contacting methods, to single devices like laser emitters or photodetectors. The latter devices are often in the form of CoS (Chip on Sub-mount), and are attached with a variety of bonding methods, from epoxy gluing to laser-induced soldering.

Listing all the required/possible steps is outside the scope of this short chapter, and the assembly of photonics devices is still in a growing phase, with new ideas, methods, and "tricks" being disclosed at every conference. A consequence of this is that in the design of a complex automated machine, the first step is always the careful analysis of the device assembly process.

14.1.4 PIC numbers, machine builders, and the flexibility versus speed issue

Production volumes obviously have a great impact on a machine's development and evolution, but when it comes to PICs numbers, accurate figures are hard to get. Some data are provided by colleagues of IMEC [5] in reference to CMOS 12″-equivalent wafers, with 1000 chips per wafer:

- 25 million wafers were produced in 2014 for conventional CMOS chips;
- 25 thousand CMOS PIC wafers per year are required, based on the needs of the datacom market;
- 250 thousand PIC wafers per year could be reached by 2025.

More detailed information is contained in the cited references, but it is likely that the large semicon machine manufacturers have hardly seen a "photonics beep" on their radar screen, and that PICs assembly is still the domain of few, highly specialized, and fairly small, automated equipment manufacturing companies. The variety of packaging solutions, coupled with a lack of packaging standards, demands that today's machines should prioritize flexibility rather than speed, allowing the end users to quickly adapt their equipment to process changing, moving on from early prototypes/limited series, to production-optimized devices.

As PIC production volumes will increase, the design of the machines will likely change, targeting speed rather than flexibility, resulting in "arrays" or lines of simpler machines, interconnected by suitable transport mechanisms for the devices being assembled, and allowing to distribute/parallelise/balance the time duration of specific steps in the whole process. Addressing speed will also address the issue of capital equipment cost and ROI, with a better definition of assembly cost per device. Foot-print, and space requirement in the clean-room, will also become more important, as well as running costs of the assembly machines.

14.1.5 Photonics materials and optical fibers pig-tailing

While the quest for the ideal photonics materials goes on, and while it is clear that in terms of volumes existing CMOS foundries are and will play a major role in the development of photonics, the average machine manufacturer remains reasonably agnostic in the debate over indium phosphide, silicon on insulator, lithium niobate, etc.

A certain degree of "mix and match" of photonics elements based on different materials is likely to be around for quite a while, and getting fibers pig-tailed to the package will also remain, even if a suitable method for passive-only alignment should be developed.

14.1.6 Addressing different markets

This book is about optical interconnects for data centers, but photonics covers many other fields and so does the assembly of PICs. The development of automated assembly processes will benefit from the variety of different application fields and different packaging requirements. A non-exhaustive list is given below:

- High performance computers interconnect with photonics technologies that share most of the needs of data center interconnects;
- Life sciences, with lab-on-chip devices adopting photonics technologies;
- Image sensing, 2D and 3D, for the smart phones and consumer markets, where the assembly of the CCM (Compact Camera Modules) is reaching overall sizes and complexities requiring PIC-like assembly processing;
- Sensing in general, especially based on laser interferometry and multi-spectral analysis, where whole set ups, occupying in the past whole optical breadboards, in the labs are being squeezed down at chip sizes;
- Sensing for automotive, addressing the needs of autonomous vehicles, and combining visible and IR image sensing, LIDAR, motion and gesture recognition sensors. V2V (vehicle to vehicle) and V2I (vehicle to infrastructure) and dynamic map/info updating will also contribute to increase the demand for overall data flows, and hence more data centers;
- Photonics and micro-fluidics, addressing different needs from bubble-based cross-matrix switching for datacom, to bio-medical and life sciences applications.

14.1.7 The more that can be done at wafer level, the better

Most of today's PIC-based devices are assembled/packaged at single die level, usually on pre-tested chips. Another effect of the increase in production volumes will be an effort to perform assembly operations at wafer level, i.e., before dicing.

Conventional semiconductor manufacturing is also moving more and more to 3D assemblies—with some issues on thermal dissipation and increasing numbers of electrical interconnects—and some technologies will spill out to photonics.

Complete wafer-to-wafer bonding is unlikely to occur for photonics devices, at least for the time being, and as long as individual elements need to be added to the PIC substrate, but the mounting of photonics components, including micro-optics elements, onto a complete PIC wafer, prior to dicing, is already being done. Wafer level testing/in-line testing (see Section 14.8) will become more relevant in this context.

14.2 Positional accuracy and the debate on passive/active alignment

There is a substantial difference between accuracy, repeatability, and resolution of the multi-axis mechanics that are used to position and align the components of photonics devices. While the combination of conventional motor-driven mechanics and piezo-actuated elements can easily provide resolutions of a few tens of nanometers, achieving bidirectional repeatability to sub-micron level in industrial-grade, robust, and reasonably priced equipment is still challenging. However, it can be safely stated that mechanical sub-micron repeatability of 0.5 or even 0.05 μm is now achievable by higher class multi-DOF positioners. The repeatability just stated also corresponds to the large majority of user-defined requirements, in order to minimize coupling losses, especially when fibers are aligned to either vertical grating or edge-coupled structures.

Before discussing further let us define what is passive and what is active:

- Passive alignment refers to using machine vision to identify the geometric boundary of a given component (single fiber, fiber array, laser emitter, photo-detector, micro-optical element), and place it on the photonics chip surface. It is assumed that a geometric calibration has been performed between the pixels of image space and the actual encoders of the mechanical axis.
- Active alignment refers to actually "switching on" the component(s), ensuring that a continuous light path exists between a light source (typically a laser) and a detector. The detector can be an optical power meter with a calibrated power output, or a simpler photo-detector connected via a TIA to a data acquisition analog input board. The controllers of the mechanical axis and the data acquisition electronics must share a common clock—or some other form of synchronization—in order to acquire data points that are linked to specific positions in space, and hence determine the best coupling alignment.

Machine vision will reach its limits, mainly due to diffraction or MTF limits of the sensor-optics combination, but it is often used to limit the peak search area of the active alignment. This two-step approach has been also referred to as coarse + fine alignment (see Fig. 14.9 in Section 14.1.3).

Passive versus active is often hotly debated during workshops, as it is considered fairly complex to implement and especially time-consuming. While it is likely that development of pig-tailing methods that are more position/alignment tolerant—one example is the use of micro-lenses to defocus/refocus the light path from PIC to fibers—will continue and mature, it is unlikely that active alignment will disappear any time soon from the requirements of the PICs assembly machines. With today's technology, active alignment can be reduced to a few seconds, and it is worth remembering that it is only one step in a more complex assembly process.

Active alignment has been around in some form or another for a long time—Fig. 14.10 is from 2002—and it is today extremely robust and reliable. The user interface panel in the figure shows the alignment of a fiber-array to a fiber-array performed acquiring the power signal on the first and last fibers of the array (blue and red in the 3D graph). The search area is quite wide by today's standards, and 4000 data points were acquired along an Archimedes spiral motion trajectory (the mechanical interpolation inaccuracies are clearly visible) in some 60 seconds. An important issue of that period was to overcome any latency in the communication between the motion control board and the analog acquisition board, a problem completely settled today by fast CPUs and fast computer busses [6]. At the end of the search, the fibers will be positioned over the maximum peak and monitored during the epoxy bonding curing time, effectively providing also a "bond tracking function".

Fast alignment can be achieved on both a conventional and a piezo-driven axis, and it is today possible to contain active alignment time down to a few seconds. However, adopting a combination of conventional and piezo motion integrated on the same mechanics and completed with a dual loop controller, would allow fast

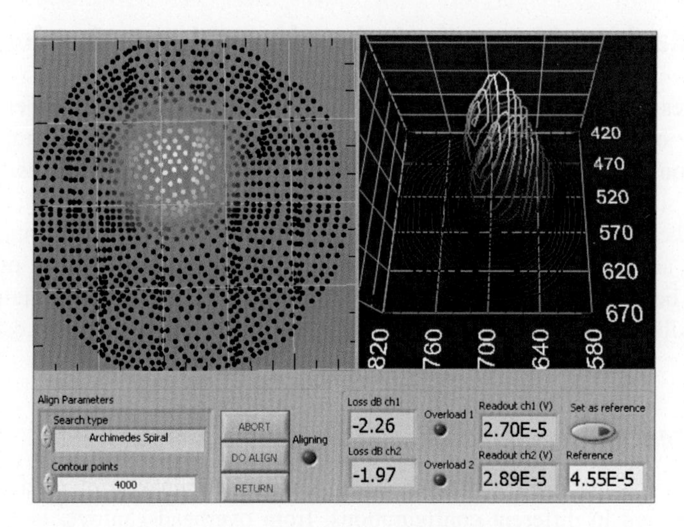

Figure 14.10 An early example of dual-fibers active alignment (2002).

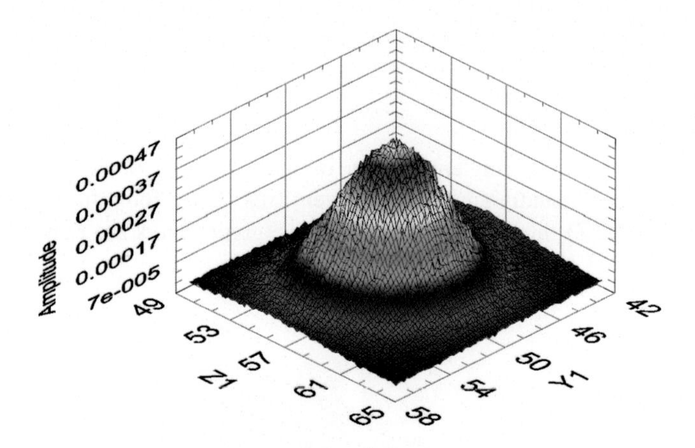

Figure 14.11 Fast alignment check on an area of approximately 18 × 18 μm.

and large travel ranges to be combined with the resolution and repeatability of short travel piezo actuated flexure stages.

The active search area can also be greatly reduced by guiding the fibers (or whatever component needs to be aligned, like a micro-optical lens) with machine vision to a position that ensures that ensures a continuous path from the light source and the detector, usually referred to as "first light".

High-speed alignment algorithms coupled with dual sided alignment set ups are particularly relevant for PICs testing (see also Section 14.8).

The 3D plot in Fig. 14.11 depicts a quasi-Gaussian fast peak search over an area of approximately 18 × 18 μm. Some measurement noise is evident on the plot floor.

14.3 Machine "ingredients" and machine technologies

What makes PICs automated assembly machines complex and thus rather costly is the variety of different devices and technologies that need to be integrated. Ideas and solutions can be mutuated from other industrial automation fields, but they need to be scaled-down to the size and positioning accuracies of the photonics components. The following paragraphs are a non-exhaustive list of the main ingredients. Successful assembly equipment suppliers have managed to turn the mix of ingredients listed below into a set of interlocking modules, preserving both a high level of design flexibility, as well as reducing NRE (non recursive engineering) costs.

14.3.1 Mechanical positioners/manipulators

The core of the machines is heavily based on the combination of motion axis or stages, arranged in different configurations, from overhead gantries, to *x-y* tables, rotation wafer chucks, and multi degree-of-freedoms (DOF) assemblies. No attempt is made here to delve into details such as bearings, guides, shaft-couplings, and the myriad of stages sub-components.

14.3.1.1 Stack of axis

The earliest and simplest implementation of multi-DOF is obtained by "bolting together" a number of individual motor-driven axes. Thus, for a full 6 DOF stack (X, Y, Z, θX, θY, θZ) three linear axes and three goniometers are required, as can be seen in Fig. 14.12. Great care has to be taken in the assembly to ensure

Figure 14.12 A 6-DOF stack of axis.
Source: Courtesy of ficonTEC.

orthogonality and centering of the axes, with some metrology performed on the assembly using interferometry measurements.

A concept that greatly simplifies the operation of multi-DOF devices (and common to the world of industrial robotics) is that of "pivot point", where the known geometry of the mechanical stack of axis is combined with the inverse kinematic functionality of the motion controller to define a point in space (it could be the tip of a single fiber or a convenient point/feature of an active photonics element) that also takes into account the geometry of the gripper and of the device to be aligned.

An alternative to a stack of axis is presented in the paragraph that follows, though the adoption of these devices has been somewhat hampered mainly by the cost of these devices, and, sometimes by the limited travel range or by the robustness in the field.

14.3.1.2 Multi DoF/6 axis devices: hexapods and SpaceFabs (tripods)

The 6-DOF devices shown in Fig. 14.13 share a common concept: a top platform/flange that can be moved in an *X-Y-Z* space, including the rotation around these axes. These intrinsic 6-DOF devices are fully integrated with a dedicated controller that also provides the "pivot point" functionality and exhibits very good mechanical characteristics in terms of foot-print and compactness, overall accuracy, and stiffness.

Six legs/actuators are used in a Hexapod, while a SpaceFab can be considered a tripod, its three legs being moved by three *x-y* stages. Like their stack-of-axis counterparts, these devices are available in a range of sizes and payloads. The choice of one or the other is really application-specific, the authors having a slight penchant for the SpaceFabs, that have a squatter geometry and tend to merge better in the overall design of a machine.

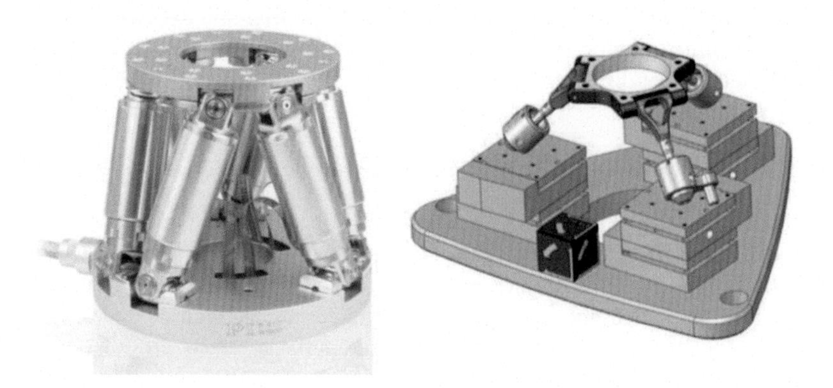

Figure 14.13 A 6-DOF Hexapod and a SpaceFab. Not to scale.
Source: Courtesy of PI.

14.3.1.3 Motors/actuators, feedback sensors, and controllers

Electrical motors fall into two basic categories: stepper motors and servo-motors (the latter by now of the brush-less variety), selected for specific application requirements, such as speed and torque. Both types can be found in their linear form for the implementation of the long stages required by the geometry of gantry structures. A linear motor can be visualized as the "unrolling" of the stator and rotor of a conventional motor, thus transforming a torque into a direct linear translation. A feedback sensor—an encoder—is essential for the operation of a servo-motor, while a stepper can be operated also in open-loop, though in precision stages it is preferable to associate encoders also with stepper motors.

Piezo actuators differ substantially from conventional motors, and are based on the property of certain materials to convert an applied voltage to mechanical force. They are mostly used for short motion ranges with nanometers resolution, often combined with flexure-based stages.

Backlash and bi-directional repeatability, especially long-term effects, are the nightmares of stage designers. Encoders—the most common being incremental optical or magnetic encoders—exist in both rotary or linear versions, the latter being used to minimize the effect of backlash in precision stages.

Most piezo/flexure stages resort to capacitive sensors for both accuracy and dimensional constraints.

It is not uncommon to include external high accuracy positioning sensors when measuring distance or planarity, based on a variety of methods, from simple laser triangulation to white light or laser-based interferometry. It is interesting to note that photonics integration will also contribute dramatically to the cost and size of interferometry-based sensors, allowing their direct integration in the mechanical stages.

The generic name of "controller" is usually indicating both the positioning electronics (also known as indexing) and the power electronics driving the motors. Multi-axis interpolation and trajectory control is based on real time computation executed by an embedded computer, and application-level interfacing and programing is made easier by the presence of firmware interpreting and executing high level commands. Direct analog inputs and analog-digital converters facilitate active alignment and accept direct voltage signals from power meters or other detectors.

14.3.1.4 Conventional industrial robotics for pick and place

Industrial robots are extremely robust and reliable COTS (Commercial Off The Shelf) positioners with sufficient resolution, accuracy, and repeatability to solve a number of complex pick and place and other ancillary jobs, like epoxy dispensing and curing, in PICs assembly automation. They offer a very high degree of flexibility and can be easily reprogrammed for different tasks. Two different types of industrial robots are shown in Fig. 14.14. The small EPSON 6-axis C3 model with a payload of 1 kg (max payload is 3 kg) exhibits a repeatability of $\pm 20\ \mu m$ and a cycle time of 0.37 seconds, while the two Mitsubishi 4-axis, expressly designed for

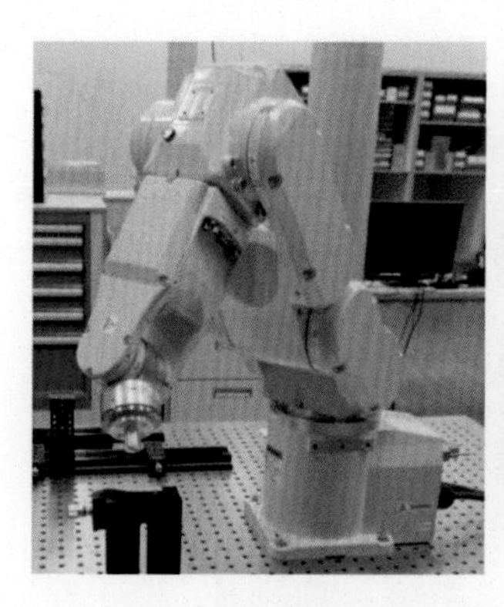

Figure 14.14 A 6 axis EPSON C3 and two Mitsubishi parallel robots (see text).

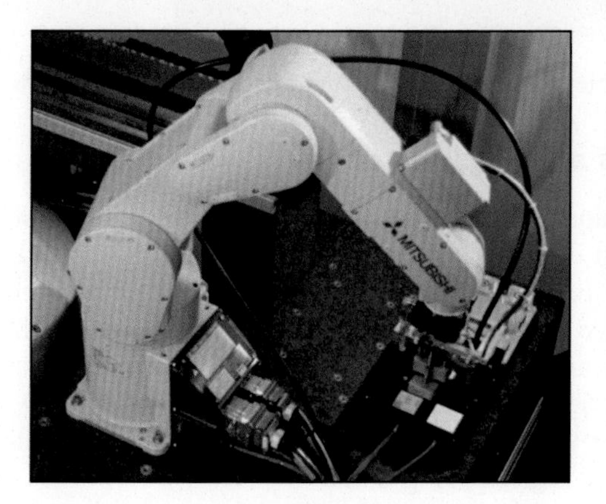

Figure 14.15 A 6 axis Mitsubishi robot and details showing the tilting multi-tool flange. *Source*: Courtesy CNIT-Inphotec.

SMT and micro-electronics assembly offer a payload of 1 kg in a work-space of 150×105 mm with a repeatability of ± 5 μm and a cycle time of 0.28 seconds for the RP-1AH model, while this value becomes 210×148 mm, ± 8 μm, and 0.33 seconds for the RP-3AH version.

In terms of cost, industrial robot arms are highly competitive when compared with a conventional multi-axis configuration, and are available in "clean room" versions, complete with controller and high-level programming software. It is envisaged that more industrial robot will be engaged in fast pick and place in complete PICs assembly lines, where the transferring of the device being assembled across dedicated specific-function stations will also be required.

However, absolute accuracy is not sufficient for alignment tasks—either passive of active—and industrial robots should be seen as "ancillary" motion equipment in the context of PICs automated assembly. In Fig. 14.15 a 6-axis Mitsubishi robot can be seeing, with a multi-tool flange that can be tilted to load two different components, dispense epoxy, and perform UV curing.

Figure 14.16 A hefty granite optical table mounted in an enclosure with isolation support pads.

14.3.2 Optical grade tables and enclosures

All the mechanical positioning equipment has to be mounted and referenced to an optical-grade table surface. The most common platforms used are optical breadboards, with hollow honey-combed internal structure and pre-spaced tapped holes, or monolithic fine-grained granite slabs with tapped metal inserts, usually made-to-measure. While the breadboards are more suited to an experimental laboratory environment, granite is almost universally used for industry-grade automated machines. Passive or active vibration damping is used to support the optical tables, and this can be coupled with self-leveling functionality. An example is given in Fig. 14.16.

Granite, with a specific weight of approximately 3.000 kg/m^3 adds considerably to the overall weight of an automated assembly machine, and the typical weight of a granite table of a machine with an overall footprint of 2×1.5 m can easily exceed 1 ton.

Unlike laboratory equipment, fully automated assembly machines are completed with a fully closed outer enclosure with sliding or lifting doors. During operation the doors are closed and interlocked, also providing protection to the operators, in accordance with machine safety laws and regulations. Although designed for clean room use, the machine's enclosure can be completed with filter units and/or controlled atmosphere flow.

14.3.3 Bonding techniques and equipment

Once a specific component has been positioned and carefully aligned to the PIC substrate, some form of permanent fixing is required. A number of options are available and are listed below:

- Epoxy UV bonding, requiring accurate metering of the correct amount, dispensing to a precise location, and flash-curing by means of UV light. UV can be generated locally by means of UV LEDs, or conveyed by means of large diameter optical fibers connected to a UV source. Exposure time depends on the materials/devices being bonded, and whether they are transparent or not to the UV. Symmetric dispensing and curing is important, as well as epoxy shrinkage, to avoid unwanted displacement of the component being bonded, and active tracking is possible with the same equipment used for active alignment or other means of monitoring. A large variety of epoxy glues are available, with different optical properties, viscosity/fluidity, curing time, and shrinkage, together with different dispensing equipment, from metering pumps and needles, to bubble-jet type dispensing. Small uniform gaps between component and substrate ensure good epoxy distribution due to the capillary effect. Spurious dropping of epoxy, or blockages due to unwanted curing in the dispensing needles/nozzles are some of the worries of fully automated bonding cycles. Tens of seconds or even minutes are not uncommon in epoxy bonding processes (Fig. 14.17).

- Epoxy thermal bonding, is similar to the above, but relies on thermal transfer to start the epoxy curing. It requires local heating, usually resistive with a controlled electrical current, and thermal insulation to avoid thermal spreading and unwanted dilatation effect of surrounding equipment.
- Post-curing usually refers to the long-term stabilization of the epoxy bonding process, but it is usually done in batches heating up trays of components to moderate temperatures for rather long periods—could be hours—in off-line controlled ovens.
- Laser welding's main advantage is the short time required for this type of bonding, known for its long-term reliability and lack of outgassing as currently occurs from epoxies. It is usually confined to components with some metal-clad or metallic structures, such as metal-coated/metal-flanged fibers, alignment collects of TOSA/ROSA packages and the like. Using simultaneous symmetric welding reduces thermal displacement of the component to be welded, and it is illustrated in Fig. 14.18, a combined image that shows the 3D CAD drawing of the three laser beams, a self-leveling chuck, and a detailed of the welded ferrule. The small dimension of the welding spots and placement accuracy requires the use of machine-vision guided positioning equipment.
- Laser-induced soldering is particularly suited to complex multi-element assembly, especially when associated with flip-chip bonding. It can be easily extended to large area

Figure 14.17 UV curing of a ferruled fiber inside a PIC butterfly package.

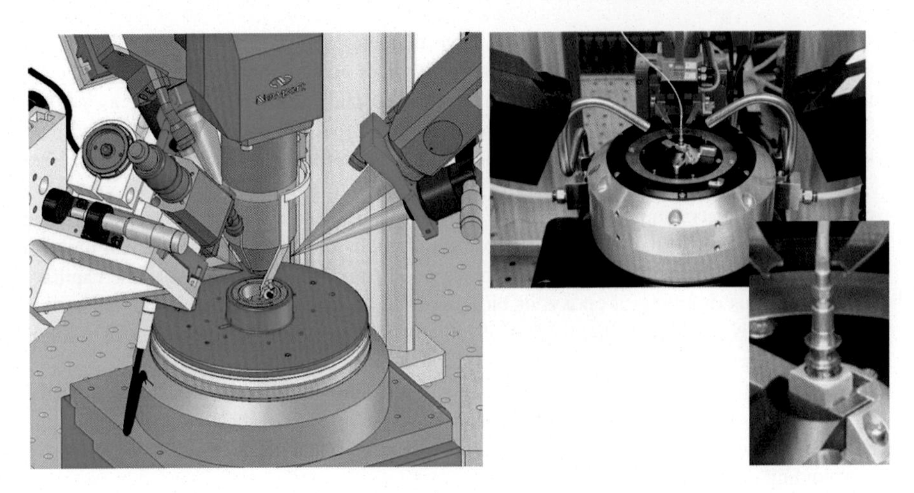

Figure 14.18 Simultaneous laser welding with three focused beams spaced 120° apart.

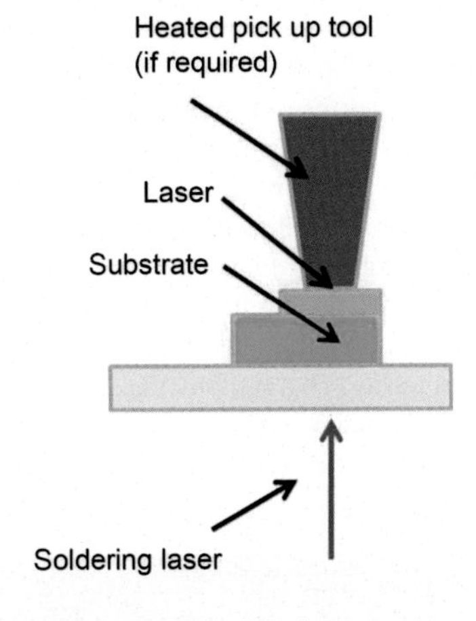

Figure 14.19 The principle of laser induced soldering.

substrates up to wafer level assemblies, as heating is applied only locally. It relies on positioning and focusing an IR laser beam via the back face of a silicon substrate/wafer to induce sufficient heating to melt previously dispensed pads of soldering alloys (Au/Sn, Ag/Sn, etc.). The associated equipment is known as fairly complex and expensive, but offers a high degree of flexibility and speed. Modern machine set ups can reduce this complexity by intelligent chuck design and software architecture (Fig. 14.19).

- Thermal compression bonding refers to a flip-chip interconnect technique, that uses micro-pillars or other micro-bumps, and relies on the application of heat and pressure to achieve simultaneous bonding of large arrays of these fine-pitch electrical contacts. Some 10–20 g of compressive force need to be applied for each bump, and the application of this technology for PICs automated assembly is limited to a limited number of interconnects. Full WLP (wafer level packaging) of wafers on wafers requires a massive amount of pressure and accurate thermal control over large areas, thus requiring the use of specialized machines.

14.3.4 Pickers, grippers, tweezers technology, electrical probing and chucks

An assembly process necessarily entails the picking/transport of a multitude of devices with different dimensions and characteristics, and delivery in a number of different packages, from pre-diced wafers to gel-packs. A PUT (pick-up-tool) is almost always "tipped" or terminated with one of these devices. A mixed list of devices/techniques follows:

- Vacuum pickers are the simplest devices, as they rely on a depression to pick up a device. While they can exercise a considerable vertical force, they are not suitable for large mechanical loads. Several vacuum nozzles can be combined to pick up larger devices, and top surface contact is normally required. Fig. 14.20 shows a vacuum picker in operation with a simultaneous ejector pushing upwards to facilitate the detachment of a small component from a blue-tape.
- Grippers or tweezers come in a variety of forms and sizes. They can be designed and implemented also for side gripping and to withstand higher loads than vacuum pickers. Actuation can make use of miniature pneumatic cylinders, electrical coils, miniature motors, piezo elements, or memory-effect metal alloy devices. An example of a gripper holding/aligning a $400 \times 200 \times 80$ μm lens in front of a laser emitter is given Fig. 14.21.
- A distinction is usually made between the actuators and their "fingers", the latter being usually interchangeable and designed for a specific component size and shape. Micromachining is often used to manufacture these mechanical elements.
- Vacuum pickers lend themselves to easily detect the presence/absence of a component, by simply measuring a vacuum/pressure level. Feedback on the operation of grippers and tweezers, including the actual position of the component being held, is usually provided via machine vision (via top, bottom, or side cameras). With the miniaturization of imaging sensors and the low cost of CMOS imagers and micro-optics, it will not be long before grippers are equipped with built-in cameras.

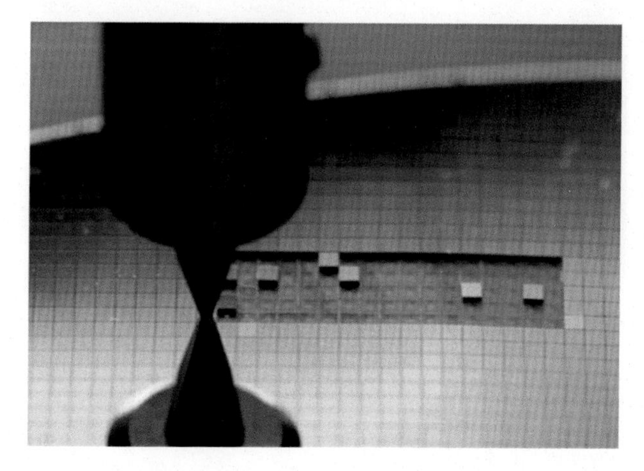

Figure 14.20 A vacuum picker lifting a 500×500 μm chip from a pre-diced wafer on blue-tape.

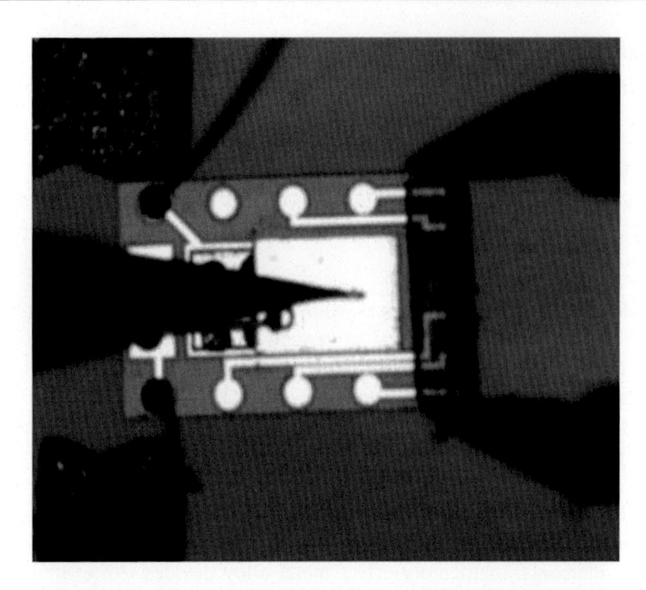

Figure 14.21 A gripper holding/aligning a micro lens in front of a laser emitter.

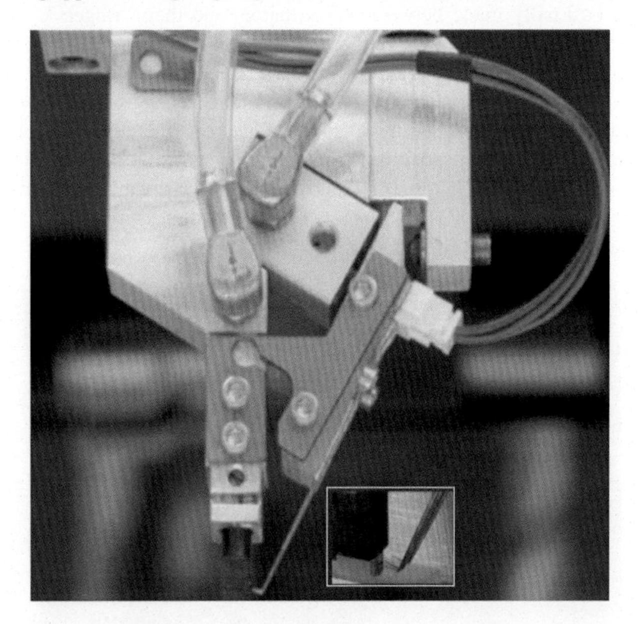

Figure 14.22 Electrical probing of active components.

- Electrical probing of photonics devices, such as laser diodes or photo-detectors, is required for active alignment, and grippers need to be equipped with electrical probing contacts, as shown in Fig. 14.22.
- Complex, multi-element PICs assembly processes require the rapid interchangeability of pick up tools. This is achieved by standardized flanges that are equipped with

pneumatic, vacuum, and electrical connections, and are also designed for accurate mechanical locking. Different tools can be housed in a "carousel" and exchanged automatically.

- Pick up tools and probes are by necessity based on rather fragile mechanical structures and are subject to a degree of "wear and tear", as well as being the first casualties of the operator's errors in the commissioning and programming of the machine sequence of operations. Field interchangeable spares are usually supplied to the end users.

The photonic device being assembled and completed with a number of different active or passive optical components is itself held in what is commonly termed a "chuck". Again, the chuck design and size are specific for a given component, and almost invariably are interchangeable to guarantee flexible operation of the assembly machine.

While all the different elements described above do not represent the highest complexity and cost of the whole machine, they occupy a key position in its continuous and successful automated operation. A refined and detailed know-how of both mechanical design and material selection, including surface treatment and hardening, of these tiny machine parts is essential, and can have a considerable impact on the overall assembly equipment up-time.

14.4 Machine vision

Machine vision has already been mentioned in the context of passive/active alignment, but it is also used for a variety of other tasks and it represent an essential part of an assembly machine. It is ubiquitously found in pick and place operations, positioning and alignment of parts, automated optical inspection, parts identification via OCR (optical character recognition), and code reading, as well as beam characterization, both in the visible and in the IR range.

Cameras have been used as visual aids since the early days of photonics packaging, but purely to display enlarged images of the devices being manipulated and assembled. Machine vision refers instead to the use of image processing algorithms to automatically extract features and information from an acquired image. In most instances, this information is then correlated to motion equipment positional data and exploited to perform automated operations.

These few paragraphs are not sufficient to provide a "primer" on machine vision for PICs assembly, but a few concepts and examples are given, with the help when appropriate of a few images (it would be difficult to discuss machine vision without images!).

The image in Fig. 14.23 dates back to 2001, and was acquired with a digital camera of 1300×1030 pixel resolution and 8-bit dynamic range (then state-of-the-art), but it exhibits both poor contrast, possibly due to a bad illumination, and poor focusing. Nevertheless, the edge-detection based algorithm correctly identifies the edges of the V-groove, the outer circumference of the fiber and its center, and the red dot

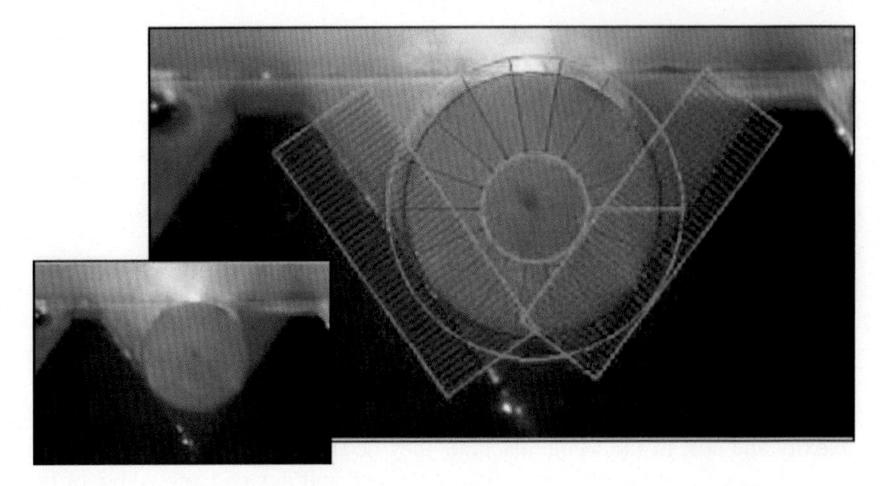

Figure 14.23 A digital camera image (from 2001) of a FO epoxied in a V-groove.

that lies within the "smudge" that corresponds to the mono-mode fiber. A colored overlay graphic helps the user to understand how the image processing is performed:

- The two green rectangles and the green annulus define three ROIs (Region Of Interest) where the image processing is applied.
- The blue lines show where an edge-detection algorithm (a function that looks at pixel intensity variations along a line) is repetitively performed.
- The red overlays show the identified results: the left and right edge of the groove, and the circumference and center of the fiber.

The same image can be used to introduce one of the major limits of machine vision in photonics assembly automation: optical resolution. The image was acquired with a resolution of approximately 0.32 μ/pixel and covers a FOV (field of view) of 968×656 pixel, corresponding to an area of 0.312×0.211 mm. This is further illustrated in Fig. 14.24, which shows the image of a 10 μm/line reticule on a glass plate, acquired via a Leitz microscope with a $20\times$ objective and $1.25\times$ relay lens on the same digital camera. A loss of sharpness and some artifacts (a "shadow" on the right of each line) are noticeable. The intensity of each pixel along the line in the zoomed image is plotted in the "line profile" graph. The distance between two adjacent peaks is approximately 40 pixels, corresponding to 250 nm/pixel. A better concept in terms of image quality is that of MTF (Modulation Transfer Function) that combines both spatial resolution and image contrast.

It can be seen in the line profile plot that out of the 8-bit dynamic range (255 levels) only about 1/3 is actually used.

In practical use the overall quality of the lenses, the illumination that can be implemented, the actual morphology of the markers and features we can find on a photonics chip, and the need to have reasonably large FOVs make it difficult to go below 1 μm or 0.5 μm/pixel resolution.

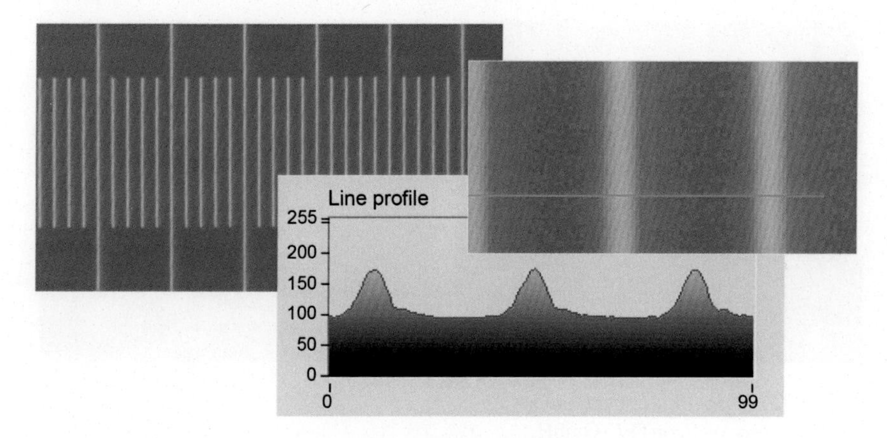

Figure 14.24 Onset of diffraction limits and MTF.

Large sensors are now available and a 29 Mpixel (with 6576 × 4384 pixel sensors) camera's cost is today approximately half of that of a 1 Mpixel camera in 2001, and allows acquisition of a full 6.5 × 4.3 mm area with a 1 μm/pixel resolution. This would greatly reduce the need to move the camera and allow the use of a single focal lens (zoomable lenses are not of great use in photonics assembly as they have poor optical/mechanical stability). Sub-sampling, selective ROI, and software zooming and panning can be used to minimize the amount of acquired data.

Although imaging algorithms can be developed from scratch, and ample literature, dedicated "recipe cook-books", and free software sources are available, it is felt that robust, well documented, and extensive image processing libraries are available on the market from several reputable vendors, and this should be the preferred choice in the implementation of industry-grade assembly equipment.

Combining imaging and motion also requires the correlation of the dimensionless world of pixels with that of the mechanical axis and their accurate encoders. These might include also the relative calibration of separate top and bottom cameras, often used in flip-chip type of components placement and other alignment routines. Dedicated calibration targets are required, either fixed or inserted into the cameras optical paths when required. Automated software calibration routines are than executed and can be repeated during machine commissioning, periodically, or whenever maintenance requires it. In the set up of Fig. 14.25 a top and bottom camera calibrated with the motion equipment also allow for correction of misalignment after picking.

Calibration procedures and other guiding/alignment algorithms can also be based on markers purposefully etched on the photonic chip to be assembled; aligned or exploiting geometric features already existing that are intrinsically accurate, being based on the lithography masks of the chip foundry processes. An example is given in Fig. 14.26, where a multiple grating structure provides alignment information for a v-groove fiber array. The optical connection between the outer ports provides angular correction along the major axis of the array, while the geometry of a single

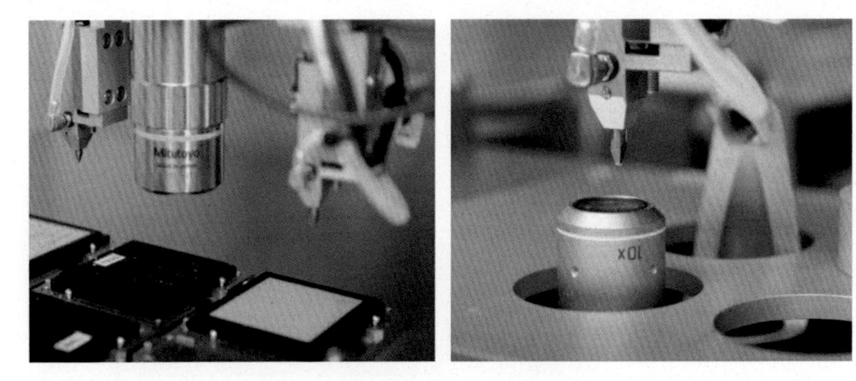

Figure 14.25 A top–bottom camera set up.
Source: Courtesy of ficonTEC GmbH.

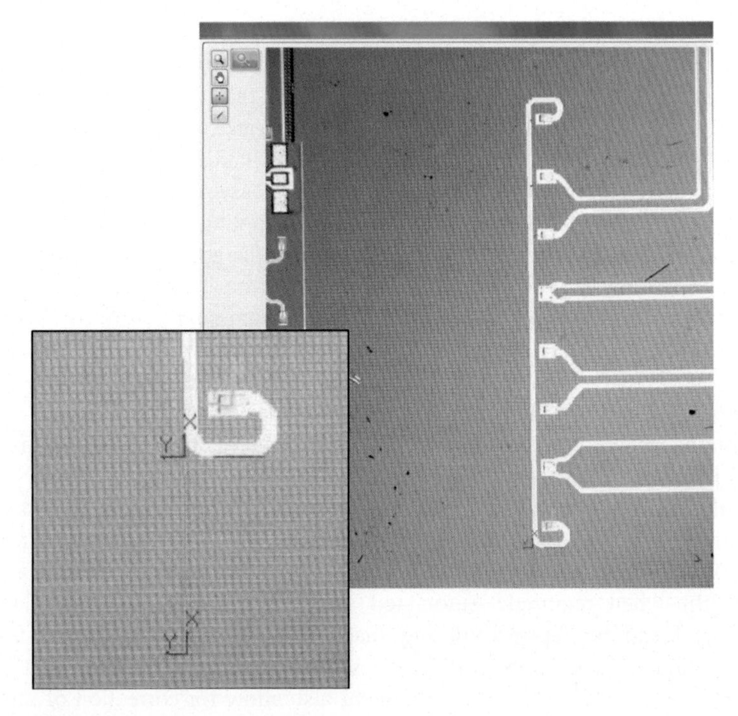

Figure 14.26 Using on chip existing features.
Source: Courtesy of ImagingLab/ST Microelectronics.

grating provides the coordinates that ensure "first light" for successive active alignment. The outer port connection is effectively an optical short circuit that allows the use of an external laser source and an external power meter to perform the active alignment.

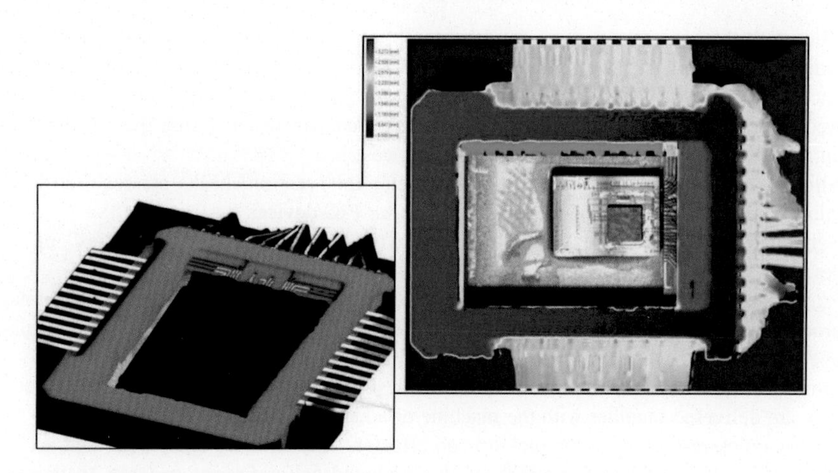

Figure 14.27 A 3D image of a photonics chip with a calibrated Z-scale.

To conclude this section of the chapter, a few words will be spent on 2D versus 3D imaging. The images shown above, and indeed most of the imaging that is today used in assembly equipment, is based on gray-scale 2D imaging. Assembly is essentially a 3D process and would certainly benefit from 3D machine vision. The *X-Y* matrix of intensity values in 2D imaging is replaced in a 3D image by a COP (cloud of points), where each point represents a triplet of *X-Y-Z* values in space. Three-dimensional cameras with sufficient *Z* axis resolution are not yet widely available, especially when speed and cost are taken into account, but different techniques are being developed, together with the corresponding software for 3D data processing.

A 3D test image acquired with a fast line scan camera and based on auto-correlation methods is given in Fig. 14.27: the false-color scale of the right-most image is actually a calibrated *Z* measurement.

14.5 The role of software and HMI in the optimization of new processes

Complex machines like the one outlined in the previous paragraphs can only be operated via a massive amount of software, and it is ultimately the software that qualifies and determine the success of a new piece of assembly equipment on the factory floor.

The role and main tasks of the machine application specific software can be briefly explained as follows:

- Provides a functional and user friendly HMI (Human Machine Interface) with a modern GUI (Graphical User Interface), hiding the underlying complexity of the machines, and possibly separating different classes of users, from the operator on the factory floor, to the process engineer responsible to configure a complete packaging production sequence.

- Allows all the parameters of a complex assembly to be described and set in a sequence of individual steps that can be individually tested/debugged prior to a fully-automated production cycle.
- Accommodates easily all the inevitable changes that will occur during the development/optimization phase of a new PIC packaging process.
- Allows semi-automated operation when operator intervention/checks are required in the initial ramp-up to production of new or unfamiliar PIC devices.
- Generates editable "recipes" to allow the automated packaging of "families" of PIC devices.
- Provides also an easy user interface to advanced machine vision, hiding complex image processing algorithms tuning via simple parameter settings and clear visual image overlays.
- Provides interface access to a wide pallet of test and measurement instruments, whether they are delivered together with the machine or added at a later stage.
- Allows quick re-tooling of the pick-up end effectors and handling equipment, and accommodates hardware changes, possibly with automated calibration procedures, to preserve the capital equipment investment over time.
- Provides production data down to the single assembled device and interface to the manufacturers databases for yields and production monitoring.
- Provides self-diagnosing/power-up tests when the machine has to undergo a "cold start".
- Allows remote connection (under end user control) for both remote maintenance/support, and for remote software upgrades.

It is likely that a machine manufacturer will use a collection of different software programming environments for different parts of the machine (like C, C++, Python, LabVIEW, etc.) to program the motion control subsystem, off-the-shelf machine vision libraries, specific graphical user interface development tools, etc. As long as the various software elements are well integrated and seamlessly presented via the HMI interface, this should not constitute a worry for the end user. It is likely that the machine will be run with a conventional PC, either in a conventional or industrial ruggedized form, and it should similarly be possible to mix both real time operating systems and conventional OS (like MS Windows).

The key issue is that the end user should not need to provide any programming efforts or skills, but only learn how to configure the machine to optimize a given process, as well as integrating the machine in the IT infrastructure of the manufacturing plant if/when required.

A recurring discussion is also on whether the software provided with the machine should be given in a closed (run-time/executable only) or in an open (access to source code) form. Opening the software to the end user carries two major problems: first, the manufacturer is liable for the correct functioning of the machine, expected life time of the major components, safety of operation, etc., and this is highly dependent on the software implementation, and second, a lot of intellectual property is built in/accumulated in the machine software. An easier way out is to provide a specific high-level interface layer that allows the end user to add specific software "tools" at HMI level, via a defined protocol for handshaking and variables exchange. This specific point should perhaps have been added to the list at the beginning of the paragraph.

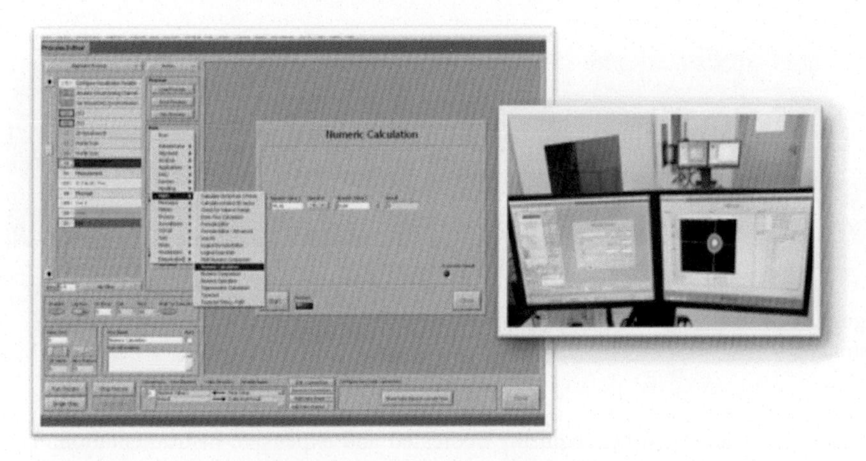

Figure 14.28 A high-level HMI interface developed at ficonTEC.

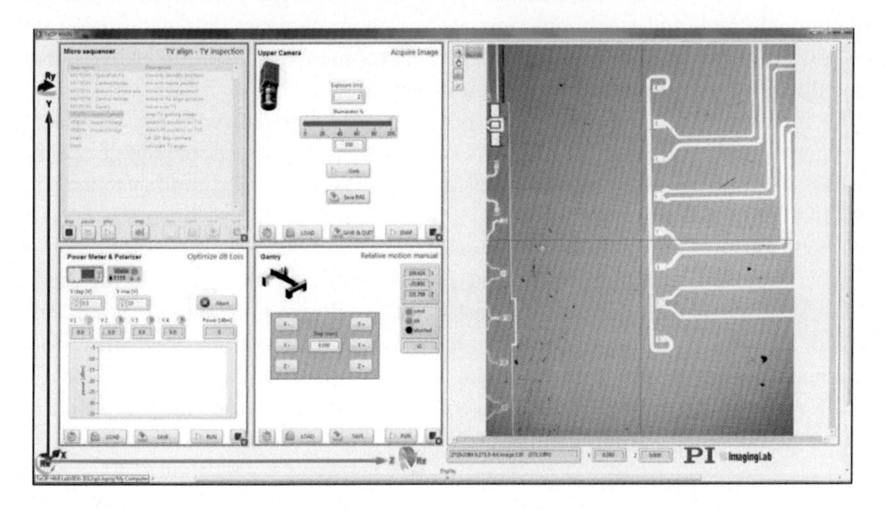

Figure 14.29 A high-level HMI interface developed by ImagingLab for PI.

Interestingly, the two examples of HMI software interfaces provided in Figs. 14.28 and 14.29, and developed separately by two different manufacturers, share most of the concepts just outlined and are both based on LabVIEW.

14.6 Test and measurement instrumentation

As the assembly of PICs often entails different kinds of measurements (voltages, currents, temperatures, light intensity and power, light frequency, beam characteristics, monitoring of laser sources, RF measurements, etc.) the integration of a vast array of instrumentation is also required. Full characterization of a PIC device is

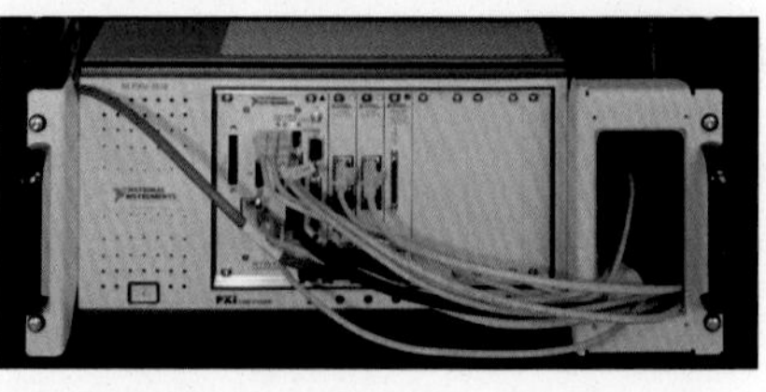

Figure 14.30 A PXI-based data acquisition system compared to a stack of conventional instruments.

usually performed off-line, requiring dedicated and expensive instrumentation. The testing required for assembly is better referred as in-line testing, and might not require the same class of instrumentation. A distinction can also be made between conventional "rack and stack" instruments and more compact data acquisition systems, offering also a lower foot-print in a clean room environment, as well as likely cost reductions, with a comparison given in Fig. 14.30.

The instrumentation cost issue will become more apparent with the increase in PIC manufacturing volumes, and might well lead to the development of specific low-cost modules targeting optical measurements but exploiting multi-vendor standards like PXI.

In both software application/HMI examples given in Section 14.5, software interfacing is included at a high level, and is performed via an open interface using LabVIEW (a trademark of National Instruments), almost a de facto standard also adopted in semiconductor and other electronics manufacturing industries. Most instrument vendors offer free interface software, referred as "drivers", that is LabVIEW compatible. This facilitates both the instrumentation integration during machine assembly and production, as well as later on at the customer premises.

14.7 Design for automated assembly/testing and standardization

Photonics packaging is still at a rather early stage, with a number of technologies being studied and developed to reduce its complexity and cost. It is also characterized by a remarkable lack of standards, and while the two things just mentioned are certainly related, some easy steps could also be discussed and implemented, and few are listed:

- Define some preferential layout whereby electrical and optical interconnect ports occupy specific sides and chip area.
- Define classes of sizes, as this will reduce the number of different chucks, grippers, and various holders, and thus cost, of the assembly machines.

- Remember that the placement and spacing of discrete components like laser sources, CoS (chip on sub-mount) components, and free-space micro-optics elements might require some physical space for the pickers to maneuver; the same is valid for dispensing glue.
- Keep these devices away from wire-bonding areas if they need to be assembled after wire-bonding.
- Avoid high-walled bath-tub types of packaging, as it makes it more difficult to enter the package with grippers and probes (electrical and optical).

The list could continue, but what is really required is better communication among all the players in the long chain of PICs manufacturing, starting from the providers of photonics design tools all the way to wafer production and packaging process experts, keeping the assembly equipment manufacturers in the loop also.

14.8 Automated testing

Reworking during the assembly of a complex PIC layout is almost impossible, and even when it is would require costly manual intervention. Testing is therefore required in-line (i.e., during assembly) and off-line, at either single die, bar, or full wafer level.

A broad distinction can already be made for devices adopting top grating coupling and those adopting edge coupling, the latter requiring either dicing before testing or some other approaches, like etching or cutting trenches in order to reach the edge coupling structures. A couple of novel technologies are being explored at the University of Southampton, UK [7], and at the Polytechnic of Milan, Italy [8], and more might be in progress not known to the authors.

Full wafer-level testing will require some creative approaches to mixing electrical probing and optical probing in a fast and cost-effective manner.

A distinction could also be made between full characterization and expensive, top-class instrumentation, and that dedicated to in-line testing during assembly and/or full wafer-level testing. For the latter, some low-cost modular optical instrumentation front-end, possibly adopting existing test and measurements platforms, has not yet emerged.

14.9 Conclusions

A more detailed description of the processes and the machines dedicated to PICs automated assembly would probably require a full book rather than a single chapter, but it is hoped that a good and sufficiently clear overview was provided in the preceding paragraphs. It is felt that more work will be required to link the design of PICs devices to full volume manufacturing processes, and that growing volumes will also promote changes in the design of the machines. Some of today's flexibility will probably be sacrificed for speed, and full production lines will be based on a number of simplified and single-task stations, with some parallelism from the slower steps of the full assembly process. Shuffling partially assembled devices and handling all the incoming parts could be done by robotics arms, and much shorter fiber ribbons in some standardized delivery packaging will have to be devised to allow fully automated loading/unloading. Machine manufacturers will need to monitor advances in micro-optics and multi-port miniaturized optical connectors, that could greatly simplify fibers pig-tailing, and, at least partially, remove some active alignment requirements. The increase in volumes, and hence more investment in capital equipment and multiple machines single purchases, will also allow machine manufacturers to allocate some NRE (Non Recurrent Engineering) efforts to reduce costs and speed up production cycles.

Plenty of automation is already available today adopting best-in-class equipment and technologies, also borrowing existing concepts from adjacent large-volume manufacturing industrial segments.

A number of changes are likely to occur, and adaption to these changes will be essential for both PICs manufacturing companies and machine suppliers.

To conclude with few words for all those involved in PICs packaging: design for automated assembly, design for testing, design for speed of both assembly and testing, and minimize/limit packaging differences by promoting standards!

References

[1] Offrein B. System-level integration aspects of silicon photonics. In: EPIC/PhoxTroT symposium on optical interconnect in data centers, Berlin (Germany); 18–19 March 2014.

[2] Doe P. Silicon photonics looks for 2.5D assembly at OSATs. 3D Packaging 2013;(26).

[3] Preve G.B. Silicon photonics packaging automation: problems, challenges, and considerations, Chapter 8.2, pp 240–246. In: Pavesi L, Lockwood DJ, editors. Silicon photonics III. Berlin and Heidelberg: Springer; 2016.

[4] Lee M, Yoo M, Cho J, Lee S, Kim J, Lee C, et al. Study of interconnection process for fine pitch flip chip. In: ECTC. Amkor Technologies; 2009.

[5] Khanna A, Bode D, Das C, Absil P, Becker S. CMOS cost-volume paradigm and silicon photonics production, Chapter 9.3.4. In: Pavesi L, Lockwood DJ, editors. Silicon photonics III. Berlin and Heidelberg: Springer; 2016.

[6] Piacentini I. Tight integration of imaging, motion and data acquisition techniques for the fiber optic alignment in single and arrayed photonics devices. In: 4th international conference on photonics, devices, systems, Prague (Czech Republic); 26–29 May 2002.

[7] Topley R, Martinez-Jimenez G, O'Faolain L, Healy N, Mailis S, Thomson DJ, et al. Locally erasable couplers for optical device testing in silicon on insulator. J Lightwave Technol 2014;32(12).

[8] Morichetti Francesco, Grillanda Stefano, Carminati Marco, Ferrari Giorgio, Sampietro Marco, Strain Michael J, et al. Non-invasive on-chip light observation by contactless waveguide conductivity monitoring. IEEE J Sel Top Quantum Electron 2014;20(4).

Part IV

Using Optical Interconnects to Improve Network Architectures in Data Centers

The role of optical interconnects in the design of data center architectures

A. Siokis[1,2], K. Christodoulopoulos[1,2] and E. Varvarigos[1,2]
[1]University of Patras, Patras, Greece, [2]Computer Technology Institute and Press – Diophantus, Patras, Greece

15.1 Introduction

Data centers (DCs) experience exponential increases in traffic volumes both in their connection to the end-user and also in the internal communication among servers. This trend is due to both evolution in processor technology and the expansion of the Internet, amplified by the ever-increasing use of wireless and cellular networks and the related information-centric services and applications for these platforms. A particularly interesting observation is that the majority of the DC traffic (76%) stays within the DC [1]. The traditional fat-tree topology built out of electronic switches [2] presents several problems, since it scales superlinearly to the number of servers and servers' rate, leading to increased requirements for switching equipment and power consumption.

Optical technology is a promising, energy-efficient solution for satisfying the increased bandwidth requirements for both telecoms and datacoms. In telecoms, optical fibers that were widely used in long-haul networks have now replaced most of the copper technology in WAN and MAN, and are gradually finding their way to datacom networks *inside* the DCs [3]. Several optical interconnection solutions have been proposed for specific parts of the DC network, while more ambitious holistic all-optical architectures are being researched. The first applications of optics are for rack-to-rack or server-to-switches communications. Since optics can once more be the solution to the bandwidth and energy problems for next generation DCs, their adaptation at the lower layers of the packaging hierarchy, including board-to-board, on-board, and on-chip, is impending. A promising technology are the Optical Printed Circuit Boards (OPCBs): boards with integrated optical waveguides can be used at the on-board and board-to-board (backplane-boards) levels to interconnect optically-enabled modules, such as opto-electronic chips packed with vertical-cavity surface-emitting laser (VCSEL), and photodiodes (PD). Laying out topologies via optical waveguides on-OPCBs presents a number of issues that have to be addressed when designing architectures for DCs.

In this chapter we outline the similarities and differences between layout models for electrical interconnects and optical waveguided communications, in order to

Optical Interconnects for Data Centers. DOI: http://dx.doi.org/10.1016/B978-0-08-100512-5.00015-2

understand the peculiarities of the latter. Taking these into account, we then present a layout model suitable for optical interconnects. We also describe strategies that can be used for laying out logical topologies on-OPCBs, assuming the aforementioned model.

15.2 Overview of optical interconnects technologies

Optical networks have been widely used in the long-haul and metropolitan telecom networks (MAN) providing low power and latency, and increased throughput. Initially, the copper channels were replaced by fiber for point-to-point communication to form fat circuit pipes. In this case, opto-electro-optical regeneration took place at every network node. Long-haul and MAN networks have now evolved to all-optical (but still circuit switched) approaches to avoid power hungry conversions from the electrical to the optical domain, and vice versa. The use of optics has been extended to cover smaller distances in LAN as well as DC networks. Currently, optics have replaced electrical links between Top-of-Rack switches, to achieve higher bandwidth, reducing the power consumption and latency somewhat. Even so, power consumption of data communication is still daunting. In order to cope with both the energy and bandwidth limitations of the electrical interconnects, optical technologies have to be deployed at even shorter distances in the near future: optics are gradually becoming more cost-effective for board-to-board, on-board, and even on-chip communications.

This new era brings an entirely new technology portfolio of network modules for short distance communication. These include Optical Printed Circuit Boards (OPCBs) printed with multi-mode (usually polymer) or single-mode (polymer or glass) waveguides, chip-to-board coupling technologies, optical transceiver chips (equipped e.g., with VCSELs—Vertical Cavity Surface-Emitting Laser for Tx, and PDs—PhotoDiodes for Rx), photonic switching and routing elements, (de)multiplexing elements, Wavelength Selective Switches (WSS), Arrayed Waveguide Gratings (AWGs), and optical RAMs, among others.

15.3 Applications of optical interconnects in the individual layers of the packaging hierarchy

In this section we review optical interconnection network architectures presented in the literature for all the layers of the packaging hierarchy: rack-to-rack, on-board, board-to-board, and on-chip.

15.3.1 Rack-to-rack interconnections

The proposed rack-to-rack optical interconnects architectures for DCs fall into two categories: (1) hybrid approaches that enhance the legacy DC architecture with optical interconnects, and (2) optical switch architectures targeting higher radices in order to lead to flatter DC architectures (such as fat trees with fewer tiers).

Regarding the first approach, two well-known hybrid architectures have been proposed that rely on both electrical (commodity) packet switches and optical circuit switches, as proposed by Farrington et al. [4] and Wang et al. [5]. In the approach of Wang et al. [5], the Top-of-Rack switches are connected both to an electrical packet-based network (based on commodity switches), and to an optical circuit-based network. The optical switch must be configured so as to connect pairs of racks with high bandwidth demands through this optical switch. In Farrington et al. [4], a similar approach is followed, but wavelength division multiplexed (WDM) links are used for the optical circuits. The electrical packet switches are used for all-to-all communication of the pod switches, while the optical circuit switches are used for high bandwidth, slowly changing (and usually long lived) communication between the pod switches.

A number of optical switch architectures have been proposed for all-optical (non-hybrid) communication between racks in the DCs. Singla et al. [6] proposed an architecture based on Wavelength Selective Switches (WSS) and Micro-Electro-Mechanical Systems Switches (MEMS). Each port has several optical transceivers operating at different wavelengths. The optical wavelengths are combined using a multiplexer, the output port of which is routed to a WSS. The outputs of the WSS are connected in the MEMS optical switch. At the output stage (after the MEMS), all of the wavelengths are demultiplexed and routed to the optical transceiver in the output port. The architecture proposed by Luijten et al. [7] is based on wavelength- and space-division multiplexing taking place in two different stages. In the first stage, multiple wavelengths are multiplexed in a common WDM line and are broadcast to all the modules of the second stage through a coupler. The second stage uses SOAs (Semiconductor Optical Amplifiers) as fiber-selector gates to select the wavelength that will be forwarded to the output. Another approach based on the combination of wavelength- and space-division multiplexing is proposed by Castoldi et al. [8], in which a Space-Time (ST) switched architecture is also proposed. Shacham and Bergman [9] proposed an architecture based on 2×2 SOA-based switches that can be scaled efficiently in tree-based topologies. Zhu et al. [10] proposed an optical switching platform, the key idea of which is the combination of Wavelength Division Multiplexing (WDM) and Space Division Multiplexing (SDM) utilizing an $N \times 1$ Wavelength Selective Switch (WSS). A number of high radix optical switch architectures have been proposed based on Arrayed Waveguide Grating Routing (AWGR) elements. Xia et al. [11] proposed a switch architecture based on a three-stage AWGR-based Clos network, using tunable wavelength converters. The architecture proposed by Ye et al. [12] consists of an array of tunable wavelength converters (one TWC for each node), an AWGR, and a loopback shared buffer. Each port can connect with any other port through the AWGR by configuring the transmitting wavelength of the tunable wavelength converter. Gripp et al. [13] proposed a three-stage architecture using AWGR. The first stage is a space switch based on AWGRs that distributes the packets uniformly across the ports of the second stage. The second stage, a time switch, holds the packets until the third stage; another round-robin space switch based on AWGRs provides a path to the output port. Proietti et al. [14] investigated the scalability issues in the AWGR-based interconnect architectures and pursued an active AWGR switch architecture assuming a distributed control plane.

15.3.2 On-board and board-to-board interconnections

Optical interconnects for on-OPCB and OPCB-to-OPCB levels of the packaging hierarchy is an active research field, which includes research on a wide range of technologies such as optical waveguides (single-mode and/or multi-mode) of various materials (polymer, glass, . . .), optical transmitters (e.g., VCSELs), various coupling techniques, etc. Architecture-wise a number of (mainly passive) interconnection architectures have been proposed for use in optical backplanes such as large parallel waveguide arrays [15], a waveguide-based optical bus structure [16], meshed waveguide architectures [17] and [18], a shared optical bus [19], and a regenerative bus structure of 40 Gbps [20]. On-OPCB architecture is a research area that will probably attract more interest in the future in order to define next generation optical DC architectures spanning from the higher/rack packaging levels to the lower ones. OPCB is the packaging level on which we will focus in the remaining of this chapter, beginning with Section 15.4.

15.3.3 Networks-on-Chip

Networks-on-Chip (NoC) constitute the lowest layer of the network system, which can benefit greatly from the introduction of photonics [21]. A great deal of research effort has been put into this area. A number of traditional topologies have been implemented for NoC environments, such as buses [22–24], crossbars [25,26], Butterfly and variants [27–29], torus [21], mesh [30], Clos [31,32], and fat-tree [33] topologies.

15.4 On-optical printed circuit boards (OPCB) layout strategies

In this section we focus on the on-OPCB layer of packaging hierarchy. First, in Section 15.4.1, we briefly discuss models presented in the literature for laying out logical topologies on electrical PCBs. In Section 15.4.2 we outline the differences between these models and the characteristics of optical waveguided communication, in order to adjust the former to a model suitable for laying-out topologies on OPCBs. Based on these modifications, we present simple layout strategies that can be used for both point-to-point as well as multi-point networks. In Section 15.4.3 we give some examples using the presented layout techniques for a number of topologies.

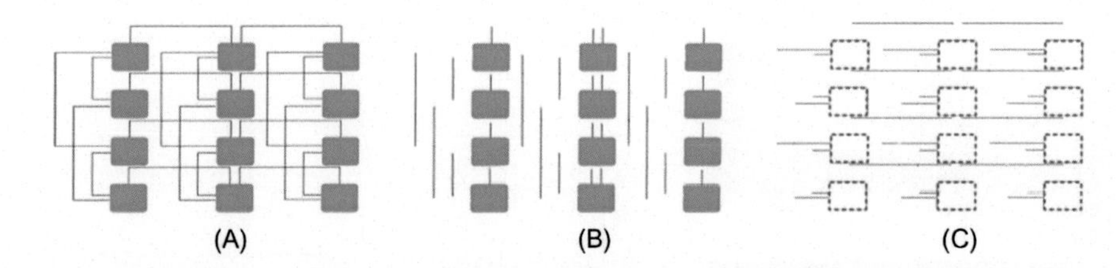

Figure 15.1 2D (3 × 4) Lay-out of a 3 × 2 × 2 mesh using the Thompson model. (A) Both layers. (B) Layer 1 (chips and vertical wiring). (C) Layer 2 (horizontal wiring).

15.4.1 Layout strategies for electrical interconnections on boards

Classic layout models for electrical interconnection networks rely on the Thompson model and variants [34]. In the Thompson model the interconnection network is modeled as a graph whose nodes represent processing elements and whose edges represent wires/links. The graph is mapped on a 2D grid. The wires run either horizontally or vertically along the grid lines, called tracks. The Thompson model assumes two wiring layers: one layer is used to lay out the horizontal segments of the wires and the other one the vertical segments. When a wire makes a turn (a 90 degree bend), the horizontal and vertical segments in the two layers are interconnected using inter-layer connectors (vias). The required area is the area of the smallest rectangle in the 2D grid containing all nodes and wires. An example of a layout using the Thompson model is depicted in Fig. 15.1. The Thompson model has been generalized from two wiring layers to the multilayer (layer > 2) 2D grid model. In the multilayer 2D grid model all the nodes are located in a single layer (active layer), while this first layer and the remaining layers contain only wiring. The multilayer 2D grid model has been further extended to the multilayer 3D grid model where the nodes of the network are embedded in more than one layer.

Yeh et al. [34] and [35] present a variety of layouts for various topologies based on the aforementioned models. In the following section we will examine (and adjust for OPCBs) two such network topology layouts: collinear and 2D. In the former, all network nodes are placed along a line, while in the latter nodes are placed along rows and columns, forming a 2D grid array. Note that in both cases the wires run on 2D grid lines. Fig. 15.2B depicts an example of a 3 × 2 × 2 mesh, laid out in a 2D grid of 3 × 4 nodes, with wires also laid out in a 2D grid. Note that the wiring, although depicted in one layer, is done in two (or more) layers. 2D layouts are constructed using collinear layouts along the rows and columns. A single row of the 2D layout in Fig. 15.2A is a collinear layout of three nodes, requiring one wiring track. A single column of the 2D layout is a collinear layout of four nodes (2 × 2), requiring three wiring tracks.

15.4.2 Layout strategies for OPCB

In this section we present layout strategies for point-to-point and multi-point interconnection networks on OPCBs. A node of the network topology could be either

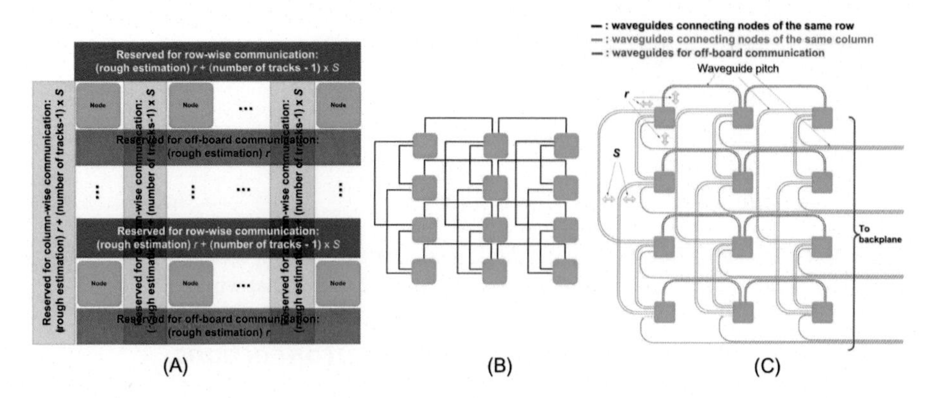

Figure 15.2 (A) Layout design rules on 2D grid for OPCBs. Space reserved for row-wise, column-wise and off-board communication. (B) 2D (3 × 4) Lay-out of a 3 × 2 × 2 mesh. (C) The same layout on an OPCB, following the strategy shown in (A).

a single chip or a group of chips, e.g., a number of optical host chips connected in an optical/opto-electronic router forming a star network (see Siokis et al. [36]). Assuming only collinear layouts, the nodes could also be assumed to be the coupling points of linecards and the 2D grid surface, the optical backplane. In the text that follows we will view a node as a square of size d (the layout strategies can be easily generalized for nodes in the shape of rectangles). We will examine layouts both for point-to-point topologies and multi-point topologies, and we will also discuss how WDM can be used in order to implement point-to-point topologies using multi-point layouts. All the layouts will be presented without length matching. If length matching is required (to meet the requirements of the chip-to-chip protocol in timing skew) when more than one waveguide connects two nodes, additional small S-bends can be used in the shorter waveguide to even up the length of the waveguides (see also discussion in Siokis et al. [36]).

15.4.2.1 Layouts for point-to-point topologies

The main differences between optical waveguided communication and the models described for copper interconnects in Section 15.4.1, from the layout point of view, are:

1. Waveguide bends require a (non-sharp) bending radius r in order to allow the propagation of light. Smaller r means more losses. In electrical interconnects a bend with $r \approx 0$ is possible (in the Thompson model it is implemented as an inter-layer connection though a via).
2. Crossings are allowed in the same layer (a crossing angle of 90 degree is preferable due to losses and crosstalk). Crossings in the same layer are not possible in the electrical interconnects, since this would lead to a closed circuit.

The layout strategies described in Section 15.4.1 can be applied on OPCBs with the following modifications. We assume two layers for waveguide routing, each for one direction of communication between nodes: so for each communicating nodes the first layer implements the Tx→Rx connection, and the second (almost identical) layer the Rx→Tx connection. This two-layer approach does not impose important restrictions on the placement of the Tx and Rx elements on the node. For example, assuming separate arrays of Tx and Rx elements on-chip, the Tx and Rx arrays could be placed either side-by-side, or the Tx array right behind or in front of the Rx array. Given the collinear layout of nodes (remember that 2D layouts are constructed from row- and column-wise collinear layouts), at each layer the links are laid out in a 2D grid, bends have a given radius, and crossings are allowed to occur. Alternatively, a single layer can be used to accommodate both directions of communication (Tx→Rx links and Rx→Tx links) side-by-side in a single "waveguide track" or bundle (see below). This approach lends itself to an alternating TxRx|TxRx|... pinout placement of the Tx and Rx elements on the chip.

In a case where more than one link is needed between two nodes, and since bends are (space and loss) expensive, to save on area we route multi-waveguide links together, as bundles, in a single "waveguide track". The distance of waveguides within a track can be as low as 250 μm, considered as the standard pitch in our study, or higher as preferred. Since the bending radius r and the chips' sizes are at least two orders of magnitude larger than the standard pitch, we neglect track width in our calculations. The first track parallel to the collinear layout direction of nodes is placed at space r from the node, while the space S left between the following tracks is related to the desired waveguide crossing angle θ and the bending radius r as follows:

$$S = (1 - \cos\theta)r \tag{15.1}$$

Thus, according to Eq. (15.1), if 90 degree crossings are used, the track spacing equals the bending radius ($S = r$). Smaller bending radii and smaller crossing angles lead to less required area, but to higher losses. Since crossings are allowed in the same layer, even only one layer would suffice if the worst case losses (due to bends, crossing, and distance) allow that (assuming also an alternating TxRxTxRx... pin placement on the nodes).

Also note that in the adopted strategy the bends and crossings appear in a specific and deterministic order: for every waveguide, an initial bend (or bends) takes place, followed by all the crossings, followed by a final bend (or bends).

To layout a topology on an OPCB we reserve an area for row-, column-wise, and off-board communication. Our generalized approach for 2D grid layouts is depicted in Fig. 15.2A. It assumes that network nodes have pinouts from two of their sides for inter-node interconnection. For the communication of the nodes in the same row, we reserve the area above the nodes (black area in Fig. 15.2A). The required area depends on the number of waveguide tracks, which is determined by the row-wise collinear topology. For the communication of the nodes in the same column, we reserve the space left to the nodes (green space in Fig. 15.2A), again depending on the required tracks. Finally, for off-board communication we reserve the space beneath the nodes (red space in in Fig. 15.2A) that has a width equal to r, since we assume that all off-board waveguides from all nodes at the same row are routed in parallel with standard pitch (or the pitch preferred) between them, at distance r from the nodes. If nodes use a single side for pinout, instead of the two sides assumed above, then the required area for waveguides will be the same, but more bends will be required. For simple collinear layouts, the proposed strategy is that of a single row of 2D, as depicted in Fig. 15.2A, but because no column-wise communication takes place, the required distance between nodes is $2r-d$ if $r \geq d/2$ (assuming that the waveguides originate from about the center of the chip) or 0 otherwise (nodes are positioned as close as possible next to each other). Fig. 15.2A also gives an estimation of the total required area. In Fig. 15.2C a 2D (3×4) layout of a $3 \times 2 \times 2$ mesh is depicted (equivalent to the network of Fig. 15.2B). Two waveguides form a bundle and are used within column and row tracks, while one waveguide/node is used for off-board communication. The off-board waveguide tracks can be omitted completely if the off-board communication takes place via vertical cabling. In the latter case, off-board routing is implemented using fiber optics (such as in Hasharoni et al. [37]). However, in racks containing a large number of boards with a large number of on-board modules, optical fibers across boards could lead to a cabling mess. Furthermore, the incorporation of on-OPCB waveguides for off-board communication would result in pluggable boards offering ease of installation.

The layout strategies outlined above based on the Thompson model follow, by definition, a X−Y routing approach (or Manhattan routing). A more general routing approach can be used by applying λ-geometry, where λ represents the number of possible routing directions and π/λ the routing angles allowed [38]. $\lambda = 2$, 3, and 4 correspond to the Manhattan architecture, Y-architecture, X-architecture, respectively. In the Manhattan architecture there are only vertical or horizontal routing options as described above (0, 90, 180, and 270 degrees). In Y-architecture (or hexagonal routing) and X-architecture (or octagonal routing) the routing options vary by 60 and 45 degrees respectively. These approaches are depicted in Fig. 15.3A−C. λ-geometry routing approaches with $\lambda > 2$ lead to alternative mesh architectures with higher connectivity degrees. Fig. 15.3D, E−F depict meshes based on these approaches, adjusted for OPCBs, assuming nodes with pinout from all four sides and non-unit side size. The four-side pinout allows the nodes of the regular 4×4 mesh to be placed as close as possible to each other (Fig. 15.3D). In Fig. 15.3E we present a generalization of the Y-mesh for arbitrary routing angles (normally $\theta = 60$ degree in 3-geometry) and in Fig. 15.3F a generalization of the X-mesh for arbitrary crossing angles (normally 90 degree crossings are present in 4-geometry). The required layout area can be estimated approximately using basic geometric shapes: isosceles triangles for the generalized Y-mesh and rectangles for the generalized X-mesh.

Redefining the routing grid in order to allow crossings of various crossing angles could lead to various such "extended" mesh architectures with even higher connectivity degrees. These extended architectures (using $\lambda > 2$) would require more area if implemented using Manhattan routing. For example, Fig. 15.3G depicts a X-mesh implemented using X-Y routing and 90 degree crossing angles.

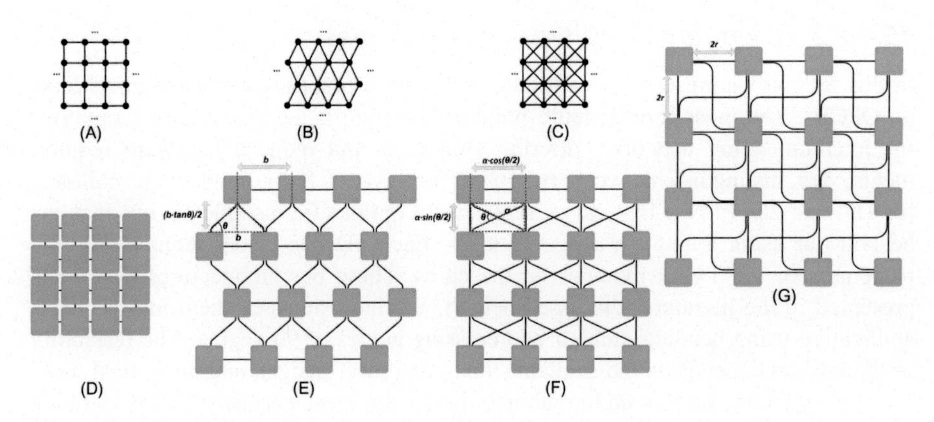

Figure 15.3 λ-geometry with: (A) $\lambda = 2$ (Manhattan routing), (B) $\lambda = 3$ (Y-routing), (C) $\lambda = 4$ (X-routing). Adjustment of λ-geometry approaches for OPCBs for 16 nodes with pinout from all four sides and non-unit side size: (D) 4×4 Mesh ($\lambda = 2$). (E) 4×4 generalized Y-Mesh (normally $\theta = 60$ degree in the Y-routing approach with $\lambda = 3$). (F) 4×4 generalized X-Mesh (normally $\theta = 90$ degree in the X-routing approach with $\lambda = 4$). (G) a X-Mesh implemented using Manhattan routing and crossing angles 90 degree.

If we set for simplicity $2r = d$, then this topology would require a $7d \times 7d$ area. Setting, for a fair comparison, crossing angle $\theta = 90$ degree (as in the $\lambda = 4$ routing approach) and $\alpha \cdot \sin(45 \text{ degree}) = d$ in the layout approach of Fig. 15.3F, the latter would require an area equal to $4d \times 7d$ (in the vertical dimension the nodes would be placed as close as possible next to each other). In principle, the allowance of crossings in the same layer and the reduced link-to-link separation (waveguide pitch), compared to electrical interconnects, allow denser integration and reduction of PCB thickness (layer count). However, a potential issue is crosstalk with respect to the crossing angle, for angles less than 90 degree . To the best of our knowledge there is not yet a design rule/analytical formula for crosstalk as a function of the crossing angle. Measurements for crosstalk can be found in the work of Bamiedakis et al. [20], but only for the examined bus architecture. Another manufacturing issue for OPCBs is that the performance of waveguide components depends on the launch conditions at the component input, e.g., whether light enters the waveguide using multi-mode MMF, or SMF, or mirrors for chip-to-board coupling. Furthermore, in multi-point topologies where splitters/combiners are used, it is possible for the light entering the first splitter along a multi-mode waveguide path to resemble light input from a well-aligned MMF, while the light entering the following splitters on the waveguide path resembles light input from a displaced MMF toward the bent output of the splitter [20]. In principle, splitters and combiners are the most expensive components, followed next by bends and finally by crossings. The layout strategies presented for both point-to-point and multi-point topologies in this chapter are detailed, requiring specific numbers of waveguide components, appearing in a specific order on the waveguide paths, while at the same time they are general enough, abstracting implementation details, thus offering flexibility to the designers.

15.4.2.2 Layouts for multi-point topologies

In this section we present layout strategies for multi-point interconnection networks on OPCBs. The most popular multi-point architecture is the bus, a legacy topology for interconnection networks, offering simplicity and reduced hardware requirements. We distinguish between two types of layouts for a single bus: collinear (or 1D) and 2D. In Fig. 15.4 we present several options for a single 1D bus that can be laid out using a single waveguide layer. Each 1D bus layout requires specific placement of the Tx/Rx modules on the chips. These bus architectures have been presented in the literature (discussed below). We have adjusted them for on-OPCB application using bending radius *r* and crossing angles of 90 degree. The feasibility of the layouts depends on the available area, the power budget, and the optical modules losses (with splitters and combiners being the most expensive). Regeneration units placed in strategic points can be used in order to render a layout that is infeasible, due to losses, feasible. Note that in Fig. 15.4 the bus layouts are presented without using any regeneration. In the following we briefly discuss the depicted architectures.

The architecture depicted in Fig. 15.4A is based on the bus architecture presented by Dou et al. [19]. It is a bidirectional bus consisting of two waveguides with splitting/combining occurring at the Tx and Rx points of the nodes (with the exception of the Tx of the first node and the Rx of the last node). It assumes that the transmitters and receivers are located on opposite sides of the node. The bus architecture in Fig. 15.4B consists of two separate multi-point channels, one for every communication direction [39,40], which therefore is called a dual bus. It assumes that the transmitters and the receivers for the first link are located at the same side of the node, with the transmitters and receivers of the second link at the opposite side of the node in reverse order. It also assumes that the separation distance between a Tx element and an Rx element in a single side of the node is *r*. An alternative layout of the same architecture where the distance between the Tx and Rx elements is the used waveguide pitch is depicted in Fig. 15.4C. In this case more bends and layout area are required. The architecture in Fig. 15.4D is

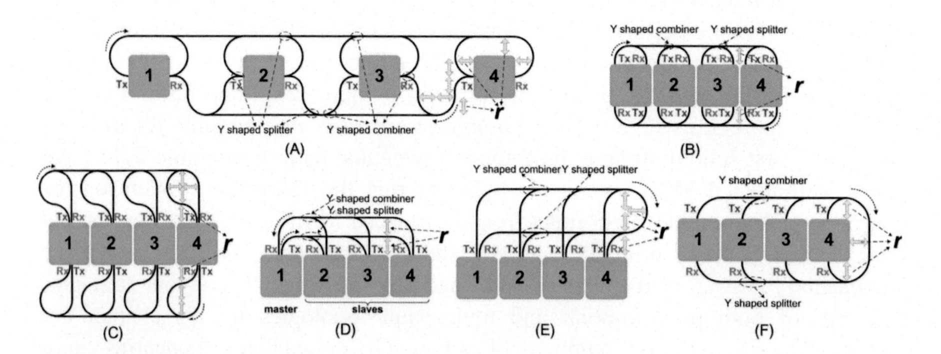

Figure 15.4 (A) Bi-dir bus, (B) dual bus, (C) dual bus (alternative), (D) master-slave bus, (E) folded bus 1, (F) folded bus 2.

Table 15.1 Comparison of the 1D bus layouts assuming N nodes

	Width	Height	Split.	Comb.	Bends	Cross.
Bi-dir bus	$N \cdot (d+2r)+(N-1) \cdot 2r$	$4r$	$N-1$	$N-1$	4	–
Dual bus	$N \cdot d$	$d+2r$	$N-1$	$N-1$	2	–
Dualb (alt.)	$\begin{cases} N \cdot d, d \geq 4r \\ N \cdot d+2r-d/2, d<4r \end{cases}$	$d+6r$	$N-1$	$N-1$	4	–
Master-slave bus	$\begin{cases} N \cdot d, d \geq 2r \\ d+2r \cdot (N-1), d<2r \end{cases}$	$d+2r$	$N-2$	$N-2$	2	$N-2$
Folded bus 1	$N \quad d+r$	$d+3r$	$N-1$	$N-1$	4	$N-1$
Folded bus 2	$N \quad d+r$	$d+2r$	$N-1$	$N-1$	4	–

a master-slave parallel optical bus [41] consisting of two parallel buses. The master node broadcasts signals on the bus using the first waveguide, where any slave node can receive them and send data back to the master using the second waveguide. The bus layouts in Fig. 15.4E and F are folded buses using a single waveguide [39,42]. The first folded bus layout assumes that the Tx and Rx elements are located at the same side of the node, separated by a distance equal to waveguide pitch. The second folded bus layout assumes that the Tx and Rx elements are located at the opposite side of the nodes.

Table 15.1 summarizes the characteristics of the bus layouts presented above in terms of area (width, height) as well as number of splitters, combiners, crossings, and bends in the worst case (for a single waveguide channel). We count each S-bend as two waveguide bends. The dual bus options need twice the number of Tx and Rx modules than the other ones. Splitters and combiners are both present in all the layout approaches (thus there are $2(N-1)$ splitting/combining elements in the "worst-case waveguide"), with the exception of the master-slave bus where only splitters or combiners are present in a single waveguide.

All the aforementioned bus layouts can be extended using multiple waveguide layers and identical waveguide routing in every layer to increase aggregate bandwidth. Alternatively, more waveguides can be added using the same waveguide layer (or a combination of both approaches). Fig. 15.5 depicts how the bus layouts can be extended in the same layer using more waveguides.

Adding bus waveguides in the same layer for the bi-directional bus presents problems due to the presence of splitters/combiners at the Tx and Rx points of the nodes (using only 90 degree crossing angles and bending radiuses equal to r). The addition of a single extra bus waveguide increases the layout height by $2r$ (for all bus layouts). It also increases the required width by r in the folded bus approaches. The worst case for the number of splitters, combiners, and bends remains the same. The worst case for the number of crossings assuming W bus waveguides occurs for the waveguides that are located closest to the nodes. Table 15.2 summarizes the total (worst case) crossings assuming W bus waveguides in the same layer.

The 1D bus layouts considered above may be restricting in terms of area since they require a lot of area for their width. In Fig. 15.6 we provide two serpentine

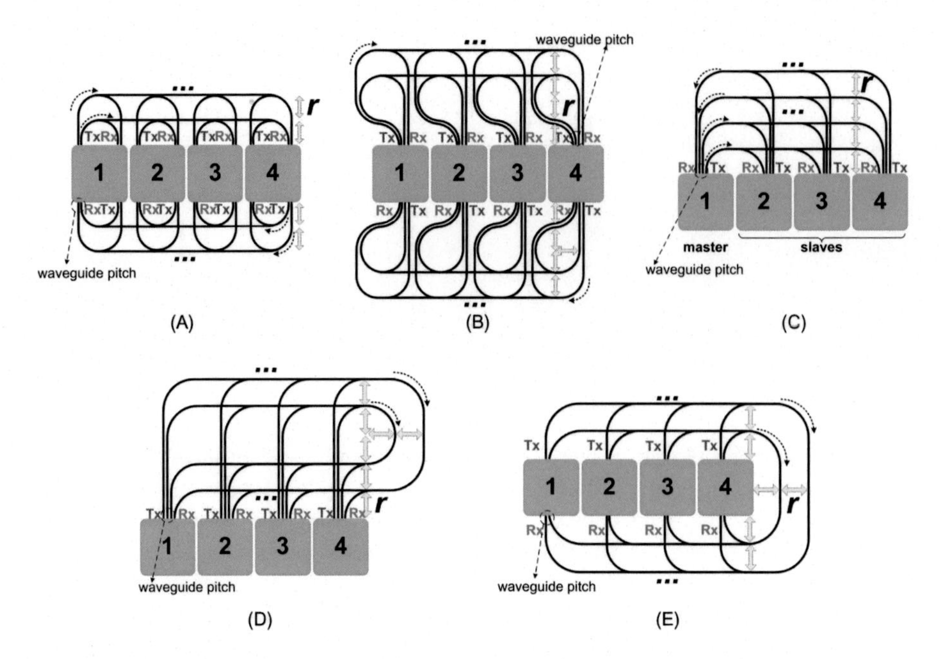

Figure 15.5 Additional bus waveguides in the same layer for increased aggregate bandwidth. (A) Dual bus, (B) Dual bus (alternative), (C) Master-slave bus, (D) Folded bus 1, (E) Folded bus 2.

Table 15.2 Number of crossings for the 1D bus lay-outs assuming W bus channels

Dual bus	Dual bus (alt.)	Master-slave bus	Folded bus 1	Folded bus 2
$(2(N-2)+2)\cdot W$	$(2(N-2)+2)\cdot W$	$(N-2)\cdot(2W-1)$	$(N-1)\cdot(2W-1)$	$2(N-1)\cdot(W-1)$

2D layout approaches (requiring a single layer) for a dual bus and a folded bus, allowing better balancing between the required height and the required width.

Finally, there could be combinations between point-to-point and multi-point architectures such as mesh of buses [43]. A 2-layer layout of a 4 × 4 mesh of buses is depicted in Fig. 15.7.

15.4.2.3 WDM for point-to-point topologies using multi-point layouts

WDM (Wavelength Division Multiplexing) is an important advantage of optical technology, giving the ability for a single waveguide to support multiple optical channels simultaneously using different wavelengths. This allows the implementation of

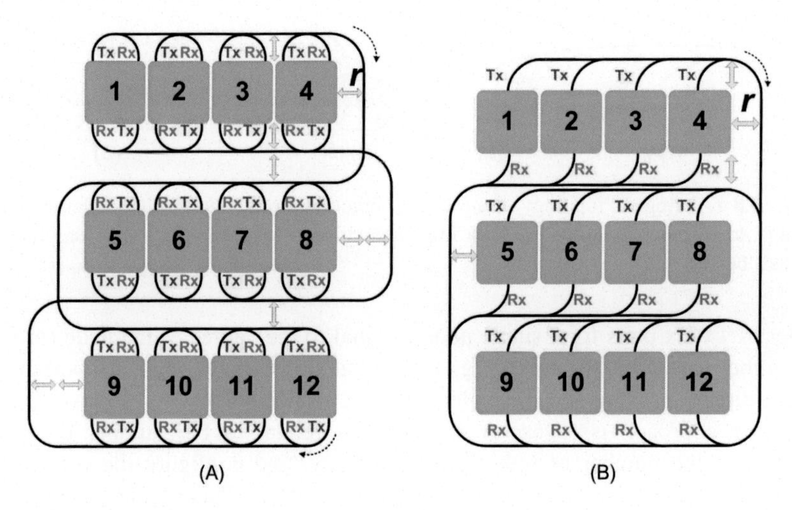

Figure 15.6 Serpentine 2D layouts for: (A) a dual bus, (B) a folded bus.

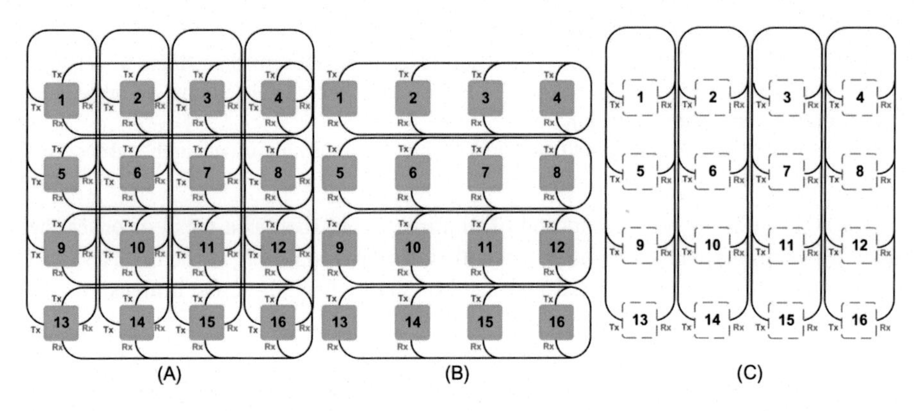

Figure 15.7 (A) 2D 4×4 mesh of buses topology (2 layers): (B) Layer 1, (C) Layer 2.

many point-to-point connections over physical waveguides laid out as busses. For example, a fully-connected network could be implemented using a single bus layout, and a mesh of fully-connected networks could be implemented as a mesh of buses. The multiplexers/demultiplexers required to create the WDM signal on a waveguide can take place either on the chip, or on the OPCB. In the latter case it can be realized by using $1 \times N$ splitters/combiners on the OPCB to combine different signals (of different wavelengths) in a single waveguide.

A simple approach to implement a point-to-point topology using an optical bus would be to use as many wavelengths as the number of uni-directional links of the topology. For example, a uni-directional ring of N nodes has N links, while an equivalent bi-directional ring has $2N$ links. Thus, for their implementation using a bus architecture, N and $2N$ wavelengths would be needed respectively. The

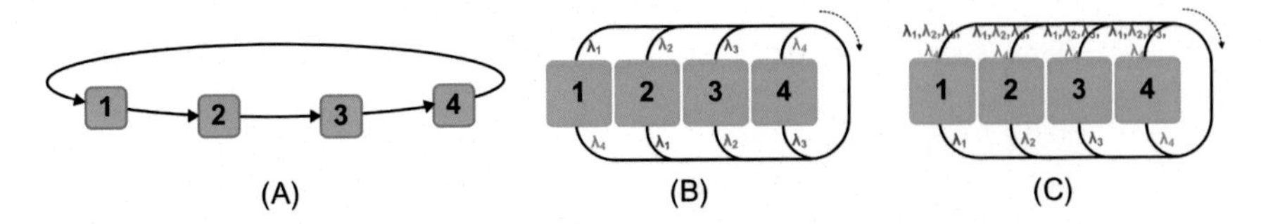

Figure 15.8 (A) Logical topology of a four uni-directional ring, and (B) its implementation using a (folded) bus. (C) An *N*-Tx, 1-Rx implementation for point-to-point connections using a (folded) bus layout.

number of Tx/Rx pairs for a single node is equal to the degree of the node (or twice that number for the dual buses). Fig. 15.8A and B depict the logical topology of a four uni-directional ring and its implementation, respectively.

Another approach is to use N wavelengths (equal to the number of nodes), smaller than the number of links of the topology, and configure the connectivity dynamically using a wavelength assignment algorithm. This would require every node to have:

- A Tunable transmitter and a burst mode receiver, or
- N separate Tx elements, N separate Rx elements, or
- N Rx elements, 1 Rx element—see Fig. 15.8C (each node transmits in a single wavelength determined by the wavelength assignment algorithm in order to ensure that no other node transmits in the same wavelength), or
- One Tx element, N Rx elements (each node transmits in a single wavelength and receives all wavelengths).

Note that in a topology composed of multiple buses such as a mesh of buses the same wavelengths can be re-used (in both the horizontal and vertical buses in a mesh of buses), since there is a different set of Tx/Rx for the second dimension.

15.4.3 Applying the proposed on-OPCB layout strategies: illustrative examples

In this section we apply the layout approaches described in the previous section for four logical topologies: a $3 \times 2 \times 2$ mesh, a 4×4 torus, a 9-fully-connected network, and a 9×9 mesh of fully-connected networks (a topology resembling a 9×9 mesh where every row and column is a fully-connected network instead of a linear array as in mesh networks). We omit the waveguide tracks for off-board communication in all cases for simplicity. We assume that at most two sides of the node can be used pinout. For the point-to-point networks we assume that two waveguide layers are used. Fig. 15.9 depicts: (A) a collinear layout for a $3 \times 2 \times 2$ mesh network, (B) a 2D (4×4) lay-out for a 4×4 torus network, and (C) a collinear layout for a 9-fully-connected network using the strategies described by Yeh et al. [34] and modified as described in Section 15.4.2 using bending radius r and crossing angles equal to 90 degree.

Table 15.3 presents the required area for all four topologies, as well as the number of crossings, using various layout approaches, assuming $d = 50$ mm and $r = 10$ mm.

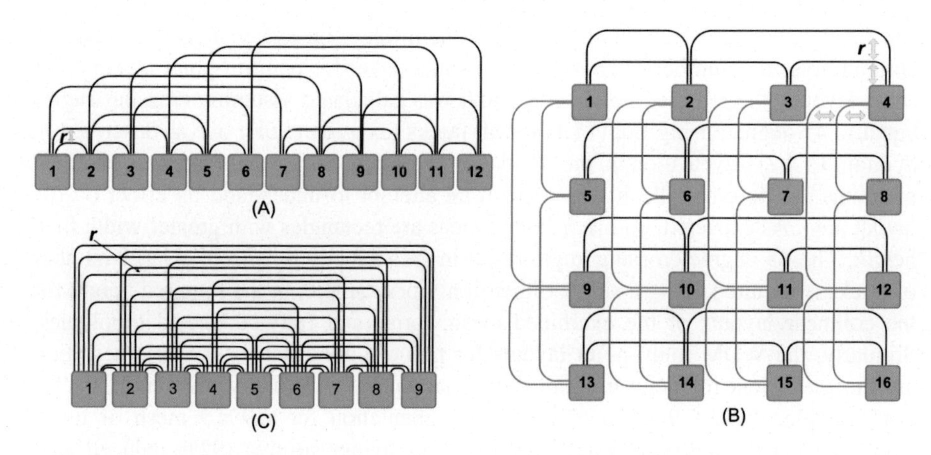

Figure 15.9 (A) Collinear layout for a 3 × 2 × 2 mesh network. (B) 2D (4 × 4) layout for 4 × 4 torus networks. (C) Collinear layout for a 9-fully-connected network.

Table 15.3 Area requirements, number of crossings and bends using the point-to-point and multi-point layout techniques for various topologies (d = 50 mm, r = 10 mm)

		Width (in mm)	Height (in mm)	Crossings
3 × 2 × 2 mesh	collinear layout	820	150	9
	collinear layout (θ = 45 degree)	820	86	9
	3 × 4 2D layout	240	240	3
	3 × 4 2D layout (θ = 45 degree)	189	240	3
	2 × 6 2D layout	410	180	5
	dual bus layout	600	70	–
	folded bus layout	610	70	–
	2D 3 × 4 folded bus layout	220	240	–
4 × 4 torus	collinear layout	1100	150	8
	collinear layout (θ = 45 degree)	1100	86	8
	4 × 4 2D layout	280	280	6
	4 × 4 2D layout (θ = 45 degree)	252	252	6
	4 × 4 mesh of buses layout	280	280	–
	dual bus layout	800	70	–
	folded bus layout	810	70	–
	2D 4 × 4 folded bus layout	220	280	–
9 fully-connected	collinear layout	610	250	12
	collinear layout (θ = 45 degree)	610	116	12
	dual bus layout	450	70	–
	folded bus layout	460	70	–
	2D 5 × 4 folded bus layout	270	140	–
9×9 mesh of fully-connected networks	2D layout	2250	2250	160
	2D mesh of buses layout	630	630	–

The bus layouts need only one layer (excluding mesh of buses layouts). The crossings column gives the number of crossings in a single layer for point-to-point layouts. For some layouts, the required height and width was calculated assuming crossing angles equal to 45 degree, using Eq. (15.1) for all tracks (except the first one as described in Section 15.4.2). Even though the 45 degree crossing angle could present practical problems due to crosstalk, it was used in an attempt to understand its effect on the layout area. As expected, collinear layout areas are rectangles with greater width than height. The 45 degree crossing angle result in area savings whenever a large number of tracks is required along a single dimension. For example, in the required height for the collinear layouts of the examined mesh, torus, and fully-connected topologies. Similarly, the WDM multi-point layouts for point-to-point networks implementation are more efficient for point-to-point topologies that require many waveguide tracks. For example, the 9×9 mesh of buses implementation for a 9×9 mesh of fully-connected networks would result in impressive savings in area (92% reduced area requirements by using the former). Similarly, the bus implementations for the 9 fully-connected network lead to significantly smaller lay-out areas (e.g, 79% less area is needed if the folded bus lay-out is used).

15.5 Conclusion

Optics have already found their way inside the DC for rack-to-rack and server-to-rack connections. In order to cope with both the energy and bandwidth requirements, and to overcome the limitations of the electrical interconnects, DCs will have to deploy optical technologies, if possible, in all packaging levels of their architecture. As short distance optical interconnects and nano photonics mature, new architectures for DCs using these building blocks will be proposed in order to maximally exploit the benefits of optics.

A large number of architectures for optical switches (for rack-to-rack communication) as well as Networks-on-Chip have already been presented, and new architectures will continue to be proposed. Implementing and laying out complex topologies via optical waveguides on the on-OPCB level presents a number of issues that have to be considered when designing architectures for DCs. To this end we have outlined layout strategies for both point-to-point and multi-point topologies for OPCBs, general enough to be easily applied by designers.

Acknowledgments

This work was supported by the European Union (European Social Fund—ESF) and Greek national funds through the Operational Program "Education and Lifelong Learning" of the National Strategic Reference Framework (NSRF) – Research Funding Program: Thales Investing in knowledge society through the European Social Fund and by the European Commission through the FP7 ICT-PHOXTROT (ICT 318240) project.

References

[1] Cisco Networks White Paper. Cisco global cloud index: forecast and methodology, 2013–2018, Cisco Systems. [Online] Available from: http://www.cisco.com/c/en/us/solutions/collateral/service-provider/global-cloud-index-gci/Cloud_Index_White_Paper.html; 2013 [accessed 01.09.15].

[2] Al-Fares M, Loukissas A, Vahdat A. A scalable, commodity data center network architecture. ACM SIGCOMM Comput Commun Rev 2008;38(4):63–74.

[3] Taubenblatt MA. Optical interconnects for high-performance computing. J Lightwave Technol 2012;30(4):448–57.

[4] Farrington N, et al. Helios: a hybrid electrical/optical switch architecture for modular data centers. ACM SIGCOMM Comput Commun Rev 2011;41(4):339–50.

[5] Wang G, et al. c-Through: Part-time optics in data centers. ACM SIGCOMM Comput Commun Rev 2011;41(4):327–38.

[6] Singla A, et al. Proteus: a topology malleable data center network. In: Proc. 9th ACM SIGCOMM workshop on hot topics in networks; 2010, p. 8.

[7] Luijten R, et al. Optical interconnection networks: the OSMOSIS project. In: The 17th annual meeting of the IEEE lasers and electro-optics society; 2004.

[8] Castoldi P, et al. Energy efficiency and scalability of multi-plane optical interconnection networks for computing platforms and data centers. In: Optical fiber communication conference, Optical Society of America; 2012, p. OW3J-4.

[9] Shacham A, Bergman K. An experimental validation of a wavelength-striped, packet switched, optical interconnection network. J Lightwave Technol 2009;27(7):841–50.

[10] Zhu Z, et al. Fully programmable and scalable optical switching fabric for petabyte data center. Opt Express 2015;23(3):3563–80.

[11] Xia K, et al. Petabit optical switch for data center networks. Polytechnic Institute of New York University, New York, Tech. Rep; 2010.

[12] Ye X, et al. DOS: a scalable optical switch for datacenters. In: Proc. 6th ACM/IEEE symposium on architectures for networking and communications systems; 2010, p. 24.

[13] Gripp J, et al. Photonic terabit routers: the IRIS project. In: Optical fiber communication conference, Optical Society of America; 2010, p. OThP3.

[14] Proietti R, et al. Scalable optical interconnect architecture using AWGR-based TONAK LION switch with limited number of wavelengths. J Lightwave Technol 2013;31(24): 4087–97.

[15] Schmidtke K, et al. 960 Gb/s optical backplane ecosystem using embedded polymer waveguides and demonstration in a 12G SAS storage array. J Lightwave Technol 2013;31(24):3970–5.

[16] Chen RT, et al. Fully embedded board-level guided-wave optoelectronic interconnects. Proc IEEE 2000;88(6):780–93.

[17] Pitwon RC, et al. FirstLight: pluggable optical interconnect technologies for polymeric electro-optical printed circuit boards in data centers. J Lightwave Technol 2012;30(21): 3316–29.

[18] Beals IV J, et al. A terabit capacity passive polymer optical backplane based on a novel meshed waveguide architecture. Appl PhysA 2009;95(4):983–8.

[19] Dou X, et al. Optical bus waveguide metallic hard mold fabrication with opposite 45 micro-mirrors. In: OPTO, International Society for Optics and Photonics; 2010, p. 76070P-76070P.

[20] Bamiedakis N, et al. A 40 Gb/s optical bus for optical backplane interconnections. J Lightwave Technol 2014;32(8):1526–37.

[21] Shacham A, Bergman K, Carloni LP. Photonic networks-on-chip for future generations of chip multiprocessors. IEEE Trans Comput 2008;57(9):1246–60.

[22] Vantrease D, et al. Corona: system implications of emerging nanophotonic technology. ACM SIGARCH Comput Arch News 2008;36(3):153–64.

[23] Beamer S, et al. Re-architecting DRAM memory systems with monolithically integrated silicon photonics. ACM SIGARCH Comput Arch News 2010;38(3):129–40.

[24] Pan Y, Kim J, Memik G. Flexishare: channel sharing for an energy-efficient nanophotonic crossbar. In 2010 IEEE 16th international symposium on high performance computer architecture (HPCA); 2010, p. 1–12.

[25] Kirman N, et al. Leveraging optical technology in future bus-based chip multiprocessors. In Proc. 39th annual IEEE/ACM international symposium on microarchitecture; 2006, p. 492–503.

[26] Kurian, et al. ATAC: a 1000-core cache-coherent processor with on-chip optical network. In: Proc. 19th international conference on parallel architectures and compilation techniques; 2010, p. 477-488.

[27] Batten C, et al. Building manycore processor-to-dram networks with monolithic silicon photonics. In: 16th IEEE symposium on high performance interconnects, HOTI'08; 2008, p. 21–30.

[28] Morris R, Kodi AK. Exploring the design of 64-and 256-core power efficient nanophotonic interconnect. IEEE J Sel Top Quantum Electron 2010;16(5):1386–93.

[29] Koka P, et al. Silicon-photonic network architectures for scalable, power-efficient multi-chip systems. ACM SIGARCH ComputArch News 2010;38(3):117–28.

[30] Cianchetti MJ, Kerekes JC, Albonesi DH. Phastlane: a rapid transit optical routing network. ACM SIGARCH Comput Arch News 2009;37(3):441–50.

[31] Joshi A, et al. Silicon-photonic clos networks for global on-chip communication. In: Proc. 2009 3rd ACM/IEEE international symposium on Networks-on-Chip; 2009, p. 124–33.

[32] Pan Y, et al. Firefly: illuminating future network-on-chip with nanophotonics. ACM SIGARCH Comput Arch News 2009;37(3):429–40.

[33] Gu H, Xu J, Zhang W. A low-power fat tree-based optical network-on-chip for multiprocessor system-on-chip. In: Proc. conference on design, automation and test in Europe; 2009, p. 3–8.

[34] Yeh CH, Varvarigos E, Parhami B. Multilayer VLSI layout for interconnection networks. In: Proc. 2000 international conference on parallel processing; 2000, p. 33–40.

[35] Yeh CH, et al. VLSI layout and packaging of butterfly networks. In: Proc. twelfth annual ACM symposium on parallel algorithms and architectures; 2000, p. 196–205.

[36] Siokis A, Christodoulopoulos K, Varvarigos E. Laying out interconnects on optical printed circuit boards. In Proc. tenth ACM/IEEE symposium on architectures for networking and communications systems; 2014, p. 101–12.

[37] Hasharoni K, et al. A high end routing platform for core and edge applications based on chip to chip optical interconnect. In: Optical fiber communication conference. Optical Society of America; 2013, p. OTu3H-2.

[38] Chen H, et al. The Y-architecture: yet another on-chip interconnect solution. In: Proc. ASP-DAC 2003. Asia and South Pacific design automation conference. IEEE; 2003, p. 840–6.

[39] Li K, Pan Y, Zheng SQ. Parallel computing using optical interconnections. Boston, MA: Springer Science & Business Media; 1998.

[40] Guo Z, et al. Pipelined communications in optically interconnected arrays. J Parallel Distributed Comput 1991;12(3):269–82.
[41] Tan M, et al. A high-speed optical multi-drop bus for computer interconnections. Appl Phys A 2009;95(4):945–53.
[42] Melham RG, Chiarulli DM, Levitan SP. Space multiplexing of waveguides in optically interconnected multiprocessor systems. Comput J 1989;32(4):362–9.
[43] Iwama K, Miyano E, Kambayashi Y. Routing problems on the mesh of buses. In: Algorithms and computation. Berlin: Springer; 1992, p. 155–64.

Index

Note: Page numbers followed by "*f*" and "*t*" refer to figures and tables, respectively.

A

Abrication limited Q-factor, 130–131
ACID (Atomicity, Consistency, Isolation, Durability), 30–31
Active alignment, 346, 346*f*, 351–353
  dual-fibers, 353*f*
Active optical cable (AOC), 38, 44–45, 50–53, 219, 219*f*
Address Resolution Protocol (ARP), 20
ADS, 251
Agile data center architecture, 22
Aluminum quaternary (Al-Q) aperture layer, 101–102
ANSI (American National Standards Institute), 9
Application architectures, 10–14
Application program interfaces (APIs), 12
Applications of optical interconnects
  in individual layers of packaging hierarchy, 376–378
    Networks-on-Chip (NoC), 378
    on-board and board-to-board interconnections, 378
    rack-to-rack interconnections, 376–378
Architectural requirements, of data centers, 5–6
Architectures, data center, 1
  application architectures, 10–14
  classifications, 9–10
  cloud data center architectures, 14–16
  design considerations, 21–32
    agility and elasticity, 21–22
    availability and reliability, 29–31
    flattened, converged networks, 22–25
    network security, 31–32
    network subscription level, 27–29
    scalability, 26–27
    virtualization and latency, 25–26
  environment considerations, 4–8
  next generation data center architectures, 32–36
  optical interconnects, 36–39
  physical architecture, 17–21
    Layer 2 and Layer 3 network architectures, 18–21
Arrayed Waveguide Grating Routing (AWGR) elements, 377–378
Arrayed waveguide gratings (AWGs), 376
Availability, of data center network, 29–31

B

Ball grid array (BGA), 294, 303
Bandwidth reallocation, 233–234
BASE (Basically Available, Soft State, Eventual Consistency), 30
Bend insensitive fibers, 335
Bit error rate (BER), 144, 191, 191*f*, 218*f*
BitTorrent, 11–12
Blade server, 4, 5*f*, 23–24, 79
Blades and boards, optical interconnect between, 85–88
Board-level optical interconnection, 288*f*
Board-to-board interconnections, 378
Board-to-board interconnects, 48–49, 49*f*
Bond tracking, 352
Border Gateway Protocol (BGP), 19–20
Branch, 204*t*
Brief taxonomy of data centers, 78–79
Bring-your-own-device (BYOD), 21–22
Buried oxide (BOX), 206–207
Busy Write Set (BWS), 231–232
Butterworth/maximally flat response, 257–258
Butt-joint regrowth, 99

C

Cabling, reduced size and weight of, 198
Cadence Virtuoso, 251
Carrier depletion, 148–149, 148*f*
Center for Integrated Access Networks (CIAN), 37–38
Channel waveguide, 204*t*

Chiplets, 88, 91*f*
Chip-level optical interconnection, 53–54
Chip-Multiprocessor technology (CMP), 65
Chips and dies, optical interconnect between, 88–90
Chip-scale silicon photonics transceiver. *See* Optical I/O core
Chip-to-chip coupling, 269–270
Chip-to-chip optical interconnects, 63–64, 171
Chucks, 361–363
Classifications, data center, 9–10
Client-server architectures, 11*f*, 34
Clock and data recovery (CDR), 247–248
Clos networks, 18–19
Cloud data center architectures, 14–16
Cloud data center networks, 15
Coarse alignment, 349, 352
Colocation, 16
Colocs. *See* Modern colocation centers
Colos. *See* Modern colocation centers
Complementary metal oxide semiconductor (CMOS), 53–54, 57–58, 62–63, 87, 112–113, 223–224
  CMOS ICs, 207–210
Consortium for Onboard Optics (COBO), 38
Conventional data center architectures, 35–36
Conventional networks, 32–33
Converged Enhanced Ethernet (CEE), 24
Converged networks, 29
Corning Clearcurve LX multi-mode fiber (MMF), 217–218
CoS (Chip on sub-mount), 349, 371
Cost of energy to power data center equipment, 7–8
Cost-efficient scaling, 26–27
Coupler, 204*t*
Crochzalski (CZ) silicon, 135–136
Current-mode logic (CML), 207–209, 247–248
Cut-back method, 161

D

Damping materials, 329–330
Data center (DC) architectures
  on-optical printed circuit boards (on-OPCB), 378–390
    applying the proposed on-OPCB layout strategies, 388–390
    layout strategies, 379–388
  optical interconnects technologies, 376
  packaging hierarchy, optical interconnects in individual layers of, 376–378
    board-to-board interconnections, 378
    networks-on-Chip (NoC), 378
    on-board interconnections, 378
    rack-to-rack interconnections, 376–378
Data Center Bridging (DCB), 24
Data Center Bridging Exchange protocol (DCBx), 24
Data center raised floor, 6*f*
Data communication requirements, 82–83
Data Deluge, 46, 75–76, 92
Data growth vs. Moore's Law, 76*f*
Data servers, 81, 84, 87–88
  storage hierarchy in, 82*t*
Data traffic, 17
DC phase shifter, 204*t*
Deep Ultra Violet (DUV) Photolithography, 127–128
Demilitarized zone (DMZ), 31–32
Deming Cycle, 12
Dense wavelength division multiplexing (DWDM), 328
Design considerations, data center, 21–32
  agility and elasticity, 21–22
  availability and reliability, 29–31
  flattened, converged networks, 22–25
  network security, 31–32
  network subscription level, 27–29
  scalability, 26–27
  virtualization and latency, 25–26
DevOps, 12
Diesel-fueled onsite generators, 7–8
Diffractive grating, 269–270, 270*f*
Diluted-waveguide spot-size converter (SSC), 101–102
Diphenyl sulfide (DPS), 175–176
Disaggregated data center, 35–36
Dispersion adapted (DA) cavity, 128
Dispersion adaption, 127–128
Dispersion relation, 122–123
Distributed Bragg reflectors (DBR), 271–272
Distributed feedback laser (DFBL), 101–102

Distributed management architectures, 34
Double-corrugated gratings, 270f
DRAM (Dynamic Random Access Memory), 65–66, 82–83, 142
Dual bus, 384–385
Dual quantum-well integration platform, 99–100
Dumb terminals, 10
DVFS (Dynamic Voltage and Frequency Scaling), 81
Dynamic Host Configuration Protocol (DHCP), 20

E

EAM-modulated laser PIC (EMLs), 116
E-beam lithography, 112
Electric filed, Maxwell's equation for, 122
Electrical and photonic off-chip interconnection and system integration, 263
  Moore's law and off-chip I/O trends, 265
  packaging in retrospect, 265–267
  "silicon bridging" concept, large-scale interconnected system using, 272–282
    3D stacked logic, photonics, and thermal challenges, 278–282
    electrical flexible interconnect integration with photonic interconnects, 276–277
    self-alignment devices and performance, 274–275
    system overview, 273
  silicon interposer technology, 267–269
  silicon photonics and optical coupling, 269–272
Electrical motors, 356
Electrical probing, 361–363, 362f
  of photonics devices, 362
Electro-absorption modulation, 61
Electro-absorption modulator (EAM), 101–102
Electro-absorption-modulated lasers (EMLs), 102
Electroluminescence, 136–137, 137f
Electron beam lithography (EBL), 126
Electronic drivers/TIAs, 247
  electronics and photonics, co-design and co-simulation of, 250–251
  rationale, 247–250
  transimpedance amplifiers, 256–261
  VCSEL drivers, 252–256
Electronics and photonics
  co-design and co-simulation of, 250–251
Electro-optic circuits, 97–98
Electro-optic modulation, 142–143
  high-speed, 144–149
Electro-optical circuit boards (EOCBs), 287
  basic concept of, 289–290
  classification of, 288f
  integrated glass waveguide based EOCBs, 297–304
    fiber-to-board and chip-to-board coupling interface, 302–303
    glass waveguide panel fabrication, 298–299
    integration of glass waveguide panels in PCB stack-ups, 300–302
    optical backplane demonstration, 303–304
  integrated planar polymer waveguides, manufacturing of, 289–297
    electro-optical PCB, manufacturing of, 290–291
    planar polymer waveguides, integration of, 293–294
    planar polymer waveguides, manufacturing of, 291–292
    polymer waveguides on flexible substrates, 296–297
    technology demonstrator, 294–296
  mass production and reliability, 304–305
  motivation and classification of optical interconnects at board level, 287–289
  requirements for, 287
Embedded multi-die interconnect bridge (EMIB), 267, 268f
Embedded Optic Modules (EOM), 51–52
EMI immunity, 197–198
Encircled angular flux (EAF), 333–334, 334f
Encircled flux (EF), 216, 332–333
Energy and data centers, 79–81
Energy Efficient Ethernet (EEE) capability, 54
Energy Star rating, 7–8

Enhanced Transmission Selection (ETS), 24
Enterprise class computing, 3
Enterprise data center networks, 28
Environment considerations, data center, 4–8
Epoxy bonding, 346–347, 359
Epoxy thermal bonding, 359
Epoxy UV bonding, 358–359
Equal Cost Multi-Pathing (ECMP), 14, 19*f*, 20*f*
Equilibrium modal distribution (EMD), 332
Erbium Doped Fiber Amplifiers (EDFAs), 59–60
Ethernet architecture, 18
Ethernet data networks, 17
Ethernet LANs, 17–18
Ethernet networks, 17–18, 22–23
Ethernet switches, 8
EU FP7 Phoxtrot project, 255
Exaflop computer, 80
Expectations for optical interconnection, 197–198
  EMI immunity, 197–198
  high density, 197
  reduced size and weight of cabling, 198
  wide bandwidth over long distances, 197

F

Fabricated optical devices, 204*t*
Fabrication, 126–127
Fabrics, 23–24
Fabry–Perot laser diode (LD), 214–216
Fano effect, 129
Far-field pattern (FFP), 210–214
  measurements, 182
Feed-forward equalization (FFE), 255
Fiber cabling solutions, 51
Fiber optic cables, 7–8
Fiber optic circuit, laminated, 311–317
  fiber coating, 317
    coating material, 317
    coating process, 317
  fiber laying technology, 313
  fiber optic circuit laminates manufacturing process, 313–316
    product requirements, 314–316
  fiber optic flexplane technologies, future outlook for, 317
Fiber-array, 344, 345*f*, 346–347, 352
Fiber-to-chip coupling, 270
Fibre Channel protocol, 17–18
Fibre Channel to Fibre Channel over Ethernet (FCoE), 24–25
Figures of Merit (FoM), 84
Fine alignment, 349, 352
Finite impulse response filter (FIR), 255
Finite-difference time-domain (FDTD), 210–214
Fire suppression system, 6–7
First light, 353
Flat network, 23–24
Flattened, converged networks, 22–25
Flip-chip bonding, 343, 349
Floquet–Bloch theorem, 122–123
Fly Src routing algorithm, 236–237
Franz-Keldysh effect, 99
Free space optics, 312

G

Glass waveguide OPCBs, 321–324
Glass waveguide panel fabrication, 298–299
Glass waveguides, 289
  planar, 311–312
Global Internet traffic, 76
"Glueless" socket architectures, 48
Graded index (GI) MMF, 199–201
Graded index circular-core waveguide fabrication method, 178*f*
Graded index polymer waveguides, 160–161, 161*f*
Graded index-core polymer optical, fabrication process of, 175*f*
Graded index-core polymer waveguides, 192, 192*f*
Graded-index core, 173–174
Grating coupler (GC), 204*t*, 206–207, 206*f*, 210–214
Grippers, 361–363
GUI (Graphical User Interface), 367

H

Hadoop processing, 38–39
Health Insurance Portability and Accountability Act (HIPPA), 16
Heterogeneous integration, 62, 114–116
Hexapods, 355, 355*f*
High performance computing (HPC) applications, 13–14

High Performance Computing (HPC) centers, 78–79
High speed (GHz) electro-optic modulation, 144–149
High-speed connectors, 265–266
High-speed electronic circuits, 247
HMI (Human Machine Interface), 367
Hops, 22–23
Human Internet, 75–76
Hybrid-Tree-Benes architecture, 109–110
Hydrogen-treated photonic crystal nanocavities, photoluminescence of, 135*f*

I

IBM, 64
IBM Portable Modular Data Center (PMDC), 8, 8*f*
ICT (Information and Communication Technology), 75
IEEE 802.1Qbg standard, 26
Indexing, 356
Indium gallium zinc oxide (IGZO), 227
Indium phosphide (InP) for optical interconnects, 95
  future trends, 114–117
    heterogeneous integration, 114–116
    inexpensive and innovative integrated photonics, 116–117
    optoelectronic integrated circuits (OEICs), 114
  InP and Si photonics, comparison of, 111–114
  InP photonic integration platforms, 98–100
  InP PICs for optical switching, 108–111
  transceiver InP PICs by vertical integration, 101–107
    receiver PICs, 105–107
    transmitter PICs, 101–105
  transceivers, 107–108
Inductive-capacitive (LC) lines, 45
Industrial manufacturing processes for OPCB, 312–324
  fiber optic circuit, laminated, 312–317
    fiber coating, 317
    fiber laying technology, 313
    fiber optic circuit laminates manufacturing process, 313–316
    fiber optic flexplane technologies, future outlook for, 317
  glass waveguide OPCBs, 321–324
  polymeric planar embedded optical waveguides, 318–321
Inexpensive and innovative integrated photonics, 116–117
InfiniBand Trade Association (IBTA), 17–18
Infrastructure-as-a-service, 15
InGaAsP passive waveguide (PWG), 101–102
In-package photonics, 89, 90*f*
In-plane coupling, 167–168
Integrated glass waveguide based EOCBs, 297–304
  fiber-to-board and chip-to-board coupling interface, 302–303
  glass waveguide panel fabrication, 298–299
  integration of glass waveguide panels in PCB stack-ups, 300–302
  optical backplane demonstration, 303–304
Integrated planar polymer waveguides, manufacturing of, 289–297
  electro-optical PCB, manufacturing of, 290–291
  integration, 293–294
  polymer waveguides on flexible substrates, 296–297
  technology demonstrator, 294–296
Intel Rack-Scale Architectures (RSA), 49–50
Intel's embedded multi-die interconnect bridge (EMIB), 267, 268*f*
International Electrotechnical Commission (IEC), 324–325
International Technology Roadmap for Semiconductors (ITRS), 46, 223–224
Internet, 75
Internet of Things (IoT), 36, 75–77
  and cyber physical systems, 77
Inter-switch links (ISLs), 19–20
Intersymbol interference (ISI), 250
Intrusion detection and prevention systems (IDS/IPS), 31–32
I/O bottleneck, 197
Ion-exchange waveguide process flow, 298–299, 299*f*

J

Jevon's Law, 37–38

K

Kerr effect, 142–143
Kramers–Kronig relation, 142–143
k-vector distribution, 139–141

L

Laser diode, 85–86
Laser soldering, 348
Laser welding, 348, 359, 360*f*
Laser-induced soldering, 344, 349, 359–360, 360*f*
Latency, virtualization and, 25–26
Layer 2 and Layer 3 network architectures, 18–21
Leaf-spine architecture, 18–19
Light emission, 131–137
  electroluminescence, 136–137
  photoluminescence, 132–136
Limited combustibility, 6–7
Link Aggregation Control Protocol (LACP), 19–20
Link aggregation groups (LAGs), 22–23
Local area networks (LANs), 10–11
Location of data center, 5
Lossless Ethernet, 24, 37–38

M

Machine "ingredients" and machine technologies, 354–363
  bonding techniques and equipment, 358–360
  chucks, 361–363
  electrical probing, 361–363
  grippers, 361–363
  mechanical positioners/manipulators, 354–357
    controllers, 356
    conventional industrial robotics for pick and place, 356–357
    feedback sensors, 356
    motors/actuators, 356
    multi DoF/6 axis devices, 355
    stack of axis, 354–355
  optical grade tables and enclosures, 358
  pickers, 361–363
  tweezers technology, 361–363
Machine vision, 345, 352, 363–367
Mach-Zehnder interferometer (MZI) modulator, 204–205
Mach-Zehnder Interferometer, 108–109, 142–143
Mach-Zehnder modulators (MZMs), 61
Malicious attacks, 32
Mechanically flexible interconnects (MFIs), 273
Media access control (MAC), 20
Membrane photonic crystal, 127
Micro-Electro-Mechanical Systems Switches (MEMS), 377–378
Micromachining, 361
Micro-ring resonators (MRRs), 224
Mid-board optical engines, 53–54
Minicomputers, 75
Modern colocation centers, 16
Modulated sources, 328
Modulation schemes, comparison of, 249, 249*f*
Modulators, 60–61
Moore's law, 76–77, 197
  data growth vs., 76*f*
  and off-chip I/O trends, 265
More Than Moore technologies, 88
Mosquito method, 174–175, 177–179, 189
MOTOR device, 110
MTF (Modulation Transfer Function), 364, 365*f*
Multi DoF/6 axis devices, 355
Multi-axis interpolation, 356
Multi-channel devices, 248
Multi-element PICs, 362–363
Multi-Guide Vertical Integration (MGVI) platform, 101
Multimode fiber (MMF), 86, 171–172, 193–194, 199–201
  connection loss with, 185
Multi-mode glass waveguides, 299
Multi-mode polymer waveguides, 157, 171
  characterization, 179–191
    connection loss with MMF, 185
    inter-channel crosstalk, 185–189
    multichannel operation, 189–191
    propagation loss, 183–185
    waveguide structure and refractive index profile, 179–183
  connectors and coupling, 166–168
    in-plane coupling, 167–168

out-of-plane coupling, 168
experimental setup for, 190*f*
fabrication method, 174–179
Mosquito method, 177–179
photo-addressing method, 176–177
soft lithography method, 175–176
numerical aperture (NA) of, 172
polymer optical waveguide circuit for optical PCB, 192–193
polynorbornene, 159–163
waveguide fabrication, 160–161
waveguide performance, 161–163
silicones, 163–166
structure of, 172–174, 173*f*
graded-index core, 173–174
step-index core, 172
Multi-mode wiring, 198–201
basic concept, 198–199
with a wavelength of 1.3 μm, 199–201
Multipath routing, 20–21, 21*f*
Multiple bandgaps, in quantum-well intermixing, 100
Multiple-Fiber Push-On/Pull-Off (MPO) connector, 52–53
Multiple-Write-Single-Read (MWSR) photonic channels, 227–231
Multi-point topologies, 384–386
Multisource agreements (MSAs), 38

N

Nanophotonics, 38, 137–138, 227–229, 283
National Science Foundation (NSF), 37
Near-field pattern (NFP), 210–214
Near-parabolic refractive index profiles, 176, 181–182
Network access control (NAC), 31–32
Network edge, 26–27
Network Equipment Building Standard (NEBS), 5–6
Network security, 31–32
Network subscription level, 27–29
Network tree, 18–19
Network-on-Chips (NoCs), 223–224, 378
Next generation data center architectures, 32–36
Non Volatile Memories (NVMs), 82–83
Non-oversubscribed network, 27–28
Non-Return-to-Zero (NRZ), 161, 249–250, 253, 255

O

Off-chip photonics, 90*f*
Offset quantum-well integration platform, 99
On-board interconnections, 378
On-board interconnects, 47–48, 47*f*
On-Demand Self-Service, 14–15
OneChip Photonics, 00004#p0065., 101–102, 107–108, 113–114
On-optical printed circuit boards (on-OPCB), 378
illustrative examples, 388–390
layout strategies, 378–390
for electrical interconnections on boards, 379
for multi-point topologies, 384–386
for OPCB, 379–388
for point-to-point topologies, 380–383
WDM for point-to-point topologies using multi-point layouts, 386–388
Open Compute Project, 17, 35–36, 55
Open Data Center Alliance, 17
Open Data Center Interoperable Network (ODIN), 17
Open Shortest Path First (OSPF), 19–20
OpenStack, 35
Optical backplane, 64, 303–304, 304*f*
Optical bus, 378
Optical coupling, 210–216, 290
Optical die, 227
Optical energy and loss model, 240–241
Optical engines, 168
mid-board, 53–54
Optical fiber pig-tailing, 343, 346–347, 347*f*, 350
Optical fibers, 83–86
Optical I/O core, 198–199, 201–216, 219–220
assembly process for, 215*f*
basic concept of, 199*f*
CMOS ICs, 207–210
constituent elements, 202–203
design concept, 201–202
fundamental photonics devices, 203–207
optical coupling structure, 210–216
packaging, 214–216
Optical interconnects, fundamentals of, 43
active optical cables, 52–53
classes of, 46–51

Optical interconnects, fundamentals of (*Continued*)
board-to-board interconnects, 48–49, 49*f*
on-board interconnects, 47–48, 47*f*
rack-to-rack interconnects and AOCs, 49–51, 50*f*
for data centers, 36–39
driver behind future data centers, 43–46
mid-board optical engines, 53–54
photonic key enabling technologies, 56–66
modulators, 60–61
optical memory elements, 64–66
optical PCBs, 63–64
optochips and 3D integration, 62–63
photo-detectors, 61–62
photonic integrated circuit technologies, 57–58
III–V on SOI active devices, 58–60
techno-economic requirements of future optical interconnects, 54–56
cost reduction, 55
energy efficiency, 54–55
longer reach, 56
Optical loss, calculating, 150–151
Optical memory elements, 64–66
Optical modulation, 142–149
high speed (GHz) electro-optic modulation, 144–149
Optical modulator, 204*t*
Optical power reduction, 242–243
Optical printed circuit board (OPCB) technologies, 63–64, 172, 192, 326–339
coordinate AAA: source characteristics, 338
coordinate BBB(b1): launch conditions, 339
coordinate CCC: input coupling conditions, 339
coordinate DDD: output coupling conditions, 339
coordinate EEE: capturing conditions, 339
diversity in OPCB types, 311–312
fiber optic circuit, laminated, 311
free space optics, 312
planar glass waveguides, 311–312
polymer waveguides, 311
target applications, 312
fabrication, 310
industrial manufacturing processes for, 312–324
fiber optic circuit, laminated, 312–317
glass waveguide OPCBs, 321–324
polymeric planar embedded optical waveguides, 318–321
international standardization of, 324–325
International Electrotechnical Commission (IEC), 324–325
polymer MT connector standard, 325
measurement definition system criteria, 326–331
capturing conditions, 331
input coupling conditions, 329–330
launch conditions, 329
output coupling conditions, 330
requirements, 327
source characteristics, 328–329
measurement identification code, 338
recommended measurement launch conditions, 331–338
mode filters and conditioners, 334–335
recommended launch conditions, 335
recommended multimode fiber launch measurement setup, 335–338
recommended single-mode fiber launch measurement setup, 335
standardized optical modal profiles, 332–334
standardized definition of waveguide position, 331
Optical probing, 371
Optical resolution, 364
Optical resonators, 121
Optical Society of America (OSA), 37
Optical technology, 44–45, 65, 375
Optical waveguides, 97–98, 171, 310, 375
planar, 323–324
polymeric planar embedded, 318–321
polymer optical waveguide, 192–193
Optical/Electrical (O/E) and Electrical/Optical (E/O) conversion, 236
Optochips and 3D integration, 62–63
Optoelectronic-integrated circuit (OEIC), 114, 117

Optoelectronics Industry Association (OIDA), 37, 37*f*
ORION 2.0, 240
Out-of-plane coupling, 168
Oversubscribed switch, 27–28

P

Packaging, 265–267
 of optical I/O cores, 214–216
 of PICs, 343–349
PAM4, 249–250, 249*f*, 253*f*, 254*f*, 255
Parallel-optical transceivers, 189
PARSEC and SPEC CPU2006 applications, 237–239
Passive alignment, 168*f*, 352
Passive waveguide (PWG), 101–102, 157–159
p-doped cladding, 99
Peer-to-peer architectures, 11–12, 11*f*
Photo detector (PD), 61–62, 204*t*, 207
 arrays, 52–53
Photo-addressing, 174–177
 fabrication technique of, 177*f*
Photo-detection, 149–152
Photodiode (PD), 250–251
Photolithography, 127–128, 165–166
Photoluminescence, 132–136
 silicon band-diagram, 133*f*
Photonic and three-dimensional interconnects, 224–225
Photonic band gap (PBG), 121–122, 124
Photonic components, 57, 66
Photonic crystal cavities, 121, 123–126
 cross-section of, 125*f*
 fiber coupling problem, 137–142
 light emission, 131–137
  electroluminescence, 136–137
  photoluminescence, 132–136
 mass production, 127–131
 optical modulation, 142–149
  high speed (GHz) electro-optic modulation, 144–149
 photo-detection, 149–152
 photonic crystal background, 121–127
  fabrication, 126–127
  theory, 122–123
 SEM image of, 146*f*
Photonic crystal-based structures, 59–60
Photonic integrated circuits, 57, 87–88, 108–109, 140–141, 311
Photonic key enabling technologies, 56–66
 III–V on SOI active devices, 58–60
 modulators, 60–61
 optical memory elements, 64–66
 optical PCBs, 63–64
 optochips and 3D integration, 62–63
 photo-detectors, 61–62
 photonic integrated circuit technologies, 57–58
Photonic switches, 108
Photonics, 43–44, 57
 chip assembly, 344*f*
 in-package, 90*f*
 off-chip, 90*f*
 silicon, 83–84, 87–90
Photonics materials and optical fibers pigtailing, 350
PhoxTroT project, 252–253, 294–296
Physical architecture, 17–21
 Layer 2 and Layer 3 network architectures, 18–21
PIC packaging
 addressing different markets, 350–351
 automated testing, 371
 automation of, 343
 complexity levels in, 344–348
 design for automated assembly/testing and standardization, 370–371
 example, 343–344
 flexibility versus speed issue, 350
 machine builders, 350
 machine vision, 363–367
 machine “ingredients” and machine technologies, 354–363
  bonding techniques and equipment, 358–360
  chucks, 361–363
  electrical probing, 361–363
  grippers, 361–363
  mechanical positioners/manipulators, 354–357
  optical grade tables and enclosures, 358
  pickers, 361–363
  tweezers technology, 361–363
 main steps of, 348–349

PIC packaging (*Continued*)
photonics materials and optical fibers pigtailing, 350
PIC numbers, 350
PICs assembly machines, 352
positional accuracy and passive/active alignment, 351–353
software and HMI role in optimization of new processes, 367–369
test and measurement instrumentation, 369–370
wafer level, assembly operations at, 351
Pick and place, 356–357, 363
Pick up tools, 361–363
Pickers, 361–363
Piezo actuators, 356
Pivot point, 355
Planar glass waveguides, 311–312
Plasma dispersion, 60
Plasma Enhanced Chemical Vapor Deposition, 138–139
Plasmonic cavities, 121
Plasmonic organic hybrid (POH) devices, 61
Platform-as-a-service, 12
Plenum rated cable, 6–7
PMOS transistors, 258–259
pn junction, 145
Pockels effect, 142–143
Point-to-point on-OPCB topology, 379–383
Point-to-point topologies, 380–383
Polarization dependent input coupling conditions, 330
Poly dimethyl siloxane (PDMS), 175–176
Poly methyl methacrylate (PMMA), 175–176
Polymer MT connector standard, 325
Polymer waveguide layers, 288–289
Polymer waveguides (PWG), 157, 159, 311
Polymeric planar embedded optical waveguides, 318–321
Polynorbornene, 159–163
waveguide fabrication, 160–161
waveguide performance, 161–163
Portable Modular Data Center (PMDC), 8
Positive self-alignment structures (PSAS), 273–274
Positive-intrinsic-negative (PIN) photodiode, 250–251
Post-curing, 359
Power reduction technique (PRT), 235–236
Power reduction technique, 235–236
Power usage effectiveness (PUE), 7–8
Printed circuit board (PCB), 55, 157, 287–288, 309
Priority-based Flow Control (PFC), 24
Probe cards, 344–345
Professional electronic IC design tools, 251
Pseudo-random binary sequence (PRBS), 191, 259
Purcell factor, 132–133

Q
Q-factor, 125, 138–139
QSFP (Quad Small Form-factor Pluggable) connector, 50–51
Quad Small Form Factor Pluggable (QSFP), 38, 52–53
Quantum-Confined Stark Effect (QCSE), 113
Quantum-well intermixing, 100

R
Rack and stack instruments, 369–370
Rack disaggregation, 81
Rack mounted server, 4*f*
Racks, optical interconnect between, 85
Rack-Scale Architecture (RSA), 49–50, 56
Rack-to-rack interconnections, 376–378
Rack-to-rack interconnects and AOCs, 49–51, 50*f*
Radiation directionality, 270–271
Rapid and Elastic Resource Provisioning, 14–15
RDMA over converged Ethernet standards (RoCE), 24–25
Reactive ion etching, 126–127
Receiver photonic integrated circuits (PICs), 105–107
Reconfiguration controller (RC), 235
Reduced Brillouin zone, 123
Refractive index matching, 329–330
Refractive index profile, 175–176, 179–183, 180*f*
Reliability, availability, and scalability (RAS), 3
Reliability, of data center network, 29–31
Requirements for optical interconnects within data centers, 75

blades and boards, optical interconnect between, 85–88
chips and dies, optical interconnect between, 88–90
data centers, optical interconnect between, 84–85
data communication requirements, 82–83
Data Deluge, 75–77
IoT and cyber physical systems, 77
racks, optical interconnect between, 85
Resistance temperature detectors (RTDs), 281
Resistive-capacitive (RC) lines, 45
Resonant modulators, 144–145
Resonant scattering, 128, 135
Resonant Si ring modulators, 60–61
Resource Pooling, 14–15
RF phase shifter, 204*t*
Riser rated cable, 6–7
Router microarchitecture, 231–232, 232*f*

S

Scalability, of data center network, 26–27
Scalable three-dimensional optical interconnects, 223
energy comparison, 240–243
optical energy and loss model, 240–241
optical power reduction, 242–243
future directions, 244
performance evaluation, 236–243
photonic and three-dimensional interconnects, 224–225
reconfiguration, 232–236
bandwidth reallocation, 233–234
dynamic reconfiguration technique, 234–235
power reduction technique, 235–236
simulation set up, 236–240
PARSEC and SPEC CPU2006 applications, 237–239
Splash-2 applications, 237
synthetic traffic, 239–240
three-dimensional-NoC, 225–232
corona, firefly and, 227–229
intra and inter-group, 229–231
proposed implementation, 227
router microarchitecture, 231–232
Search-based applications, 223
Selective area growth approach, 99
Self-biased shunt-feedback TIA implementation, 258, 258*f*
Servo-motors, 356
Shannon limit, 45
Shortest path bridging (SPB), 20–21
Shunt–shunt feedback, 259
Silica based micro-toroids, 121
Silicon based micro-ring resonators, 121
"Silicon bridging" concept, large-scale interconnected system using, 272–282
3D stacked logic, photonics, and thermal challenges, 278–282
electrical flexible interconnect integration with photonic interconnects, 276–277
self-alignment devices and performance, 274–275
system overview, 273
Silicon dioxide, 97–98, 127
Silicon interposer technology, 267–269
Silicon nitride, 138–139
Silicon Organic Hybrid technology, 61
Silicon photonic crystal, 124*f*, 126, 138
Silicon photonics, 38, 83–84, 87–90, 126, 138
and optical coupling, 269–272
technology, 198
Silicon photonics for multi-mode transmission, 197
application, 219–220
system LSI with optical I/Os, 219–220
transceivers, 219
evaluation, 216–218
optical coupling, 216
transmission characteristics, 217–218
expectations for optical interconnection, 197–198
EMI immunity, 197–198
high density, 197
reduced size and weight of cabling, 198
wide bandwidth over long distances, 197
multi-mode wiring, 198–201
basic concept, 198–199
with a wavelength of 1.3 μm, 199–201
Optical I/O core, 201–216
CMOS ICs, 207–210

Silicon photonics for multi-mode transmission (*Continued*)
constituent elements, 202–203
design concept, 201–202
fundamental photonics devices, 203–207
optical coupling structure, 210–216
Silicon waveguide-based optical modulators, 144
Silicon-based photonic crystal cavities, 121
Silicone polymer waveguides, 166
Silicones, 163–166
Silicon-on-insulator (SOI), 126, 142
platforms, 111–112
wafer, 203–204, 207
Silicon-photonic WDM components, 87–88
SIMOX (Separation by Implantation of Oxygen) 3D sculpting, 224–225
Single-Mode Fiber (SMF), 86
Single-mode waveguides, 299
Single-Write-Multiple-Read (SWMR) channel, 227–231
Si-on-insulator (SOI), 271
64 cores implementations, 237–239
Soft lithography, 174–176
Software defined data center (SDDC) architecture, 34–35, 35*f*
Software defined networking (SDN), 34–35, 54–55
Software-as-a-service, 12
Space Division Multiplexing (SDM), 377–378
SpaceFabs, 355
Space-Time (ST) switched architecture, 377–378
S-parameter measurements, 251
Spin-on-glass (SOG) process, 138–139
Splash-2 applications, 237
Spot size converters, 137–138
Spot-size convert (SSC), 204*t*
Spot-size converter (SSC), 101–102
Step-index core (SI), 172
Stepper motors, 356
Storage area networks (SANs), 17–18
Storage hierarchy in a data center, 83*f*
Stringers, 5–6
Subscription rate, 27–29
Substrate-corrugated gratings, 270*f*
Surface Mount Technology (SMT), 86
Surface-corrugated gratings, 270*f*
Surface-relief grating, 269–270
Synthetic traffic, 239–240
System LSI with optical I/Os, 219–220

T

TC-86 JWG-9, 159
Techno-economic requirements of future optical interconnects, 54–56
cost reduction, 55
energy efficiency, 54–55
longer reach, 56
Teflon-based jackets, 6–7
Telecom networks, 18
Telecommunications Industry Association (TIA), 9
Thermal compression bonding, 360
Thermo-optic effects, 142–143
Three-dimensional interconnects, 224–225
Three-dimensional-NoC (3D-NoC), 225–232
corona, firefly and, 227–229
intra and inter-group, 229–231
proposed implementation, 227
router microarchitecture, 231–232
3D packaging technology, 90
Through silicon vias (TSVs), 62–63, 88, 278
Tier 1 data center, 9
Tier 2 data center, 9–10
Tier 3 data center, 9–10
Tier 4 data center, 9–10
Top of rack (TOR) design, 18
Total Cost of Ownership (TCO), 48
TPIR-202 monomer, 175–176
Traditional data center network architecture, 13*f*
"Traffic trombone" effect, 33–34
Transceiver circuits for optical interconnects, 249
Transceivers, 107–108, 219
Transimpedance amplifier (TIA), 117, 207, 209–210, 210*f*, 247–248, 256–261, 258*f*
Transmitter photonic integrated circuits (PICs), 101–105
Transparent interconnect for lots of links (TRILL), 20–21
Tweezers, 361

25 Gb/second VCSEL links, 249–250
2D photonic crystal, 122, 124, 124*f*
2D versus 3D imaging, 367
2.5D technology, 88–89, 89*f*
256-core implementations, 239–240
Two tier data center network architecture, 24*f*
"Typical" data center, 3

U

Uninterruptable power supplies (UPS), 7–8
University of California at Santa Barbara (UCSB), 58–59

V

Vacuum pickers, 361
Vertical coupling technique, 138, 144
Vertical integration platform, 100
Vertical-cavity surface-emitting laser (VCSEL), 52–53, 58–59, 85–86, 171, 185–186, 235–236, 252–256, 252*f*, 375–376
Vertically-coupled photonic crystals, 147, 149*f*
Virtual Ethernet bridge (VEB), 26
Virtual Link Aggregation Group (VLAG), 19–20
Virtual local area networks (VLANs), 12–13
Virtual machines (VMs), 4
Virtual Router Redundancy Protocol (VRRP), 19–20
Virtual switch (vSwitch) model, 26
Virtualization, 78–79
  and latency, 25–26
Volume grating, 269–270

W

W1 photonic crystal waveguide, 124
Wafer level, assembly operations at, 351
Wafer-level testing, 371
Warehouse-scale computers (WSC), 223
Wavelength division multiplexed sources, 328–329
Wavelength Division Multiplexing (WDM), 86, 267, 377–378
Wavelength Selective Switch (WSS), 110, 376–378
WDM (Wavelength Division Multiplexing), 386–387
  for point-to-point topologies using multi-point layouts, 386–388
Wide area networks (WANs), 31
Wide bandwidth over long distances, 197

Printed in the United States
By Bookmasters